HAZARDOUS WASTE HANDBOOK FOR HEALTH AND SAFETY

HAZARDOUS WASTE HANDBOOK FOR HEALTH AND SAFETY

Second Edition

William F. Martin
John M. Lippitt
Timothy G. Prothero

Butterworth–Heinemann
Boston London Oxford Singapore Sydney Toronto Wellington

Library of Congress Cataloging-in-Publication Data

Martin, William F.
  Hazardous waste handbook for health and safety /
William F. Martin, John M. Lippitt, Timothy G.
Prothero. — 2nd ed.
      p.   cm.
  Includes bibliographical references and index.
  ISBN 0-7506-9235-9 (case bound)
  1. Hazardous waste sites—Safety measures—
Handbooks, manuals, etc. 2. Hazardous waste
sites—Health aspects—Handbooks, manuals, etc.
3. Environmental health—Handbooks, manuals, etc.
I. Lippitt, John M. II. Prothero, Timothy G.
III. Title.
TD1052.M38   1992
628.4′4—dc20   91-38367
                                              CIP

British Library Cataloguing in Publication Data

Martin, William F.
Hazardous waste handbook for health and
safety. — 2nd ed
I. Title II. Lippitt, John M.
III. Prothero, Timothy G.
363.728068

ISBN 0-7506-9235-9

Butterworth–Heinemann
80 Montvale Avenue
Stoneham, MA 02180

10 9 8 7 6 5 4 3 2 1

Printed in the United States of America

CONTENTS

ACKNOWLEDGMENTS

Many individuals and organizations contributed substantially to the development of the original four agency occupational safety and health documents by providing technical information, technical review, and editorial and production assistance. They are gratefully acknowledged again for their valuable contribution to the original documents from which this handbook was condensed. This second edition was published partially due to some excellent recommended changes from James D. Kirk, director of training, Ecological Safety Services, Houston, and Joseph A. Gispanski, Jr. and William R. Gourdie, III, Hygiene Safety and Training, Inc., Pittsburgh.

This practical hazardous waste health and safety handbook would not be possible without the previous work of many individuals, companies, and government agencies. During the past 10 years, the authors have worked with a host of highly qualified professionals in the nation's efforts to contain hazardous waste spills, clean up abandoned landfills, and control hazardous chemical threats to the environment and public health.

The authors take this opportunity to acknowledge a portion of those individuals and organizations that contributed to the vast array of publications, lectures, and training programs that served as the basis for this publication.

Mazen Y. Anastas
Linda R. Anku
Robert Arnott
Barrett Benson
Robert J. Bicknell
Linda Bochert
Nancy J. Bollinger
Thomas Burke
Barry Burrus
George A. Carson
Byron R. Chadwick

Roger A. Clark
Leslie Cole
Jan Connery
Rory Connolly
Richard J. Costello
Robert D. Cox
William DeVille
Max Eisenberg
John Farthing
James S. Ferguson
Patrick Ford

Anthony A. Fuscaldo
Maurice Georgevich
Charles L. Geraci
Ralph F. Goldman
Austin Henschell
Al Hines
Dean Y. Ikeda
Chris Jennings
William J. Keffer
Richard Kent
Gail F. Kleiner
Jay C. Klemme
James W. Lake
Mike Larsen
Steven P. Levine
Brana Lobel
Donald Mackenzie
Karen L. Mann
John B. Miles
Ronald Miller
Frank L. Mitchell
Edward Morris
Charles M. Nenadic
Lawrence J. Partridge

George Pettigrew
B. Jim Porter
Gerald P. Reidy
Stanley J. Reno
James A. Rodgers
Richard M. Ronk
Walter E. Ruch
Gilbert J. Saulter
Heidi Schultz
Tom Sell
James J. Severns
James Solyst
Alexander Stanrunas
Frank L. Strahselm
Wesley Straub
Rod Turpin
Leslie J. Ungers
Leonard Vance
Lynn P. Wallace
James B. Walters
David Weitzman
Charles Whilhelm
Mary K. White

Recognition is given to the U.S. Public Health Service, the Occupational Safety and Health Administration (OSHA), the U.S. Environmental Protection Agency (EPA), and the U.S. Coast Guard for their efforts under Superfund to gather, develop, and make publicly available health and safety publications and contractor reports.

The authors give special recognition and greatly appreciate the editorial and manuscript assembly assistance of Virginia T. Kiefert, M.A.

1

INTRODUCTION

*I*n the past decade, industry, government, and the general public have become increasingly aware of the need to respond to the hazardous waste problem, which has grown steadily over the past 100 years. In 1980, Congress passed the Comprehensive Environmental Response, Compensation, and Liability Act (CERCLA)—the Superfund law—to provide for liability, compensation, cleanup, and emergency response for hazardous substances released into the environment and the cleanup of abandoned and uncontrolled hazardous waste disposal sites. The Superfund Amendments and Reauthorization Act (SARA) of 1986 extended CERCLA and added new authorities under Title III of SARA that included Emergency Planning, Community Right-to-Know, and Toxic Chemical Release Reporting. The Resource Conservation and Recovery Act (RCRA) of 1976 sets the standards for waste handling, storage, and disposal. The 1975 Hazardous Materials Transportation Act provides regulation of hazardous materials labeling, packaging, placarding, manifesting, and transporting.

This handbook is a guidance document for supervisors responsible for occupational safety and health programs at hazardous waste sites. It was developed to give site supervisors specific instructions and guidelines on how to protect the safety and health of workers. A second goal of this handbook is to improve hazardous waste operations efficiency through knowledge and training of the work force. A third goal is to reduce the cost of hazardous waste cleanups through reduced lawsuits and liability losses of employers and individuals.

This manual is intended for individuals who have direct responsibility to carry out hazardous waste site cleanup and their subcontractors. It can be used as

- a planning tool
- a management tool
- an educational tool
- a reference document

It also serves as an applied industrial hygiene handbook for hazardous waste activities and is a valuable sourcebook on hazardous waste occupational safety and health. It should be used as a preliminary basis for developing a specific health and safety program. Consult other sources and experienced individuals as necessary for the details needed to design and implement occupational safety and health plans at specific hazardous waste sites.

Although this manual cites federal regulations, it is not a definitive legal document and should not be taken as such. Individuals who are responsible for the health and safety of workers at hazardous waste sites should obtain and comply with the most recent federal, state, and local regulations.

Several of the key hazardous waste–related regulations are briefly summarized in this chapter.

The Codes of Federal Regulations (CFRs) provide the complete text of current regulations. Some of the CFRs of direct application to hazardous waste operations are as follows: 40 CFR 300, 29 CFR 1910, 40 CFR 260–265, 30 CFR 11, and 49 CFR 100–199. These federal publications can be located at major public libraries, university libraries, and most major federal and state offices. Many data bases will provide access to these regulations. Two of these are the Computer-Aided Environmental Legislative Data System (CELDS) and LEXIS. CELDS contains abstracts of environmental regulations and is designed for use in environmental impact analysis and environmental quality management. The abstracts are written in an informative narrative style, with excessive verbiage removed. Characteristics of this system are as follows:

1. Legislative information is indexed to a hierarchical keyword thesaurus, in addition to being indexed to a set of environmental keywords.
2. Information can be obtained for federal and state environmental regulations, as well as regulatory requirements related to the keywords.
3. Appropriate reference documents, such as enactment/effective date, legislative reference, administrative agency, and bibliographical reference, are provided.

The system is structured to satisfy the user agency's (U.S. Department of Defense) specific needs for environmental regulations; consequently, the needs of other agencies may not be completely satisfied by this system. Augmentations to the system include regulations of concern to the U.S. Environmental Protection Agency (EPA).

LEXIS is a full-text system from Mead Data Central. It is a data base with a family of files that contain the full text of the following:

1. *United States Code*—a codification by major title of the body of U.S. statutes
2. *Code of Federal Regulations*—a codification by major title of current effective administrative agency regulations
3. *Federal Register*—July 1980 to present
4. Supreme Court decisions since 1960
5. State court decisions—courts of last resort, intermediary courts, lower courts

FEDERAL REGULATIONS

SUPERFUND AMENDMENTS AND REAUTHORIZATION ACT (SARA) (42 U.S.C. 11001 ET SEQ.)

Basic Objective   This act revises and extends CERCLA (superfund authorization). CERCLA is extended by the addition of new authorities known as the Emergency Planning and Community Right-to-Know Act of 1986 (also known as Title III of SARA). Title III of SARA provides for "emergency planning and preparedness, community right-to-know reporting, and toxic chemical release reporting."

Key Provisions   There are key provisions which apply when a hazardous substance is handled, and when an actual release has occurred. Even before any emergency has arisen, certain information must be made available to state and local authorities, and to the general public upon request. Facility owners and operators are obligated to provide information pertaining to any regulated substance present on the facility to the appropriate state or local authorities (Subtitle A). Three types of information are to be reported to the appropriate state and local authorities (Subtitle B):

1. Material Safety Data Sheets (MSDSs), which are prepared by the chemical manufacturer of any hazardous chemical and are retained by the facility owner or operator (or if confidentiality is a concern, a list of hazardous chemicals for which MSDSs are retained can be made available). These sheets contain general information on a hazardous chemical and provide an initial notice to the state and local authorities.
2. Emergency and Hazardous Chemical Inventory Forms, which are submitted annually to the state and local authorities. Tier 1 information includes the maximum amount of a hazardous chemical which may be present at

any time during the reporting year, and the average daily amount present during the year prior to the reporting year. Also included is the "general location of hazardous chemicals in each category." This information is available to the general public upon request. Tier II information is reported only if requested by an emergency entity or fire department. This information provides a more detailed description of the chemicals, the average amounts handled, the precise location, storage procedures, and whether the information is to be made available to the general public (allowing for the protection of confidential information).

3. Toxic chemical release reporting, which releases general information about effluents and emissions of any "toxic chemicals."

In the event a release of a hazardous substance does occur, a facility owner or operator must notify the authorities. This notification must identify the hazardous chemical involved; amounts released; time, duration, environmental fate; and suggested action.

A multilayer emergency planning and response network on the state and local government levels is to be established (also providing a notification scheme for the event of a release).

Enforcement Responsibilities; Federal-State Relationship   Local emergency planning committees or an emergency response commission appointed by the governor of the state are responsible for the response scheme. The primary drafters of the local response plans are local committees, which are also responsible for initiating the response procedure in the event of an emergency. Each state commission will supervise the local activities.

Accomplishments and Impacts   SARA legislation to promote emergency planning and to provide citizen information at the local level was a response to the Bhopal, India, disaster. A major intent is to reassure U.S. citizens that a similar tragedy will not occur in this country, and thus have a calming effect. The standardization of reporting and record keeping should produce long-term benefits and well-designed response plans. Whether a high-quality emergency response involvement can be maintained indefinitely at the local level remains a question.

COMPREHENSIVE ENVIRONMENTAL RESPONSE, COMPENSATION, AND LIABILITY ACT (42 U.S.C. 9601 ET SEQ.)

Basic Objective   The act, known as CERCLA or "Superfund," has four objectives. These are:

1. To provide the enforcement agency the authority to respond to releases of hazardous wastes (as defined in

the Clean Water Act, Clean Air Act, Toxic Substances Control Act, Solid Waste Disposal Act, and by the administrator of the enforcement agency) from "inactive" hazardous waste sites which endanger public health and the environment.
2. To establish a Hazardous Substance Superfund.
3. To establish regulations controlling inactive hazardous waste sites.
4. To provide liability for releases of hazardous wastes from such inactive sites.

The act amends the Solid Waste Disposal Act. It provides for an inventory of inactive hazardous waste sites and for the appropriate action to protect the public from the dangers possible from such sites. It is a response to the concern for the dangers of negligent hazardous waste disposal practices.

Key Provisions Key provisions of this act are:

1. The establishment of a Hazardous Substance Superfund based on fees from industry and federal appropriations to finance response actions.
2. The establishment of liability to recover costs of response from liable parties, and to induce the cleanup of sites by responsible persons.
3. The determination of the number of inactive hazardous waste sites by conducting a national inventory. This inventory shall include coordination by the Agency for Toxic Substances and Disease Registry within the Public Health Service for the purpose of implementing the health-related authorities in the act.
4. The provision of the authority for the EPA to act when there is a release or threat of release of a pollutant from a site which may endanger public health. Such action may include "removal, remedy and remedial action."
5. The revision, within 180 days of enactment of the act, of the National Contingency Plan for the Removal of Oil and Hazardous Substances (40 CFR, Part 300). This plan must include a section to establish procedures and standards for responding to releases of hazardous substances, pollutants, and contaminants and abatement actions necessary to offset imminent dangers.

Enforcement Responsibilities; Federal-State Relationship The EPA has responsibility for enforcement of the act as it pertains to the inventory, liability, and response provisions. The EPA is also responsible for claims against the Hazardous Substance Superfund, which is administered by the president. The EPA is responsible for promulgating regulations to designate hazardous substances, reportable quantities, and procedures for response. The National Response Center, established by the Clean Water Act, is responsible for notifying appropriate government agencies of any release.

The following Department of Transportation agencies also have responsibilities under the act:

1. U.S. Coast Guard—responses to releases from vessels.
2. Federal Aviation Administration—responses to releases from aircraft.
3. Federal Highway Administration—responses to releases from motor carriers.
4. Federal Railway Administration—responses to releases from rolling stock.

States are encouraged by the act to participate in response actions. The act authorizes the EPA to enter into contracts or cooperative agreements with states to take response actions. The fund can be used to defray costs to the states. The EPA must first approve an agreement with the state, based on the commitment by the state to provide funding for remedial implementation. Before undertaking any remedial action as part of a response, the EPA must consult with the affected state(s).

Accomplishments and Impacts On July 16, 1982, the EPA published the final regulations pursuant to Section 105 of the act, revising the National Contingency Plan for Oil and Hazardous Substances under the Clean Water Act, reflecting new responsibilities and powers created by CERCLA. The plan establishes an effective response program. Because the act requires a national inventory of inactive hazardous waste sites, the intent is to identify potential danger areas and effect a cleanup or remedial actions to avoid or mitigate public health and environmental dangers. In studying a sampling of these sites, the House Committee on Interstate and Foreign Commerce (House Report No. 96-1016) found four dangerous characteristics common to all the sites. These characteristics are:

1. Large quantities of hazardous wastes
2. Unsafe design of the sites and unsafe disposal practices
3. Substantial environmental danger from the wastes
4. The potential for major health problems to people living and working in the area of the sites

The intent of the act is to eliminate the above problems by dealing with the vast quantities of hazardous and toxic wastes in unsafe disposal sites in the country. The immediate impact of the act has been the identification of the worst sites where the environmental and health dangers are imminent. This priority list will be used to spend the money available in the Hazardous Waste Response Fund in the most effective way to eliminate the imminent dangers. The long-term impact of the act will be to eliminate and clean up all the identified inactive sites and develop practices and procedures to prevent future hazards in such sites, whether active or inactive. Another accomplishment of the act is to establish liability for the cost of cleanup to discourage unsafe design and disposal practices. The act has armed the

EPA with the authority to pursue an active program of cost recovery for cleanup from responsible parties.

RESOURCE CONSERVATION AND RECOVERY ACT (RCRA) (42 U.S.C. 6901 ET SEQ.)

Basic Objective RCRA, as it exists now, is the culmination of a long series of pieces of legislation, dating back to the passage of the Solid Waste Disposal Act of 1965, which address the problem of waste disposal. It began with the attempt to control solid waste disposal and eventually evolved into an expression of the national concern with the safe and proper disposal of hazardous waste. Establishing alternatives to existing methods of land disposal and to conversion of solid wastes into energy are two important needs noted by the act.

The RCRA of 1976 gives EPA broad authority to regulate the disposal of hazardous wastes; encourages the development of solid waste management plans and nonhazardous waste regulatory programs by states; prohibits open dumping of wastes; regulates underground storage tanks; and provides for a national research, development, and demonstration program for improved solid waste management and resource conservation techniques.

The control of hazardous wastes will be undertaken by identifying and tracking hazardous wastes as they are generated, ensuring that hazardous wastes are properly contained and transported, and regulating the storage, disposal, or treatment of hazardous wastes.

A major objective of the RCRA is to protect the environment and conserve resources through the development and implementation of solid-waste plans by the states. The act recognizes the need to develop and demonstrate waste-management practices that are not only environmentally sound and economical but also conserve resources. The act requires EPA to undertake a number of special studies on subjects such as resource recovery from glass and plastic waste and managing the disposal of sludge and tires. An Interagency Resource Conservation Committee has been established to report to the president and the Congress on the economic, social, and environmental consequences of present and alternative resource conservation and resource recovery techniques.

Key Provisions Some of the significant elements of the Act follow.

Hazardous wastes are identified by definition and publication. Four classes or definitions of hazardous waste have been identified—ignitability, reactivity, corrosivity, and toxicity. The chemicals that fall into these classes are regulated primarily because of the dangerous situations they can cause when landfilled with typical municipal refuse. Four lists, containing approximately 1000 distinct chemical compounds, have been published. (These lists are revised as new chemicals become available.) These lists include waste chemicals from nonspecific sources, by-products of specific industrial processes, and pure or off-specification commercial chem-

ical products. These classes of chemicals are regulated primarily to protect groundwater from contamination by toxic products and by-products.

The act requires tracking of hazardous wastes from generation, to transportation, to storage, to disposal or treatment. Generators, transporters, and operators of facilities that dispose of solid wastes must comply with a system of record keeping, labeling, and manufacturing to ensure that all hazardous waste is designated only for authorized treatment, storage, or disposal facilities. The EPA must issue permits for these facilities, and they must comply with standards issued by the EPA.

The states must develop hazardous waste management plans which must be EPA-approved. These programs will regulate hazardous wastes in the states and will control the issuance of permits. If a state does not develop such a program, the EPA, based upon the federal program, will do so.

Solid waste disposal sites are to be inventoried to determine compliance with sanitary landfill regulations issued by the EPA. Open dumps are to be closed or upgraded within five years of the inventory. As with hazardous waste management, states must develop management plans to control the disposal of solid waste and to regulate disposal sites. EPA has issued guidelines to assist states in developing their programs.

As of 1983, experience and a variety of studies dating back to the initial passage of the RCRA legislation found that an estimated 40 million metric tons of hazardous waste escaped control annually through loopholes in the legislative and regulatory framework. Subsequently, Congress was forced to reevaluate RCRA, and in doing so found that RCRA fell short of its legislative intent by failing to regulate a significant number of small-quantity generators, regulate waste oil, ensure the environmentally sound operation of land disposal facilities, and realize the need to control the contamination of groundwater caused by leaking underground storage tanks.

Major amendments were enacted in 1984 in order to address the shortcomings of RCRA. Key provisions of the 1984 amendments include:

- Notification of underground tank data and regulations for detection, prevention, and correction of releases
- Incorporation of small-quantity generators (which generate between 100 and 1000 kg of hazardous waste per month) into the regulatory scheme
- Restriction of land disposal of a variety of wastes unless EPA determines that land disposal is safe from human health and environmental points of views
- Requirement of corrective action by treatment, storage, and disposal facilities for all releases of hazardous waste regardless of when the waste was placed in the unit
- Requirement of EPA to inspect government-owned facilities (which handle hazardous waste) annually, and other

permitted hazardous waste facilities at least every other year

- Regulation of facilities which burn wastes and oils in boilers and industrial furnaces.

Enforcement Responsibilities; Federal-State Relationship Subtitle C of the Solid Waste Disposal Act, as amended by the RCRA of 1976, directs the EPA to promulgate regulations for the management of hazardous wastes.

The hazardous waste regulations initially published in May 1980 from the RCRA control the treatment, storage, transport, and disposal of waste chemicals that may be hazardous if landfilled in the traditional way. These regulations (40 CFR 261–265) identify hazardous chemicals in two ways, by listing and by definition. A chemical substance that appears on any of the lists or meets any one of the definitions must be handled as a hazardous waste.

Like other environmental legislation, RCRA enforcement responsibilities for hazardous waste management will eventually be handled by each state, with federal approval. Each state must submit a program for the control of hazardous waste. These programs must be approved by the EPA before the state can accept enforcement responsibilities.

The state programs will pass through three phases before final approval will be given. The first phase is the interim phase, during which the federal program will be in effect. The states will begin submitting their programs for the control of hazardous wastes. The second-phase programs will address permitting procedures. A final phase will provide federal guidance for design and operation of hazardous waste disposal facilities. Many states have chosen to allow the federal programs to suffice as the state program to avoid the expense of designing and enforcing the program.

It should also be noted that the Department of Transportation has enforcement responsibilities for the transportation of hazardous wastes and for the manifest system involved in transporting.

Accomplishments and Impacts The 1980 regulations for the control of hazardous wastes were a response to the national concern over hazardous waste disposal. States have begun to discover their own "Love Canals," and the impacts of unregulated disposal of hazardous wastes on their communities. While the "Superfund" legislation provides funds for the cleanup of such sites, RCRA attempts to avoid future "Love Canals."

TOXIC SUBSTANCES CONTROL ACT (TSCA) (15 U.S.C. 2601 ET SEQ.)

Basic Objective This act sets up the toxic substances program which is administered by the EPA. If the EPA finds that a chemical substance may present an unreasonable risk to health or to the environment and that there is insufficient data to predict the effects of the substance, manufacturers may be required to conduct tests to evaluate the characteristics of the substance, such as persistence, acute toxicity, or carcinogenic effects. Also, the act establishes a committee to develop a list of chemical substances by priority to be tested. The committee may list up to 50 chemicals which must be tested within one year. However, the EPA may require testing for chemicals not on the priority list.

Manufacturers must notify the EPA of the intention to manufacture a new chemical substance. The EPA may then determine if the data available are inadequate to assess the health and environmental effects of the new chemical. If the data are determined to be inadequate, the EPA will require testing. Most importantly, the EPA may prohibit the manufacture, sale, use, or disposal of a new or existing chemical substance if it finds the chemical presents an unreasonable risk to health or the environment. The EPA can also limit the amount of the chemical that can be manufactured and used and the manner in which the chemical can be used.

The act also regulates the labeling and disposal of polychlorinated biphenyls (PCBs) and prohibited their production and distribution after July 1979.

In 1986, Title II, "Asbestos Hazard Emergency Response," was added to address issues of inspection and removal of asbestos products in public schools and to study the extent of (and response to) the public health danger posed by asbestos in public and commercial buildings.

Key Provisions Testing is required on chemical substances meeting certain criteria to develop data with respect to the health and environmental effects for which there are insufficient data relevant to the determination that the chemical substance does or does not present an unreasonable risk of injury to health or the environment.

Testing shall include identification of the chemical and standards for test data. Testing is required from the following:

1. Manufacturers of a chemical meeting certain criteria
2. Processors of a chemical meeting certain criteria
3. Distributors or persons involved in disposal of chemicals meeting certain criteria

Test data required by the act must be submitted to the EPA, identifying the chemical, listing the uses or intended uses, and listing the information required by the applicable standards for the development of test data.

EPA will establish a priority list of chemical substances for regulation. Priority is given to substances known to cause or contribute to cancer, gene mutations, or birth defects. The list is revised and updated as needed.

A new chemical may not be manufactured without notifying the EPA at least 90 days before manufacturing begins. The notification must include test data showing that the manufacture, processing, use, and disposal of the chemical will not present an unreasonable risk of injury to health or the environment. Chemical manufacturers must keep records for submission to the EPA as required. The EPA will use these reports to compile an inventory of chemical substances manufactured or processed in the United States.

The EPA can prohibit the manufacture of a chemical found to present an unreasonable risk of injury to health or the environment or otherwise restrict a chemical. The act also regulates the disposal and use and prohibits the future manufacture of PCBs, and requires the EPA to engage, through various means, in research, development, collection, dissemination, and utilization of data relevant to chemical substances.

Enforcement Responsibilities; Federal-State Relationship The EPA has enforcement responsibilities for the act, but the act makes provisions for consultations with other federal agencies involved in health and environmental issues, such as OSHA and the Department of Health and Human Services. Initially, the states could receive EPA grants to aid them in establishing programs at the state level to prevent or eliminate unreasonable risks to health or the environment related to chemical substances.

Accomplishments and Impacts TSCA has provided a framework for establishing that chemical manufacturers take responsibility for the testing of chemical substances as related to their health and environmental effects. It places the burden of proof on the manufacturer to establish the safety of a chemical, yet still gives the EPA the final authority to prohibit or severely restrict chemicals in commerce. Thus, it is an attempt at the onset of a chemical to prevent significant health and environmental problems that may surface later on. The fact that when this legislation was initially passed, PCB effects were such an issue because of their widespread and uncontrolled use is reflective of public concerns over the number of other possible chemicals commonly used which could be carcinogenic. Public concern was so visible that an immediate need was perceived to regulate PCBs. Thus, PCBs are controlled and specifically prohibited by TSCA rather than RCRA.

NATIONAL ENVIRONMENTAL POLICY ACT (NEPA) (42 U.S.C. 4341 ET SEQ.)

Enforcement Responsibilities; Federal-State Relationship The President's Council on Environmental Quality (CEQ) has the main responsibility for overseeing federal efforts to comply with NEPA. In 1978, CEQ issued regulations to comply with the procedural provisions of NEPA. Other provisions of NEPA apply to major federal actions significantly affecting the quality of the human environment.

Accomplishments and Impacts The enactment of this act has added a new dimension to the planning and decision-making process of federal agencies in the United States. This act requires federal agencies to assess the environmental impact of implementing their major programs and actions early in the planning process. For those projects or

actions which are either expected to have a significant effect on the quality of the human environment or are expected to be controversial on environmental grounds, the proponent agency is required to file a formal environmental impact statement (EIS). Other accomplishments and impacts of the act are:

- It has provided a systematic means of dealing with environmental concerns and including environmental costs in the decision-making process.
- It has opened governmental activities and projects to public scrutiny and public participation.
- Some projects have been delayed because of the time required to comply with the NEPA requirements.
- Many projects have been modified or abandoned to balance environmental costs with other benefits.
- It has served to accomplish the four purposes of the act as stated in its text.

FEDERAL INSECTICIDE, FUNGICIDE, AND RODENTICIDE ACT (FIFRA) (7 U.S.C. 136 ET SEQ.)

Basic Objective FIFRA is designed to regulate the use and safety of pesticide products within the United States (which is in excess of one billion pounds). The 1972 amendments (a major restructuring which established the contemporary regulatory structure) are intended to ensure that the environmental harm resulting from the use of pesticides does not outweigh the benefits.

Key Provisions Key provisions of FIFRA include:

- The evaluation of risks posed by pesticides (requiring registration with the EPA)
- The classification and certification of pesticides by specific use (as a way to control exposure)
- The restriction of the use of pesticides which are harmful to the environment (or suspending or canceling the use of the pesticide)
- The enforcement of the above requirements through inspections, labeling, notices, and state regulation

Enforcement Responsibilities; Federal-State Relationship The EPA is allowed to establish regulations concerning registration, inspection, fines, and criminal penalties, and to stop the sales of pesticides. Primary enforcement responsibility, however, has been assumed by almost every state. Federal law only specifies that each state must have adequate laws and enforcement procedures to assume primary authority.

As in the case for almost any federal law, FIFRA preempts state law to the extent that it addresses the pesticide problem. Thus, a state cannot adopt a law or regulation that counters a provision of FIFRA, but can be more stringent.

Accomplishments and Impacts While the volume of pesticides and related information is enormous, FIFRA has enabled the EPA to acquire much information for analysis of risk and environmental degradation that results from the use of pesticides. This information has been, and will continue to be, generally invaluable in such analyses. However, Congress continues to struggle with the balancing of benefits and detriments of the use of pesticides in its attempt to deal with the economic, scientific, and environmental issues which are involved in the regulation of pesticides.

FEDERAL AND STATE REGULATORY AGENCIES

The following federal agencies and their parallel state agencies can be contacted for the latest regulations, training materials, and technical updates.

National Institute for Occupational Safety and Health (NIOSH)
4676 Columbia Parkway
Cincinnati, OH 45226

Occupational Safety and Health Administration (OSHA)
200 Constitution Avenue, NW
Washington, DC 20210

Environmental Protection Agency (EPA)
401 M Street, SW
Washington, DC 20460

Federal Emergency Management Administration (FEMA)
500 C Street, SW
Washington, DC 20472

U.S. Coast Guard (USCG)
2100 Second Street, SW
Washington, DC 20593

Agency for Toxic Substances and Disease Registry
1600 Clifton Road, NE
Atlanta, GA 30333

Although this handbook was designed to assist supervisors at abandoned or uncontrolled hazardous waste sites, the information can be used in planning for and responding to emergencies involving hazardous materials.

A short bibliography is provided at the end of each chapter to provide some additional sources of more technical information.

BIBLIOGRAPHY

American Conference of Governmental Industrial Hygienists (ACGIH). *Guidelines for the Selection of Chemical Protective Clothing.* Cincinnati: ACGIH, 1983.

Bierlein, L. *Red Book on Transportation of Hazardous Materials.* Kahners Books International, 1977.

Bretherick, L. *Handbook of Reactive Chemical Hazards.* 3d ed. Boston: Butterworth-Heinemann, 1985.

Cheremisinoff, P.N. *Hazardous Materials: Emergency Response Pocket Handbook.* Lancaster, Penn.: Technomic, 1988.

Clayton, G.D., and F.E. Clayton, eds. *Patty's Industrial Hygiene and Toxicology.* Vol. 1, 3d ed. New York: John Wiley & Sons, 1978.

Corbitt, R.A. *Standard Handbook of Environmental Engineering.* New York: McGraw-Hill, 1989.

Government Institutes, Inc. *Hazardous Material Spills—Conference Proceedings.* Rockville, Md.: Government Institutes, Inc., yearly.

———. *Management of Uncontrolled Hazardous Waste Sites—Conference Proceedings.* Rockville, Md.: Government Institutes, Inc., annually.

Hackman, E. III. *Toxic Organic Chemicals: Destruction and Waste Treatment.* Parkridge, N.J.: Noyes Data Corporation, 1978.

Levine, S.P., and W.F. Martin, eds. *Protecting Personnel at Hazardous Waste Sites.* Boston: Butterworth-Heinemann, 1985.

Lindgren, G.F. *Guide to Managing Industrial Hazardous Waste.* Boston: Butterworth-Heinemann, 1983.

Malina, E.J.F., Jr. *Environmental Engineering. Proceedings of the 1989 Specialty Conference.* New York: ASCE, 1989.

The Merck Index. 11th ed. Rahway, N.J.: Merck & Company, 1989.

National Institute for Occupational Safety and Health (NIOSH). *Occupational Safety and Health Guidance Manual for Hazardous Waste Site Activities.* U.S. Department of Health and Human Services Publication No. 85-115. Washington, D.C.: GPO, October 1985.

———. *Registry of Toxic Effects of Chemical Substances.* Cincinnati: NIOSH, 1990.

Noyes Data Corporation. *Hazardous Chemicals Data Book.* 2d ed. Parkridge, N.J.: Noyes Data Corporation, 1986.

Sullivan, T.F.P. *Directory of Environmental Information Sources.* 3d ed. Rockville, Md.: Government Institutes, Inc., 1990.

U.S. Department of Transportation (DOT). *Hazardous Materials Emergency Response Guide Book.* DOT Publication No. 5800.5. Washington, D.C.: DOT, 1990.

U.S. Environmental Protection Agency (EPA). *Everybody's Problem: Hazardous Waste.* Publication No. SW-826. Washington, D.C.: EPA, 1980.

Zajic, J.E., and W.A. Himmelman. *Highly Hazardous Materials Spills and Emergency Planning.* Marcel-Dekker, 1978.

2
HAZARDS

*H*azardous waste sites pose a multitude of health and safety risks, any one of which could result in serious injury or death (see Table 2.1). These hazards are due to the physical and chemical nature of the site as well as the work being performed. They include the following:

- chemical exposure
- fire and explosion
- oxygen deficiency
- ionizing radiation
- biologic (etiologic) hazards
- physical safety hazards
- electrical hazards
- heat stress
- cold exposure
- noise

Interaction among the substances may produce additional compounds not originally deposited at the site. Workers are subject to dangers posed by the disorderly physical environment of uncontrolled sites. The stress of working in protective clothing adds its own risk. Selection of protective equipment often is overly conservative due to many unknowns (see Chapter 6, Personal Protective Equipment).

In approaching a site, it is prudent to assume that all these hazards are present until site characterization has shown otherwise. A site health and safety plan must provide protection against the potential hazards and specific protection against individual known hazards. The safety plan must be continuously updated with new information and changing site conditions.

CHEMICAL EXPOSURE

Preventing exposure to toxic chemicals is a primary concern at hazardous waste sites. Most sites contain a variety of chemical substances in gaseous, liquid, or solid form. These substances can enter the unprotected body by inhalation, direct skin contact, ingestion, or through a puncture wound (injection). A contaminant can cause damage at the point of contact or act systemically by causing a toxic effect at other points in the body.

Chemical exposures are generally divided into two categories, acute and chronic. Symptoms resulting from *acute exposures* usually occur during or shortly after exposure to a sufficiently high concentration of a contaminant. The concentration required to produce such effects varies widely from chemical to chemical. The term *chronic exposure* generally refers to exposures to low concentrations of a contaminant over a long period of time. The concentrations required to produce symptoms of chronic exposure depend on the chemical, the duration of each exposure, and the number of exposures. For a given contaminant, the symptoms of an acute exposure may be completely different from those resulting from chronic exposure.

For either chronic or acute exposure, the toxic effect may be temporary and reversible or permanent (disability or death). Some chemicals may cause obvious symptoms such as burning, coughing, nausea, tearing eyes, or rashes. Other chemicals may cause health damage without any such warning signs. Health effects such as cancer or respiratory disease may not become manifest for several years or decades after exposure. In addition, some toxic chemicals may be colorless and/or odorless, may dull the sense of smell, or may not produce any immediate discomfort that could act as a warning that toxic chemicals are present. Thus, the ability of a worker to notice an exposure by his senses or feelings of discomfort cannot be relied upon to determine whether a worker is at risk of toxic exposure to these chemicals.

The effects of exposure are not just dependent on the chemical, its concentration, the route of entry, and the duration of exposure. A given response to a toxic chemical also may be influenced by a number of personal factors, such as the individual's smoking habits, alcohol consumption, medication use, nutrition, age, and sex (see Chapter 10, Medical Monitoring Program).

A primary exposure route of concern on a hazardous waste site is *inhalation*. The lungs are extremely vulnerable to chemical agents. Even substances that do not directly affect the lungs may pass through lung tissue into the bloodstream. Chemicals also can enter the respiratory tract through a punctured eardrum.

Direct contact of the skin and eyes by gaseous, liquid, or solid substances is another important route of exposure.

TABLE 2.1 Overview of Typical Hazards Encountered at Hazardous Waste Sites

Hazard Type	Exposure Route or Cause	Symptom or Effect	Measurement or Measuring Device	Prevention	Personal Protection	Additional Comments
Chemical exposure	Inhalation, eye/skin contact, ingestion, puncture	Headaches, nausea, rashes, burning, coughing, cancer, liver damage, kidney damage, convulsions, coma, death	TLV, PEL, $IDLH$, LD_{50}, etc. Devices: OVA, HNU, air sampling, detection tubes, personal monitoring, field GC	Follow SOP and safety procedures; extra caution when working in hot zones. Use remote control devices whenever possible.	Protective clothing Respiratory protection	Remove person immediately and decontaminate if exposed to a chemical.
Ionizing radiation	Molecular degradation releases gamma, beta, and alpha radiation; gamma is most serious; alpha is most hazardous in the case of ingestion	Radiation burns, mutagenicity, death	Radiation detection meters, Geiger-Müller detector, gamma scintillation meter	Do radiation survey early in investigation. Gamma is detectable through thin metal (e.g., drums), so survey can be done without disturbing chemical wastes.	Protective clothing and dust masks will protect against alpha and help limit beta. Only limited protection against gamma is available.	Consult health physicist if measurements are above 10 mR/hr. At 1 mR/hr, back off and map the 1 mR/hr area.
Fire and explosion	Unstable chemicals; incompatible reactions; shock-sensitive chemicals; vapor buildup in enclosed spaces or low-lying areas with sparks, open flames, or static electricity	Burns, concussion, shock, dismemberment, death	Flash point: <100°F—flammable material 100°F–200°F—fire hazard Explosimeter: >25% LEL—Withdraw and reevaluate situation; will not read accurately in oxygen-deficient environment. >10% LEL—Stop all spark-producing operations	Ventilate to prevent vapor buildup; use only nonsparking tools; use explosion-proof or intrinsically safe instruments.	Nomex Fire proximity suits Blast suits SCBAs	Always test chemicals before mixing: in enclosed areas, use a nitrogen blanket. Generation of toxic vapors and fumes may result in chemical exposures.
Oxygen deficiency	In *enclosed* spaces (e.g., buildings, tanks, manholes) or in *low* areas (e.g., trenches), oxygen is replaced by other gases; generally they are dangerous toxic gases and vapors	Inattention, impaired judgment, reduced coordination, altered breathing and heart rate, nausea, brain damage, unconsciousness, death	21% O_2 is normal. <19.5% O_2 is of concern at HW sites. <16% O_2 is dangerous to humans. When O_2 is displaced in air, N_2 is displaced also. They are in a 4:1 ratio. Therefore, if	Monitor enclosed and low areas before entry; use mechanical ventilation.	SCBAs, supplied air Safety lines	When O_2 is below 19.5% at HW site, assume O_2 has been replaced by toxic gases or vapors. Toxic gases will likely require protective clothing as well.

Hazard	Cause	Effects	Monitoring	Control	Protection	Comments
			O_2 is reduced from 21% to 19.5%, a total of 1.5% O_2 was displaced. Since N_2 is in a 4:1 ratio with O_2, 6.0% of N_2 was displaced, for a total of 7.5% air; this is equal to 75,000 ppm, a potentially deadly concentration for many chemicals to which one could be exposed.			
Physical safety hazards	Sharp objects, slippery surfaces, steep grades, uneven terrain, fogged eyewear, bulky protective clothing	Slip, trip, or fall resulting in cuts, broken bones, bruises, concussions, torn protective clothing	Visual inspection and monitoring	Identify physical hazards. Correct those that can be corrected and rope or fence off others.	Lighter protective clothing, better-fitting clothing, hard hats, boots with good gripping soles. Rubber or other nonconducting gloves, handgrips, etc.	
Electrical hazards	Exposed skin	Electrical shock	Electricity, Ohmmeter, circuit tester, visual inspection	Lock out and tag circuit breakers, switches, and controls.	Rubber and other nonconducting gloves, handgrips, ladders, etc.	Particular care is necessary when using large equipment where overhead electrical wires are present. Also need to check underground utilities if excavation is involved.
Heat stress	Caused by difficult work done in clothing designed to protect against chemicals but not against weather conditions	Inattention, impaired judgment, tiredness, exhaustion, fatigue, stroke, death	Temperature, both ambient and body	Frequent rest breaks, monitor body temperature and condition, drink fluids, appropriate clothing for weather conditions	Cool packs or vests in hot conditions; warm clothing under cold conditions. Conditioned air if supplied air suits being used	Beware of cryogenic chemicals. Fogging of face piece is a common problem. Electrolytes in body fluids should be replaced in heat stress conditions.
Cold exposure	Caused by difficult work done in clothing designed to protect against chemicals but not against weather conditions	Frostbite, pulled muscles, reduced coordination, hypothermia, death	Temperature, both ambient and body	Frequent rest breaks, monitor body temperature and condition, drink fluids, appropriate clothing for weather conditions	Warm clothing under cold conditions	Beware of cryogenic chemicals.

(continued)

TABLE 2.1 (Continued)

Hazard Type	Exposure Route or Cause	Symptom or Effect	Measurement or Measuring Device	Prevention	Personal Protection	Additional Comments
Biologic (etiologic) hazards	Wastes from hospitals and research facilities	Fever, disease, death	Swab, swipe, and grab; samples and high-volume air samplers using liquid impingement media. Requires laboratory incubation and identification. *No direct measurement available.*	Decontaminate with disinfectant and use good personal hygiene. Immunize if agent is known.	Gloves, respirators, protective clothing	Biohazard label should result in presumption of etiologic agents present until additional data are obtained. Hospital wastes (e.g., needles, bloody bandages, surgical tubing) should be considered suspect.
Noise	Compressors, machinery, large equipment	Temporary or permanent hearing loss, aural pain, nausea, reduced muscular control (when exposures are severe), distraction, interference with communication	Sound-level meter and octave-band analyzer	Shielding or enclosure of source Distance/isolation Substitution of equipment/machines generating less noise	Earmuffs, earplugs, noise-insulating earphones	Use of earphones with communication built in can improve coordination and warnings. Use of earplugs must include consideration of potential indirect exposures if earplugs become contaminated.

Note: See Appendix K for additional information on a selection of 380 chemicals and their hazards, IDLH, PEL, symptoms, and personal protective equipment.

Some chemicals directly injure the skin. Some pass through the skin into the bloodstream, where they are transported to vulnerable organs. Skin absorption is enhanced by abrasions, cuts, heat, and moisture. The eye is particularly vulnerable because airborne chemicals can dissolve onto its moist surface and be carried to the rest of the body through the bloodstream (capillaries are very close to the surface of the eye). Do not wear contact lenses when wearing protective equipment, since they can trap chemicals against the eye surface. Keeping hands away from the face, minimizing contact with liquid and solid chemicals, and using protective clothing and eye wear will protect against skin and eye exposure to hazardous substances.

Although *ingestion* should be the least significant route of exposure at a site, it is important to be aware of ways in which this type of exposure can occur. Deliberate ingestion of chemicals is unlikely, but personal habits such as chewing gum or tobacco, drinking, eating, or smoking cigarettes on-site may provide a route of entry for chemicals.

Chemical exposure by *injection* must be prevented. Chemicals can be introduced into the body through puncture wounds by stepping or tripping and falling onto contaminated sharp objects. Safety shoes with steel shanks are important protective measures against injection.

FIRE AND EXPLOSION

There are many potential causes of fires and explosions at hazardous waste sites:

- chemical reactions that produce explosion, fire, or heat
- ignition of explosive or flammable chemicals
- ignition of materials due to oxygen enrichment
- irritation of shock- or friction-sensitive compounds
- sudden release of materials under pressure

Explosions and fires may arise spontaneously. More commonly, however, they result from site activities such as moving drums, accidentally mixing incompatible chemicals, or introducing an ignition source, such as a spark from equipment. At hazardous waste sites, explosions and fires not only pose the obvious hazards of intense heat, open flame, smoke inhalation, and flying objects, but they also may cause the release of toxic chemicals into the environment. Such releases can threaten both personnel on-site and members of the general public living or working nearby. To protect against the hazard, monitor for explosive atmospheres and flammable vapors, keep all potential ignition sources away from a fire or explosive area, use nonsparking explosion-proof equipment, and follow the work practice instructions when performing any hazardous task such as bulking or mixing chemicals.

OXYGEN DEFICIENCY

The oxygen content of normal air at sea level is approximately 21%. Physiological effects of oxygen deficiency in humans is readily apparent when the oxygen concentration in the air decreases to 16%. These effects include impaired attention, judgment, and coordination and increased breathing and heart rate. Oxygen concentrations lower than 16% can result in nausea and vomiting, brain damage, heart damage, unconsciousness, and death. To take into account individual physiological responses and errors in measurement, oxygen concentrations of 19.5% or lower are considered to be indicative of oxygen deficiency.

Oxygen deficiency may result from the displacement of oxygen by another gas or the consumption of oxygen by a chemical reaction. Confined spaces or low-lying areas are particularly vulnerable to oxygen deficiency. Field personnel must monitor oxygen levels and should use supplied air respiratory equipment when oxygen concentrations drop below 19.5% by volume. Any decrease in the oxygen level of the breathing zone should be considered a potential immediately dangerous to life or health (IDLH) atmosphere if the gases or vapors are unknown.

IONIZING RADIATION

Radioactive materials emit one or more of three types of harmful radiation: alpha, beta, and gamma. Alpha radiation has limited penetration ability and is usually stopped by clothing and the outer layers of the skin. Alpha radiation poses little threat outside the body but can be hazardous if materials that emit alpha radiation are inhaled or ingested. Beta radiation can cause harmful beta burns to the skin and damage the subsurface blood system. Beta radiation also can be hazardous if materials that emit beta radiation are inhaled or ingested. Use of protective clothing, coupled with good personal hygiene and decontamination, affords protection against alpha and beta radiation.

Gamma radiation easily passes through clothing and human tissue and can cause serious and permanent damage to the body. Chemical protective clothing (CPC) affords no protection against gamma radiation, but use of respiratory and protective equipment can help keep radiation-emitting materials from entering the body by inhalation, injection, or skin contact.

If you discover levels of gamma radiation slightly above natural background, consult a health physicist. At high levels of any type of radiation, cease activities until the site has been examined and assessed by health physicists.

BIOLOGIC (ETIOLOGIC) HAZARDS

Wastes from hospitals and research facilities may contain disease-causing bacteria and viruses that could infect site personnel. Like chemical hazards, etiologic agents may be dispersed in the environment via water and wind. Other biologic hazards that may be present at a hazardous waste site include poisonous plants, insects, animals, and indigenous pathogens. Protective clothing and respiratory equipment can help reduce the chances of exposure. Thoroughly wash any exposed body parts and equipment to help protect against infection.

PHYSICAL SAFETY HAZARDS

Hazardous waste sites may contain numerous physical hazards such as the following:

- holes or ditches
- precariously positioned objects such as drums or boards that may fall
- sharp objects such as nails, metal shards, or broken glass
- slippery surfaces
- steep grades
- uneven terrain
- unstable surfaces, such as walls that may cave in or flooring that may give way

Some physical hazards are a function of the work itself. For example, heavy equipment creates an additional hazard for workers in the vicinity of the operating equipment. Protective equipment impairs a worker's agility, hearing, and vision, thus increasing the chance of an accident.

Accidents involving physical hazards can result in direct injury to workers. Accidents also can create additional hazards; for example, increased chemical exposure can result from damaged protective equipment, or a danger of explosion may be caused by the mixing of chemicals. Site personnel should be constantly on the lookout for potential safety hazards and should immediately inform their supervisors of any new hazards so that mitigative action can be taken.

ELECTRICAL HAZARDS

Overhead power lines, downed electrical wires, and buried cables all pose a danger of shock or electrocution if workers contact or sever them during site operations. Electrical equipment used on-site also may pose a hazard to workers. Use low-voltage equipment with ground-fault interrupters and watertight, corrosion-resistant connecting cables to help minimize this hazard. In addition, lightning is a hazard during outdoor operations, particularly for workers handling metal containers or equipment. Monitor weather conditions and suspend work during electrical storms to eliminate this hazard.

HEAT STRESS

Heat stress can be a major hazard, especially for workers wearing protective clothing (see Table 2.2). The same protective materials that shield the body from chemical exposure limit the dissipation of body heat and moisture. Protective clothing can, therefore, create a hazardous condition. Depending on the ambient temperature and the work being performed, heat stress can occur very rapidly—within as little as 15 minutes. It can pose as great a danger to worker health as chemical exposure. In its early stages, heat stress can cause rashes, cramps, discomfort, and drowsiness,

resulting in impaired functional ability that threatens the safety of both the individual and her coworkers. Continued heat stress can lead to heatstroke and death. Avoid overprotection, carefully train and monitor personnel wearing protective clothing, judiciously schedule work and rest periods, and replace fluids frequently to protect against this hazard. For further information on heat stress, see Chapter 6, Personal Protective Equipment, and that chapter's section on heat stress and other physiological factors.

HEATSTROKE

The classical description of heatstroke includes (1) a major disruption of central nervous function (unconsciousness or convulsions); (2) a lack of sweating; and (3) a rectal temperature in excess of 41 °C (105.8 °F). The 41 °C rectal temperature is an arbitrary value for hyperpyrexia because the disorder has not been produced experimentally in humans, so observations are made only after the admission of patients to hospitals, which may vary in time from about 30 minutes to several hours after the event. In some heatstroke cases, sweating may be present. The local circumstances of metabolic and environmental heat loads that give rise to the disorder are highly variable and are often difficult or impossible to reconstruct with accuracy. The period between the occurrence of the event and admission to a hospital may result in a quite different medical outcome from one patient to another depending on the knowledge, understanding, skill, and facilities available to those who render first aid in the intervening period. Recently, the sequence of biologic events in some fatal heatstroke cases has been described.

Heatstroke is a *medical emergency,* and any procedure from the moment of onset that will cool the patient improves the prognosis. Placing the patient in a shady area, removing outer clothing and wetting the skin, and increasing air movement to enhance evaporative cooling are all urgently needed until professional methods of cooling and assessment of the degree of the disorder are available. Frequently, by the time a patient is admitted to a hospital, the disorder has progressed to a multisystem lesion affecting virtually all tissues and organs. In the typical clinical presentation, the central nervous system is disorganized, and there is commonly evidence of fragility of small blood vessels, possibly coupled with the loss of integrity of cellular membranes in many tissues. The blood-clotting mechanism is often severely disturbed, as are liver and kidney functions. It is not clear, however, whether these events are present at the onset of the disorder or whether their development requires a combination of a given degree of elevated body temperature and a certain period for tissue or cellular damage to occur. Postmortem evaluation indicates there are few tissues that escape pathological involvement. Early recognition of the disorder or its impending onset, associated with appropriate treatment, considerably reduces the death rate and the extent of organ and tissue involvement. An ill worker should not be sent home or left unattended without a physician's specific order.

HEAT EXHAUSTION

Heat exhaustion is a mild form of heat disorder that readily yields to prompt treatment. This disorder has been encountered frequently in experimental assessment of heat tolerance. Characteristically, it is sometimes but not always accompanied by a small increase in body temperature (38 °C to 39 °C or 100.4 °F to 102.2 °F). The symptoms of headache, nausea, vertigo, weakness, thirst, and giddiness are common to both heat exhaustion and the early stages of heatstroke. There is a wide interindividual variation in the ability to tolerate an increased body temperature; some individuals cannot tolerate rectal temperatures of 38 °C to 39 °C, and others continue to perform well at even higher rectal temperatures.

There are, of course, many variants in the development of heat disorders. Failure to replace water may predispose the individual to one or more of the heat disorders and may complicate an already complex situation. Therefore, cases of hyperpyrexia can be precipitated by hypohydration. It is unlikely that there is only one cause of hyperpyrexia without some influence from another. Recent data suggest that cases of heat exhaustion can be expected to occur some 10 times more frequently than cases of heatstroke.

HEAT CRAMPS

Heat cramps are common in individuals who work hard in the heat. They are attributable to a continued loss of salt in the sweat, accompanied by a copious intake of water without appropriate replacement of salt. Other electrolytes such as Mg^{++}, Ca^{++}, and K^+ also may be involved. Cramps often occur in the muscles principally used during work and can be readily alleviated by rest, the ingestion of water, and the correction of any body fluid electrolyte imbalance.

HEAT RASHES

The most common heat rash is prickly heat (miliaria rubra), which appears as red papules, usually in areas where the clothing is restrictive, and gives rise to a prickling sensation, particularly as sweating increases. It occurs in skin that is persistently wetted by unevaporated sweat, apparently because the keratinous layers of the skin absorb water, swell, and mechanically obstruct the sweat ducts. The papules may become infected unless they are treated.

Another skin disorder (miliaria crystallina) appears with the onset of sweating in skin previously injured at the surface, commonly in sunburned areas. The damage prevents the escape of sweat with the formation of small to large watery vesicles, which rapidly subside once sweating stops, and the problem ceases to exist once the damaged skin is sloughed.

Miliaria profunda occurs when the blockage of sweat ducts is below the skin surface. This rash also occurs following sunburn injury but has been reported to occur without clear evidence of previous skin injury. Discrete and pale elevations of the skin resembling gooseflesh are present.

In most cases, the rashes disappear when the individuals are returned to cool environments. It seems likely that none of the rashes occurs (or if it does, certainly with greatly diminished frequency) when a substantial part of the day is spent in cool and/or dry areas so that the skin surface can dry.

Although these heat rashes are not dangerous in themselves, each of them carries the possibility of resulting patchy areas that are anhydrotic and thereby adversely affects evaporative heat loss and thermoregulation. In experimentally induced miliaria rubra, sweating capacity recovers within 3 to 4 weeks. Wet and/or damaged skin could absorb toxic chemicals more readily than dry, unbroken skin.

COLD EXPOSURE

Cold injury (hypothermia and frostbite) and impaired ability to work are dangers at low temperatures and at extreme windchill factors. To guard against them, wear appropriate clothing, have warm shelter readily available, carefully schedule work and rest periods, and monitor workers' physical conditions. Learn to recognize warning symptoms, such as reduced coordination, drowsiness, impaired judgment, fatigue, and numbing of toes and fingers.

HYPOTHERMIA

Hypothermia is the result of the body losing heat faster than it can produce it.

Risk Factors

    vasodilators

Stages

    shivering

    apathy

    unconsciousness

    freezing

    death

Treatment

    Get the victim out of the cold.

    Remove wet clothes and dry the victim's skin.

    Rewarm the victim by active or passive means.

FROSTBITE

Frostbite is caused by the freezing of tissues.

Risk Factors

    vasoconstrictors

Types

    frost nip

    superficial frostbite

    deep frostbite

TABLE 2.2 Classification, Medical Aspects, and Prevention of Heat Illness

Category and Clinical Features	Predisposing Factors	Underlying Physiologic Disturbance	Treatment	Prevention
1. Temperature regulation heatstroke				
Heatstroke: (1) Hot, dry skin usually red, mottled, or cyanotic; (2) t_{re}, 40.5°C (104°F) and over; (3) confusion, loss of consciousness, convulsions, t_{re} continues to rise; fatal if treatment delayed	(1) Sustained exertion in heat by unacclimatized workers; (2) Lack of physical fitness and obesity; (3) Recent alcohol intake; (4) Dehydration; (5) Individual susceptibility; (6) Chronic cardiovascular disease	Failure of the central drive for sweating (cause unknown), leading to loss of evaporative cooling and an uncontrolled accelerating rise in t_{re}; there may be partial rather than complete failure of sweating	Immediate and rapid cooling by immersion in chilled water with massage or by wrapping in wet sheet with vigorous fanning with cool, dry air; avoid overcooling; treat shock if present	Medical screening of workers, selection based on health and physical fitness; acclimatization for 5 to 7 days by graded work and heat exposure; monitoring workers during sustained work in severe heat
2. Circulatory hypostasis heat syncope				
Fainting while standing erect and immobile in heat	Lack of acclimatization	Pooling of blood in dilated vessels of skin and lower parts of body	Remove to cooler area; rest in recumbent position; recovery prompt and complete	Acclimatization; intermittent activity to assist venous return to heart
3. Water and/or salt depletion				
(a) Heat exhaustion				
(1) Fatigue, nausea, headache, giddiness; (2) Skin clammy and moist; complexion pale, muddy, or hectic flush; (3) May faint on standing with rapid, thready pulse and low blood pressure; (4) Oral temperature normal or low but rectal temperature usually elevated (37.5°C–38.5°C) (99.5°F–101.3°F); water restriction type: urine volume small, highly concentrated; salt restriction type: urine less concentrated, chlorides less than 3 g/L	(1) Sustained exertion in heat; (2) Lack of acclimatization; (3) Failure to replace water lost in sweat	(1) Dehydration from deficiency of water; (2) Depletion of circulating blood volume; (3) Circulatory strain from competing demands for blood flow to skin and to active muscles	Remove to cooler environment; rest in recumbent position; administer fluids by mouth; keep at rest until urine volume indicates that water balances have been restored	Acclimatize workers using a breaking-in schedule for 5 to 7 days; supplement dietary salt only during acclimatization; ample drinking water to be available at all times and to be taken frequently during workday

Heat disorder		Cause	Treatment	Prevention
(b) Heat cramps Painful spasms of muscles used during work (arms, legs, or abdominal); onset during or after work hours	(1) Heavy sweating during hot work; (2) Drinking large volumes of water without replacing salt loss	Loss of body salt in sweat; water intake dilutes electrolytes; water enters muscles, causing spasm	Salted liquids by mouth or more prompt relief by IV infusion	Adequate salt intake with meals; in unacclimatized workers supplement salt intake at meals
4. Skin eruptions (a) Heat rash (miliaria rubra; "prickly heat") Profuse tiny raised red vesicles (blisterlike) on affected areas, prickling sensations during heat exposure	Unrelieved exposure to humid heat with skin continuously wet with unevaporated sweat	Plugging of sweat gland ducts with retention of sweat and inflammatory reaction	Mild drying lotions, skin cleanliness to prevent infection	Cool sleeping quarters to allow skin to dry between heat exposures
(b) Anhydrotic heat exhaustion (miliaria profunda) Extensive areas of skin do not sweat on heat exposure but present gooseflesh appearance, which subsides with cool environments; associated with incapacitation in heat	Weeks or months of constant exposure to climatic heat with previous history of extensive heat rash and sunburn	Skin trauma (heat rash, sunburn) causes sweat retention deep in skin; reduced evaporative cooling causes heat intolerance	No effective treatment available for anhydrotic areas of skin; recovery of sweating occurs gradually on return to cooler climate	Treat heat rash and avoid further skin trauma by sunburn; periodic relief from sustained heat
5. Behavioral disorders (a) Heat fatigue—transient Impaired performance of skilled sensorimotor, mental, or vigilance tasks in heat	Performance decrement in unacclimatized and unskilled worker	Discomfort and physiologic strain	Not indicated unless accompanied by other heat illness	Acclimatization and training for work in the heat
(b) Heat fatigue—chronic Reduced performance capacity, lowering of self-imposed standards of social behavior (e.g., alcoholic overindulgence), inability to concentrate, etc.	Workers at risk come from temperate climates for long residence in tropical latitudes	Psychosocial stresses probably as important as heat stress; may involve hormonal imbalance, but no positive evidence	Medical treatment for serious cases; speedy relief of symptoms on returning home	Orientation on life in hot regions (customs, climate, living conditions, etc.)

Treatment

Get the victim out of the cold.

Remove clothing.

Soak the victim in warm water (102°F to 106°F).

NOISE

Work around large equipment often creates excessive noise. The effects of noise can include the following:

* psychological effects from workers being startled, annoyed, or distracted
* physiological effects, including physical damage, pain, temporary and/or permanent hearing loss, or reduced muscular control (when exposure is severe)
* communication interference that may increase potential hazards due to the inability to warn of danger or to communicate safety precautions

Permissible noise exposures are listed in OSHA regulations 29 CFR 1910.95. Earmuffs, earplugs, or other noise attenuators can be used for hearing protection.

BIBLIOGRAPHY

Dahlstrom, D.L. "Occupational Health and Safety Programs for Hazardous Waste Workers." In *Protecting Personnel at Hazardous Waste Sites,* S.P. Levine and W.F. Martin, eds. Boston: Butterworth-Heinemann, 1985.

Employee Health Risks at Toxic Waste Sites. Publication No. 91-955P. Serial No. 101-83. Washington, D.C.: GPO, September 1990.

Hill, V.H. "Control of Noise." In *The Industrial Environment—Its Evaluation and Control.* Cincinnati: National Institute for Occupational Safety and Health, 1973.

Meyer, E. *Chemistry of Hazardous Materials.* Englewood Cliffs, N.J.: Prentice-Hall, 1977.

National Fire Protection Association. *Fire Protection Guide on Hazardous Materials.* 7th ed. Quincy, Mass.: National Fire Protection Association, 1978.

National Institute for Occupational Safety and Health (NIOSH). *Hot Environments.* U.S. Department of Health and Human Services Publication No. 80-132. Cincinnati: NIOSH, July 1980.

———. *Occupational Exposure to Hot Environments.* U.S. Department of Health and Human Services Publication No. 86–113. Cincinnati: NIOSH, April 1986.

Noyes Data Corporation. *Hazardous Chemicals Data Book.* 2d ed. Parkridge, N.J.: Noyes Data Corporation, 1986.

Occupational Safety and Health Administration (OSHA). *Control of Hazardous Energy (Lockout-Tagout).* U.S. Department of Labor Publication No. 3120. Washington, D.C.: OSHA, 1991.

Sax, N.I. *Cancer Causing Chemicals.* New York: Van Nostrand Reinhold, 1981.

———. *Dangerous Properties of Industrial Materials.* 7th ed. New York: Van Nostrand Reinhold, 1988.

Verschueren, K. *Handbook of Environmental Data on Organic Chemicals.* New York: Van Nostrand Reinhold, 1983.

3

PLANNING AND ORGANIZATION

*P*lanning is the first step in hazardous waste site response activities. By anticipating and taking steps to prevent potential health and safety hazards, work at a waste site can proceed with minimum risk to workers and the public.

Planning can be organized into three phases: developing an organizational structure for site operations, establishing a work plan that considers each specific phase of the operation, and developing and implementing a health and safety plan.

The organizational structure should identify the personnel needed for the operation, establish the chain of command, and specify the responsibilities of each employee. The work plan should establish the objectives of site operations and the logistics and resources required to achieve the goals. The health and safety plan should deter-

mine the health and safety concerns for each phase of the operation and describe the procedures for worker and public protection.

Coordinating with the existing response organizations is required and will give you access to a wide range of experienced individuals. A national response organization was established by Congress under the National Contingency Plan to coordinate response actions to releases of hazardous substances. National Contingency Plan response teams are composed of representatives of federal, state, and local agencies. The EPA has designated individuals responsible for coordinating federal activities related to site cleanup.

Planning should be viewed as an ongoing process. The cleanup activities and health and safety plans must be continuously adapted to new site conditions and new information.

PERSONNEL AND RESPONSIBILITIES

An organizational structure and personnel requirements should be developed in the first phase of planning. This structure should do the following:

- identify a leader who has the authority to direct all response activities

- identify the other personnel needed for the project and assign their general functions and responsibilities

- show lines of authority, responsibility, and communication

- identify the contact points and relationships with other response agencies

The organizational structure and responsibilities may require adjustments as new information is gained and site conditions change. Any changes to the organizational structure should be recorded in the work or safety plans and communicated to all parties involved.

The following list of responsibilities must be assigned to designated individuals:

- Provide the necessary facilities, equipment, and money.

- Provide adequate personnel and time resources to conduct activities safely.

- Support the efforts of on-site management.

- Provide appropriate disciplinary actions when unsafe acts or practices occur.

- Provide advice on the design of the work plan and the health and safety plan.

- Become familiar with the types of materials on-site and the potential for worker exposure; recommend the medical program for the site.

- Provide emergency treatment and decontamination procedures for the specific type of exposures that may occur at the site. Obtain special drugs, equipment, or supplies necessary to treat such exposures.

- Provide emergency treatment procedures appropriate to the hazards on-site.

- Prepare and organize the background review of the situation, the work plan, and the field team.

- Obtain permission for site access and coordinate activities with appropriate officials.
- Ensure that the work plan is completed on schedule.
- Brief the field teams on their specific assignments.
- Prepare the final report and support files on response activities.
- Serve as the liaison with public officials.
- Choose protective clothing and equipment.
- Periodically inspect protective equipment.
- Ensure that protective clothing and equipment are properly stored and maintained.
- Control entry and exit at the access control points.
- Coordinate safety program activities with the scientific advisers.
- Confirm each team member's suitability for work based on a physician's recommendation.
- Monitor the work parties for signs of stress, such as cold exposure, heat stress, and fatigue.
- Monitor on-site hazards and conditions.
- Conduct periodic inspections to determine whether the health and safety plan is being followed.
- Enforce the buddy system.
- Know emergency procedures, evacuation routes, and the telephone numbers of the ambulance, local hospital, poison control center, fire department, and police department.
- Notify, when necessary, local public emergency officers.
- Coordinate emergency medical care.
- Manage field operations.
- Execute the work plan and schedule.
- Enforce safety procedures.
- Enforce site control.
- Document field activities and sample collection.
- Serve as a liaison with public officials.
- Notify emergency response personnel by telephone or radio in the event of an emergency.
- Assist the site safety officer (SSO) in a rescue operation.
- Maintain a log of communication and site activities.
- Assist other field team members in the clean areas, as needed.
- Maintain line-of-sight and communication contact with the work parties via walkie-talkies, air horns, or other means.
- Set up decontamination lines and the decontamination solutions appropriate for the type of chemical contamination on-site.
- Control the decontamination of all equipment, personnel, and samples from the contaminated area.
- Assist in the disposal of contaminated clothing and materials.
- Ensure that all required equipment is available.
- Advise medical personnel of potential exposures and consequences.
- Stand by, partially dressed in protective gear, near hazardous waste areas.
- Rescue any workers whose health or safety is endangered.
- Notify the SSO or supervisor of unsafe conditions.
- Plan and mobilize the facilities, materials, and personnel required for the response.
- Photograph site conditions.
- Archive photographs.
- Provide financial and contractual support.
- Release information to the news media and the public concerning site activities.
- Manage site security.
- Maintain the official records of site activities.
- Advise on the properties of the materials on-site.
- Advise on contaminant control methods.
- Advise on the dangers of chemical mixtures that may result from site activities.
- Provide immediate advice to those at the scene of a chemical-related emergency.
- Provide communication to the public in the event of an emergency.
- Predict the movement of released hazardous materials through the atmospheric, geologic, and hydrologic environment.
- Assess the effect of this movement on air, groundwater, and surface water quality.
- Predict the exposure of people and the ecosystem to the materials.
- Help plan for public evacuation.
- Mobilize transit equipment.
- Assist in public evacuation.
- Respond to fires that occur on-site.
- Provide meteorological information.
- Control access to the site.
- Advise on methods of handling explosive materials.
- Assist in safely detonating or disposing of explosive materials.

- Conduct health hazard assessments.
- Advise on adequate health protection.
- Conduct monitoring tests to determine worker exposures to hazardous substances.
- Advise on toxicological properties and health effects of substances on-site.
- Provide recommendations on the protection of worker health.

This list is intended to illustrate the scope of responsibilities and functions that must be covered. One individual may perform one or several of the functions described depending on the size of the operation and the training and experience of the individual. Use Appendix J, Health and Safety Checklist, as a tool for ensuring that all the tasks are being done and the responsibilities are adequately assigned.

Regardless of the size of the effort, all response teams should include an individual responsible for implementing health and safety requirements. The designated safety person should have access to other occupational health and safety professionals. Once an organizational system has been developed, all individuals should be identified and their respective authorities clearly explained to all members of the response team.

One of the critical elements in worker safety is the attitude of all levels of project management. This attitude sets the tone for the entire operation. The SSO and the supervisor or team leader must have the clear support of senior management for establishing, implementing, and enforcing safety programs. The importance of management's attitude toward safety throughout the project cannot be overemphasized, since site personnel are more likely to cooperate with these programs if they sense a genuine concern on the part of management.

The following organizational factors are good indicators of successful worker safety programs:

- strong management commitment to safety
- close contact and interaction among workers, supervisors, and management, enabling open communication on safety as well as other job-related matters
- a work force subject to less turnover, including a core of workers with significant lengths of service in their jobs
- a high level of housekeeping, orderly workplace conditions, and effective environmental quality control
- well-developed selection, job placement, and advancement procedures
- training practices emphasizing early indoctrination and follow-up instruction in job safety procedures

The most effective industrial safety programs are often identified by their success in dealing with people. Open communication among workers, supervisors, and management concerning work site safety is essential.

The effective management of response actions at hazardous waste sites requires a commitment to the health and safety of the general public as well as to the on-site personnel. Prevention and containment of contaminant release into the surrounding community should be addressed in the planning stages of a project. Not only must the public be protected, but they also must be made aware of the health and safety program and have confidence in it. To accomplish these goals, the project team leader or public information officer, under the supervision of the project team leader, should establish community liaison well before any response action is begun and be in continuous contact with community leaders.

WORK PLAN

A work plan describing anticipated cleanup activities must be developed before beginning on-site response actions. The work plan should be periodically reexamined and updated as new information about site conditions is obtained.

The following steps should be taken in formulating a work plan:

- Review available information:
  site records
  waste inventories
  generator and transporter manifests
  previous sampling and monitoring data
  site photos
  state and local environmental and health agency records
- Define work objectives.
- Determine methods of accomplishing the objectives (e.g., sampling plan, inventory, disposal techniques).
- Determine personnel requirements.
- Determine the need for additional training of personnel. Evaluate their current knowledge/skill level against the tasks they will perform and the situations they may encounter (see Chapter 11, Training).
- Determine equipment requirements. Evaluate the need for special equipment or services, such as the subcontracting of drilling equipment or heavy equipment and operators.

Preparation of the work plan often requires a multidisciplinary approach. Input from all levels of management and outside consultants may improve the plan and prevent oversights.

HEALTH AND SAFETY PLAN

The health and safety plan must provide measures to minimize accidents and injuries that may occur during normal daily activities or adverse conditions such as hot or cold weather. This section describes the planning process for health and safety during normal site operations—that is, nonemergency situations. Chapter 9, Site Emergencies, describes planning and response to emergencies.

Development of a written health and safety plan helps ensure that all safety aspects of site operations are thoroughly examined prior to commencing fieldwork. The health and safety plan may need updating as site cleanup progresses.

Planning requires information; thus, planning and site characterization should be coordinated. An initial health and safety plan should be developed so that the preliminary site assessment can proceed in a safe manner. The information from this assessment can then be used to refine the health and safety plan so that further site activities can proceed safely.

At a minimum, the plan should do the following:

- name key personnel and alternates responsible for site safety
- describe the safety and health risks or hazards associated with each site operation conducted (see Chapter 4, Site Characterization)
- confirm that personnel are adequately trained to perform their job responsibilities and to handle the specific hazardous situations they may encounter (see Chapter 11, Training)
- describe the protective clothing and equipment to be worn by personnel during various site operations (see Chapter 6, Personal Protective Equipment)
- describe the program for periodic air monitoring, personnel monitoring, and environmental sampling (see Chapter 4, Site Characterization, and Chapter 7, Site Control and Work Practices)
- describe the actions to be taken to mitigate existing hazards
- define site control measures and include a site map (see Chapter 7, Site Control and Work Practices)
- establish decontamination procedures for personnel and equipment (see Chapter 8, Decontamination)
- set forth the site's standard operating procedures (SOPs)
- ensure that all employees have completed all medical monitoring requirements
- set forth a contingency plan for safe and effective responses to emergencies
- establish confined-space entry procedures
- establish a spill-containment program

Appendix E provides a sample site safety plan for a fairly complex hazardous waste site cleanup operation. The sample plan can be used as a guide, *not a standard,* for designing a safety plan.

SAFETY MEETINGS AND INSPECTIONS

To ensure that the health and safety plan is being followed, the safety officer should conduct a safety meeting prior to initiating any site activity and before each workday. These safety meetings serve the following purposes:

- describe the assigned tasks and their potential hazards
- review the SOPs for the planned assignments
- coordinate activities
- identify methods and precautions to prevent injuries
- plan for emergencies
- describe any changes in the safety plan
- get worker feedback on safety conditions
- get worker feedback on how well the safety plan is working

The SSO also should conduct frequent inspections of site conditions, facilities, equipment, and activities to determine whether the health and safety plan is adequate and being followed.

At a hazardous waste site, risks to workers can change quickly and dramatically when there are changes in the following:

- the actions of other people
- the state of degradation of containers and containment structures
- the state of equipment maintenance
- weather conditions
- the work being done
- the workers assigned to the site

The following safety inspection guidelines should be observed:

- Develop a checklist for each site, listing the items that should be inspected. See Appendix J.
- Review the results of these inspections with supervisors and workers.
- Reinspect any identified problems to ensure that they have been corrected.
- Document all inspections and subsequent follow-up actions. Retain these records until site activities are completed and as long as required by regulatory agencies.

The frequency at which inspections should occur varies with the characteristics of the site and the equipment used on-site. The frequency of inspections will depend on the following:

- the severity of risk on-site
- regulatory requirements
- operation and maintenance requirements
- the expected lifetime of clothing, equipment, vehicles, and other items
- recommendations based on professional judgment, laboratory test results, and field experience

BIBLIOGRAPHY

Gere Engineers, Inc. *Hazardous Waste Site Remediation: The Engineer's Perspective.* New York: Van Nostrand Reinhold, 1988.

National Institute for Occupational Safety and Health (NIOSH). *Safety Program Practices in Record-Holding Plants.* U.S. Department of Health, Education, and Welfare Publication No. 79–136. Cincinnati: NIOSH, 1979.

National Oil and Hazardous Substances Pollution Contingency Plan. 40 CFR Part 300.

Superfund Innovative Cleanup Technologies. Publication No. 86–199P. Washington, D.C.: GPO, 1986.

U.S. Environmental Protection Agency (EPA). Office of Emergency and Remedial Response. Hazard Response Support Division. *Standard Operating Safety Guides.* Washington, D.C.: EPA, November 1986.

World Health Organization. "Planning." In *Hazardous Waste Management: Health Aspects of Chemical Safety.* Interim Document 7. Copenhagen: World Health Organization, 1982.

4

SITE CHARACTERIZATION

*S*ite characterization is directly related to worker protection. The more accurate, detailed, and comprehensive the information about a site, the more the protective measures can be tailored to the hazards workers may encounter.

At each phase of site characterization, first obtain information and then evaluate it to define the hazards the site may pose to field personnel. Then use this assessment to develop work and safety plans that define the scope and limits of the next phase of investigation.

The personnel with primary responsibility for site characterization and assessment are the site safety officer and the project team leader. In addition, outside experts, such as chemists, health physicists, industrial hygienists, and toxicologists, may be needed to interpret all the available information on site conditions.

Site characterization generally proceeds in three phases:

1. Conduct off-site surveys to gather all information prior to site entry.

2. Conduct on-site surveys. During this phase, restrict site entry to reconnaissance personnel.
3. Once the site has been determined safe, perform ongoing monitoring to provide a continuous source of information about site conditions.

It is important to recognize that site characterization is a continuous process. In addition to the formal information gathering that takes place during the phases of site characterization described here, all site personnel should be constantly alert for new information about site conditions.

This chapter details the three phases of site characterization and provides a general guide that should be adapted to meet the specific situation. Within each phase of information gathering, determine the most appropriate sequence of steps, particularly if time or budget considerations limit the scope of the investigation. Wherever possible, pursue all information sources.

OFF-SITE CHARACTERIZATION

Obtain as much information as possible before site entry so that the hazards can be evaluated and preliminary controls instituted to protect initial entry personnel. Focus initial information-gathering missions on identifying all potential or suspected IDLH conditions. Some indicators of potential IDLH conditions are listed in Table 4.1.

Information can be obtained off-site in two ways: remote resources and off-site reconnaissance.

REMOTE RESOURCES

Collect as much data as possible before any personnel go on-site. Obtain the following information where possible:

- exact location of the site
- detailed description of the activity that occurred at the site
- duration of the activity
- meteorological data—current weather and forecast, pre-

TABLE 4.1 Indicators of Potential IDLH Conditions

Large containers or tanks that must be entered

Enclosed spaces (such as buildings or trenches) that must be entered

Potentially explosive or flammable situations (indicated by bulging drums, effervescence, gas generation, or instrument readings)

Extremely toxic materials (such as cyanide or phosgene)

Visible vapor clouds

Presence of uncontained wastes (such as standing pools of liquids or severely discolored soil)

Areas where biological indicators (such as dead animals or vegetation) are located

---

vailing wind direction, precipitation levels, temperature profiles

- terrain—historical and current site maps, site photographs, aerial photographs, U.S. Geological Survey topographic quadrangle maps, land use maps, land cover maps

- utility company records
- geologic and hydrologic data
- habitation—population centers, population at risk
- accessibility by air and roads
- paths of dispersion
- present status of response and who has responded
- hazardous substances involved and their chemical and physical properties; some sources of this information follow:

    company records, receipts, logbooks, and ledgers
    records from state and federal pollution-control regulatory/enforcement agencies, state attorney general's office, state occupational safety and health agencies, and state fire marshall's office
    waste storage inventories and manifests or shipping papers
    interviews with personnel and their families (verify all information from interviews)
    records of generators and transporters
    water department and sewage district records
    interviews with neighbors (note possible site-related medical problems and verify all information from interviews)
    local fire and police department records
    court records
    utility company records
    media reports (verify all information from the media)
    interviews with nearby residents (verify all information from interviews)

- previous surveying (including soil, ground-penetrating radar, and magnetometer surveys), sampling, and monitoring data

OFF-SITE RECONNAISSANCE

At a site in which the hazards are largely unknown or there is no need to go on-site immediately, make visual observations, monitor atmospheric concentrations of airborne pollutants at the site perimeter (see Chapter 5, Air Monitoring), and collect samples near the site. Samples taken off-site are not definite indications of on-site conditions but do assist in the preliminary evaluation. Off-site reconnaissance should involve the following actions:

- Develop a preliminary site map, with the locations of buildings, containers, impoundments, pits, ponds, and tanks.
- Review historical and current aerial photographs. Note the following:

    disappearance of natural depressions, quarries, or pits
    variation in revegetation of disturbed areas
    mounding or uplift in disturbed areas or paved surfaces

    modifications in grade
    changes in vegetation around buildings and storage areas
    changes in traffic patterns at site

- Note any labels, markings, or placards on containers or vehicles.
- Note the amount of deterioration or damage of containers or vehicles.
- Note any biological indicators, such as dead animals or plants.
- Note any unusual conditions, such as clouds, discolored liquids, oil slicks, vapors, or other suspicious substances.
- Monitor the ambient air at the site perimeter for the following materials:

    toxic substances
    combustible gases
    inorganic gases, particulates, and vapors
    organic gases, particulates, and vapors
    oxygen deficiency
    ionizing radiation
    specific materials, if known

- Note any unusual odors.
- Collect and analyze off-site samples, including the following:

    soil
    drinking water
    groundwater
    site runoff
    surface water

PROTECTION OF ENTRY PERSONNEL

The selection of protective equipment for the initial site survey should be based on (1) the information from remote resources and off-site reconnaissance and (2) the proposed work to be accomplished. For example, if the purpose of the survey is to inspect on-site conditions, count containers, measure the ambient air for hot spots (areas with high concentrations of toxic chemicals), and generally get acquainted with the site, the level of protection will be less stringent than if containers are to be opened and samples taken. (Chapter 6, Personal Protective Equipment, provides more detail on the selection of protective items.)

The ensemble of clothing and equipment referred to as Level B protection is generally the minimum level recommended by the EPA for an initial entry until the site hazards have been identified and the most appropriate protective clothing and equipment chosen. Level B is preferred over Level A whenever there is no reason to believe in the presence of skin-penetrating toxins, such as cyanide gases or rocket fuels, because of the physical risks posed by the lack of visibility, dexterity, and heat stress from the full encapsulating suit in Level A protection. Those physical risks must be carefully weighted against the chemical risks of the site. Level B equipment is described in Chapter 6.

ON-SITE INFORMATION GATHERING
SITE SURVEY

The purpose of an on-site survey is to verify and supplement information from the off-site characterization. Prior to going on-site, use the off-site characterization to develop a safety plan for site entry, addressing the work to be accomplished and prescribing the procedures to protect the health and safety of the entry team. Establish priorities for monitoring and investigation after carefully evaluating probable conditions. Because team members are entering an unknown environment, caution and conservative actions are appropriate. The composition of the entry team depends on the site characteristics but should consist of four persons—two who will enter the site and two outside support people, outfitted with personal protective equipment (PPE) and prepared to enter the site in case of emergency.

During the site survey, you should do the following:

- Monitor the air for IDLH conditions (combustible or explosive atmospheres, oxygen deficiency, toxic substances).
- Monitor for ionizing radiation. Survey for gamma and beta radiation with a Geiger-Müller (GM) detection tube or a gamma scintillation tube. If alpha radiation is expected, use a proportional counter.
- Visually observe for signs of IDLH or potential IDLH conditions, including the presence of the following:
    explosives
    Class A poisons
    percutaneous (through the skin) agents
    radioactive materials

Exercise extreme caution in continuing the site survey when IDLH hazards are indicated. Tables 4.1 and 4.2 provide some guidelines for decision making. If IDLH conditions are not present, or if proper precautions can be taken, continue the survey.

TABLE 4.2 Atmospheric Hazard Guidelines

Hazard	Monitoring Equipment	Ambient Level	Action
Explosive atmosphere	Combustible gas indicator	<10% LEL[a]	Continue investigation; discontinue all sparking activities.
		10% to 25% LEL	Continue on-site monitoring with extreme caution as higher levels are encountered.
		<25% LEL	Explosion hazard; withdraw from area immediately.
Oxygen	Oxygen meter	<19.5%	Monitor SCBA. *Note:* Combustible gas readings are not valid in atmospheres with less than 19.5% oxygen. Determine what gases are displacing oxygen.
		19.5% to 25%	Continue investigation with caution. Deviation from normal level may be due to oxygen displacement. SCBA may not be needed based on oxygen content only; look at other indicators.
		>25.0%	Fire hazard potential; discontinue inspection and consult fire department or other fire specialist.
Radiation	Radiation survey instruments	>1 mR/hr[b]	Radiation above background levels (normal background = 0.02 mR/hr) signifies the possible presence of radiation sources. Continue investiga-

(continued)

TABLE 4.2 (Continued)

Hazard	Monitoring Equipment	Ambient Level	Action
Radiation (continued)			tion. Perform thorough monitoring. Consult with a health physicist.
		>10 mR/hr	Potential radiation hazard; evacuate site. Continue monitoring only upon advice of a health physicist.
Inorganic and organic gases and vapors	Colorimetric tubes	Depends on species	Consult standard reference manuals for air concentration/toxicity data. Action level dependent on TLV/PEL.[c]
	HNU photoionizer	Depends on species	Consult standard reference manuals for air concentration/toxicity data. Action level dependent on TLV/PEL.
Organic gases and vapors	Organic vapor analyzer 1. Operated in GC mode 2. Operated in survey mode	Depends on species	Consult standard reference manuals for air concentration/toxicity data. Action level dependent on TLV/PEL.

[a]LEL = lower explosive limit.
[b]mR/hr = milliroentgen per hour.
[c]TLV = threshold limit value; PEL = permissible exposure limit (see Table 6.3).

- Conduct further air monitoring as necessary.
- Note the types of containers, impoundments, or other storage systems:
    paper or wood packages
    metal or plastic barrels or drums
    underground tanks
    aboveground tanks
    compressed-gas cylinders
    pits, ponds, or lagoons
- Note the condition of waste containers and storage systems:
    sound (undamaged)
    visibly rusted or corroded
    leaking
    bulging
    types and quantities of material in containers
    labels on containers indicating corrosive, explosive, flammable, or toxic materials
- Note the physical condition of the materials:
    gas, liquid, or solid
    color and turbidity
    behavior—corroding, foaming, or vaporizing
    conditions conducive to splash or contact
- Identify natural wind barriers:

    buildings
    hills
    tanks
    trees
- Determine the potential paths of dispersion:
    air
    biological routes, such as food chains and animals
    groundwater
    land surface
    surface water
- Note any indicators of potential chemical exposure:
    dead fish, animals, or vegetation
    dust or spray in the air
    fissures or cracks in solid surfaces that expose deep waste layers
    pools of liquid
    foam or oil on liquid surfaces
    gas generation or effervescence
    deteriorating containers
    cleared land areas or possible landfill areas
- Note any physical hazards:
    conditions of site structures
    obstacles to entry and exit
    trenches, pits, or abandoned wells

soil stability, slopes, mud, and dust
stability of stacked material
electrical wires (overhead or underground)
noise sources

- Identify any reactive, incompatible, or highly corrosive wastes.

- Note land features.

- Note the presence of any natural dermatitis agents:
  poison ivy
  poison oak
  poison sumac

- Note any identification tags, labels, markings, or other indicators of material.

- Take samples:
  air (see Chapter 5, Air Monitoring)
  drainage ditches (water and sediment)
  soil (surface and subsurface)
  surface water
  storage containers
  streams and ponds (water and sediment)
  groundwater (upgradient, beneath site, downgradient)

- Sample for or identify the following:
  biological or pathological hazards
  disease-carrying animals or insects

- If necessary, use one or more of the following remote sensing or subsurface investigative methods to locate buried wastes or contaminant plumes:
  electromagnetic conductivity
  magnetometry
  metal detection
  ground-penetrating radar

INFORMATION DOCUMENTATION

Accurate, current, and readily accessible information about site conditions and activities is essential for assessing hazards, reviewing plans, and making decisions in emergency situations. However, action may be required before all the highly desirable information is available. Thus, any plan for action should provide for the continuous input of information. Documentation may become crucial in the event of any litigation. If litigation is likely, develop a document-tracking plan with the assistance of your attorneys.

Record all information pertinent to field activities, sample analysis, and site conditions in one of several forms:

- logbooks
- field data records
- graphs
- photographs
- sample labels
- chain-of-custody forms
- analytical records

These documents must be controlled to ensure that they are all accounted for when the project is completed. Assign the task of document control to one individual on the project team and specify the following responsibilities:

- Number each document (including sample labels) with a serial number.

- List each document in a document inventory.

- Record the whereabouts of each document in a separate document register so that it can be readily located. In particular, record the name and location of site personnel who have documents in their possession.

- Collect all documents at the end of each work period.

- Make sure that all document entries are made in waterproof ink.

- File all documents in a central file at the completion of the site response.

Field personnel should record all activities and observations while on-site in a field logbook, a bound book with consecutively numbered pages. Entries should be made during or just after completing a task to ensure thoroughness and accuracy. The following should be recorded during sampling:

- date and time of entry

- purpose of sampling

- name, address, and affiliation of personnel performing sampling

- name and address of the material's producer, if known

- type of material (e.g., sludge or wastewater)

- description of material container

- description of sample

- chemical components and concentrations, if known

- number and size of samples taken

- description and location of sampling point

- date and time of sample collection

- difficulties experienced in obtaining sample

- visual references such as maps or photographs of sampling site

- field observations (including weather and site conditions)

- field measurements of the materials (e.g., explosiveness, flammability, and pH)

- whether chain-of-custody forms have been filled out for the samples

- whether chain-of-custody seals have been used

Sometimes, because of the team structure, number of individuals involved, or need for a specific type of paper, notes might be made on separate sheets. When that is necessary, the field notes, data records, graphs, and other records must be initialed, listed in the logbook, and stapled into it.

Photographs can be an accurate, objective addition to a fieldworker's written observations. For each photograph taken, record the following in the field logbook:

- date, time, and name of site
- name and signature of photographer
- location of subject within the site
- general compass direction of the photograph's orientation
- general description of subject
- sequential number of the photograph and film roll number
- camera and lens type used

Assign serially numbered sample labels or tags to sampling team personnel and record them in the field logbook. Note lost, voided, or damaged labels in the logbook. Firmly affix the labels to the sample containers using either gummed labels or tags attached by string or wire. Record on the tag, in waterproof ink, information such as the following:

- sample log number
- date and time sample collected
- source of sample (e.g., name, location, and type of sample)
- preservative used
- analysis required
- name of collector
- pertinent field data

In addition to supporting litigation, written records of sample collection, transfer, storage, analysis, and destruction help ensure the proper interpretation of analytical test results. Record information describing the chain of custody on a form that accompanies the sample from collection to destruction. See Figure 4.1 for a sample chain-of-custody record.

Hazardous Materials Laboratory — Collector's Sample No. ________

CHAIN-OF-CUSTODY RECORD
Hazardous Materials

Location of Sampling: ____ Producer ____ Hauler ____ Disposal Site ____ Other: ________

Company's Name ________ Telephone (____) ________

Address ________
number street city state zip

Collector's Name ________ Telephone (____) ________
signature

Date Sampled ________ Time Sampled ________ hours

Type of Process Producing Waste ________

Waste Type Code ________ Other ________

Field Information ________

Sample Allocation:

1. ________ name of organization
2. ________ name of organization
3. ________ name of organization

Chain of Possession

1. ________ signature ________ title ________ inclusive dates
2. ________ signature ________ title ________ inclusive dates
3. ________ signature ________ title ________ inclusive dates

Figure 4.1. Example of chain-of-custody record

TABLE 4.3 Guidelines for Assessing Chemical and Physical Hazards

Hazard	Guideline[a]	Explanation	Sources for Values[b]
Chemical			
Airborne concentration	TLV-TWA	The time-weighted average concentration for a normal 8-hour workday and 40-hour workweek to which nearly all workers may be repeatedly exposed without adverse effects; should be used as an exposure guide rather than an absolute threshold	ACGIH
	TLV-STEL	A 15-minute time-weighted average exposure that should not be exceeded during any 15-minute period	ACGIH
	TLV-C	The concentration that should not be exceeded even instantaneously	ACGIH/OSHA
	PEL	Time-weighted averages and ceiling concentrations similar to (in many cases derived from) the ACGIH TELs (OSHA enforceable)	29 CFR 1910.1000
	IDLH	The maximum level from which a worker could escape within 30 minutes without any escape-impairing symptoms or any irreversible health effects	NIOSH/OSHA
	REL	8- or 10-hour time-weighted average	NIOSH
Dermal contact	Permissible concentration	Concentrations of airborne chemicals that represent acceptable levels of dermal exposure over an 8-hour workday; not to be used for assessing respiratory hazards	EPA
Dermal absorption through air-borne or direct contact	Skin-S	Indicates that a substance may be readily absorbed through or damage the skin; provides no threshold for safe exposure; direct contact with a substance designated "skin" should be avoided	ACGIH/OSHA
Dusts	TLV	Same as TLV-TWA, TLV-STEL, and TLV-C above	ACGIH
Carcinogens	TLV, CA	Some human carcinogens have an assigned TLV; for others, no contact is recommended by the ACGIH	ACGIH/NIOSH
Noise	Permissible noise exposure	Threshold levels for acceptable noise exposure	29 CFR 1910.95
	TLV	Sound pressure levels and durations of exposure that represent conditions to which nearly all workers may be repeatedly exposed without adverse effects on their ability to hear and understand normal speech	ACGIH
Ionizing radiation	TLV		NCRP
Explosion	LEL	The minimum concentration of vapor in air below which propagation of a flame will not occur in the presence of an ignition source	NFPA
Fire	Flash point	The lowest temperature at which the vapor of a combustible liquid can be made to ignite momentarily in air	NFPA

[a]TLV = threshold limit value; TWA = time-weighted average; STEL = short-term exposure limit; TLV-C = threshold limit value = ceiling; PEL = permissible exposure limit; IDLH = immediately dangerous to life or health; REL = recommended exposure limit; CA = potential carcinogen; LEL = lower explosive limit.
[b]Sources: TLV, ACGIH; PEL, OSHA; REL, NIOSH.

HAZARD ASSESSMENT

Once the presence and concentration of a specific chemical or class of chemicals have been established, the health hazards associated with that chemical or class must be determined. This is done by referring to standard data reference sources and sometimes original toxicological studies. For many of the more commonly encountered chemicals, several government agencies have specified acceptable exposure limits. The agencies and their exposure limit terminology are as follows:

- Occupational Safety and Health Administration (OSHA)—permissible exposure limit (PEL)
- National Institute of Occupational Safety and Health (NIOSH)—recommended exposure limit (REL)
- American Conference of Governmental Industrial Hygienists (ACGIH)—threshold limit value (TLV)

OSHA also considers RELs and TLVs as PELs, but in this case PEL stands for published exposure limit. Table 4.3 lists guidelines for assessing chemical and physical hazards.

HAZARDOUS SUBSTANCE DATA SHEET

The information on the chemical, physical, and toxicological properties of each compound known or expected to occur on-site should be recorded on a hazardous substance data sheet (see Appendix G). Response personnel will then have the necessary health and safety information in one place, and new personnel can be quickly briefed. Use as many reference sources as possible to fill out the sheets, as information on the same property of a compound may vary from one source to another.

Using the data gathered off-site and on-site, determine whether the site is or can be made safe for the entry of cleanup personnel or whether additional information is needed to define the necessary protective measures. Conduct further site surveys as necessary.

MONITORING

Because site activities and weather conditions change, a continuous monitoring program must be implemented after characterization has determined that the site is safe for the commencement of other operations.

Perform ongoing monitoring of atmospheric and chemical hazards using a combination of stationary sampling equipment, personnel monitoring devices, and periodic area monitoring with direct-reading instruments (see Chapter 5, Air Monitoring). Use data obtained during off-site and on-site surveys to develop a plan that details the procedures to be used for monitoring ambient conditions during cleanup operations. Where possible, monitor routes of exposure other than inhalation. For example, skin swipe tests or personal monitoring with a passive dosimeter may be used to determine the effectiveness of personal protective clothing (PPC). Depending on the physical properties and toxicity of the on-site materials, community exposure resulting from remedial action operations may need to be assessed.

Monitoring also includes continual evaluation of any changes in site conditions or work activities that could affect worker safety. When a significant change occurs, reassess the hazards. Following are some indicators of the need for reassessment:

- commencement of a new work phase, such as the start of drum sampling
- change in job tasks during a work phase
- change of season
- drastic change in weather
- change in ambient levels of contaminants

BIBLIOGRAPHY

American Conference of Governmental Industrial Hygienists (ACGIH). *Threshold Limit Values for Chemical Substances and Physical Agents.* Cincinnati: ACGIH.

California Department of Health Services. *Samplers and Sampling Procedures for Hazardous Waste Streams.* Publication No. 80–135353. California Department of Health Services, January 1980.

Centers for Disease Control (CDC). *A System for Prevention, Assessment, and Control of Exposures and Health Effects from Hazardous Waste Sites (SPACE for Health).* Atlanta: CDC, January 1984.

National Institute for Occupational Safety and Health (NIOSH) and Occupational Safety and Health Administration (OSHA). *Pocket Guide to Chemical Hazards.* U.S. Department of Health and Human Services (DHHS) Publication No. 90–117. Washington, D.C.: DHHS, June 1990.

Occupational Safety and Health Administration. *Air Contaminants—Permissible Exposure Limits.* 29 CFR 1910.1000. Washington, D.C.: U.S. Department of Labor, 1989.

5

AIR MONITORING

*A*irborne contaminants can present a significant threat to worker health and safety. Thus, identification and quantification of these contaminants through air monitoring is an essential component of a health and safety program at a hazardous waste site. Reliable measurements of airborne contaminants are necessary to

- *determine the level of PPE required*
- *define the areas where protection is needed*
- *assess the potential health hazards of exposure*
- *determine whether continual exposure is occurring, which may indicate the need for medical monitoring*

If the identification of the major airborne contaminants is incomplete, any assessment of the health and safety conditions at the site will be incomplete.

This chapter identifies the factors to consider when conducting air monitoring at a hazardous waste site by (1) presenting strategies for assessing inhalation exposure to chemicals at hazardous waste sites and (2) describing instruments and methods for measuring exposures. The information presented here stems from several studies conducted by the EPA and NIOSH and is derived from experience gained from Superfund operations at operational hazardous waste treatment facilities, hazardous waste sites, emergency response sites, and landfills.

MEASURING INSTRUMENTS

The purpose of air monitoring is to identify and quantify all airborne contaminants so that you can determine the level of worker protection needed. Identification is often a qualitative event—that is, the contaminant or the class to which it belongs is demonstrated to be present, but the determination of its concentration (quantification) must await subsequent testing. Two principal approaches toward identification and quantification of airborne contaminants are available, both derived from NIOSH and industrial hygiene standards: (1) the on-site use of direct-reading air survey instruments and (2) laboratory analyses of air samples. These methods are meant to identify the air contaminants to consider when establishing the level of protection for site entry. They are not a substitute for industrial hygiene sampling during site entry.

DIRECT-READING INSTRUMENTS

Direct-reading instruments were developed as early warning devices for use in industrial settings where a leak or an accident could release a high concentration of a known chemical into the ambient atmosphere. Generally, they detect and/or measure high exposure levels that are likely to cause acute reactions. Unlike conventional instruments, which take samples that must be subsequently analyzed in a laboratory, direct-reading instruments provide information in real time, enabling immediate decision making.

Direct-reading instruments may be used to detect IDLH conditions or to demonstrate the presence of specific chemicals or classes of chemicals. They are the primary tools of initial site characterization. The information provided by

direct-reading instruments can be used to institute appropriate protective measures (i.e., PPE), to determine the most appropriate equipment for further monitoring, and to develop optimum sampling and analytical protocols.

All direct-reading instruments have inherent constraints in their ability to detect hazards:

- The measurements are relative to specific chemicals.
- Generally, the instruments are not designed to measure and/or detect low (minute) airborne concentrations.
- Many of the direct-reading instruments that have been designed to detect one particular substance also detect other substances (cross-sensitivity), so they can give false readings.

For all these reasons, it is imperative that these instruments be operated and their data interpreted by qualified individuals who are thoroughly familiar with the particular device's operating principles and limitations and who have obtained the device's latest operating instructions and calibration curves. At hazardous waste sites, where unknown and multiple contaminants are the rule rather than the exception, be sure to interpret the readings of these instruments conservatively. The following guidelines may facilitate accurate recording and interpretation:

- Develop response curves for pollutants not designated on the instrument.
- Remember that the instrument's readings have no value where contaminants are unknown. When recording read-

ings of unknown contaminants, report them as *needle deflection* or *positive instrument response* rather than specific concentrations (i.e., parts per million, or ppm). Conduct additional monitoring at any location where a positive response occurs.

- Report a reading of zero as *no instrument response* rather than *clean* because there may be quantities of chemicals present that are not detectable by the instrument.

- Repeat the survey with several detection systems to maximize the number of chemicals detected.

Tables 5.1 and 5.2 list several direct-reading instruments and the conditions and/or substances they measure. The flame ionization detector (FID) and the photoionization detector (PID) (see Table 5.1) are commonly used at hazardous waste sites. However, these ionization devices do not detect two particularly toxic agents: (1) particulate cyanides (or any particulates) and (2) hydrogen sulfide and other inorganic gases that have no achievable ionization potentials. Thus, the use of ionization devices must be supplemented with detector tubes for these agents and with laboratory analysis.

Using a range of colorimetric detector tubes, one can distinguish between the following classes of contaminants:

- acid-reacting substances
- alcohols
- amines or reactive nitrogen materials
- aromatic hydrocarbons
- halogenated hydrocarbons
- ketones

Results are valuable as a rapid, qualitative indicator of contaminant classes but cannot provide quantitative pollutant levels without identification of specific compounds present.

LABORATORY ANALYSIS

Direct-reading personal monitors are available for only a few specific substances and are rarely sensitive enough to measure the minute quantities of contaminants that, over years of accumulative exposure, may induce chronic health changes. To detect relatively low-level concentrations of contaminants, long-term personal air samples, also called full-shift samples, must be analyzed in a laboratory. These samples are usually collected after preliminary monitoring with direct-reading instruments indicates that no IDLH conditions are present. Full-shift air samples may be collected passively or by means of a pump that draws air through a filter or sorbent. Table 5.3 lists some sampling and analytical techniques used by NIOSH at hazardous waste sites.

When the concentration of contaminants is being determined, the correct measurement apparatus depends on the physical state of the contaminants. For example, chemical

hazards such as polychlorinated biphenyls (PCBs) and polynuclear aromatic hydrocarbons (PNAs) occur as both vapors and particulate-bound contaminants. A dual-media system is needed to measure both forms of these substances. The volatile component is collected on a solid adsorbent, and the nonvolatile component is collected on a filter. More than two dozen dual-media sampling techniques have been developed by NIOSH.

A major disadvantage of long-term air monitoring is the time required to obtain data. The time lag between sampling and obtaining the analysis results may be a matter of hours if an on-site laboratory is available or days, weeks, or even months if a remote laboratory is involved. This can be a significant problem if the situation requires immediate decisions concerning worker safety. Also, by the time samples are returned from a remote laboratory, the waste site cleanup may have progressed to a different stage or to a location where different contaminants or different concentrations may exist. Careful planning and/or the use of an on-site laboratory may alleviate these problems.

Based on this concept, NIOSH has developed a mobile operational laboratory for use at hazardous waste sites. Several cleanup contractors also have mobile laboratories available for lease, hire, or contract arrangement. The mobile laboratory is generally a trailer that houses analytical instruments capable of classifying contaminants in nearly real time by a variety of techniques. Typical instruments include gas chromatographs, ion chromatographs, and X-ray fluorescence spectrometers. When not in use in the mobile laboratory, these devices can be relocated to fixed-base facilities.

Usually only a few of the field samples collected are analyzed on-site to provide rapid estimates of the concentration of airborne contaminants. These data can be used to determine the initial level of worker personal protection, to modify field sampling procedures, and to guide the fixed-base laboratory analysis. If necessary, samples screened in the mobile laboratory can be subsequently reanalyzed in sophisticated fixed-base laboratories. The mobile laboratory also provides storage space, countertop staging areas for industrial hygiene equipment, and facilities for recharging self-contained breathing apparatuses (SCBAs).

Air monitoring is conducted in various stages and categories, including the following:

- IDLH monitoring to identify acutely hazardous conditions during initial site entry and for high-risk activities such as confined-area entry
- general monitoring to identify major sources, classes, and concentrations of airborne contaminants
- perimeter monitoring to provide fixed-location monitoring to measure and ensure appropriate designations of clean versus contaminated areas
- periodic monitoring to identify potential variations in air

TABLE 5.1 Direct-Reading Instruments for General Survey

	Example	Hazard Monitored	Application	Detection Method	Limitations	Ease of Operation	Care and Maintenance	Operating Time
Combustible gas indicator (CGI)	MSA 260 (O₂/LEL meter)	Combustible gases and vapors	Measures the concentration of a combustible gas or vapor	A filament, usually made of platinum, is heated by the burning of the combustible gas or vapor.	Accuracy is, in part, dependent on the difference between the calibration and sampling temperatures. Sensitivity is a function of the differences in the chemical and physical properties between the calibration gas and the unknown. The filament can be damaged by certain compounds, such as silicones, halides, and tetraethyl lead. Accuracy can be affected by oxygen-deficient atmospheres.	Effective use requires that the operator understand the operating principles and procedures.	Recharge or replace battery. Calibrate immediately before use.	
Flame ionization detector (FID)	Organic vapor analyzer (Century Systems OVA or equivalent)	Many organic gases and vapors	In survey mode, detects the total concentrations of many organic gases and vapors; in gas chromatography (GC) mode, identifies and measures specific compounds In survey mode, all the organic	Gases and vapors are ionized in a flame. A current is produced in proportion to the number of carbon atoms present.	Does not detect inorganic gases and vapors or some synthetics. Sensitivity is dependent on the compound. Should not be used at temperatures less than 40°F.	Requires experience to interpret data correctly, especially in the GC mode.	Recharge or replace battery. Monitor fuel and/or combustion air supply gauges. Perform routine maintenance as described in the manual. Check for leaks.	8 hours; 3 hours with strip chart recorder

(continued)

TABLE 5.1 (Continued)

Example	Hazard Monitored	Application	Detection Method	Limitations	Ease of Operation	Care and Maintenance	Operating Time
		compounds are ionized and detected at the same time; in GC mode, volatile species are separated		Difficult to identify compounds absolutely

High concentrations of contaminants or oxygen-deficient atmospheres require system modification.

Reduced reliability in high humidity due to incomplete combustion | | | |
| Ultraviolet (UV) photoionization detector (PID)

HNU or equivalent | Many organic and some inorganic gases and vapors | Detects total concentrations of many organic and some inorganic gases and vapors; some identification of compounds is possible if more than one probe is used | Molecules are ionized using UV radiation, and a current is produced that is directly proportional to the number of ions. | Does not detect methane

Does not detect a compound if the probe used has a lower energy than the compound's ionization potential. Response may change when gases are mixed.

Other voltage sources may interfere with measurements.

Does not readily ionize fully chlorinated materials.

High humidity affects readings.

Low humidity affects operation. | Effective use requires that the operator understand the operating principles and procedures and be competent in calibrating, reading, and interpreting the instrument. | Recharge or replace battery.

Regularly clean lamp window.

Regularly clean and maintain the instrument and accessories. | 10 hours; 5 hours with strip chart recorder |

Infrared (IR) spectrophotometer	Miniature IR spectrophotometer (Miran or equivalent)	Many gases and vapors	Measures concentration of many gases and vapors in air; designed to quantify one- or two-component mixture	Different frequencies of IR light are passed through the sample. The frequencies absorbed are specific for each compound.	Response is sensitive to dust or moisture on the lamp. In the field, Miran must make repeated passes to achieve reliable results. Requires 115-volt AC power Not approved for use in a hazardous location Requires personnel with extensive experience in IR spectrophotometry

TABLE 5.2 Direct-Reading Instruments for Specific Survey

	Hazard Monitored	Application	Detection Method	Limitations	Ease of Operation	Care and Maintenance	Operating Time
Direct-reading colorimetric indicator tube	Specific gases and vapors	Measures concentrations of specific gases and vapors	The compound reacts with the indicator chemical in the tube, producing a stain whose length is proportional to the compound's concentration.	The measured concentration of the same compound may vary among different manufacturers' tubes. Many similar chemicals interfere. Affected by temperature, pressure, and humidity. NIOSH certification program has been discontinued.	Greatest sources of error are how the operator judges the stain's end point and the ±25% cross-sensitivity of the instrument.	Do not use a previously opened tube even if the indicator chemical is not stained. Check for leaks. Refrigerater prior to use to maintain shelf life of about 2 years. Calibrate with a volumetric test quarterly.	Essentially instantaneous
Oxygen meter[a]	Oxygen (O_2)	Measures the percentage of O_2	The meter uses an electrochemical sensor to measure the partial pressure of O_2 in the air and converts that reading to O_2 concentration.	Must be calibrated prior to use to compensate for altitude and barometric pressure. Certain gases, especially oxidants, can affect readings. Carbon dioxide poisons the detector cell.	Effective use requires that the operator understand the operating principles and procedures.	Protect with a disposable cover. Change detector solution according to manufacturer's recommendations. Recharge or replace batteries prior to expiration of the specified interval. If the ambient air is more than 0.5% carbon dioxide, replace or rejuvenate the O_2 detector cell frequently.	8 to 12 hours

[a]Oxygen measurements are most informative when paired with combustible gas measurements. Several combination units package a combustible gas indicator (CGI) with an oxygen meter.

TABLE 5.3 Sample Collection and Analytical Methods

Substance	Collection Device	Analytical Method	Typical Limit of Detection (μg)
Anions	Prewashed silica gel tube	Ion chromatography	
Chloride			5
Nitrate			10
Bromide			10
Fluoride			5
Sulfate			10
Phosphate			20
Aliphatic amines	Silica gel	GC/NPD	10
Asbestos	AA filter	PCM	4500*
Metals	AA filter	ICP-AES	0.5
Organics I	Charcoal tube	GC/MS	10
Nitrosamines	Thermosorb/N	GC-HECD	0.01
Particle size Distribution	Personal Cascade impactor	Gravimetric	
PCBs	GC filter and florisil tube		0.05
Pesticides	13 mm GF filter and Chromosorb 102 tube	GC/MS	0.05

*Fibers per filter
GC/NPD = gas chromatography and nitrogen-phosphorus detector; PCM = phase contrast microscopy; ICP- AES = inductively coupled plasma atomic emission spectrometry; GC/MS = gas chromatography and mass spectrometry; GC-HECD = gas chromatography using a Hall Electrical Conductivity Detector; GF = glass fiber.

contamination concentration and patterns based on changes in site conditions, weather, and activities

- personnel monitoring to check worker exposures based on samples collected in the workers' breathing zones during site operations and on analysis of air-purifying respirator (APR) cartridges and canisters

SITE CHARACTERIZATION

Site characterization begins with the initial off-site investigation into the current circumstances, the origin, and the history of the wastes at the site (see Chapter 4). Establish priorities for air monitoring based on this information. With the selected monitoring equipment in hand, and wearing the selected PPE as deemed necessary by the off-site characterization, proceed to monitor the background air quality on-site in the following manner.

IDLH MONITORING

First, conduct air monitoring to identify any IDLH conditions, such as flammable or explosive atmospheres, oxygen-deficient environments, and high levels of airborne toxic substances. Monitoring instruments normally include combustible gas indicators, oxygen meters, colorimetric tubes, and organic vapor monitors. This should be done as soon as possible after the site has been designated as hazardous. Exercise extreme caution in continuing the site survey when atmospheric hazards are indicated. Be aware that conditions can change suddenly from nonhazardous to hazardous.

Acutely hazardous concentrations may persist in confined and low-lying spaces for long periods of time. Consider whether the suspected contaminants are lighter or heavier than air. Look for any natural or artificial barriers, such as hills, tall buildings, or tanks, behind which air might stand still, allowing concentrations to build up. Examine any confined spaces, such as cargo holds, mine shafts, silos, storage tanks, boxcars, buildings, bulk tanks, and sumps, where chemical exposures capable of causing acute health effects are likely to accumulate. Low-lying areas, such as hollows and trenches, also are suspect. Monitor these spaces for IDLH conditions.

In open spaces, toxic materials tend to be emitted into the atmosphere, transported away from the source, and dispersed. Acutely hazardous conditions are not likely to persist in open spaces for extended periods of time unless there is a very large (and, hence, readily identifiable) source, such as an overturned tank car. Therefore, open spaces are generally a lower monitoring priority.

GENERAL MONITORING

Conduct air monitoring using a variety of media to identify the major classes of airborne contaminants and their concentrations. The following sampling pattern can be used as a guideline.

First, after visually identifying a possible source of contaminants, pull air samples downwind from the source along the axis of the wind direction. When downwind monitoring is complete, work upwind until reaching or getting as close as possible to the source. Be sure to wear an SCBA during all sampling operations. Once the greatest concentrations are determined, workers must suit up to the appropriate level of PPE. Additional PPE must be available in case the level of protection must be upgraded.

After reaching the source, or finding the highest concentration, sample across the axis of the wind direction to determine the degree of dispersion. Smoke plumes or plumes of instrument-detectable airborne substances may be released as an aid in this assessment. To ensure that there is no background interference and that the detected substances are originating at the identified source, also pull air samples from upwind of the source.

PERIMETER MONITORING

Fixed-location monitoring along the perimeter, where PPE is no longer required, measures contaminant migration away from the site and enables the SSO to evaluate the integrity of the site's clean areas. Since the fixed-location samples may reflect exposures either upwind or downwind from the site, wind direction and velocity data are needed to interpret the sample results. Designation of the perimeter must be based on sample results, not on fences or property boundaries.

PERIODIC MONITORING

Site conditions may change following the initial characterization. Monitoring should be repeated periodically, especially when

- work begins on a different portion of the site
- different contaminants are being handled
- a markedly different type of operation is initiated (e.g., barrel opening as opposed to exploratory well drilling)

    Monitoring may be discontinued if

- it can be demonstrated that site contaminants are homogeneously distributed and consistent exposures are anticipated
- sampling in maximally contaminated areas has produced negative results

MONITORING PERSONNEL

The selective daily monitoring of high-risk workers—that is, those who are closest to the source of contaminant generation—is highly recommended. This approach is based on the rationale that the probability of significant exposure varies directly with the distance from the source. If workers closest to the source are not significantly exposed, then all other workers are, presumably, also not significantly exposed and probably do not need to be monitored. Monitoring the high-risk workers first conserves resources.

Since occupational exposures are linked closely with active material handling, personal air sampling should not be necessary until site mitigation has begun. Collect personal monitoring samples in the breathing zone and, if workers are wearing respiratory protective equipment, outside the face piece. These samples represent the potential inhalation exposure of workers who are not wearing respiratory protection. It is best to use flow rate–controlled pumps to collect samples, since it is difficult to observe and adjust pumps while wearing gloves, respirators, and other PPE. Protect the pumps with disposable coverings, such as small plastic bags, to make decontamination procedures easier.

Personal monitoring is performed using a variety of sampling media. Unfortunately, single workers cannot carry multiple sampling media because of the added strain and because it is not usually possible to draw air through different sampling media using a single portable, battery-operated pump. Consequently, several days are required to measure the exposure of a specific individual using each of the media. Alternatively, if workers are in teams, a different monitoring device can be assigned to each team member. Another method is to place multiple sampling devices on pieces of heavy equipment. While these are not, technically, personal samples, they are collected very close to the breathing zone of the heavy equipment operator and thus would be reasonably representative of personal exposure. These multimedia samples can yield as much information as several personal samples.

VARIABLES OF HAZARDOUS WASTE SITE EXPOSURE

Complex, multisubstance environments such as hazardous waste sites pose significant challenges to accurate and safe assessment of airborne contaminants. Several independent and uncontrollable variables, most notably temperature and weather conditions, can affect airborne concentrations. These factors must be considered when developing an air-monitoring safety program and when analyzing data. Some demonstrated variables and variances include the following:

- *Temperature.* A 10 °C increase in temperature can more than double the vapor pressure of PCBs.
- *Wind speed.* An increase from 3.7 meters per second to 6.9 meters per second at 20 °C can double the PCB vapor concentration near a free-liquid surface. Dust and particulate-bound contaminants also are affected.
- *Rainfall.* Water from rainfall can essentially cap or plug vapor emission routes from open or closed containers, saturated soil, or lagoons, thereby reducing airborne emissions of certain substances.
- *Moisture.* Dust, including finely divided hazardous solids, is highly sensitive to moisture content, which can vary significantly with respect to location and time.
- *Vapor emissions.* The physical displacement of saturated vapors can produce short-term, relatively high-vapor

concentrations. Continuing diffusion and/or evaporation may be important long-term, low-concentration phenomena involving large areas.

- *Work activities.* Remedial action can lead to mechanical disturbance of contaminated materials, thereby changing the concentration and composition of airborne contaminants at any one time or place at the site.

FIELD STUDY RESULTS

NIOSH has collected more than 500 air samples at hazardous waste sites. About half of these were collected at remedial action sites where drums were being handled. The other half were collected at a municipal landfill that accepted limited types of hazardous waste and at a hazardous waste treatment facility that accepted virtually all categories of waste.

Of the 500 samples, no contaminant exceeded 10% of the OSHA 8-hour permissible time-weighted average (TWA) concentration. The substances analyzed included acids, bases, cyanides, heavy metals, organic vapors, PCBs, and pesticides. The metals scan typically included 30 elements from aluminum to zinc. The NIOSH study found evidence that particulate-bound PNAs were responsible for eye and respiratory irritation among some workers.

BIBLIOGRAPHY

Costello, R.J., and J. Melius. "Inhalation Exposures at Hazardous Waste Sites." Paper presented at the AIHC, Las Vegas, May 1985.

Hill, R.H., and J.E. Arnold. "A Personal Air Sampler for Pesticides." *Arch. Environ. Contam. Toxicol.* 8 (1979):621–628.

Leichnitz, K. "Qualitative Detection of Substances by Means of Draeger Detector Tube Ethyl Acetate 200/a." *Draeger Review* 46 (1980):13–21.

National Institute for Occupational Safety and Health (NIOSH). *NIOSH Manual of Analytical Methods.* 4th ed. Cincinnati: NIOSH, 1984.

6

PERSONAL PROTECTIVE EQUIPMENT

*A*nyone entering an uncontrolled hazardous waste site must be protected against any potential hazards they may encounter. The purpose of PPC and PPE is to shield or isolate individuals from the chemical, physical, and biological hazards presented by a waste site. Careful selection of adequate PPE should protect the

* *respiratory system*
* *skin and body*
* *face and eyes*
* *feet and hands*
* *head*
* *hearing*

This chapter describes the various types of PPE appropriate for use on uncontrolled hazardous waste sites and provides guidance in their selection and use. The final section discusses heat stress and other key physiological factors that must be considered in connection with PPE use. More detailed guidance on PPE selection and use can be found in Personal Protective Equipment for Hazardous Materials Incidents: A Selection Guide *(NIOSH, 1984).*

Based on the Superfund Amendments and Reauthorization Act (SARA) of 1986, the U.S. Congress tasked OSHA with establishing requirements for worker protection on hazardous waste sites. In accordance with requirements in Section 126 of Title I of SARA, OSHA established the Hazardous Waste Operations and Emergency Response rule in 29 CFR 1910.120. The final rule

became effective March 6, 1990. Pursuant to 29 CFR 1910.120(b), employers engaged in covered activities are required to establish a safety and health program for each site. Within the program, employers are required to prepare a written PPE program applicable to each task identified. This program identifies requirements for PPE selection, use, and maintenance. It must comply with requirements in 29 CFR 1910.120(g) (3), (4), and (5). The written PPE program also must comply with established OSHA standards for PPE, also found in 29 CFR 1910. Additional respiratory regulations, passed pursuant to the Mine Safety Act, can be found in 30 CFR Part 11. Table 6.1 provides a list of OSHA standards that should be addressed in the PPE program.

No one piece of PPE, nor any single combination of equipment and clothing, is capable of protecting against all threats. In fact, no PPE is capable of providing protection against even one threat for a prolonged period of time. PPE should be considered the last option for protection. It should be used in conjunction with other protective methods, such as safety procedures, alternative remedial actions, and/or engineering controls.

The use of PPE can create significant worker hazards, such as heat stress, physiological stress, and impaired visibility, mobility, and communication. The greater the level of PPE protection, the greater are the associated risks. For any given situation, equipment and clothing should be selected to provide an adequate level of protection. Overprotection is hazardous and should be avoided.

DEVELOPING A PPE PROGRAM

Two basic objectives must be achieved in a PPE program: (1) to prevent overexposure of workers to safety and health hazards and (2) to prevent injury to workers resulting from incorrect use or malfunction of the required PPE. To accomplish these goals, a comprehensive PPE program must be established. In accordance with the OSHA 29 CFR 1910.120 rule, the PPE program must be in writing as part of the site-specific safety and health plan. No specific format has been required, but the following elements should be included:

* design and use of engineering controls and site work practices established for protection of worker safety and health (required to the extent feasible to reduce and control potential worker exposures to safety and health hazards)
* designation of required levels of protection for established work zones (consistent with engineering controls and work practices referenced above)
* requirements for selection, use, and maintenance of respiratory equipment

TABLE 6.1 OSHA Regulations and Sources for Use of Protective Clothing and Equipment

Type of Protection	Regulation	Source
General	29 CFR 1910.132	41 CFR 50–204.7 (General Requirements for Personal Protective Equipment)
Eye and face	29 CFR 1910.133(a)	ANSI Z87.1–1968 (Eye and Face Protection)
Noise	29 CFR 1910.95	
Respiratory	29 CFR 1910.134	ANSI Z88.2–1969 (Standard Practice for Respiratory Protection)
Head	29 CFR 1910.135	ANSI Z89.1–1969 (Safety Requirements for Industrial Head Protection)
Foot	29 CFR 1910.136	ANSI Z41.1–1967 (Men's Safety Toe Footwear)

Source: American National Standards Institute (ANSI), New York.

- requirements for selection, use, and maintenance of PPC
- establishment of required PPE ensembles to provide the necessary level of worker protection for the designated work zones
- procedures and requirements for employee training and supervision of equipment use to address and control the following:
    duration of activities for each task
    limitations of PPE and ensemble use
    donning of PPE (i.e., dressing)
    respirator fit testing and evaluation during use
    monitoring of environmental conditions
    monitoring adequacy of PPE ensemble and performance criteria/limitations for existing conditions
    doffing PPE (i.e., undressing)
    decontamination of PPE and personnel
    inspection and monitoring of the PPE's condition and procedures for repair or replacement
    storage and access to PPE
    PPE program maintenance

Other elements required in the site-specific safety and health plan provide necessary input into the PPE program. These elements include the following:

- identification of hazardous materials and conditions (Chapter 4)
- environmental sampling and surveillance (Chapters 4 and 5)
- site controls and work practices (Chapter 7)
- decontamination (Chapter 8)
- response to site emergencies (Chapter 9)
- medical surveillance and monitoring in the field (Chapter 10)
- employee training requirements (Chapter 11)

The PPE program should be reviewed at least annually and when new hazards are identified, site conditions change, or after any incident involving PPE use that results in, or could have resulted in, injury. When reviewing the PPE program, the following issues should be considered:

- the number of person-hours required for PPE ensembles at the various levels of protection
- accident and illness experiences
- the levels of exposures encountered
- the appropriateness of PPE, engineering controls, and standard work safety practices and procedures
- the adequacy of operational guidelines for conducting site activities
- the effectiveness of PPE decontamination, cleaning, inspection, maintenance, and storage practices
- coordination of PPE information and needs with site safety and health program requirements
- program costs
- the maintenance of necessary records (e.g., PPE use and assignment, employee training, and medical surveillance)
- an evaluation of PPE program goals and accomplishments
- recommended PPE program improvements and modifications

The results of the PPE program review and evaluations should be made available to the affected employees and presented to upper management for implementation.

SELECTION OF RESPIRATORY EQUIPMENT

Respiratory protection is of primary importance, as the lungs present the body's greatest exposed surface area. Respiratory protective devices (respirators) consist of a face piece connected to an air or oxygen source. The three major categories of respirators differ with respect to the air or oxygen source:

- Self-contained breathing apparatuses (SCBAs) supply air from a source carried by the user.
- Air-line respirators (ALRs) supply air from a source located some distance away and connected to the user by a hose, sometimes called an umbilical cord.

TABLE 6.2 Relative Advantages and Disadvantages of Respiratory Protective Equipment

Type of Respirator	Advantages	Disadvantages
Self-contained breathing apparatus (SCBA)	Operated in positive-pressure mode, provides the highest available level of protection against most airborne contaminants Mobile Protection factor of 10,000	Bulky and heavy (up to 35 pounds) Finite air supply limits work duration Less suitable for strenuous work or work in confined spaces
Air-line respirator (ALR)	Enables longer work periods than a SCBA Less bulky and lighter than an SCBA; weighs less than 5 pounds (or around 15 pounds if escape protection is included) Protects against most airborne contaminants	Not approved for use in IDLH atmospheres or for initial site entry unless equipped with an emergency backup SCBA escape respirator for emergency use in the event of air-line failure; emergency SCBA shall be rated for at least 5 minutes of continuous use; enabling the worker to escape from the hazardous environment Impaired mobility; NIOSH limits the length of the supply air hose to 300 feet; auxiliary piping may extend this distance provided no decrease in air pressure occurs within the auxiliary piping
Air-purifying respirator (APR)	Enhanced mobility Lighter than an SCBA (generally 2 pounds or less) Less expensive than an SCBA or ALR	Air supply hose vulnerable to damage, chemical contamination, and degradation; decontamination of hoses may be difficult Worker must retrace steps to leave work area Requires supervision/monitoring of the air supply line Cannot be used in IDLH or oxygen-deficient (less than 19.5% oxygen) atmospheres Limited duration of protection; may be hard to gauge safe operating time in field conditions Protects only against specific chemicals and up to specific concentrations Requires continuous monitoring of contaminant and oxygen levels Should be used only against organic vapors with adequate warning properties (taste, odor, irritation, etc.)

- Air-purifying respirators (APRs) enable the user to inhale purified ambient air.

Because SCBAs and ALRs both supply air to the user, they are sometimes categorized together as *supplied-air respirators*. Table 6.2 lists the relative advantages and disadvantages of SCBAs, ALRs, and APRs.

Respirators are further differentiated by the type of airflow supplied to the face piece:

- Negative-pressure respirators (also referred to as demand respirators) draw air into the face piece via the negative pressure created by user inhalation. The disadvantage of demand respirators is if any leaks develop in the system (e.g., a crack in the hose or an ill-fitting face piece), the user draws contaminated air into the face piece during inhalation.

- Positive pressure respirators (also referred to as pressure-demand respirators) maintain a slight positive pressure in the face piece during both inhalation and exhalation. A pressure regulator and an exhalation valve on the mask maintain the mask's positive pressure at all times. If a leak develops, the regulator sends a continuous flow of clean air into the face piece, preventing penetration of contaminated ambient air. Only positive-pressure respirators are recommended for work at hazardous waste sites.

- Continuous-flow respirators send a continuous stream of air into the face piece at all times. Continuous airflow prevents infiltration by ambient air but exhausts the air supply much more rapidly than positive-pressure or negative-pressure respirators.

Different types of face pieces are available for the various types of respirators:

- Full masks cover the face from the hairline to below the chin. They are recommended for use on uncontrolled sites, since they provide eye as well as respiratory protection.

- Half masks cover the face from below the chin to over the nose. They can be used when the airborne contaminants have been identified and are judged unlikely to irritate the eyes.

- Quarter masks cover the nose and mouth but do not cover the chin. Their face piece–to–face seal is not as good as that of half and full face masks. NIOSH does not recommend their use on hazardous waste sites, but OSHA does accept them for industrial use.

Federal regulations require use of approved respirators. Approval numbers are clearly written on all approved respiratory equipment. Respirators are tested by NIOSH, and if they pass the OSHA requirements specified in 30 CFR 11, they are jointly approved by NIOSH and the Mine Safety and Health Administration (MSHA). Testing procedures are described in 30 CFR 11. Periodically, NIOSH publishes the *NIOSH Certified Equipment List*, which iden-

tifies all approved respirators and respiratory components. Certification is negated if parts are modified or changed to modify the equipment as listed.

PROTECTION FACTOR

The protection factor, described by a number, denotes the overall level of protection provided by the respirator. The protection factor is determined by the fit and, with APRs, by the filtering ability of the respirator. The number indicates the relative difference in concentrations of substances outside and inside the face piece that can be maintained by the respirator. For example, the protection factor for full-mask APRs is 100 according to the American National Standards Institute (ANSI). This means, theoretically, that workers wearing these respirators should be protected in atmospheres containing chemicals at concentrations that are 100 times higher than their safe levels. Protection factors are determined by quantitative analytical tests. The protection factors for various types of SCBAs, ALRs, and APRs are listed in Table 6.3.

To determine whether an SCBA or ALR provides adequate protection in a given situation, multiply the protection factor by the threshold limit value (TLV) or permissible exposure limit (PEL) for the chemical(s) in the atmosphere (see Chapter 4, Site Characterization). Adequate protection is provided against a particular chemical if the product is greater than the measured ambient concentration of the chemical. (For APRs, the maximum safe use concentration for specific chemicals is designated on the respirator.)

TABLE 6.3 Respirator Protection Factors

	Face Piece Pressure	Protection Factor	
		ANSI[a]	NIOSH[b]
Self-Contained Breathing Apparatus[c]			
Open-circuit, positive-pressure full face piece	+	10,000+	10,000
Open-circuit, negative-pressure full face piece	−	100	50
Closed-circuit, negative-pressure, oxygen tank–type full face piece	−	100	50
Air-Line Respirator			
Positive-pressure full face piece with escape provision	+	10,000+	—
Positive-pressure full face piece without escape provision	+	—	2000
Negative-pressure full face piece	−	100	50
Negative-pressure half mask	−	10	10
Continuous-flow full face piece or hood	+	—	2000
Continuous-flow full face piece with escape provision	+	10,000+	—
Continuous-flow half mask	+	—	1000
Air-Purifying Respirator			
Negative-pressure gas- and vapor-removing full face piece	−	100	50
Negative-pressure gas- and vapor-removing half mask	−	10	10

[a]ANSI Standard Z88.2-1980 (New York: American National Standards Institute).
[b]*A Guide to Industrial Respiratory Protection*, National Institute of Occupational Safety and Health, June 1976.
[c]See Table 6.5 for definitions of open- and closed-circuit face pieces.

Bear in mind that a respirator's protection factor can be compromised in several situations, most notably (1) if a worker has a high respiration rate or (2) if the ambient temperature is high or low.

If a worker's respiration rate exceeds 67 liters per minute, many positive-pressure respirators will not maintain positive pressure during peak inhalation. This *inboard face piece leakage* occurs in ALRs and SCBAs. Also, at high work rates, exhalation valves may leak. Consequently, positive-pressure respirators working at volumes of 67 liters per minute or greater offer no more protection than a similarly equipped negative-pressure respirator. In terms of the numerical protection factor, this is a reduction from 10,000 to approximately 50. (Actual protection reduction is difficult to estimate and varies from unit to unit.)

A similar reduction in the protection factor may be occasioned by high or low ambient temperatures. Table 6.4 summarizes possible effects of temperature on respirator function. As a general precaution, consider that protection factors may decrease at temperatures below 60°F or above 90°F. Note that the temperature inside a fully encapsulating suit, within which an SCBA might be worn, may exceed 90°F.

SELF-CONTAINED BREATHING APPARATUSES

An SCBA consists of a face piece connected by a hose and a regulator to an air source (compressed air, compressed oxygen, or an oxygen-generating chemical) carried by the wearer. SCBAs are the only respirators approved for entry into IDLH atmospheres. Because SCBAs use an independent rather than an ambient air source, they offer protection against most types and levels of airborne contaminants. However, the duration of the air supply is limited, based on the amount of oxygen or air and the rate of air consumption, and this is an important planning factor in SCBA use. SCBAs also are bulky and heavy. They increase the likelihood of heat stress and may impair movement in confined spaces. Generally, only workers handling hazardous materials or operating in contaminated zones are equipped with SCBAs. By mandate of 30 CFR 11.70(a), SCBAs may be approved (1) for escape only or (2) for both entry into and escape from a hazardous atmosphere. The types of SCBAs and their relative advantages and disadvantages are described in Table 6.5.

Escape-only SCBAs are frequently continuous-flow devices with hoods that can be placed directly over other respiratory face pieces to provide immediate emergency protection. Employers must provide and ensure that employees carry an escape respirator where exposure to extremely toxic substances may occur (an extremely toxic substance is defined as a gas or vapor having an LC_{50} of less than 10 ppm).

Entry-and-escape SCBAs give workers untethered access to nearly all portions of the work site but decrease worker mobility, particularly in confined areas, due to the units'

TABLE 6.4 Possible Temperature-Related Effects That May Compromise Respiratory Protection

High temperatures	Excessive sweat may break the face piece–to–face seal. Exhalation valve and regulator diaphragm may malfunction due to softening and increased flexibility.
Low temperatures	Exhalation valve may leak. Exhalation valve may become clogged with ice due to moisture of breath.

bulk and weight. Their use is particularly advisable when dealing with unidentified and unquantified airborne contaminants. There are two types of entry-and-escape SCBAs, open-circuit and closed-circuit. In an *open-circuit SCBA,* air is exhaled directly into the ambient atmosphere. In a *closed-circuit SCBA,* exhaled air is recycled by removing the carbon dioxide with an alkaline scrubber and replenishing the consumed oxygen with oxygen from a liquid or gaseous source or from an oxygen-generating chemical.

Any *compressed-air cylinder* used with an SCBA must meet U.S. Department of Transportation (DOT) General Requirements for Shipment and Packaging (49 CFR Part 173) and Shipping Container Specifications (49 CFR Part 178). Breathing air quality must meet the requirements of Grade D breathing air as described by the Compressed Gas Association:

- Oxygen content must be between 19.5% and 23.5%, with the remainder mainly nitrogen.

- Hydrocarbon concentrations must not exceed 5 milligrams per cubic meter (mg/m^3).

- Carbon monoxide concentrations must not exceed 20 ppm.

- Carbon dioxide concentrations must not exceed 1000 ppm.

- There must not be any pronounced odor.

Following are some key questions to ask when considering whether an SCBA is appropriate:

- Is the atmosphere immediately dangerous to life or health? If so, you must use an SCBA.

- Is the duration of air supply sufficient for accomplishing the necessary tasks? If not, use a larger cylinder or modify the work plan.

- Will the bulk and weight of the SCBA interfere with task performance or cause unnecessary stress? If so, consider using an ALR if conditions permit.

- Will temperature effects compromise respirator effectiveness or pose added stress to the worker? If so, shorten the work period or postpone the mission until the temperature changes.

TABLE 6.5 Self-Contained Breathing Apparatus (SCBA) Types

Types	Description	Advantages	Disadvantages	Comments
Entry and escape SCBA Open-circuit SCBA	Supplies clean air or oxygen to the wearer from an air cylinder; wearer exhales air directly to the atmosphere	Operated in a positive-pressure mode, provides highest respiratory protection currently available; warning alarm signals when only 25% of the air supply remains	Shorter operating time (30 to 60 minutes) and heavier weight (up to 35 pounds) than a closed-circuit SCBA Vulnerable to flame from spark impingement	Operating time may vary depending on the size of the air tank and the work rate of the individual
Closed-circuit SCBA (rebreather)	Recycles exhaled gases (CO, CO_2, O_2, and nitrogen) by removing CO_2 with an alkaline scrubber and replenishing the consumed O_2 with O_2 from a liquid or gaseous source or from an O_2-generating chemical	Longer operating time (15 minutes to 4 hours) and lighter weight (21 to 30 pounds) than open-circuit apparatus Warning alarm signals when only 25% of the air supply remains O_2 supply is depleted before CO_2 sorbent scrubber supply, thereby protecting the wearer from CO_2 breakthrough	Approved by NIOSH only as negative-pressure respirator, although pressure-demand units are available If the release valve fails or the face seal leaks, the suit may fill with O_2, making the wearer vulnerable to flame from spark impingement At cold temperatures, scrubber efficiency is markedly reduced and CO_2 breakthrough can occur, rendering the wearer unconscious without warning Retains heat normally exchanged in exhalation and generates heat in the CO_2 scrubbing operation, adding to the danger of heat stress; auxiliary cooling devices may be required When worn outside an encapsulating suit, breathing bag may be permeated by chemicals, contaminating the breathing apparatus and the respirable air; decontamination of breathing bag may be difficult	Once initiated, O_2-generating devices cannot be turned off, and the regenerative chemical is a safety hazard
Escape-only SCBA	Supplies clean air to the wearer from an air cylinder or an O_2-generating chemical; approved for escape purposes only	Lightweight (10 pounds or less) and easy to carry Available in positive-pressure or continuous-flow mode	Provides only 5 to 15 minutes of respiratory protection; does not carry safety features necessary for longer work periods	Used for hazardous substances fieldwork as a backup device to replace breathing apparatus that has failed or run out of air

AIR-LINE RESPIRATORS

ALRs (or supplied-air respirators) supply air, not oxygen, to a face piece via a supply line from a stationary source. ALRs are available in positive-pressure, negative-pressure, and continuous-flow systems. Positive-pressure and continuous-flow ALRs with escape provisions provide the highest level of protection (among ALRs) and are the only ALRs recommended for use in remedial actions involving hazardous materials. ALRs are not approved for entry into IDLH areas (30 CFR 11) or for initial site entry in areas requiring respiratory protection (29 CFR 1910.120) unless the apparatus is equipped with an emergency bailout bottle or the worker also wears an escape SCBA rated for at least 5 minutes' duration.

The air source for ALRs may be a compressor, which purifies and pumps ambient air to the face piece, or compressed-air cylinders. ALRs suitable for use with compressed-air cylinders are classified as Type C supplied-air respirators. All ALR couplings must be incompatible with the outlets of other gas systems used on-site to prevent a worker from connecting to a hazardous compressed-gas source. This incompatibility is generally standardized.

ALRs enable longer work periods than SCBAs and are not as bulky. However, the air-line hose impairs worker mobility and requires workers to retrace their steps when leaving the area. Also, the air-line hose is vulnerable to puncture from rough or sharp surfaces, chemical permeation, damage from being run over by heavy equipment, and obstruction from falling drums. To maintain safe conditions, all such hazards should be removed prior to use. When in use, air lines should be kept as short as possible, and other workers and vehicles should be kept away from the air lines.

The use of air compressors as the air source for an ALR at a hazardous waste site is severely limited by the same concern that requires workers to wear respirators: the questionable quality of the ambient air. On-site compressor use is recommended only when contaminants can be identified and readily removed by sorbents in the compressor's air-purification system. Even in these conditions, an air compressor's purification system shares the same constraints as an APR—that is, effective filters and/or sorbents capable of removing these contaminants must be available.

Following are some key questions to ask when considering ALR use:

- Is the atmosphere immediately dangerous to life or health? If so, use an ALR-SCBA combination or an SCBA.
- Will the hose significantly impair worker mobility? If so, modify the task or use other respiratory protection.
- Is there a danger of the air-line hose being damaged or obstructed (e.g., by heavy equipment, falling drums, rough terrain, or sharp objects) or permeated or degraded by chemicals? If so, remove the hazard or use other respiratory protection.

- If a compressor is the air source, is it possible for air-borne contaminants to enter the air system? If so, have the contaminants been identified, and are efficient filters and/or sorbents capable of removing those contaminants available? It not, use cylinders as the air source or use another form of respiratory protection.
- Can other workers and vehicles be kept away from the area where the line has been laid? If not, consider another form of respiratory protection.

AIR-PURIFYING RESPIRATORS

APRs consist of a face piece and an air-purifying device that is a removable component of the face piece or is worn on a body harness and attached to the face piece by a corrugated breathing hose. APRs selectively remove specific air-borne contaminants (particulates, gases, vapors, and fumes) from ambient air by filtration, absorption, or adsorption and are colorcoded (Table 6.6). They are approved for use in atmospheres containing specific chemicals up to designated concentrations but not for IDLH atmospheres. APRs have limited use in remedial actions involving hazardous materials and can be used only when the ambient atmosphere contains sufficient oxygen (19.5%) to support life at high work rates. Table 6.7 lists the conditions precluding APR use.

APRs usually operate only in the negative-pressure mode. Blower-powered devices maintain a positive face piece pressure, but they generally remove only dust, fumes, and particulates, not gases or vapors.

Three types of purifying devices exist: (1) particulate filters, which remove particulates; (2) cartridges and canisters, which contain sorbents for specific chemicals, gases, and vapors; (3) combination devices. Their efficiency varies considerably even for closely related materials.

Cartridges attach directly to the face piece. The larger-volume canisters attach to the chin of the face piece or are carried with a harness and attached to the face piece by a breathing tube. Single cartridges and canisters remove only one chemical or one class of chemical. They have designated maximum concentration limits above which they should not be used (30 CFR Part II, Section 150). Combination canisters and cartridges (also known as Type N, all-service, universal, or all-purpose canisters) contain layers of different sorbent materials and remove multiple chemicals or classes of chemicals from the ambient air. Though approved for use against more than one substance, these canisters are tested independently against single substances. Thus, their effectiveness against two or more substances has not been demonstrated.

A number of standard cartridges and canisters are commercially available. They are color-coded to indicate the general chemicals or classes of chemicals against which they are effective. 29 CFR 1910.134 lists the OSHA-approved color coding used on air-purifying canisters.

TABLE 6.6 Contaminants to Be Protected Against and Assigned Color Codes

Atmospheric Contaminants to Be Protected Against	Colors Assigned*
Acid gases	White
Hydrocyanic acid gas	White with ½-inch green stripe completely around the canister near the bottom
Chlorine gas	White with ½-inch yellow stripe completely around the canister near the bottom
Organic vapors	Black
Ammonia gas	Green
Acid gases and ammonia gas	Green with ½-inch white stripe completely around the canister near the bottom
Carbon monoxide	Blue
Acid gases and organic vapors	Yellow
Hydrocyanic acid gas and chloropicrin vapor	Yellow with ½-inch blue stripe completely around the canister near the bottom
Acid gases, organic vapors, and ammonia gases	Brown
Radioactive materials, except tritium and noble gases	Purple (magenta)
Particulates (dust, fumes, mist, fog, or smoke) in combination with any of the above gases or vapors	Canister color for contaminant, as designated above, with ½-inch gray stripe completely around the canister near the top
All of the above atmospheric contaminants	Red with ½-inch gray stripe completely around the canister near the top

*Gray shall not be assigned as the main color for a canister designed to remove acids or vapors. Orange shall be used as a complete body or stripe color to represent gases not included in this table. The user will need to refer to the canister label to determine the degree of protection the canister will afford.

Note: Filters may be combined with cartridges to provide additional protection against particulates.

NIOSH has granted approval for complete assemblies of APRs (i.e., a gas mask with a canister or cartridge) against a limited number of specific chemicals, including ammonia, chlorine, formaldehyde, hydrogen chloride, methyl iodide, monomethylamine, and sulfur dioxide. Certain respirators have received special approval against vinyl chloride under carefully controlled conditions, while

TABLE 6.7 Conditions That Exclude the Use of Air-Purifying Respirators

Oxygen deficiency

IDLH concentration

Entry into an unventilated or confined area

Fire fighting

Situation requiring a protection factor greater than 50

Presence of unidentified contaminants

Contaminant concentrations unknown or exceeding designated maximum use concentration

Identified chemicals with inadequate or no warning properties

One or more shock-sensitive airborne contaminants that may be sorbed

Presence of two or more incompatible contaminants on-site that might react in the cartridge or canister to produce a toxic or hazardous condition

Relative humidity greater than 65%

others have been approved for use against carbon monoxide.

Respirators should be used only against those substances for which they have been approved. A sorbent should not be used when there is reason to suspect that it does not provide adequate sorption efficiency against a specific contaminant. In addition, approval testing is performed at a given temperature and over a narrow range of flow rates and relative humidities. Thus, protection may be compromised in nonstandard conditions. The assembly that has been approved by NIOSH to protect against organic vapors is tested against only a single challenge substance, carbon tetrachloride. Therefore, its effectiveness for protecting against other vapors has not been demonstrated.

Chemical sorbent cartridges and canisters have an expiration date. They may be used up to that date as long as they were not opened previously. Once opened, they begin to sorb humidity and air contaminants whether or not they are in use, and their efficiency and service life decrease. Discard cartridges after use.

Where a canister or cartridge is being used against gases or vapors, the device should be used only if the chemical(s) have ''adequate warning properties'' (30 CFR 11, Section 150). The regulation also prohibits the use of APRs against organic vapors that have poor warning properties. A substance is considered to have adequate warning properties when

- its odor, taste, or irritant effects are detectable and persistent at concentrations below the PEL (see Chapter 4)

- its odor or irritation threshold is somewhat above the PEL, but no ceiling limits exist and no serious or irreversible health effects occur within this concentration range

A substance is considered to have poor warning properties when

- its odor or irritation threshold is three or more times the PEL

- its odor or irritation threshold is extremely low in relation to the PEL (i.e., the worker can detect the substance(s) even when the respirator is working properly)

These warning properties are essential to safe use of APRs, since they allow detection of contaminant breakthrough. Warning properties are not foolproof because they rely on human senses, but they do provide some indication of possible sorbent exhaustion, poor face piece fit, and other malfunctions.

Following are some key questions to ask when considering APR use:

- Does the ambient atmosphere contain at least 19.5% oxygen? If not, do not use an APR.

- Is the atmosphere immediately dangerous to life or health? If so, do not use an APR.

- Are the airborne contaminants identified and their concentrations known? If not, do not use an APR.

- Is a cartridge, canister, or filter that has been approved for the chemical or chemicals present in the atmosphere available? It not, do not use an APR.

- Is the ambient chemical concentration below the designated maximum use concentration? If not, do not use an APR.

- Is the ambient air monitored periodically to ensure that workers are not being exposed to dangerous levels of toxic chemicals? It not, institute such monitoring.

- Are there any dangerous conditions that could change without warning? If so, equip workers with escape SCBAs.

- Could the contaminants cause skin or eye irritation or penetrate the skin? If so, use a full face piece on the respirator.

- Does the face piece provide a good seal against the user's face as demonstrated by fit testing (see ''PPE Use'')? Do not use respirators that do not provide a good face piece-to–face seal.

SELECTION OF PPC

For this book, clothing is considered to be any article offering skin and/or body protection. It includes the following:

- fully encapsulating suits
- nonencapsulating suits
- aprons, leggings, and sleeve protectors
- gloves

- fire fighters' protective clothing
- proximity, or approach, garments
- blast and fragmentation suits
- cooling garments
- antiradiation suits

Table 6.8 describes the various types of PPC, details the types of protection they offer, and lists the factors to consider in their selection and use. This table also describes a number of accessories that might be used in conjunction with a PPE ensemble:

- knife
- flashlight or lantern
- personal locator beacon
- personal dosimeter
- two-way radio
- safety belts and lifelines

Each type of PPC has a specific purpose; many, but not all, are designed to protect against chemical exposure. Chemical protective clothing (CPC) is available in a variety of materials that offer a range of resistances to different chemicals. The most appropriate clothing material will depend on the chemicals present. Ideally, the chosen material resists *degradation* (a chemical reaction between the chemical and the material resulting in damage to the material) and *permeation* (the seepage of a chemical substance through the material). No material protects against all chemicals and combinations of chemicals, and no material is an effective barrier to prolonged chemical exposure.

Selection is a complex task and should be performed by personnel with training and experience. PPE selection generally occurs under the following three circumstances: (1) the contaminants are identified or classified; (2) the contaminants are unknown; (3) special hazards exist that require specialized PPE. Under all conditions, PPC is selected by evaluating the performance characteristics of the clothing against the requirements and limitations of the site- and task-specific conditions.

CONTAMINANTS IDENTIFIED OR CLASSIFIED

The selection of CPC depends greatly on the type and physical state of the contaminants. This information is determined during the site characterization (Chapter 4). Once the chemicals have been identified, consult available information sources to determine a suit's resistance to permeation and degradation by the known chemicals and its heat-transfer characteristics as described below. Use this information to narrow down the suitable options. Then, if possible, physically inspect representative garments before purchase and discuss use and performance factors with someone having previous experience with the clothing under consideration.

TABLE 6.8 Protective Clothing and Accessories

Body Part Protected	Type of Clothing	Description	Type of Protection	Use Considerations
Full body	Fully encapsulating suit	One-piece garment; boots and gloves may be integral, attached and replaceable, or separate	Protects against splashes, dust, gases, and vapors	Does not allow body heat to escape; may cause heat stress in wearer, particularly if worn in conjunction with a closed-circuit SCBA; consider the need for a cooling garment; impairs worker mobility, visibility, and communication
	Nonencapsulating suit	Jacket, hood, pants, bib overalls, or one-piece coveralls	Protects against splashes, dust, and other materials but not against gases and vapors; does not protect head or neck	Do not use where gas-tight or pervasive splashing protection is required; may cause heat stress in wearer; tape-seal connections between pant cuffs and boots and between gloves and sleeves
	Aprons, leggings, and sleeve protectors	Fully sleeved and gloved apron; separate coverings for arms and legs; commonly worn over nonencapsulating suit	Provides additional splash protection of chest, forearms, and legs	When worn over nonencapsulating suit, results in less heat stress and more comfort than fully encapsulating suit; useful for sampling, labeling, and analysis operations; use partial coverings only when the possibility of total body contact with contaminants is minimal
	Fire fighters' protective clothing	Boots, gloves, helmet [National Fire Protection Administration (NFPA) Standard 1972–1979)], running or bunker coat, and running or bunker pants	Protects against heat, hot water, and some particles; does not protect against gases and vapors or chemical permeation or degradation; NFPA Standard 1971–1981 specifies that a three-piece garment consisting of an outer shell, an inner liner, and an intermediate vapor barrier have a minimum water penetration of 25 pounds per square inch to prevent the passage of hot water	Decontamination is difficult; should not be worn in areas where protection against chemical splashes or permeation is required
	Proximity garment (approach suit)	One- or two-piece overgarment with boot covers, gloves, and hood of aluminized nylon or cotton fabric; normally worn over other protective clothing such as chemical protective clothing, fire fighters' bunker gear, or flame-retardant coveralls	Protects against brief exposures to radiant heat; does not protect against chemical permeation or degradation; can be custom-manufactured to protect against some chemical contaminants	If the wearer may be exposed to a toxic atmosphere or needs more than 2 or 3 minutes of protection, use auxiliary cooling and an SCBA

	Equipment	Description	Notes/Limitations
	Blast and fragmentation suit	Blast and fragmentation vests and clothing, bomb blankets, and bomb carriers	Provides some protection against very small detonations (the equivalent of less than 2 ounces of TNT at distances of 20 feet or more); bomb blankets and baskets can help redirect a blast	In confined spaces, "safe" distances may need to be doubled
	Antiradiation suit	Protects against alpha and beta particles; does not protect against gamma radiation	Designed to prevent ingestion, inhalation, or skin contamination; if radiation is detected on-site, consult an experienced radiation expert and evacuate personnel until radiation levels can be reduced to background level
	Flame/fire-retardant coveralls	Normally worn as an undergarment	Provide protection against flash fires
	Flotation gear	Life jackets or work vests; commonly worn underneath chemical protective clothing to prevent degradation	Adds 15.5 to 25 pounds of buoyancy to personnel working in or around water	Adds bulk and restricts mobility; must meet U.S. Coast Guard (USCG) standards 46 CFR 160
	Cooling garment	One of three methods: 1. A pump circulates cool, dry air throughout the suit or portions of it via an air line. Cooling may be enhanced by use of a vortex cooler, refrigeration coils, or a heat exchanger. 2. Packets of ice are inserted into the pockets of a jacket or vest. 3. A pump circulates chilled water from a water/ice reservoir and through circulating tubes, which cover part of the body (generally the upper torso).	Removes excess heat generated by worker activity, the equipment, or the environment	1. Requires 10 to 20 cubic feet of respirable air per minute, so it is often uneconomical for use at a waste site 2. Poses ice storage and recharge problems 3. Poses ice storage problems; if pump is battery operated, adds bulk and weight
Head	Safety helmet (hard hat)	Must meet ANSI Z89.1-1969 specifications for protection	Mandatory for operations involving fire fighting, rescue, or other emergency situations; protects the head from blows	Proper adjustment of helmet is necessary to prevent it from falling off; chin straps are available but should be used only if they do not interfere with respirator-face seal
	Helmet liner		Insulates against cold; does not protect against chemical splashes

(continued)

TABLE 6.8 (Continued)

Body Part Protected	Type of Clothing	Description	Type of Protection	Use Considerations
Head (continued)	Hood	Commonly worn with a helmet, which provides protection against blows	Protects against chemical splashes, particulates, and rain	
	Protective hair covering	Particularly important for workers with long hair	Protects against chemical contamination of hair; prevents the entanglement of hair in machinery or equipment; prevents hair from interfering with the visibility or functioning of respiratory devices	
Eyes and face	Full-face respirator	Covers face from hairline to below the chin	Protects against vaporized chemicals, splashes, and dust	Face piece may fog, impairing vision, particularly in cold weather; use a nose cup, anti-fogging agents, and/or exhaust funnels
	Half mask	Covers face from below the chin to over the nose	Provides no eye protection and only partial face protection	All eye and face protection must meet OSHA Standard 29 CFR 1910
	Quarter mask	Covers face from top of chin to over the nose	Provides no eye protection and only partial face protection	Often difficult to maintain adequate respirator-face seal; not recommended for hazardous waste sites
	Face shield	Commonly worn with full-face respirator or goggles; extends from helmet to chin	Protects against chemical splashes; does not protect adequately against projectiles	Face shields and hoods must be suitably supported to prevent equipment from shifting and exposing portions of the face during work activity and to prevent equipment from obscuring visibility
	Splash hood		Provides limited protection against chemical splashes from sides; does not protect adequately against projectiles	
	Safety glasses	Glasses with safety lenses to protect from projectiles; frequently include guards on sides	Protect against large particles and projectiles	If lasers are used to survey a site, workers should wear protective lenses
	Goggles	Variety of devices providing safety lenses or eye plates with total enclosure protection around the eyes	Depending on construction, protect against vaporized chemicals, splashes, large particles, and projectiles (if constructed with impact-resistant lenses)	Selection and use should not interfere with respirator-face seal

Body part	Type	Description	Purpose	Comments
	Sweatbands		Prevents sweat-induced eye irritation	
Ears	Earplugs and earmuffs		Protect against physiological damage and psychological disturbance; required when the time-weighted average noise level is above 85 dBA; provides partial protection from entry of splashes and projectiles	Can interfere with communication; use of earplugs should be carefully reviewed by a health and safety professional because contaminants could be introduced into the ear
	Headphones	Radio headset with throat microphone	Provide some hearing, splash, and projectile protection while enabling communications	Highly desirable, particularly if emergency conditions arise
Hands and arms	Gloves and sleeves	May be integral, attached, or separate from other protective clothing	Protect hands and forearms from chemical contact	Wear jacket cuffs over glove cuffs to prevent liquid from entering gloves; tape-seal gloves to sleeves to provide additional protection
	Overgloves		Provide supplemental protection to the wearer and protect more expensive undergarments from abrasions, tears, and contamination	Decrease manual dexterity by adding bulk around fingers
	Disposable gloves		Should be used whenever possible to reduce decontamination needs	Extremely limited for prolonged contact due to permeability; should use protective outer-glove
Feet	Safety boots	Constructed of chemical-resistant material	Protect feet from contact with chemicals	All boots must meet the specifications of ANSI Z41–1981 and should provide good traction
		Constructed with some steel materials (e.g., toes, shanks)	Protect feet from compression, crushing, or puncture by falling, moving, or sharp objects	May require use of overboot to provide adequate chemical protection
		Constructed of nonconductive, spark-resistant material or coating	Protect wearer against electrical hazards and prevent ignition of combustible gases or vapors	
	Disposable shoe or boot covers	Made of variety of materials; slip over shoe or boot	Protect more expensive safety boots from contamination; protect feet from contact with chemicals	Covers may be disposed of after use, facilitating decontamination
General Accessories	Knife		Allows a person in a fully encapsulating suit to cut his or her way out of the suit in the event of an emergency or equipment failure	Caution is dictated when carrying a knife, as wearer could fall and puncture the body of the suit; user also must consider potentially lethal exposures upon breach of protective suit

(continued)

TABLE 6.8 (Continued)

Body Part Protected	Type of Clothing	Description	Type of Protection	Use Considerations
General (continued)	Accessories (continued) Flashlight or lantern		Enhances visibility in buildings, enclosed spaces, and the dark	Must be intrinsically safe for use in combustible atmospheres; seal the flashlight in a plastic bag to facilitate decontamination; only electrical equipment approved as intrinsically safe or approved for the class and group of hazard as defined in Article 500 of the National Electrical Code may be used
	Personal locator beacon	Operated by sound, radio, or light	Enables emergency personnel to locate victim	Must be intrinsically safe if used in potentially combustible atmospheres
	Personal dosimeter		Can detect oxygen-deficient atmospheres; measures worker exposure to ionizing radiation and certain chemicals	To estimate actual body exposure, place dosimeter inside fully encapsulating suit
	Two-way radio		Enables fieldworkers to communicate with personnel in the support zone	Must be intrinsically safe if used in potentially combustible atmospheres
	Safety belts and lifelines		Enable personnel to work in elevated areas or enter confined areas and prevent falls; belts may be used to carry tools and equipment	Must be constructed of spark-free hardware and chemical-resistant materials to provide proper protection; must meet OSHA Standard 29 CFR 1910; permeation of materials may require disposal of equipment after use if adequate decontamination cannot be achieved

Permeation and Degradation Appendix H provides clothing material recommendations for approximately 300 chemicals based on an evaluation of breakthrough and permeation data from vendor literature, raw material suppliers, and independent tests. Charts indicating the resistance of various clothing materials to permeation and degradation also are available from manufacturers and other sources. One good reference is *Guidelines for the Selection of Chemical Protective Clothing* (ACGIH, 1985).

When reviewing vendor literature, be aware that the data provided are of limited value for several reasons. For example, the quality of vendor test methods ranges from state-of-the-art to nonexistent; vendors often rely on the raw material manufacturers for data rather than conducting their own tests; and the data may not be updated. In no way do vendor data address the wide variety of uses and challenges to which CPC may be subjected. Most vendors strongly emphasize this point in the descriptive text that accompanies their data tables. Thus, CPC vendor recommendation tables provide only guidance in the selection of CPC—that is, they provide

- a place to begin the selection process
- a means for selecting specific CPC for further evaluation
- a means for eliminating specific CPC from consideration

Also bear in mind that the rate of permeation is a function of several factors, including material type and thickness, manufacturing method, concentration of the hazardous substance, temperature, pressure, humidity (to some extent), solubility of the chemical in the suit's material, and the diffusion coefficient of the permeating chemical in the material. The following generalizations are applicable:

- *Temperature.* Permeation rates increase and breakthrough times decrease with increasing temperatures. The degree of reduction in protective performance depends on the chemical and the material.
- *Thickness.* For a given clothing material, permeation is inversely proportional to thickness (i.e., doubling the thickness theoretically halves the permeation rate). Breakthrough time (the time from initial exposure until the hazardous material is detectable on the inside of the suit) is directly proportional to the square of the thickness (i.e., doubling the thickness theoretically quadruples the breakthrough time).

The permeation rate is a direct function of the solubility of the chemical in the material, but you must exercise caution in interpreting solubility data. Low solubilities do not necessarily correspond to low permeation rates. The diffusion coefficient is also a direct factor in the permeation rate. Gases, for example, have low solubilities but high diffusion coefficients and may permeate CPC at rates several times greater than a liquid with moderate to high solubility. Information on solubility and diffusion coefficients may be found in vendor literature or literature on permeation testing. Table 6.9 describes CPC compatibility with selected chemicals. Table 6.10 is a general description of CPC durability.

When trying to determine suit penetration or permeation rates, multicomponent liquids pose a problem, due to lack of experience in and information about permeation and degradation. Mixtures of chemicals can be significantly more aggressive toward plastics and rubbers than any one of the components alone. Even small amounts of a rapidly permeating chemical may provide a pathway that accelerates the permeation of other chemicals. Research is being performed to develop data on these effects. In the meantime, immersion and permeation testing are recommended as the best means of selecting CPC for multicomponent solutions.

Heat-Transfer Characteristics The heat-transfer characteristics of CPC are important factors in selection. Since most CPC is virtually impermeable to moisture, evaporative cooling does not occur. The thermal insulation (CLO) value of CPC influences heat loss through means other than evaporation. A CLO unit is roughly equivalent to the amount of insulation provided by clothing a person usually wears at room temperature. The larger the CLO value, the greater the insulating properties of the garment and, consequently, the lower the heat transfer. Given other equivalent protective properties, select the lowest CLO value in hot environments or for high work rates. Unfortunately, at present, CLO values are not commonly available for CPC.

CONTAMINANTS UNKNOWN

Unfortunately, contaminants are often not identified or quantified until after CPC selection and use. In lieu of knowledgeable selection, one or more suits may be preselected from those offering the widest range of protection against the chemicals one expects to encounter.

After selecting such a suit or suits, you must rule out the presence of any hazardous materials known to chemically attack (degrade) or permeate the suit(s). This is usually accomplished by using gas detector tubes to test for those chemicals known to degrade the specific protective material selected. Since permeation data are limited at present, it may not always be possible to completely rule out chemicals that can permeate.

The presence of degrading and/or permeating chemicals does not necessarily preclude use of the suit. For example, butyl rubber rapidly degrades if exposed to nitric acid for more than 20 minutes, but a butyl rubber suit may be used in the presence of nitric acid to effect a 10-minute rescue. However, the suit must be retired after this mission.

TABLE 6.9 Protective Clothing Material Compatibility with Selected Chemical Hazards*

Chemical Hazard	Natural Rubber	Neoprene	Polyvinyl Alcohol	Polyvinyl Chloride	Nitrile
Acetaldehyde	G	G	P	F	F
Acetic acid	E	E	F	G	E
Acetic anhydride	G	G	P	F	G
Acetone	E	G	F	P	P
Acetonitrile	G	G	G	G	G
Acrylonitrile	F	G	F	F	F
Alcohols	G	F	P	F	E
Ammonia (100%)	G	G	P	E	—
Amyl acetate	F	G	G	P	G
Aniline	F	G	F	G	P
Battery acid	G	E	P	E	E
Benzaldehyde	F	P	E	P	P
Benzene	P	P	E	P	P
Benzene sulfinic	P	G	P	E	—
Bromine	G	G	E	G	G
Butyric acid	F	G	F	G	—
Cadmium cyanide	G	E	G	E	G
Carbolic acid	F	E	F	E	E
Carbon tetrachloride	P	P	E	G	P
Chlorine	F	F	E	G	G
Chloroform	P	P	E	P	P
Chlorosulfonic acid	P	P	F	G	F
Chromic acid	P	F	P	G	P
Cresol	G	G	F	F	P
Cyanide solution	G	G	P	F	G
Cyclohexane	P	G	E	P	P
Dimethyl formamide	E	G	P	P	P
Dioctyl phthalate	F	G	E	P	P
Dioxane	F	F	P	P	P
Ethanolamine	F	G	P	E	P
Ethers	G	E	E	P	P
Ethyl acetate	F	G	F	P	P
Ethyl alcohol	E	E	P	G	E
Ethylene dichloride	P	P	E	P	P
Ethylene glycol	E	E	G	E	E
Formaldehyde	E	E	P	E	E
Formic acid	G	E	P	E	F
Frem	P	F	E	P	F
Fuel oil	P	G	E	G	P
Furfural	G	G	F	F	P
Gasoline (unleaded)	P	P	F	P	E
Hexane	P	G	E	P	E
Hydrobromic acid (40%)	G	E	P	E	E
Hydrochloric acid (conc)	G	E	P	E	E
Hydrofluoric acid (30%)	G	E	P	G	E
Hydrogen peroxide (30%)	G	E	P	E	E
Isopropyl alcohol	E	E	P	E	E
Kerosene	P	E	E	G	E
Mercury	G	G	P	E	E
Methyl alcohol	E	E	P	G	E
Methylene chloride	P	P	G	P	P
Methyl isobutyl ketone	F	P	F	P	P
Methyl methacrylate	P	P	G	P	P
Naphtha	P	P	E	F	E
Nitric acid (10%)	P	P	P	G	E
Nitric acid (70%)	P	G	P	P	P
Nitric acid (fuming)	P	P	P	P	P

TABLE 6.9 (Continued)

Chemical Hazard	Natural Rubber	Neoprene	Polyvinyl Alcohol	Polyvinyl Chloride	Nitrile
Nitrobenzene	P	P	E	P	P
Nitromethane	F	G	E	P	P
Oleum	P	P	P	F	P
Parathion	P	P	G	F	P
Pentane	P	E	E	P	E
Perchloric acid	F	E	P	E	E
Perchloroethylene	P	F	E	F	P
Phenol	G	G	P	G	P
Phosphoric acid	G	E	P	E	E
Pickling baths	G	G	P	G	E
Potassium hydroxide	E	E	P	E	E
Propylene oxide	F	P	G	P	P
Pyridine	F	G	G	E	F
Sodium hydroxide	E	E	P	E	E
Sodium peroxide	F	G	F	E	G
Stoddard solvent	P	G	E	P	P
Styrene	P	P	G	P	P
Sulfuric acid	P	P	P	G	P
Tannic acid	E	E	F	E	E
Tetrachlorethylene	P	P	E	P	P
Tetrahydrofuran	P	P	F	P	P
Toluene	P	P	G	P	P
Toluene di-isocyanate	F	P	G	P	P
Trichloroethylene	P	P	E	P	P
Xylene	P	P	E	P	G

*P = poor; F = fair; G = good; E = excellent.
Source: Dave Streng, *A Worker Bulletin—Occupational Safety and Health at Hazardous Waste Sites* (National Institute for Occupational Safety and Health, February 1982).

TABLE 6.10 Protective Clothing Material Wearing and Durability Characteristics*

Chemical Hazard	Natural Rubber	Neoprene	Polyvinyl Alcohol	Polyvinyl Chloride	Nitrile
Tear resistance	G	G	F	F	G
Abrasion resistance	G	G	G	G	E
Heat resistance	F	E	P	P	E
Flame resistance	P	G	P	P	P
Elongation	E	E	F	F	G

*P = poor; F = fair; G = good; E = excellent.
Note: See Appendix H for chemical protective clothing recommendations by chemical class.
Source: Dave Streng, *A Worker Bulletin—Occupational Safety and Health at Hazardous Waste Sites* (National Institute for Occupational Safety and Health, February 1982).

OTHER CONSIDERATIONS

In addition to permeation, degradation, and heat resistance, several other factors must be considered during CPC selection. These affect not only chemical resistance but also the worker's ability to perform the required task. The following list summarizes these considerations.

• *Chemical resistance.* Does the clothing have design or construction imperfections that would allow hazardous materials to penetrate (e.g., stitched seams, buttonholes, porous zippers, pinholes)?

• *Durability*
Does the material have sufficient strength to withstand the physical stress of the task(s) at hand?
Will the material resist tears, punctures, and abrasions?
Will the material withstand repeated use after contamination/decontamination?

- *Flexibility.* Will workers be able to perform their assigned tasks? (This is particularly important when evaluating gloves.)

- *Temperature effects.* Will the material maintain its protective integrity and flexibility under hot and cold extremes?

- *Ease of decontamination*
  Are decontamination procedures available on-site?
  Will the material pose any decontamination problems?
  Should disposable clothing be used?

- *Compatibility with other equipment.* Does the clothing preclude the use of another piece of protective equipment (e.g., suits that preclude hard hat use in a hard hat area)?

SELECTION OF PPC FOR SPECIAL CONDITIONS

Fire, explosion, heat, and radiation are considered special conditions requiring specialized PPE. Table 6.7 lists protection and use considerations for specialized PPE. General questions for determination of specialized needs include the following:

- *Fire fighting/explosion*
  Is structural fire fighting involved?
  Is radiant heat exposure a problem?
  Is there a potential for a flash fire (e.g., vapor pockets or rapid releases of flammable compounds)?
  Is there a potential for explosions, blasts, and fragmentation (e.g., compressed gases, confined pockets of explosive vapors or dust, or shock-sensitive compounds)?

- *Liquid impoundments or open bulk liquid containers*
  Is there a potential for workers to fall into or be immersed in liquids?
  Does the depth of the liquid, the slope of the container, or the type of container walls prevent personnel from easily getting out if they fall in?
  Are the materials likely to knock down or immobilize workers who may be immersed?
  Are there any irreversible injuries or extreme chronic risks associated with immersion-type exposures?

- *Radiation.* If radiation exposures are identified or anticipated, consult a qualified health physicist.

Specialized PPE, such as fire fighters' running or bunker coats and pants, proximity suits for radiant heat, blast and fragmentation suits, and flotation devices or safety lines for working around open bulk liquid containers or impoundments, are not designed to protect against other chemical exposures or general safety hazards. Selection of a PPE ensemble requires consideration of all available information about site condition (see Chapter 4, Site Characterization), work practices (see Chapter 7, Site Control and Work Practices), and decontamination requirements (see Chapter 8, Decontamination).

SELECTION OF A PPE ENSEMBLE
LEVEL OF PROTECTION

Up to this point, only the individual components of a PPE ensemble have been described. The final goal, however, is to assemble the necessary components into a full ensemble that both protects the worker from the site-specific hazards and minimizes the hazards and drawbacks of the PPE ensemble itself.

In developing an ensemble, note that NIOSH approves a fully encapsulated suit with either (1) an ALR, a combination ALR-SCBA, or an SCBA worn inside the suit or (2) a respirable air supply providing air directly to the fully encapsulating suit through an air inlet valve. Currently no NIOSH approvals are given for ensembles in which the respirator face piece is an integral part of the suit.

The clothing worn inside a PPE ensemble will depend on the ambient temperature, but the minimum clothing, even in hot weather, should be loose-fitting light cotton underwear. Such garments absorb perspiration and act as a wick for evaporation, thus aiding cooling. To protect the skin from contact with hot inner suit surfaces and reduce the possibility of burns in hot weather, loose-fitting, light cotton long-sleeve shirts may be needed. Under extreme temperature conditions, activities requiring use of PPE should be postponed if possible.

Table 6.11 lists the ensemble components of the widely used EPA levels of protection (Levels A, B, C, and D). In the EPA approach, a specific respirator is designated for each protection level, with the designation based primarily on airborne-contaminant characterization with total organic vapor detectors. Typical use of the EPA ensemble levels is described in Table 6.12.

Tables 6.13 through 6.16 provide additional guidance for adaptation of the EPA levels to site-specific PPE ensembles. The levels used in these tables are designated as Levels 1, 2, 3, and 4 to distinguish them from the ensembles as designated in the referenced EPA levels. The EPA designations are designed for general applications in a variety of situations and include specifications that may or may not be applicable when considering site-specific conditions. Modification of PPE ensembles must be based on data that provide sufficient characterizations of the hazards and conditions to which workers may be exposed. The PPE listed for Levels 2, 3, and 4 does not specify respiratory equipment. The selection of respiratory equipment is discussed earlier in this chapter.

Consider these lists as a starting point for ensemble creation. To provide the most appropriate level of protection, the ensembles should be tailored to the specific situation.

If work is being conducted at a very dirty (highly contaminated) site or the potential for contamination is high, it may be advisable to wear a disposable covering over the suit. Tyvek coveralls or PVC splash suits can be used. A slit must be made in the back of these suits to fit around the bulge of the encapsulating suit and SCBA.

TABLE 6.11 EPA Protective Ensembles

Level of Protection	Equipment	Protection Provided	Use Criteria[a]	Limiting Criteria[b]
Level A	Required Pressure-demand SCBA Fully encapsulating chemical-resistant suit Inner and outer chemical-resistant gloves Chemical-resistant boots with steel toe and shank Two-way radio communications Optional Coveralls Long cotton underwear Disposable protective suit, gloves, and boots	The highest level of respiratory, skin, and eye protection	Total atmospheric concentration of unidentified gases or vapors exceeds 500 ppm, as measured by some types of total organic vapor analyzer Known or potential presence of extremely hazardous chemical Known or potential presence of toxic chemical that can injure or be absorbed through the skin Oxygen-deficient atmosphere IDLH atmosphere Operations that must be conducted in confined, poorly ventilated areas until the absence of hazards requiring Level A protection is demonstrated	
Level B	Required Pressure-demand SCBA Chemical-resistant clothing (overalls and long-sleeve jacket; coveralls; hooded one- or two-piece chemical splash suit; disposable chemical-resistant coveralls) Inner and outer chemical-resistant gloves Chemical-resistant boots with steel toe and shank Hard hat Two-way radio communications Optional Coveralls Disposable outer boots Face shield	The same level of respiratory protection but less skin protection than Level A; the minimum level recommended for initial site entries until the hazards have been further identified	Oxygen-deficient, nauseating, or irritating atmosphere where the exact type of chemical present is unknown Concentrations of unidentified gases or vapors exceed 5 ppm (but are less than 500 ppm) as measured by some type of total organic vapor analyzer	Total atmospheric concentration of unidentified gases or vapors does not exceed 500 ppm Gases and vapors are not suspected of containing high levels of chemicals that can injure or be absorbed through the skin

(continued)

TABLE 6.11 (Continued)

Level of Protection	Equipment	Protection Provided	Use Criteria[a]	Limiting Criteria[b]
Level C	Required Full-face, air-purifying canister-equipped respirator Chemical-resistant clothing Chemical-resistant gloves Chemical-resistant boots with steel toe and shank Hard hat	The same level of skin protection as Level B but a lower level of respiratory protection	Danger of splashing	Atmosphere contains at least 19.5% oxygen The types of air contaminants have been identified, concentrations measured, and canister is available that can remove the contaminant Airborne contaminants possess properties that warn of exposure
Level D	Required Coveralls Chemical-resistant boots with steel toe and shank Five-minute escape mask Safety glasses or chemical splash goggles Optional Gloves Outer chemical-resistant disposable boots (required if leather boots are worn) Hard hat Face shield	No respiratory protection (except for emergency escape); minimal skin protection	Does not provide adequate protection for use in contamination exclusion zones	Atmosphere contains at least 19.5% oxygen Atmosphere contains no known hazard Work functions preclude splashes, immersion, or the potential for unexpected inhalation of any chemicals Only boots may be contaminated

[a]Use criteria are those conditions that a particular level of protection is effective against.
[b]Limiting criteria are those conditions that must be present if a particular level of protection can be used.

TABLE 6.12 Use of EPA Ensembles

Level A	Highest level of respiratory, skin, eye, and mucous membrane protection. Generally used where extremely hazardous substances are known to be present in high atmospheric concentrations, Level B splash gear does not offer adequate protection against any dermally active substances present, or materials and concentrations are unknown.
Level B	Highest level of respiratory protection but a lesser level of skin and eye protection. Generally used in situations where the chemical is known, the atmosphere is oxygen deficient, no IDLH concentrations of substances that pose a respiratory hazard are present, or dermal contact with a hazardous substance is unlikely. This level of protection is normally the minimum used for initial response or reconnaissance, unless the respiratory hazards allow a lower level of respiratory protection than an SCBA.
Level C	Generally comprising splash gear and an air-purifying respirator. Provides adequate protection when the type of airborne substance is known, its concentration is measured, the criteria for using air-purifying respirators are met, and skin exposure is unlikely. Use of this level of protection requires continuing measurement of air contaminants to ensure that IDLH concentrations do not exist and that the concentrations of the contaminants present do not exceed the service limits of the respirator.
Level D	Primarily a work uniform. Should not be worn in any hazardous environment. It is used when there is no indication of hazardous conditions and the work function precludes contact with any hazardous substance.

TABLE 6.13 Level 1 Personal Protective Clothing and Equipment

Description	Protective clothing and equipment designed to provide maximum protection and prevent contact of skin and body with hazardous substances
Conditions	Site conditions and work function involve high potential splash, immersion, or exposure to unexpected vapors, gases, or particulates
	Direct skin and eye contact with hazardous compounds or air contaminants may cause severe damage and/or irreversible effects
	Exposures above IDLH and/or TLV concentrations may be encountered
	Compounds and/or skin effects suspected or not known
Emergency entry or initial entry— contaminants unknown	Suspect highly toxic compounds on-site
	Use viton or butyl rubber
	Rule out unacceptable protective materials
	Reevaluate protective materials and level of protection as contaminants are identified/classified
Personal protective clothing and equipment[a]	Pressure-demand self-contained breathing apparatus
	Fully encapsulating chemical-resistant suit (including boots and gloves)
	Coveralls (outer)—chemical-resistant, disposable[b]
	Light, loose-fitting cotton underwear[b]

	Gloves
	    Chemical-resistant outer gloves (intrinsic to suit)
	    Chemical-resistant outer gloves (worn over glove attached to suit)[b]
	    Chemical-resistant inner gloves
	    Cloth or leather work gloves (disposed of after use)

	Boots—chemical-resistant, steel toe and shank (steel metatarsal[b]); work over or under fully encapsulating suit (depending on suit construction)
	Boot covers—chemical-resistant (disposable)[b]
	Hard hat (under suit)
	Two-way radio communication (intrinsically safe)

[a]Adapted from EPA recommendations for Level A personal protection
[b]Optional

TABLE 6.14 Level 2 Personal Protective Clothing and Equipment

Description	Protective clothing and equipment designed to minimize or prevent contact of skin and body surfaces with hazardous substances
Conditions	Direct skin and eye contact with hazardous compounds or air contaminants may cause severe damage and/or irreversible effects
	Work function precludes exposures of unprotected areas of the face above TLV concentrations
	Concentrations of sun-absorbing compounds less than TLV
Initial entry—contaminants unknown	Off-site investigations and observations do not indicate highly toxic waste present
	Use viton or butyl rubber
	Rule out inappropriate protective material
	Reevaluate protective material and level of protection as wastes are identified/classified
Personal protective clothing and equipment[a]	Protective creams and gels[b]
	Chemical-resistant clothing options
	Hooded chemical-resistant coveralls (disposable)[b]
	One- or two-piece chemical splash suit
	Hooded chemical-resistant rain suit
	Chemical-resistant leggings and/or sleeve protectors
	Chemical-resistant apron[b]
	Coveralls (outer)[b]—chemical-resistant, disposable[b]
	Light, loose-fitting cotton underwear
	Gloves
	Chemical-resistant outer gloves (extended cuff)[b]
	Chemical-resistant inner gloves
	Cloth or leather work gloves (disposed of after use)[b]
	Boots—chemical-resistant, steel toe and shank (steel metatarsal)[b]
	Boot covers—chemical-resistant (disposable)[b]
	Hard hat (face shield)[b]
	Two-way radio communication (intrinsically safe)

[a]Adapted from EPA recommendations for Level B personal protection
[b]Optional

TABLE 6.15 Level 3 Personal Protective Clothing and Equipment

Description	Protective clothing and equipment designed to minimize contact with hazardous substances
Conditions	Limited direct skin and eye contact with hazardous compounds or air contaminants will not result in severe damage and/or irreversible effects
	Work function involves potential for only minor splashes and excludes total body splashes or immersion
	Concentrations or skin-absorbing compounds less than TLV
Personal protective equipment[a]	Chemical-resistant clothing
	Hooded chemical-resistant coveralls (disposable)[b]
	Hooded two-piece chemical splash suit
	Chemical-resistant hood and apron[c]
	Leggings and/or sleeve protectors[c]
	Coveralls (outer)[b]
	Gloves
	Chemical-resistant outer gloves
	Chemical-resistant inner gloves
	Cloth or leather work gloves (disposed of after use)

TABLE 6.15 (Continued)

Personal protective equipment (continued)	Boots—chemical-resistant, steel toe and shank (steel metatarsal)[b] Boot covers—chemical-resistant (disposable)[b] Hard hat Face shield and/or chemical splash goggles (optional if respirator includes full face piece)[c] Escape mask[b] Two-way radio communication (intrinsically safe)

[a]Adapted from EPA recommendations for Level C personal protection
[b]Optional
[c]Protection from chemical spills or splashes; generally not used for protection from air contaminants

TABLE 6.16 Level 4 Personal Protective Clothing and Equipment

Description	Protective clothing and equipment designed to protect worker from common workplace hazards and minimize contact with contaminated materials
Conditions	Compounds of concern do not have adverse skin and eye effects No hazardous air pollutants measured or anticipated Work function precludes splashes, immersion, or potential for unexpected respiratory hazards No exposures anticipated above TLV levels
Personal protective equipment[a]	Coveralls (disposable)[b] Gloves—cloth, leather, and/or waterproof work gloves Boots/shoes—leather or chemical-resistant, steel toe and shank (also metatarsal if protection of top of foot required) Boots (outer)—chemical-resistant (disposable) Safety glasses or chemical splash goggles Hard hat (face shield)[b] Escape mask (air-supplied)[b] Rain suit (for precipitation)[b]

[a]Adapted from EPA recommendations for Level D personal protection
[b]Optional

The type of equipment used and the overall level of protection should be reevaluated periodically as the amount of information about the site increases and workers are required to perform different tasks. Personnel should be able to upgrade their level of protection if, after discussion with the SSO, they feel it is necessary and the SSO approves. Any downgrading of protection also must be approved by the SSO.

Reasons to upgrade include the following:

- known or suspected presence of dermal chemicals
- occurrence or likely occurrence of detonation or gas emission
- change in work task that will increase contact or potential contact with hazardous materials
- request of individual performing task

Reasons to downgrade include the following:

- new information indicating the situation is less hazardous than originally thought
- change in site conditions that decreases the hazard
- change in work task that will reduce contact with hazardous materials

PPE USE

PPE can offer the highest degree of protection only if it is used properly. This section covers the following aspects of PPE use:

- training
- work mission duration
- personal use factors

- donning an ensemble
- respirator fit testing
- in-use monitoring
- doffing an ensemble
- clothing reuse
- inspection
- storage
- maintenance

Decontamination is covered in Chapter 8. Inadequate attention to any of these areas could result in clothing and equipment that is inappropriate or unsafe for use.

TRAINING

Training in PPE use is required by OSHA regulation 29 CFR 1910. Training

- allows the user to become familiar with the equipment in a nonhazardous situation
- instills user confidence in the equipment
- increases the efficiency of work operations performed in PPE
- may increase the protective efficiency of PPE use
- reduces maintenance expenses

Training must take place at least annually and should be completed prior to actual PPE use in a hazardous environment. At a minimum, the training portion of the PPE program should delineate the user's responsibilities and explain the following (using both classroom and field training when necessary):

- the operation of the selected PPE, including its capabilities and limitations
- the nature of the hazards and the consequences of not using PPE
- the human factors influencing PPE performance
- instruction in inspecting, donning, doffing, checking, fitting, and use of PPE
- individualized quantitative PPE fitting
- use of PPE in normal air for a prolonged period and wearing PPE in a test atmosphere to evaluate its effectiveness
- the user's responsibility for decontamination, cleaning, maintenance, and repair of PPE
- how to recognize emergencies
- emergency procedures and self-rescue in the event of PPE failure
- the buddy system

- the site emergency plan and the individual's responsibilities and duties in an emergency

Since PPE use often causes discomfort and inconvenience, there is a natural resistance to wearing it. The major thrust of training must be to make the user aware of the need for PPE and to instill the motivation to wear and maintain it.

WORK MISSION DURATION

Before workers actually begin work in their PPE ensembles, the anticipated duration of the work mission should be established. Several factors limit mission length. These include the following:

- oxygen/air supply consumption
- suit/ensemble penetration
- coolant supply
- ambient temperature

Oxygen/Air Supply Consumption The duration of the air supply must be considered before planning any SCBA-assisted work activity. The anticipated operating time of an SCBA is clearly indicated on the breathing apparatus. Devices that operate for up to 4 hours are available. The designated operating time is based on a moderate work rate—some lifting, carrying, and/or heavy equipment operation. In actual operation, however, several factors can reduce the rated operating time by up to 50%. When planning an SCBA-assisted work mission, consider the following variables and adjust work actions and durations accordingly:

- *Work rate.* The actual in-use duration of SCBAs may be reduced by 30% to 50% during strenuous work (e.g., drum handling, major lifting, or any task requiring repetitive speed of motion).
- *Fitness.* Well-conditioned individuals generally use oxygen more efficiently and can extract more oxygen from a given volume of air (particularly when performing strenuous tasks) than unfit individuals, thereby increasing unit duration.
- *Body size.* Larger individuals generally consume air at a higher rate than smaller individuals, thereby decreasing unit duration.
- *Breathing patterns.* Quick, shallow, or irregular breaths use air more rapidly than deep, regularly spaced breaths. Heat-induced anxiety and lack of acclimatization may induce hyperventilation, thereby decreasing unit duration.

Suit/Ensemble Penetration The possibility of ensemble penetration during the work mission is always a matter of concern. Some possible causes of ensemble penetration follow:

- suit valve leakage, particularly under excessively hot or cold temperatures

- suit fastener leakage if the suit is not properly maintained or if the fastener becomes loose or brittle at excessively hot or cold temperatures
- exhalation valve leakage at excessively hot or cold temperatures

Also, when considering mission duration, remember that all CPC has limitations:

- No one clothing material is an effective barrier to all chemicals or all combinations of chemicals.
- No material is an effective barrier to prolonged chemical exposure.
- Commercially available clothing provides no more than 1 hour's protection against certain chemicals and chemical combinations.
- Chemical resistance estimates often refer only to resistance to degradation. Check to find out whether resistance estimates being used include resistance to permeation.

Coolant Supply The amount of coolant carried by the wearer significantly influences mission duration. If a coolant (ice or chilled air; see Table 6.8) is necessary to keep workers at a comfortable temperature, the coolant supply will affect mission duration. Adequate coolant should be provided to enable workers to complete the mission in comfort. Remote coolant supplies, such as supplied-air suits or two-stage refrigeration systems, can extend duration at the expense of a trailing umbilical cord and increased ensemble complexity.

Ambient Temperature The ambient temperature has a major influence on mission duration, as it affects both the worker and the protective integrity of the ensemble. Heat stress, which can occur even in relatively moderate temperatures, is the greatest immediate danger to an ensemble-encapsulated worker. Methods to monitor for and prevent heat stress are discussed in the final section of this chapter. Hot and cold ambient temperatures also affect the following:

- valve operation on suits and/or respirators
- the durability and flexibility of suit materials
- the integrity of suit fasteners
- the breakthrough and permeation rates of chemicals
- the carbon dioxide scrubber efficiency on closed-circuit SCBAs
- APR efficiency
- airborne-contaminant concentration levels, perhaps compromising protection factors on current-use respirators

All these factors may decrease the duration of protection provided by a given piece of clothing or respiratory equipment.

PERSONAL USE FACTORS

Certain personal features may jeopardize worker safety during equipment use. Prohibit or take precautionary measures as indicated.

Facial hair and long head hair interfere with respirator fit and visibility. Prohibit any facial hair that passes between the face and the sealing surface of the respirator. Even a few days' growth of stubble will allow excessive contaminant penetration. OSHA regulations require removal of facial hair that interferes with respirator fit tests. Long head hair must be effectively contained within protective hair coverings.

Punctured eardrums allow contaminants to enter the respiratory tract. Persons with punctured eardrums should not be allowed on-site or should be required to wear only ensembles that enclose the entire head.

Eyeglasses with conventional earpiece bars may interfere with a proper fit when using a full-face respirator. Install a spectacle kit in the face masks of workers requiring vision correction.

Contact lenses may absorb vapors and trap contaminants and/or particulates between the lens and the eye, causing irritation, damage, absorption, and an urge to remove the respirator. Prohibit contact lens use on-site. Install a spectacle kit in the face masks of workers requiring vision correction.

Facial features such as scars, hollow temples, very prominent cheekbones, deep skin creases, dentures, or missing teeth may interfere with a tight respirator fit. Full dentures should be retained when wearing a respirator; partial dentures may or may not have to be removed, depending on the possibility of swallowing them. Quarter masks may not fit properly if the wearer has full lower dentures.

Prohibit gum and tobacco chewing during respirator use.

DONNING AN ENSEMBLE

The wearer must understand all aspects of ensemble operation and its limitations. This is especially important for full-body encapsulating ensembles where misuse could result in suffocation.

The donning of a fully encapsulating ensemble is a relatively simple task, but a routine must be established and practiced frequently. It is difficult for the user to don an encapsulating suit alone. Doing so also increases the possibility of suit damage. Therefore, be sure assistance is provided for both donning and doffing.

The procedures outlined here for donning a fully encapsulating SCBA ensemble are for certain types of suits. These procedures should be modified for the particular suit or when extra gloves or boots are used. These procedures assume that the wearer has previous training in SCBA use and decontamination procedures.

When donning a suit, use a moderate amount of talcum powder to prevent chafing and to increase comfort. Talcum powder also will reduce rubber binding. Perform the following procedures in the order indicated:

1. Inspect the clothing and respiratory equipment before donning (see the section on inspection later in this chapter).
2. Adjust the hard hat or headpiece, if worn, to fit your head.
3. Open the back enclosure (used to change the air), if the suit has one, before donning the suit.
4. Standing or sitting, step into the legs of the suit. Ensure proper placement of the feet within the suit, then gather the suit around the waist.
5. Put on chemical-resistant steel-toe and -shank boots over the feet of the suit. Properly attach and affix the suit's leg over the top of the boots.
   a. If additional chemical-resistant boots are required, put these on now.
   b. Some one-piece suits have heavy-soled protective feet. With these suits, wear leather or short rubber safety boots inside the suit.
6. Put on the air tank and harness assembly of the SCBA. Don the face piece and adjust it to be secure but comfortable. Do not connect the breathing hose. Open the valve to the air tank.
7. Perform negative and positive respirator test procedures (see the section on respirator fit testing).
8. Depending on the type of suit,
   a. Put on long inner gloves (similar to surgical gloves).
   b. Secure detachable gloves to the sleeves (if not done prior to donning the suit).
   c. Additional overgloves, worn over attached suit gloves, may be donned later.
9. Put the sleeves of the suit over your arms as your assistant pulls the suit up and over the SCBA. Your assistant can adjust the suit around the SCBA and shoulders to ensure unrestricted motion.
10. Raise the hood over your head carefully so as not to dislodge your headpiece or disrupt the seal of the SCBA mask. Adjust the hood for satisfactory comfort.
11. Begin to secure the suit by closing all fasteners until there is enough room only to connect the breathing hose. Secure all belts and/or adjustable bands around the legs, head, and waist.
12. Connect the breathing hose while opening the main valve.
13. Have your assistant make sure that you are breathing properly, then make final closure of the suit.
14. Have your assistant check all closures.
15. Have your assistant observe you for a period of time to make sure you are comfortable and psychologically stable and the equipment is functioning properly.

Once you have donned the ensemble, evaluate its fit. If it is too small, worker movement is restricted, which increases the likelihood of tearing the material and may accelerate fatigue. If it is too large, the possibility of snagging the material is increased, and the worker's dexterity and coordination may be compromised. In either case, locate better-fitting clothing prior to site entry.

RESPIRATOR FIT TESTING

The fit or integrity of the face piece affects the respirator's performance. A secure fit is important with positive-pressure equipment and is essential to the safe functioning of negative-pressure equipment, such as APRs. Most face pieces fit only 60% of the population; thus, each face piece must be tested on the potential wearer to ensure a tight seal.

To conduct qualitative fit testing, place the wearer in an enclosed space, such as a plastic bag, and expose her to isoamyl acetate (a low-toxicity substance that tastes like banana oil), a sweet saccharin-water mist, or an irritant smoke. The wearer should breathe normally, move the head side to side and up and down, talk, perform exercises in an exaggerated imitation of the task(s) to be performed, and then breathe deeply, as during heavy exertion. If the wearer detects any of the test substance (bananalike smell; sweet taste; or irritation, coughing, or choking, respectively) the fit is inadequate. If the wearer does not detect any of the test substance, assign the exclusive use of that respirator to the worker or make the worker aware of the model and size respirator known to provide her with a proper fit. Conduct periodic checks to ensure that proper fit is maintained.

Fit tests have certain weaknesses. Isoamyl acetate is widely used in fit testing of organic vapor APRs, but its odor threshold varies among individuals. Also, isoamyl acetate can dull the sense of smell, thereby raising the detection threshold to very high levels. Irritant smoke also is used widely for fit testing. The irritant, commonly stannic chloride or titanium tetrachloride, comes in sealed glass tubes that are broken open at test time. Perform this test with caution, since these aerosols are highly irritating to the eyes, skin, and mucous membranes. The advantage of this test is that an involuntary reaction (coughing or sneezing) is induced. The other qualitative fit tests are subjective and rely on the wearer's reaction and honesty.

Quantitative fit testing requires expensive equipment and is generally performed by manufacturers and testing organizations. Portable testing devices are available, however. Leakage is expressed as a percentage of the test atmosphere outside the respirator and is called percent of penetration or simply penetration. Each test respirator is equipped with a sampling port to allow continual removal of an air sample from the face piece. The addition of the hole in the face piece for the quantitative test causes the NIOSH certification to become voided. Therefore, that test respirator may only be used for test purposes and may not be used in the field even if the hole is sealed. Once tested, the face piece cannot be worn in service, as the test orifice negates the approval of the respirator. Quantitative fit testing is highly recommended for work in highly toxic or IDLH atmospheres.

Each time a negative-pressure respirator is donned for use, negative- and positive-pressure tests should be performed, if these tests can be performed on that particular model. To conduct a negative-pressure test, close the inlet valve with the palm of the hand or squeeze the breathing tube so it does not pass air, then gently inhale for about 10 seconds. Any inward rushing of air indicates a poor fit. This is only a gross determination of fit; for example, a leaking face piece may be drawn tightly to the face to form a good seal, giving a false indication of adequate fit. To conduct a positive-pressure test, gently exhale (while covering the exhalation valve, if possible) to ensure that a positive pressure can be built up. Failure to build up a positive pressure indicates a poor fit. Conduct both tests carefully to prevent disrupting a good fit or damaging the valves.

IN-USE MONITORING

During equipment use, encourage workers to report any perceived problems or difficulties to their supervisor. These malfunctions include but are not limited to the following:

- discomfort
- resistance to breathing
- fatigue due to respirator use
- interference with vision or communication
- restriction of movement

If an ALR is being used, remove all hazards that might endanger the integrity of the air line prior to use. During use, keep air lines as short as possible and keep other workers and vehicles away from the area.

DOFFING AN ENSEMBLE

Exact procedures for removing fully encapsulating ensembles must be established and followed to prevent contaminant migration from the work area and transfer of contaminants to the wearer's body, the doffing assistant, and others. The following procedure assumes that appropriate decontamination procedures, commensurate with the type(s) and degree of contamination, have already occurred (see Chapter 8, Decontamination). This procedure also assumes the availability of a suitably attired helper. Throughout the procedure, avoid any contact with the outside surface of the suit.

1. Remove any extraneous or disposable clothing, boot covers, or gloves.
2. Have your assistant loosen and remove the steel-toe and -shank boots.
3. Have your assistant open the front of the suit to allow access to the SCBA regulator. Leave the breathing hose attached as long as there is sufficient pressure.

4. Have your assistant open the suit completely, lift the hood over your head, and rest the hood on top of the SCBA tank.
5. Remove your arms, one at a time, from the suit. Once your arms are free, have your assistant lift the suit up and away from the SCBA backpack, avoiding any contact between the outside surface of the suit and your body, and lay the suit out flat behind you. Leave internal gloves on.
6. Sitting, if possible, remove both legs from the suit.
7. Remove internal gloves by rolling them off your hand, inside out.
8. Proceed to the clean area and follow the procedure for doffing the SCBA.
9. Remove your internal clothing and thoroughly cleanse your body.

This procedure assumes that a sufficient air supply is available, allowing appropriate decontamination before removal. However, if the low-pressure warning alarm has sounded, signifying that approximately 5 minutes of air is remaining, follow this procedure:

1. Remove disposable clothing.
2. Quickly scrub and hose off the suit, especially around the entrance/exit zipper.
3. Open the zipper enough to allow access to the regulator and breathing hose.
4. Immediately attach an organic vapor, acid gas, dust, mist, or fume canister to the breathing hose. This provides protection against any contamination still present.
5. Follow steps 1 through 9 of the regular doffing procedure. Take extra care to avoid contaminating your assistant and yourself.

CLOTHING REUSE

Chemicals that have begun to permeate clothing during use may not be removed during decontamination and may continue to diffuse through the material toward the inside surface, presenting the hazard of direct skin contact to the next person who uses the clothing.

Where such hazards may develop, clothing should be checked inside and out for discoloration (see the next section) or, if possible, by swipe testing for chemicals before reuse. This is particularly important for full-body encapsulating suits, which are generally subject to reuse due to their cost. Note that negative (i.e., no chemical found) test results do not necessarily preclude the possibility that some absorbed chemical will reach the suit's interior.

At present, little documentation exists regarding clothing reuse. In making a decision on reuse, you must consider the known permeation rates as well as the toxicity of the contaminant(s). Unless extreme care is taken to ensure decontamination, the reuse of CPC with highly toxic chemicals is not advisable.

INSPECTION

An effective PPE inspection program will feature five different inspections:

- inspection and operational testing of equipment received from the factory or distributor
- inspection of equipment as it is issued to workers
- inspection after use or training and prior to maintenance
- periodic inspection of stored equipment
- periodic inspection when a question arises concerning the appropriateness of the selected equipment or when problems with similar equipment arise

Each inspection will cover somewhat different areas in varying degrees of depth. Detailed inspection procedures, where appropriate, are usually available from the manufacturer. The inspection checklists provided in Table 6.17 also may be helpful.

Records must be kept of all inspection procedures. Assign individual identification numbers to all reusable pieces of equipment (respirators may have ID numbers already) and maintain records by that number. In conjunction with each inspection, record as a minimum the ID number, date, inspector, and any unusual conditions or findings. A periodic review of these records may indicate an item or type of item with excessive maintenance costs or a particularly high level of downtime.

STORAGE

PPC and respirators must be stored properly to prevent damage or malfunction due to exposure to dust, moisture, sunlight, damaging chemicals, extreme temperatures, and impact. Procedures must be specified for both preissuance warehousing and, more importantly, postissuance (in-use) storage. Many equipment failures can be directly attributed to improper storage.

PPC

- Store contaminated PPC in an area separate from street clothing.
- Store contaminated PPC in a well-ventilated area with good airflow around each item, if possible.
- Do not mix different materials and types of PPC in storage.

Respirators

- Dismantle, wash, and disinfect SCBAs, ALRs, and APRs after each use.
- Store SCBAs in storage chests supplied by the manufacturer. Store APRs individually in their original cartons or carrying cases or in heat-sealed or resealable plastic bags.

MAINTENANCE

The technical depth of maintenance procedures varies. Manufacturers frequently restrict the sale of certain PPE parts to those specially trained, equipped, and authorized to purchase them. Adopt explicit procedures to ensure that maintenance is performed only by those having this specialized training and equipment. The following classification scheme is often used to divide maintenance into three levels:

- Level 1—user or wearer maintenance, requiring a few common tools or no tools at all
- Level 2—shop maintenance, which can be performed by the owner's maintenance shop
- Level 3—specialized maintenance, which can be performed only by the factory or an authorized repair depot

HEAT STRESS AND OTHER PHYSIOLOGICAL FACTORS

Wearing permeation-resistant protective clothing and equipment puts a hazardous waste worker at considerable risk for developing heat stress (the inability to release body heat), a condition that can range from transient heat fatigue and lowered work tolerance to serious illness or death. Heat stress is caused by a number of interacting factors, including environmental conditions, clothing, work load, and individual worker characteristics. Because heat stress is probably one of the most common (and potentially serious) illnesses at hazardous waste sites, regular monitoring and other preventive measures are vital.

Individuals vary in their susceptibility to heat stress. Factors that may predispose someone to heat stress include the following:

- lack of physical fitness
- lack of acclimatization
- age
- dehydration
- obesity
- alcohol and drug use
- infection
- sunburn
- diarrhea
- chronic disease

The amount and type of PPE worn is directly related to reduced work tolerance and the risk of heat stress. Protective clothing and equipment add weight and bulk and diminish or prevent liquid and vapor exchange. This (1) severely reduces the body's normal heat exchange mechanisms (evaporation, convection, and radiation) and (2) increases energy expenditure by about 1.2% for every kilogram of added weight. (A bulky suit can increase by two to four times the energy ordinarily needed to perform a task.)

TABLE 6.17 Personal Protective Equipment Inspection Checklists

Clothing

Before use:

Determine that the construction material is correct for the specified task at hand.

Visually inspect for the following:
  imperfect seams
  nonuniform coatings
  tears
  malfunctioning closures

Hold the material up to light and check for pinholes.

Flex the product and look for the following:
  cracks
  other signs of shelf deterioration

If the product has been used previously, inspect it inside and out for signs of chemical attack:
  discoloration
  swelling
  stiffness

During the work task, inspect periodically for the following:

Evidence of chemical attack such as discoloration, swelling, stiffening, and softening. Keep in mind, however, that chemical permeation can occur without any visible effects.

Closure failure

Tears

Punctures

Seam discontinuities

Gloves

Before use, pressurize glove to check for pinholes. Either blow into the glove, then roll the gauntlet toward the fingers, or inflate the glove and hold it under water. In either case, no air should escape.

Fully Encapsulating Suits

Before use:

Check the operation of pressure-release valves.

Inspect the fitting of wrists, ankles, and neck.

Check the face shield for cracks, crazing, and fogginess.

Self-Contained Breathing Apparatuses (SCBAs)

Inspect SCBAs prior to each use, weekly when in storage, and every time they are cleaned.

Check all connections for tightness.

Check the material for signs of the following:
  pliability
  deterioration
  distortion

Self-Contained Breathing Apparatuses (*Continued*)

Check for proper setting and operation of regulators, gauges, and valves (according to manufacturer's recommendations).

Check operation of alarm(s).

Check face shields and lenses for the following:
  cracks
  crazing
  fogginess

Check the air cylinder for full charge and inspect for dents, gouges, or other damage that could cause loss of cylinder integrity.

Confirm that required hydrostatic tests have been conducted in accordance with intervals prescribed by the manufacturer.

Air-Line Respirators (ALRs)

Inspect ALRs prior to each use, weekly when in storage, and every time they are cleaned.

Inspect air lines prior to each use for cracks, kinks, cuts, fraying, and weak areas.

Check for proper setting and operation of regulators and valves (according to manufacturer's recommendations).

Check operation of alarm(s).

Check all connections for tightness.

Check the material for signs of the following:
  pliability
  deterioration
  distortion

Check face shields and lenses for the following:
  cracks
  crazing
  fogginess

Air-Purifying Respirators (APRs)

Inspect APRs before each use to be sure they have been adequately cleaned, after each use, during cleaning, and monthly if in storage for emergency use.

Check the material for signs of the following:
  pliability
  deterioration
  distortion

Examine cartridges or canisters to ensure that
  they contain the proper sorbents
  the expiration date has not expired
  they have not been opened or used previously

Check face shields and lenses for the following:
  cracks
  crazing
  fogginess

When selecting PPE, carefully evaluate each item's health benefit against its potential for increasing the risk of heat stress. Once you have selected the PPE, determine the optimum length of the work period based on the following factors:

- work rate
- ambient temperature and other environmental factors
- type of protective ensemble
- individual worker characteristics

MONITORING

Because the occurrence of heat stress depends on a variety of factors, all workers, even those not wearing PPE, should be monitored.

For workers wearing permeable clothing (e.g., standard cotton or synthetic work clothes), follow the recommendations for monitoring requirements and suggested work/rest schedules in *Threshold Limit Values for Heat Stress* (ACGIH, 1982). If actual clothing worn differs from the ACGIH standard ensemble in insulation value and/or wind and vapor permeability, change the monitoring requirements and work/rest schedules accordingly.

Monitor workers wearing semipermeable or impermeable encapsulating ensembles when the ambient temperature is above 70°F. It also may be necessary to monitor them at lower temperatures if the humidity is high. (Although no protective ensemble is completely impermeable, for practical purposes, an outfit may be considered impermeable when calculating heat stress risk.)

To monitor workers, measure the following:

- *Heart rate.* Count the radial pulse during a 30-second period immediately following the end of a work period.
   If the heart rate exceeds 140 beats per minute at the end of a work period and 100 beats per minute at the end of a rest period, shorten the next work cycle by one-third or lengthen the rest period by one-third.
   If the heart rate still exceeds 140 beats per minute at the end of the next work cycle, shorten the following work cycle by one-third or lengthen the rest period by one-third.

- *Oral temperature.* Use a clinical thermometer (3 minutes under the tongue) or similar device.
   If the oral temperature exceeds 99.6°F, shorten the next work cycle by one-third or lengthen the rest period by one-third.
   If the oral temperature still exceeds 99.6°F at the end of the next work cycle, shorten the following work cycle by one-third or lengthen the rest period by one-third.
   Do not permit a worker to wear a semipermeable or impermeable garment when his oral temperature exceeds 100.6°F.

- *Skin temperature.* Measure the medial thigh temperature after the skin temperature reaches its equilibrium level (usually after just 10 minutes of work in the heat).
   If the skin temperature exceeds 96.8°F (36°C), a worker will have difficulty maintaining an acceptable heat balance. Consider the cessation of work.
   If the skin temperature exceeds 98.6°F (37°C), cease all work.

- *Body weight.* Measure body weight at the beginning and end of each workday to see whether enough fluids are being taken to prevent dehydration. Do not allow more than a 1.5% body weight loss in a workday.

The frequency of physiological monitoring depends on the air temperature adjusted for solar radiation and the level of physical work (see Table 6.18). Energy levels for work rarely exceed 250 kilo calories per hour, except for short bursts of hard labor. The length of the work cycle will be governed by the frequency of the required physiological monitoring.

PREVENTION

Proper training and preventive measures will help avert serious illness and loss of work productivity. Preventing heat stress is particularly important because once someone suffers from heatstroke or heat exhaustion, that person is predisposed to additional heat injuries. To avoid heat stress, management should take the following steps:

- Adjust work schedules.
   Modify work/rest schedules according to monitoring requirements (generally, for every hour of work, allow 30 minutes of rest).
   Mandate work slowdowns as needed.
   Rotate personnel. Alternate job functions to minimize overstress or overexertion at one task.
   Perform work during cooler hours of the day, if possible.

- Provide shelter or shaded areas to protect personnel during rest periods.

- Maintain workers' body fluids at normal levels. This is necessary to ensure that the cardiovascular system functions adequately. Daily fluid intake must approximately equal the amount of water lost in sweat (i.e., 8 ounces of water must be ingested for every 8 ounces of perspiration lost). A worker in an impermeable ensemble performing a moderately intense task at an adjusted temperature exceeding 85°F could lose as much as 1 quart of sweat per hour (2 gallons in 8 hours). The normal thirst mechanism is not sensitive enough to ensure that enough water will be drunk to replace lost sweat. When heavy sweating occurs, encourage the worker to drink more. Use these strategies:
   Maintain water temperature at 50°F to 60°F.
   Provide small disposable cups that hold about 4 ounces.
   Have workers drink two glasses (16 ounces) of fluid (preferably water or dilute drinks) before beginning work.
   Urge workers to drink 8 to 16 ounces of water every 15 to 20 minutes.
   Weigh workers before and after work to determine whether fluid replacement is adequate.

TABLE 6.18 Required Frequency of Physiological Monitoring

Adjusted Temperature[a]	Semipermeable Ensemble[b]	Impermeable Ensemble[b]
90°F or above	After each 45 minutes of work	After each 15 minutes of work
87.5°–90°F	After each 60 minutes of work	After each 30 minutes of work
82.5°–87.5°F	After each 90 minutes of work	After each 60 minutes of work
77.5°–82.5°F	After each 120 minutes of work	After each 90 minutes of work
72.5°–77.5°F	After each 150 minutes of work	After each 120 minutes of work

[a]Calculate the adjusted air temperature (ta adj) by using this equation: ta adj °F = ta °F + (13 × % sunshine). Measure air temperature (ta) with a standard mercury-in-glass thermometer, with the bulb shielded from radiant heat. Estimate % sunshine by judging what percentage of the time the sun is not covered by clouds that are thick enough to produce a shadow (100% sunshine = no cloud cover and a sharp, distinct shadow; 0% sunshine = no shadows).

[b]When semipermeable or impermeable protective clothing is worn open, raise each temperature adjustment in the left-hand column of the table by about 5% (this increases the threshold for each monitoring time). The exact adjustment depends on the level of permeability of the clothing or the extent to which an impermeable garment can be safely opened.

- See that workers maintain an optimal level of physical fitness.

    Provide an aerobic and/or other exercise program.

    Acclimatize workers to site work conditions: temperature, protective clothing, and work load.

    Urge workers to maintain normal weight levels.

    Discourage smoking and alcohol consumption during off-hours.

- Provide cooling devices to aid natural body ventilation during prolonged work or severe heat exposure:

    field showers or hose-down areas to reduce body temperature and/or cool off protective clothing

    cooling jackets, vests, or suits (see Table 6.8 for details)

    loose-fitting cotton underwear and/or clothing to help absorb moisture and protect skin from direct contact with heat-absorbing protective clothing

- Train workers to recognize and treat heat stress. As part of this training, identify the symptoms of heat stress (see Table 6.19).

OTHER FACTORS

PPE decreases worker performance. The magnitude of this effect varies considerably depending on the individual and the ensemble used. This section discusses the demonstrated physiological responses to PPE, the individual human characteristics that play a factor in these responses, and some of the precautionary and training measures that need to be taken to avoid PPE-induced injury.

The physiological factors that directly affect a worker's ability to function using PPE include the following:

- physical condition
- level of acclimatization
- age
- sex
- weight

TABLE 6.19 Symptoms of Heat Stress

Heat rash results from continuous exposure to heat or humid air.

Heat cramps are caused by heavy sweating with inadequate fluid intake. Symptoms include
  muscle spasms
  pain in the hands, feet, and abdomen

Heat exhaustion occurs when body organs attempt to keep the body cool. Symptoms include
  pale, cool, moist skin
  heavy sweating
  dizziness

Heatstroke is the most serious form of heat stress. Immediate action must be taken to cool the body before serious injury and death occur. Symptoms are
  red, hot, dry skin
  lack of perspiration
  nausea
  dizziness and confusion
  strong, rapid pulse
  coma

Physical Condition This is the most important factor in a person's ability to endure work. The higher the degree of fitness, the heavier a work load one can safely tolerate. At a given level of work, a fit person, relative to an unfit person, will have

- less physiological strain
- a lower heart rate
- a lower rectal temperature, which indicates less retained body heat (a rise in internal temperature precipitates heat injury)
- a more efficient sweating mechanism
- slightly lower air consumption
- slightly lower carbon dioxide production

Level of Acclimatization Acclimatization refers to the physiological changes occurring within an individual that reduce the strain caused by environmental heat stress. An acclimatized individual will generally have a lower heart rate and body temperature than an individual who is unaccustomed to working in the heat. The acclimatized person also begins to sweat sooner and sweats more. This enables the person to maintain a lower skin and body temperature at a given heat stress than an unacclimatized person. Sweat composition also becomes more dilute with acclimatization, so less salt is lost in the sweat.

Acclimatization can occur after just a few days of exposure to a hot environment. NIOSH recommends a progressive 6-day acclimatization period for the unacclimatized worker before allowing her to do full work on a hot job. Begin the first day with 50% of the anticipated work load and exposure time and add 10% each day through day 6. With fit or trained individuals, the acclimatization period may be shortened 2 or 3 days.

Enclosed in an impermeable suit, a fit, acclimatized individual may actually face a greater danger of heat exhaustion than an unfit, unacclimatized individual. This is because the fit, acclimatized individual sweats more than an unacclimatized individual. This higher sweat rate may contribute to rapid dehydration and an earlier onset of heat exhaustion. This can be prevented by consuming adequate quantities of water. See the previous section on prevention for additional information.

Age Generally, maximum work capacity declines with increasing age, but this is not always the case. Active, well-conditioned seniors often have performance capabilities equal to or greater than young, sedentary individuals. Some evidence, such as lower sweat rates and higher core temperatures, indicates that older individuals are less effective in compensating for a given heat stress. Older individuals also appear to become dehydrated more frequently and have a greater risk of heatstroke. At moderate thermal loads, however, the physiological responses of young and old persons are similar, and performance is not affected.

Age should not be the sole criterion for judging whether or not an individual should be subjected to moderate heat stress. Fitness level is a more important factor.

Gender The literature indicates that females tolerate heat stress at least as well as their male counterparts. Generally, a female's work capacity averages 10% to 30% less than a male's. The primary reasons for this are the greater oxygen-carrying capacity and stronger heart of the male. However, a similar situation exists as with aging: not all males have greater work capacities than all females. Therefore, maximum performance, rather than gender, is more appropriate in the selection of workers.

Weight The ability of a body to dissipate heat depends on the ratio of its surface area to its mass (surface area/weight). Heat loss (dissipation) is a function of surface area. Heat production is dependent on mass. Heat balance is determined by the ratio of the two.

Obese and stocky individuals produce a lot of heat but do not have a proportionately large surface area. Hence, they are not capable of rapidly dissipating the heat they produce and are susceptible to heat illness. In comparison, thin individuals have less weight and nearly the same amount of surface area as obese persons and are able to dissipate heat much more rapidly. Therefore, it has been suggested that those exceeding their optimum weight by 15% or more be excluded from working in a hot job. The use of height-weight tables are not recommended. More valid procedures are skin-fold measurements, anthropometry, and hydrostatic weighing.

BIBLIOGRAPHY

American Conference of Governmental Industrial Hygienists (ACGIH). *Guidelines for the Selection of Chemical Protective Clothing.* Cincinnati: ACGIH, 1987.

———. *Threshold Limit Values for Chemical Substances and Physical Agents in the Workplace Environment.* Cincinnati: ACGIH, 1991.

American National Standards Institute (ANSI). *American National Standard Protective Headwear for Industrial Workers.* Publication No. Z89.1–1981. New York: ANSI, 1981.

Dukes-Dubos, F.N., and A. Henschel, eds. *Proceedings of a NIOSH Workshop on Recommended Heat Stress Standards.* Cincinnati: U.S. Department of Health and Human Services, 1980.

Goldman, R.F. "Heat Stress in Industrial Protective Encapsulating Garments." In *Protecting Personnel at Hazardous Waste Sites,* S.P. Levine and W.F. Martin, eds. Boston: Butterworth-Heinemann, 1985.

Held, B.J., and C.A. Horter. *Effectiveness of Self-Contained Breathing Apparatus in a Fire Environment.* Livermore, Calif.: Lawrence Livermore National Laboratory, n.d.

Hyatt, E.C. *Respirator Protection Factors.* Los Alamos, N. Mex.: Los Alamos National Laboratory, n.d.

Marguglio, B.W. *Environmental Management Systems.* New York: Marcel Dekker, 1991.

National Institute for Occupational Safety and Health (NIOSH). *A Guide to Industrial Respiratory Protection.* U.S. Department of Health, Education, and Welfare Publication No. 76-189. Cincinnati: NIOSH, 1976.

———. *NIOSH Certified Equipment List.* Publication No. 83-140 350. Morgantown, W.Va.: NIOSH, 1983.

———. *NIOSH Certified Equipment List.* Publication No. 87-102. Morgantown, W.Va.: NIOSH, October 1986.

Schwope, A.D., P.P. Costas, J.O. Jackson, and D.J. Weitzman. *Guidelines for the Selection of Chemical Protective Clothing.* Prepared by A.D. Little, Inc., Cambridge, Mass., for the U.S. Environmental Protection Agency and Los Alamos National Laboratory, 1987.

———. *Guidelines for the Selection of Chemical Protective Clothing,* 2d ed. American Conference of Governmental Hygienists, Inc.: Cincinnati, 1985.

U.S. Environmental Protection Agency (EPA). Office of Emergency and Remedial Response. Hazard Response Support Division. *Interim Standard Operating Procedures.* Washington, D.C.: EPA, April 1982.

7

SITE CONTROL AND
WORK PRACTICES

*T*he purpose of site control is to minimize potential contamination of workers, protect the public from the site's chemical and physical hazards, facilitate work activities, and prevent vandalism. Site control is especially important in emergency situations to ensure communication, site access, efficient evacuation, and response. This chapter describes the basic components of a program to control the activities and movements of people, materials, and equipment at a hazardous waste site. Special attention is given to drum handling, sampling, bulking, and equipment use. Engineering controls are recommended whenever equipment and barriers can separate workers from contact with hazardous materials and reduce the risk of worker exposure.

Several site control procedures can be implemented to reduce worker and public exposure to chemical and physical hazards:

- Compile a site map.

- Prepare the site.
- Establish work zones.
- Enforce the buddy system.
- Establish and strictly enforce decontamination procedures. (See Chapter 8.)
- Establish site security measures.
- Set up communication networks.
- Enforce safe work practices.
- Establish safe drum and container management practices.

The degree of site control necessary depends on site characteristics, site size, and the surrounding community. Establish the site control program in the planning stages of a project and modify it based on new information and site assessments. Determine the appropriate sequence for implementing these measures on a site-specific basis. In many cases, it will be necessary to implement several measures simultaneously.

SITE MAP

A site map showing topographic features predominant wind direction, drainage, and the location of buildings, containers, impoundments, pits, ponds, and tanks is helpful in

- planning activities
- assigning personnel
- identifying access routes, evacuation routes, and problem areas
- identifying utilities, overhead electrical lines, and areas where equipment use is restricted based on safe clearance requirements
- identifying areas of the site that require the use of PPE
- supplementing the daily safety briefings of the field teams

Prepare the map prior to site entry and update it throughout the course of site operations to reflect the following:

- accidents
- changes in site activities
- emergencies
- hazards not previously identified
- new materials introduced on the site
- vandalism
- weather conditions

Use overlays to present information without cluttering the map.

SITE PREPARATION

Time and effort must be spent in preparing a site for the cleanup activity and eliminating obvious physical hazards. This ensures that response operations go smoothly and

TABLE 7.1 Site Preparation

Construct roadways to provide ease of access and a sound roadbed for heavy equipment and vehicles.

Arrange traffic flow patterns to facilitate efficient operations.

Eliminate physical hazards from the work area as much as possible, including the following:

- ignition sources in flammable hazard areas
- exposed or ungrounded electrical wiring and low overhead wiring that may entangle equipment
- sharp or protruding edges, such as glass, nails, and torn metal, which can puncture protective clothing and equipment and inflict puncture wounds
- debris, holes, loose steps or flooring, protruding objects, slippery surfaces, or unsecured railings, which can cause falls, slips, and trips
- unsecured objects, such as bricks and gas cylinders, near the edges of elevated surfaces, such as catwalks, rooftops, and scaffolding, which can dislodge and fall on workers
- debris and weeds that obstruct visibility

Install skid-resistant strips on slipper surfaces.

Construct operation pads for mobile facilities and temporary structures.

Construct loading docks, processing and staging areas, and decontamination pads.

Provide adequate illumination for work activities. Equip temporary lights with guards to prevent accidental contact and heavy-duty electric cords with connections and insulation maintained in good condition. Portable electric lighting shall be operated at a maximum of 12 volts.

Required levels of illumination are listed in Table H-120.1 in 29 CFR 1910.120.

Ground temporary wiring in accordance with the National Electric Code. Sheath or otherwise protect all wiring. Wherever possible, make any necessary open wiring inaccessible to unauthorized personnel. Splices should have insulation equal to that of the cable.

worker safety is protected. Site preparation should meet the needs of the work plan. Table 7.1 presents the major steps in site preparation prior to any cleanup activities.

SITE WORK ZONES

To reduce the accidental spread of hazardous substances by workers from the contaminated area to the clean area, delineate zones where different types of operations will occur and control the flow of personnel among the zones. The establishment of work zones will help to ensure that personnel are properly protected against the hazards present where they are working, confine work activities to the appropriate areas, and locate and evacuate personnel in the event of emergency. In all areas, follow good housekeeping practices at all times.

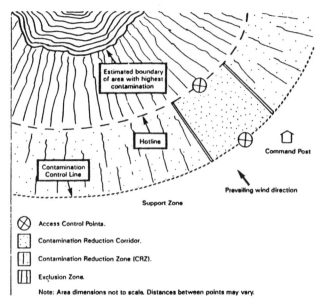

Figure 7.1. A schematic presentation of work zones

Uncontrolled hazardous substance sites are generally divided into three zones:

1. exclusion zone—the contaminated area
2. contamination reduction zone (CRZ)—the area where decontamination takes place
3. support zone—the uncontaminated area where workers should not be exposed to hazardous conditions

Base delineation of these three zones on sampling and monitoring results and on an evaluation of potential routes and amounts of contaminant dispersion in the event of a release. Limit movement of personnel and equipment among these zones through specific access control points to prevent cross-contamination from contaminated areas to clean areas. A schematic representation of the layout of work zones is given in Figure 7.1

EXCLUSION ZONE

The exclusion zone is the area where contamination does or could occur. The following primary activities are performed in the exclusion zone:

- site characterization, such as mapping, photographing, and sampling
- well installation for groundwater monitoring
- cleanup work, such as drum movement, drum staging, and material bulking

Establish the outer boundary of the exclusion zone, called the *hot line*, according to the criteria listed in Table 7.2. Clearly mark it with lines, placards, hazard tape, or

TABLE 7.2 Establishing the Hot Line

Visually survey the immediate environs of the site.

Determine the locations of the following:
   hazardous substances
   drainage, leachate, and spilled material
   visible discolorations

Evaluate the data from the initial site survey indicating the presence of the following:
   combustible gases
   organic and inorganic gases, particulates, or vapors
   ionizing radiation

Evaluate the results of soil and water sampling.

Consider the distances needed to prevent an explosion or fire from affecting personnel outside the exclusion zone.

Consider the distances that personnel must travel to and from the exclusion zone.

Consider the physical area necessary for site operations.

Consider meteorological conditions and the potential for contaminants to be blown from the area.

Secure or mark the hot line.

Modify its location, if necessary, as more information becomes available.

---

signs, or enclose it with physical barriers such as chains, fences, or ropes. Establish access control points at the periphery of the exclusion zone to regulate the flow of personnel and equipment into and out of the zone and to help verify that proper procedures for entering and exiting are followed. If feasible, establish four access control points. Provide separate entrances and exits for personnel and equipment.

The exclusion zone can be subdivided into different areas of contamination based on the known or expected types and degrees of hazard or on the incompatibility of waste streams. This allows more flexibility in operations, decontamination procedures, and resources.

The personnel requirements in the exclusion zone may include the field team leader, the work parties, and specialized personnel such as heavy equipment operators. Require all personnel within the exclusion zone to wear the level of protection described in the site health and safety plan. Within the zone, different levels of protection may be justified based on the degree of hazard presented. Specify and mark in each subarea the level of personal protection required (see Chapter 6, Personal Protective Equipment).

The required level of protection in the exclusion zone varies according to job assignment. For example, a worker who collects samples from open containers might require Level B protection, whereas one who performs walk-through ambient air monitoring might need only Level C

protection. When appropriate, assign different levels of protection within the exclusion zone to promote a more flexible, more effective, less costly operation while still maintaining a high degree of safety.

CONTAMINATION REDUCTION ZONE

The CRZ is the transition area between the contaminated area and the clean area. This zone is designed to reduce the probability that the clean support zone will become contaminated or affected by other site hazards. The distance between the exclusion and support zones, combined with decontamination, will limit the physical transfer of hazardous substances on workers and equipment.

Decontamination procedures begin at the boundary between the exclusion zone and the CRZ, called the hot line. Set up at least two lines of decontamination stations—one for personnel and one for heavy equipment. A large operation may require more than two lines. Access into and out of the CRZ from the exclusion zone is through access control points—one each for personnel and equipment entrance and one each for personnel and equipment exit, if feasible. The degree of contamination in the CRZ decreases as one moves from the hot line into the support zone, due to the distance and the decontamination procedures.

The boundary between the support zone and the CRZ, called the *contamination control line,* separates the possibly low contamination area from the clean support zone. Access to the CRZ from the support zone is through two access control points—one for personnel and another for equipment. Require personnel entering the CRZ to wear the protective clothing and equipment prescribed for working in the CRZ. When workers reenter the support zone, require them to remove any protective clothing and equipment worn in the CRZ and have them leave through the personnel exit access control point.

The personnel stationed in the CRZ are usually the SSO, a personnel decontamination station (PDS) operator, and the emergency response personnel. Additional personnel may assist the PDS operator by operating a separate mini decontamination system for samples.

The CRZ must be well laid out to facilitate the following activities:

- decontamination of equipment, PDS operators, personnel, and samples
- emergency response, including transport of injured personnel (safety harness, stretcher), first aid equipment (bandages, blankets, eyewash, splints, water), and containment equipment (absorbent, fire extinguisher)
- equipment resupply, such as air tank changes, protective clothing and equipment, sample equipment, and tools

- sample packaging and preparation for on-site or off-site laboratories

- worker temporary rest area, including toilet facilities, bench, chair, liquids, and shade (Water and other potable liquids should be clearly marked and stored properly to ensure that all glasses and cups are clean. Locate washing facilities near drinking facilities to allow employees to wash before drinking. Locate drinking, washing, and toilet facilities in a safe area where protective clothing can be removed. Inspect and clean facilities regularly.)

- drainage of water and other liquids used in the decontamination process

SUPPORT ZONE

The support zone is the location of the administrative and other support functions needed to keep the operations in the exclusion zone and CRZ running smoothly. Any function that need not or cannot be performed in a hazardous or potentially hazardous area is performed here. The personnel present in the support zone depend on the functions being performed. At a minimum, there is the command post supervisor. Others may include the project team leader and field team members who are preparing to enter or have returned from the exclusion zone.

Personnel may wear normal work clothes within this zone. Potentially contaminated clothing, equipment, and samples must remain in the CRZ until they are decontaminated.

It is the responsibility of support zone personnel to alert the proper agency in the event of an emergency. Keep all emergency telephone numbers, change for the telephone (if necessary), evacuation route maps, and vehicle keys in the support zone.

Support facilities, listed in Table 7.3, are located in the support zone. Consider the following factors when locating these facilities:

- accessibility—topography, open space available, location of highways and railroad tracks, and ease of access for emergency vehicles

- resources—adequate roads, power lines, telephones, shelter, and water

- visibility—line of sight to all activities in the exclusion zone

- wind direction—upwind of the exclusion zone, if possible

- distance—as far as possible from the exclusion zone

THE BUDDY SYSTEM

One of the most important directives for any type of fieldwork, but especially vital for work around hazardous materials, is *never work alone*. No one should ever enter a contaminated or otherwise hazardous area without a buddy who is able to

TABLE 7.3 Support Zone Activities

Facility	Function
Command post	Supervision of all field operations and field teams
	Maintenance of communications, including emergency lines of communication
	Record keeping, including the following: accident reports chain-of-custody records daily logbooks manifest directories and orders medical records personnel training records site inventories site safety map up-to-date safety manuals up-to-date site safety plans
	Interfacing with the public: government agencies, local politicians, medical personnel, the media, and other interested parties
	Monitoring work schedules and weather changes
	Maintenance of site security operations
	Maintenance of sanitary facilities
Medical station	First aid administration
	Medical emergency response
	Medical monitoring activities
	Maintenance of sanitary facilities
Equipment and supply centers	Supply, maintenance, and repair of communication, respiratory, and sampling equipment
	Maintenance and repair of vehicles
	Replacement of expendable supplies
	Storage of monitoring equipment and supplies here or in an on-site field laboratory
Administration	Sample shipment
	Interfacing with home office
	Maintenance of emergency telephone numbers, evacuation route maps, and vehicle keys
	Coordination with transporters,

	disposal sites, and appropriate federal, state, and local regulatory agencies
Field laboratory	Coordination and processing of environmental and hazardous samples
	Packaging of materials for analysis following the decontamination of the outside of the sample containers (This packaging also can be done in a designated location in the contamination reduction zone. Keep shipping papers and chain-of-custody files in the command post office.)
	Maintenance of sampling plans and procedures for quick reference
	Maintenance and storage of laboratory notebooks in designated locations in the laboratory while in use (Keep notebooks in the command post office when not in use.)

- provide his partner with assistance
- observe his partner for signs of chemical or heat exposure
- periodically check the integrity of his partner's protective clothing
- notify the command post supervisor or others if emergency help is needed

The access control point for personnel entrance to the exclusion zone is a convenient location for enforcing the buddy system for two reasons: (1) enforcement is the responsibility of the team leader, who is stationed in the CRZ; (2) all personnel who enter the contaminated area must pass through the control point.

The buddy system alone may not be sufficient to ensure that help will be provided in case of an emergency. At all times, workers in the exclusion zone should be in line-of-sight contact or communications contact with the command post supervisor or backup person in the support zone.

SITE SECURITY

Site security is necessary to

- prevent the exposure of unauthorized, unprotected people to site hazards
- avoid the increased hazards from vandals or persons seeking to dispose of other wastes on the site
- prevent theft

- avoid interference with safe working procedures

Site security can be maintained during working hours as follows:

- Maintain security in the support zone and at access control points.
- Establish an identification system to identify authorized persons and limitations to their approved activities.
- Assign enforcement authority for entry and exit requirements.
- Erect a fence or other physical barrier around the site.
- If the site is not fenced in, post signs around the perimeter and hire guards to patrol it.
- Have the project team leader approve all visitors to the site. Make sure they have a valid purpose for entering the site. Have trained site personnel accompany visitors at all times and provide them with the appropriate PPE.

Site security can be maintained during off-duty hours as follows:

- If possible, assign trained, in-house technicians for site surveillance. Such personnel should receive detailed training concerning site hazards, the nature of the work, and respiratory protection and should be qualified to provide support in the event of an emergency.
- If necessary, hire private security guards to patrol the site boundary. Such personnel should be provided a general overview of site hazards and work areas. In the event that a guard suspects or observes a problem, she can refer to written instructions identifying contacts to be notified and evacuation procedures. Under no circumstance should a private security guard enter a posted hazard area or directly handle any hazardous materials that may be released.
- Coordinate security procedures with public enforcement agencies, such as the local police and fire departments, if the site presents a significant risk to local health and safety.
- Lock the equipment behind a fence.
- Remove the equipment from the site every day if the site is not fenced.

COMMUNICATION SYSTEMS

Use *internal communication* among the work zones to

- alert team members to emergencies
- pass along safety information, such as air time left
- communicate changes in work scope
- maintain site control

Verbal communication at a site can be impeded by background noise and the use of PPE. For example, speech transmission through a respirator can be poor, and hearing can be impaired by protective hoods and respirator airflow. For effective communication, commands must be prearranged and should include audio and visual cues to help convey the message in situations where voice communication alone is not adequate.

Both a primary and a backup system are necessary. Common communication devices are listed in Table 7.4. Establish a set of signals for use only during emergencies. Sample emergency signals for each type of communication device also are given in Table 7.4.

Effective communication requires the identification of individual workers so that commands can be addressed to the right worker. Mark the worker's name on the PPE or, for long-distance identification, add color coding, numbers, or symbols. Flags may be used to help locate personnel in areas where visibility is poor due to obstructions such as accumulated drums, equipment, and waste piles.

All communication devices used in a potentially explosive atmosphere must be inherently spark-free. Check all communication systems daily to make sure they are operating.

Use *external communication* between on-site and off-site personnel to

- coordinate emergency response
- report to management
- maintain contact with essential off-site personnel

TABLE 7.4 Internal Communication

Devices	Example Emergency Signals
Radio*—citizens band or FM	Established code words
Noisemaker—bell, air horn, megaphone, or whistle	Three short blasts, pause, three short blasts: given from downrange = need assistance; given from command post = evacuate area
Visual signal—flag hand signals, lights, or signal board	Hand signals: clutching throat = personal distress; arm waved in a circle over the head given from downrange = need assistance; arm waved in a circle over head given from command post = evacuate area

*All radios used in the exclusion and contamination reduction zones must be intrinsically safe and not capable of sparking. Look for the Underwriters' Laboratories (UL) green dot.

TABLE 7.5 Example Set of Standing Orders

For personnel entering the contamination reduction zone:
  No smoking, eating, or drinking in this zone.
  Sign in at the entrance access control point before entering this zone.
  Sign out at the exit access control point before leaving this zone.

For personnel entering the exclusion zone:
  No smoking, eating, or drinking in this zone.
  Sign in at the entrance access control point before entering this zone.
  Sign out at the exit access control point before leaving this zone.
  Always have your buddy with you in this zone.
  Wear an SCBA in this zone.
  If you discover any signs of radioactivity, explosivity, or unusual conditions such as dead animals at the site, exit immediately and report this finding to your supervisor.

The primary means of external communication are telephone and radio. If telephone lines are not installed at a site, be sure all team members know the location of the nearest telephone. Make sure the correct change and necessary telephone numbers are readily available in the support zone.

SAFE WORK PRACTICES

To maintain a strong safety awareness and enforce safe procedures at a site, develop a list of standing orders stating the practices that must be followed and those that must never occur in the contaminated areas. Develop a separate set of standing orders for the CRZ and exclusion zone if the hazards are sufficiently different. An example of standing orders is given in Table 7.5. To ensure that everyone who enters the site is aware of these orders and that a high degree of familiarity with their content is maintained, the list should be

- distributed to everyone who enters the site
- posted conspicuously at the command post
- posted conspicuously at the entrance access control points into the CRZ and/or the exclusion zone
- reviewed by the field team leader or project team leader with the field crew at the beginning of each workday to inform them of any new standing orders resulting from a change in site conditions or work activities

Also prepare and post conspicuously an employee material safety data sheet (MSDS) or EPA hazardous substance data sheet that lists the names and properties of chemicals present on-site. Brief employees at the beginning of the project and all new employees on the chemical information. Hold daily or weekly safety review meetings as needed for all employees.

Working with tools and heavy equipment is a major hazard at sites. Injuries can result from equipment hitting or running over personnel, flying objects, hot objects, and damage to protective equipment such as ALRs. The following precautions will help prevent these hazards:

- Train personnel in operating procedures for each piece of equipment and develop an awareness of the equipment's weak points and frequency of failure, as well as how to inspect it for safety of operation.

- Install adequate on-site roads, signs, lights, and devices.

- Install appropriate equipment guards and engineering controls on tools and equipment. These would include emergency shutoffs, rollover protection, seat belts, safety harnesses, and backup warning lights and audio devices.

- Provide and enforce warning signs such as "Unlawful to operate this equipment within 10 feet of all power lines" on equipment such as cranes, derricks, and power shovels. Also post equipment restriction areas where utilities and overhead electrical lines are marked on the site safety map.

- Use intrinsically safe (nonsparking) tools.

- In hydraulic power tools, use fire-resistant fluid that is capable of retaining its operating characteristics at the most extreme temperatures.

- Use three-wire grounded extension cords with portable electric tools and appliances. Use ground-fault circuit interrupters on unprotected circuits to protect against line-to-ground short circuits.

- At the start of each workday, inspect brakes, hydraulic lines, light signals, fire extinguishers, fluid level, steering, and splash protection.

- Keep all nonessential people out of the work area.

- Prohibit loose-fitting clothing, loose long hair, and jewelry around moving machinery.

- Keep the cab free of all nonessential items and secure all loose items.

- Do not exceed the rated load capacity of a vehicle.

- Instruct equipment operators to report to their supervisor any abnormalities, such as equipment failure, oozing liquids, or unusual odors.

- When an equipment operator must negotiate in tight quarters or is backing up, provide a second person to act as a guide to direct the movement of the equipment and restrict the entry of persons or equipment into the area required for equipment clearance.

- Equip all on-site internal combustion engines with spark arresters that meet requirements for hazardous atmospheres. Refuel in safe areas. Do not fuel engines while the vehicle is running. Prohibit ignition sources near a fuel area.

- Lower all blades and buckets to the ground and set parking brakes before shutting off the vehicle.

- Inspect all tools and moving equipment regularly to ensure that parts are secured and intact with no evidence of cracks or weakness, that the equipment turns smoothly with no evidence of wobble, and that it is operating according to the manufacturer's specifications. Promptly repair or replace any defective items. Keep a repair log.

- Store tools in clean, secure areas so that they will not be damaged, lost, or stolen.

HANDLING DRUMS AND OTHER CONTAINERS

Accidents occur frequently during handling of drums and other containers of hazardous wastes. Hazards that may be encountered include explosions, fires, violent chemical reactions, toxic vapors, corrosive materials, and other physical injuries resulting from working around heavy equipment and deteriorated drums and containers. While these hazards are always present, workers are at an increased risk when working in the immediate vicinity of damaged or deteriorating drums and containers, opening and collecting waste samples from drums and containers, removing waste materials for treatment or repackaging, overpacking drums, reorganizing and staging materials, and loading materials for transport to off-site facilities. Proper work practices, such as minimizing handling and using equipment and procedures that isolate workers from hazardous substances, can minimize the risk to site personnel.

OSHA regulations (29 CFR 1910 and 1926) establish general requirements for maintaining chemical- and container-handling equipment and storing, containing, and handling chemicals and containers. Specific requirements for handling drums and containers during cleanup operations at hazardous waste sites are established in 29 CFR 1910.120(j). EPA hazardous waste regulations (40 CFR 264 and 265) stipulate requirements for types of containers, maintenance of containers and containment structures, and design and maintenance of storage areas. DOT regulations in 49 CFR establish requirements applicable to all shipments of hazardous wastes and hazardous materials. Consistent with DOT regulations, the EPA regulations under 40 CFR 263 establish specific requirements for hazardous waste transporters.

INSPECTING DRUMS

The appropriate procedures for handling drums depend on the drum contents. Thus, prior to any handling, visually inspect the drums to gain as much information as possible about their contents. Look for

- symbols, words, or other marks on the drum indicating that its contents are hazardous (e.g., radioactive, explosive, corrosive, toxic, flammable)

- symbols, words, or other marks on a drum indicating that it contains discarded laboratory chemicals, reagents, or other potentially dangerous materials in small-volume individual containers

- signs of deterioration, such as corrosion, rust, and leaks
- signs that the drum is under pressure, such as swelling and bulging
- drum type (see Table 7.6)
- configuration of the drumhead (see Table 7.7)

Conditions in the immediate vicinity of the drums may provide information about drum contents and their associated hazards. Monitor around the drums using a gamma radiation detector, drum surface wipe samples, organic vapor analyzer, and combustible gas meter (check instruments).

Use the results of this survey to classify the drums into preliminary hazard categories such as the following:

- radioactive
- leaking/deteriorated
- bulging
- explosive/shock-sensitive
- contains small-volume individual containers of laboratory wastes or other dangerous materials

Assume that unlabeled drums contain hazardous materials until their contents are characterized. Bear in mind that

TABLE 7.6 Drum Types and Their Associated Hazards

Type	Hazards
Polyethylene or PVC-lined drums	Often contain strong acids or bases. If the lining is punctured, the substance usually corrodes the steel quickly, resulting in a significant leak or spill.
Exotic metal drums (e.g., aluminum, nickel, or stainless steel)	Very expensive drums that usually contain an extremely dangerous material.
Single-walled drums used as pressure vessels	These drums have fittings for both product filling and placement of an inert gas, such as nitrogen. May contain reactive, flammable, or explosive substances.
Laboratory packs	Used for disposal of expired chemicals and process samples from university laboratories, hospitals, and similar institutions. Many of these bottles contain incompatible materials and are not packed in absorbent material. They may contain radioisotopes or shock-sensitive, highly volatile, highly corrosive, or highly toxic exotic chemicals. Laboratory packs are the primary ignition source for fires at most hazardous waste sites.

TABLE 7.7 Information Provided by Drumhead Configuration

Configuration	Information
Whole lid removable	Designed to contain solid material
Has a bung	Designed to contain a liquid
Contains a liner	May contain a highly corrosive or otherwise hazardous material

drums are frequently mislabeled, particularly drums that are reused. Thus, a drum's label may not accurately describe its contents.

If buried drums are suspected, use ground-penetrating systems, such as electromagnetic waves, electrical resistivity, ground-penetrating radar, magnetometry, and metal detection, to estimate the location and depth of the drums.

PLANNING DRUM HANDLING

Since drum handling is fraught with danger, every step of the operation should be carefully planned, based on all the information available at the time. Use the results of the preliminary inspection to determine (1) any hazards present and the appropriate response and (2) which drums need to be moved before being opened and sampled. Develop a preliminary plan specifying the extent of handling necessary, the personnel selected for the job, and the most appropriate procedures based on the hazards associated with the probable drum contents as determined by visual inspection. Revise this plan as new information is obtained during drum handling.

HANDLING DRUMS

The purpose of handling is to (1) respond to any obvious problems that might impair worker safety, such as radioactivity, leakage, or the presence of explosive substances; (2) unstack and orient drums for sampling; and (3) if necessary, organize drums into different areas on-site to facilitate characterization and remedial action (see the section on staging later in this chapter). Handling may or may not be necessary depending on how the drums are positioned.

Since accidents occur frequently during handling, particularly initial handling, handle only if necessary. Prior to handling, warn all personnel about the hazards of handling and instruct them to minimize handling as much as possible and to avoid unnecessary handling. In all phases of handling, be alert for new information about hazards to site personnel. Respond to these hazards before continuing with more routine handling operations. Keep a spill control kit containing an adequate volume of absorbent and overpack drums near areas where minor spills may occur. Where major spills may occur, construct a berm adequate to contain the entire volume of liquid. If the drum contents spill, use personnel trained in spill-response procedures to isolate and contain the spill.

Several types of equipment can be used to move drums: a drum grappler attached to a hydraulic excavator; a small front-end loader, which can be either loaded manually or equipped with a bucket sling; a rough-terrain forklift; a roller conveyer equipped with solid rollers; drum carts designed specifically for drum handling. Drums also are sometimes moved manually.

The drum grappler is the preferred piece of equipment for drum handling. It keeps the operator removed from the drum so that there is less likelihood of injury if the drum detonates or ruptures. If a drum is leaking, the operator can stop the leak by rotating the drum and immediately placing it in an overpack container. In case of an explosion, grappler claws help protect the operator by partially deflecting the force of the explosion.

Use the following procedures to maximize worker safety during drum handling and movement:

- Train personnel in proper lifting and moving techniques to prevent back injuries.

- Make sure the vehicle selected has a sufficient rated load capacity to handle the anticipated loads and that it can operate smoothly on the available road surface.

- Air-condition the cabs of vehicles to increase operator efficiency. Protect the operator with heavy splash shields.

- Supply operators with respiratory protective equipment—either an SCBA equipped with an air-line regulator and reserve tank or an ALR and bailout bottle. This improves operator efficiency and provides protection in case the operator must abandon the equipment.

- Have overpack containers ready before any attempt is made to move drums.

- Before moving anything, determine the most appropriate sequence in which the various drums and other containers should be moved. For example, small containers may have to be removed first to permit heavy equipment to enter and move the drums.

- Do not move drums unless they are intact and tightly sealed.

- Ensure that operators have a clear view of the roadway when carrying drums. Where necessary, have ground workers available to guide the operator's movement.

Drums Containing Radioactive Waste If a container is labeled radioactive or radiation levels in a drum exceed background levels, immediately contact the state radiation officer and obtain the services of a qualified radiation health physicist to evaluate the potential hazards and appropriate storage and management requirements. Measure radiation levels to identify hazard areas (i.e., areas where radiation levels exceed background levels) and restrict all access to those areas.

Do not handle suspected radioactive materials or enter restricted areas until radiation hazards have been fully characterized, appropriate health and safety procedures have been developed, and necessary equipment and supplies are obtained. Personnel handling suspected radioactive materials must be properly trained and under the direction of a qualified radiation health physicist.

Drums That May Contain Explosive or Shock-Sensitive Waste If a drum is suspected of containing explosive or shock-sensitive waste as determined by visual inspection, seek specialized assistance before handling it. If handling is necessary, do so with extreme caution.

- When handling shock-sensitive materials, all nonessential personnel must be evacuated from the affected area.

- All material handling equipment must have explosive containment devices or protective shields to protect operators from explosions.

- Use a grappler unit constructed for explosive containment for initial handling of drums.

- Palletize the drums prior to transport. Secure them to the pallets.

- Use an audible siren signal system, similar to that used in conventional blasting operations, to signal the commencement and completion of handling activities.

- Maintain continuous contact with the SSO and/or the site control center until the operation is complete.

Bulging Drums Pressurized drums are extremely hazardous. Wherever possible, do not move drums that may be under initial pressure, as evidenced by bulging or swelling. Cool the drum, if possible, to relieve some of the internal pressure.

If a pressurized drum has to be moved, handle it with a grappler unit constructed for explosive containment. Either move the bulged drum only as far as necessary to allow seating on firm ground or carefully overpack it. Exercise extreme caution when working with or adjacent to potentially pressurized drums.

Laboratory Waste Packs Drums containing packaged laboratory wastes are one of the primary ignition sources for fires at hazardous waste sites. They sometimes contain shock-sensitive materials. Consider such containers to hold shock-sensitive wastes until they are otherwise characterized. If handling is required, follow the procedures for shock-sensitive wastes. See "Special Case Problems," the last section of this chapter, for more details concerning management of laboratory packs.

- Prior to handling or transporting lab packs, remove any nonessential personnel to a safe distance.

- Use a grappler unit constructed for explosive containment for initial handling of such drums.

- Palletize the drums prior to transport. Secure them to the pallets.

- Maintain continuous contact with the SSO and/or the site control center until the handling operation is complete.

- If the lab pack is open, have a chemist inspect and classify the individual bottles without opening them (see ''Special Case Problems'' for more details).

Leaking, Open, and Deteriorated Drums and Containers

- Provide appropriate lids or covers for open drums and containers that are otherwise in sound condition.

- If a drum or container of liquids cannot be moved without rupture, immediately transfer its contents to an overpack container using intrinsically safe equipment.

- Using a drum grappler, immediately place the following types of drums in overpack containers:
  leaking drums containing sludge or semisolids
  damaged open drums containing liquid or solid wastes
  deteriorated drums that can be moved without rupture

- If the quantity of materials or the number of containers subject to transfer or overpacking necessitates a prolonged activity, isolated holes and leaks in otherwise sound containers or drums should be patched.

- Construct containment barriers and provide a means for collecting leaking materials.

Buried Drums

- Prior to initiating subsurface excavation, use ground-penetrating systems to estimate the location and depth of the drums (see ''Inspection'').

- Remove the soil with great caution to minimize the potential for drum rupture.

- Maintain a high degree of safety control during excavation.

- Have available provisions for extinguishing chemical fires.

- Be aware of discolored soil that may indicate leaking drums.

OPENING DRUMS AND CONTAINERS

Drums are usually opened and sampled in place during site investigations. However, remedial and emergency operations may require a separate drum-opening area (see ''Staging Drum Movement''). Procedures for opening drums are the same regardless of where they are opened. To enhance the efficiency and safety of drum-opening personnel, institute the following procedures.

- If an ALR protection system is used, place a bank of air cylinders outside the work area and supply air to the operators via air lines and escape tanks. This enables workers to operate in relative comfort for extended periods of time.

- Protect personnel by keeping them at a safe distance from the drums being opened. If personnel must be located near the drums, place shielding materials, such as Plexiglas, between them and the drums to protect them from sudden releases of pressure and materials. Locate controls for opening, monitoring, and fire-suppression equipment behind the shielding.

- Place personal monitors on workers who are opening drums.

- If possible, monitor personnel continuously during opening. Place the sensors of monitoring equipment such as colorimetric tubes, dosimeters, explosion meters, organic vapor analyzers, and oxygen meters as close as possible to the source of the contaminants (i.e., at the drum opening). Monitor for alpha and beta radiation during opening to ensure that these hazards, which are undetectable when enclosed in a sealed metal container, are detected as soon as possible.

- Use the following remote-controlled devices for opening drums:
  a pneumatically operated impact wrench to remove drum bungs
  hydraulically or pneumatically operated drill piercers
  a backhoe equipped with bronze spikes for penetrating drum tops in large-scale operations

- Do not use picks, chisels, or firearms to open drums.

- Hang or balance the drum-opening equipment to minimize worker exertion.

- If the container shows signs of swelling or bulging, perform all steps slowly. Relieve excess pressure prior to opening and, if possible, from a remote location using devices such as a pneumatic impact wrench or hydraulic penetration device. If pressure must be relieved manually, place a barrier (such as Plexiglass) between the worker and the container opening to deflect any gas, liquid, or solid that may be expelled as the opening is loosened.

- Enter exotic metal drums through the bung. Enter drums lined with polyethylene or polyvinyl chloride (PVC) through the bung by removal or drilling. Exercise extreme caution when manipulating these containers.

- Do not open or sample individual drums in laboratory packs. (For more details, see ''Special Case Problems.'')

- Reseal open bungs and drill openings as soon as possible with new bungs or plugs to avoid explosions and/or vapor generation. If an open drum cannot be resealed, place the drum in an overpack container. Plug any openings in pressurized drums with pressure-venting caps set to a 5 pounds per square inch (psi) release to allow venting of vapor pressure.

- Decontaminate equipment after each use to avoid mixing of incompatible wastes.

SAMPLING DRUM CONTENTS

Drum sampling can be one of the most hazardous activities because it often involves direct contact with unidentified wastes. Prior to collecting any sample, develop a sampling plan:

- Research background information about the waste.
- Determine which drums should be sampled.
- Select the appropriate sampling device(s) and container(s).
- Develop a written sampling plan that includes the number, volume, and location of samples to be taken.
- Develop SOPs for opening drums, sampling, and sample packaging and transportation. For guidance in designing proper sampling procedures, refer to the EPA manual *Samplers and Sampling Procedures for Hazardous Waste Streams* (see de Vera et al. in bibliography).
- Have a trained health and safety professional determine, based on information about the wastes and site conditions, the appropriate personal protection to be used during sampling, decontamination, and packaging of the sample.

When manually sampling from a drum, use the following techniques:

- Keep sampling personnel at a safe distance while containers are being opened. Sample only after opening operations are complete.
- Do not lean over other containers to reach the drum being sampled unless it is absolutely necessary.
- Cover container tops with plastic sheeting or other suitable uncontaminated material to avoid excessive contact with them.
- Never stand on drums or other containers. This is extremely dangerous. Use mobile steps or another platform to achieve the height necessary to take samples.
- Obtain samples with glass rods or vacuum pumps. Do not use contaminated items such as discarded rags. Such items may contaminate the sample and may not be compatible with the waste in the drum.
- Do not draw samples into a pipette with your mouth.

CHARACTERIZING WASTES

The goal of characterization is to obtain the data necessary to determine how to package and transport the wastes for treatment and/or disposal. If wastes are bulked, they must be characterized to determine whether they can be safely combined with other wastes (see "Bulking"). As a first step

in obtaining these data, use standard tests to classify the wastes into general categories, including autoreactives, water reactives, inorganic acids, organic acids, heavy metals, pesticides, cyanides, inorganic oxidizers, and organic oxidizers (this is called *finger-printing*). In some cases, further analysis should be conducted to identify the materials more precisely.

Whenever possible, characterize materials using an on-site laboratory. This provides data as rapidly as possible and minimizes the time lag before appropriate action can be taken to handle any hazardous materials. Also, it precludes any potential problems associated with transporting samples to an off-site laboratory (e.g., sample packaging, waste incompatibility, fume generation).

If samples must be analyzed off-site, package samples on-site in accordance with DOT regulations (49 CFR) and ship them to the laboratory for analysis.

STAGING DRUM MOVEMENT

Although every attempt should be made to minimize container handling, containers sometimes must be staged (i.e., moved in an organized manner to predesignated areas) to facilitate characterization and remedial action and to protect them from potentially hazardous site conditions (e.g., movement of heavy equipment or high temperatures that might cause explosion, ignition, or pressure buildup). Staging involves a trade-off between the increased hazards associated with drum movement and the decreased hazards associated with the enhanced organization and accessibility of the waste materials.

The number of staging areas needed depends on site-specific circumstances, such as the scope of the operation, the accessibility of drums in their original positions, and the perceived hazards. Investigation usually involves little, if any, staging. Remedial and emergency operations can involve extensive staging. The extent of staging must be determined individually for each site and should be kept to a minimum. Up to four separate areas have been used:

- Locate an *opening area,* where containers are opened, sampled, and resealed, a safe distance from the original waste disposal or storage site and from all staging areas to prevent a chain reaction in case of fire and explosion.
- During large-scale remedial or emergency tasks, a separate *sampling area* may be set up at some distance from the opening area to reduce the number of people present in the opening area and to limit potential casualties in case of an explosion.
- Drums are temporarily stored in a *second staging area,* also known as a holding area, after sampling and pending characterization of their contents. Do not place unsealed drums with unknown contents in the second staging area in case they contain incompatible materials. (Either remove the contents or overpack the drum.)

- In a *final staging area,* also known as a bulking area, substances that have been characterized are bulked for transport to treatment or disposal facilities.

    Locate the final staging area as close as possible to the site's exit.

    Grade the area and cover it with plastic sheets.

    Construct approximately 1-foot-high (0.3-meter-high) dikes around the entire area.

    Segregate drums according to their basic chemical categories (acids, heavy metals, pesticides, etc.) as determined by characterization. Construct separate areas for each type of waste present to preclude the possibility of mixing incompatible chemicals when bulking.

In all staging areas, stage the drums two-wide in two rows per area and space these rows 7 to 8 feet (2 to 2.5 meters) apart to enable movement of drum-handling equipment and facilitate visual inspection and monitoring of drum conditions. Similar practices should be used for staging other types of containers.

BULKING WASTES

Wastes that have been characterized are often mixed together and placed in bulk containers such as tanks or vacuum trucks for shipment to treatment or disposal facilities. This increases the efficiency of transportation. Bulking should be performed only after thorough waste characterization by trained and experienced personnel. The preliminary tests described under "Characterization" provide only a general indication of the nature of the individual wastes. In most cases, you should conduct additional sampling and analysis to characterize the wastes further and run compatibility tests in which small quantities of different wastes are mixed together under controlled conditions and observed for signs of incompatibility, such as vapor generation and heat of reaction. Bulking is performed at the final staging area using the following procedures:

- Inspect each tank trailer and remove any residual materials from the trailer prior to transferring any bulked materials. This will prevent reactions between incompatible chemicals.

- To move hazardous liquids, use pumps that are safety rated and that have a safety relief valve with a splash shield. Make sure the pump hoses, casings, fittings, and gaskets are compatible with the material being pumped.

- Inspect hose lines before beginning work to ensure that all lines, fittings, and valves are intact and have no weak spots.

- Take special precautions when handling hoses, as they often contain residual material that can splash or spill on personnel operating the hoses. Protect personnel against accidental splashing. Protect lines from vehicular and pedestrian traffic.

- Store flammable liquids in containers approved by OSHA and the National Fire Prevention Association (NFPA).

SHIPPING WASTES

Shipping materials to off-site treatment, storage, or disposal facilities involves the entry of waste-hauling vehicles into the site. Use the following guidelines to enhance the safety of these operations:

- Locate the final staging (bulking) area as close as possible to the site exit.

- Prepare a circulation plan that minimizes conflict between cleanup teams and waste haulers. Install traffic signs, lights, and other control devices as necessary.

- Provide an adequate area for haulers to turn around. Where necessary, build or improve on-site roads.

- Equip vehicles with backup lights and bells.

- Stage haulers in a safe area until the wastes are ready to load. Make sure the drivers stay in the cab with the windows rolled up and the air conditioner off. Minimize the time that drivers spend in hazardous areas.

- Outfit the drivers with appropriate PPE.

- If drums are shipped, tightly seal them prior to loading. Overpack leaking or deteriorated drums prior to shipment. Make sure truck beds and walls are clean and smooth to prevent damage to drums. Do not double-stack drums. Secure drums to prevent shifting during transport.

- Make sure the drums are properly labeled according to EPA and DOT requirements.

- Keep bulk solids at least 6 inches (15 centimeters) below the top of the container. Cover loads with a layer of clean soil or foam and/or a tarp. Secure the load to prevent shifting or release during transport. Any free liquids, including rainfall, should be pumped off of solidified wastes prior to shipment.

- Placard vehicles for the material being hauled.

- Prepare and provide drivers with a manifest meeting federal and state requirements.

- Weigh vehicles periodically to ensure that vehicle and road weight limits are not exceeded.

- Decontaminate the vehicle's tires prior to their leaving the site to ensure that contamination is not carried onto public roads.

- Monitor the vehicles periodically to ensure that they are not releasing dust or vapor off-site.

- Develop procedures for responding quickly to off-site vehicle breakdowns and accidents to ensure minimum public impact.

- Review drivers' PPE (if any) and question their spill-response procedures. Provide drivers with an emergency phone list.

SPECIAL CASE PROBLEMS
LABORATORY PACKS

- Once a lab pack has been opened, have a chemist inspect and classify the containers within it, without opening them, according to the scheme outlined in Table 7.8, which is used by most facilities accepting such wastes. This scheme allows for the safest segregation of the lab pack's contents.

- If radioactive contents are suspected (e.g., some containers are labeled radioactive or are heavily shielded, or radiation levels from a drum exceed background), immediately close and isolate the lab pack and establish a restricted perimeter based on radiation measurements.

- Contact the state radiation officer and obtain the services of a qualified health physicist to evaluate the potential radiation hazards and appropriate storage and management requirements. Do not handle suspected radioactive materials or enter restricted radiation storage areas until hazards have been fully characterized, appropriate health and safety procedures have been developed, and the necessary equipment and supplies are obtained. Personnel handling suspected radioactive materials must be properly trained and under the direction of a qualified radiation health physicist.

- If crystalline material is noted at the neck of any bottle, handle it as a shock-sensitive waste, due to the potential presence of picric acid or another similar material. Pack these bottles, no more than five to a drum, with vermiculite and ship them to a disposal facility. Label the drums appropriately and store them remotely and securely.

TANKS AND VAULTS

- In general, when opening a tank or vault, follow the same procedures as for a sealed drum. If necessary, vent excess pressure if volatile substances are stored. Place deflecting shields between the worker and the opening to protect the worker from possible detonation of the materials and to prevent direct contamination by materials forced out by pressure.

- Guard manholes or access portals to prevent personnel from falling into a tank (29 DFR 1910, Subpart B).

- Identify the contents through sampling and analysis. If characterization indicates that the contents can be safely moved with the available equipment, pump them into a trailer for transportation to a disposal or recycling facility.

- Purge the tank or vault with an inert gas, then remove the tank or vault for scrap.

TABLE 7.8 Lab Pack Content Classification System for Disposal

Category Number	Type	Examples
1	Inorganic acids	Hydrochloric acid Sulfuric acid
2	Inorganic bases	Sodium hydroxide Potassium hydroxide
3	Strong oxidizing agents	Ammonium nitrate Barium nitrate Sodium chlorate Sodium peroxide
4	Strong reducing agents	Sodium thiosulfate Oxalic acid Sodium sulphite
5	Anhydrous organics and organometalics	Tetraethyl lead Phenylmercuric chloride
6	Anhydrous inorganics and metal hydrides	Potassium hydride Sodium hydride Sodium metal Potassium
7	Toxic organics	PCBs Insecticides Carcinogens
8	Flammable organics	Hexane Toluene Acetone
9	Inorganics	Sodium carbonate Potassium chloride
10	Inorganic cyanides	Potassium cyanide Sodium cyanide Copper cyanide
11	Organic cyanides	Cyanoacetamide
12	Toxic metals	Arsenic Cadmium Lead

- If it is necessary to enter a tank or vault for any reason (e.g., to clean off solid materials or sludge on the bottom or sides), a confined-space entry procedure must be established in the site safety and health plan. The procedure should include the following precautions:
  Prior to entry, ventilate the site thoroughly.
  Disconnect any connecting pipelines and pumps.
  Prior to entry, take air samples to ensure that safe entry can be achieved based on adequate levels of oxygen, air mixtures are not flammable, and workers can be adequately protected from any hazardous vapors that may be present.
  Equip the entry team and a safety observer with safety harnesses, ropes, SCBAs, and appropriate CPC.

Establish lifeline signals prior to entry so that workers and the safety observer can communicate by tugs on the rope.

Have an additional person available in the immediate area to call for outside assistance and provide support to the safety observer, if needed.

Do not allow anyone to enter a tank or vault if the size of the opening requires the person to squeeze through, as the worker will not be able to exit quickly in an emergency.

VACUUM TRUCKS

- Sample from the top of the vehicle.

- If possible, use mobile steps, powered platforms, or suitable scaffolding consistent with OSHA requirements in 29 CFR 1910, Subparts D and F. Avoid climbing up the ladder and walking across the tank catwalk.

- If the truck must be climbed, raise and lower equipment and samples of carriers to enable workers to use two hands while climbing.

- Wear appropriate protective clothing and equipment when opening the hatch.

- If it is necessary to sample from the drain spigot, take steps to prevent spraying of excessive substances. Have all personnel stand off to the side. Have sorbent materials on hand in the event of a spill.

ELEVATED TANKS

In general, observe the safety precautions described for vacuum trucks. In addition, follow these procedures:

- Use a safety line and harness.

- Maintain ladders and railings in accordance with OSHA requirements (29 CFR 1910, Subpart D).

COMPRESSED-GAS CYLINDERS

- Obtain expert assistance in moving and disposing of compressed-gas cylinders.

- Handle compressed-gas cylinders with extreme caution. The rupture of a cylinder may result in an explosion, and the cylinder may become a dangerous projectile.

- Record the identification numbers on the cylinders to aid in characterizing their contents.

PONDS AND LAGOONS

- Drowning is a very real danger for workers wearing PPE because the weight of the equipment increases an individual's overall density and severely impairs his swimming ability. Where there is a danger of drowning, provide necessary safety gear such as lifeboats, tag lines, railings, nets, safety belts, and flotation gear.

- Wherever possible, stay on shore. Avoid going out over the water.

- If possible, use a vertical extended platform consistent with 29 CFR 1910, Subparts D and F.

BIBLIOGRAPHY

Bierlein, L. *Red Book on Transportation of Hazardous Materials.* 2d ed. New York: Van Nostrand Reinhold, 1987.

de Vera, E.R., B.P. Simmons, R.D. Stephens, and D.L. Storm. *Samplers and Sampling Procedures for Hazardous Waste Streams.* U.S. Environmental Protection Agency (EPA) Publication No. 600/2–80–018. Cincinnati: EPA, January 1980.

Esposito, M.P., and R. Clark. *Decontamination Techniques for Buildings, Structures, and Equipment.* Parkridge, N.J.: Noyes Data Corporation, 1987.

Green, M., and A. Turck. *Safety in Working with Chemicals.* New York: McMillan Publ. Co., 1978.

Manufacturing Chemists Association. *Guide for Safety in the Chemical Laboratory.* New York: Van Nostrand Reinhold, 1972.

Mayhew, J.J., G.M. Sodear, and D.W. Carroll. *A Hazardous Waste Site Management Plan.* Washington, D.C.: Chemical Manufacturers Association, Inc., 1982.

U.S. Coast Guard Department of Transportation (DOT). *Response Methods Handbook of Chemical Hazards Response Information System (CHRIS).* Washington, D.C.: DOT, 1978.

8

DECONTAMINATION

econtamination—the process of removing and neutralizing contaminants that have accumulated on personnel and equipment—is critical to health and safety at hazardous waste sites. Decontamination protects workers from hazardous substances that may contaminate and eventually permeate the PPC, respiratory equipment, tools, vehicles, and other equipment they use on-site. It also protects all site personnel by minimizing the transfer of harmful materials into clean areas, and it helps to prevent mixing of incompatible chemicals.

This chapter describes the types of contami-

nation that workers may encounter at a waste site, the factors that influence the extent of contamination, and methods for preventing or reducing contamination. In addition, this chapter provides general guidelines for designing and selecting decontamination procedures at a site, and it presents a decision logic for evaluating the health and safety aspects of decontamination methods. The chapter does not cover decontamination of radioactively contaminated personnel and equipment. Procedures for radiological decontamination should be prepared and supervised by a qualified radiation health physicist.

DECONTAMINATION PLAN

A decontamination program must be developed and set up before any personnel or equipment enter areas where the potential for exposure to hazardous substances exists. All aspects of the decontamination program must be organized and documented in a written plan that should do the following:

- establish methods and procedures to minimize worker contact with contaminants during removal of PPE

- establish procedures to prevent contamination of clean areas

- determine appropriate decontamination methods

- determine the number and layout of decontamination stations

- identify incompatible wastes requiring separate decontamination stations

- determine the decontamination equipment needed

- establish methods for disposing of clothing and equipment that are not completely decontaminated

- establish the target level of decontamination (A quick test of the final rinse solution may be used to determine the effectiveness of the decontamination.)

The plan should be revised whenever the type of protective clothing or equipment changes, the site conditions change, or the site hazards are reassessed based on new information.

PREVENTING CONTAMINATION

The first step in decontamination is to establish SOPs that minimize contact with wastes and thus the potential for contamination. Through training, make all site personnel aware of the importance of minimizing contact and enforce the appropriate practices and procedures throughout site operations.

- Stress work practices that minimize contact with hazardous substances (e.g., do not walk through areas of obvious contamination; do not directly touch potentially hazardous substances).

- Use remote sampling, handling, and container-opening techniques (e.g., drum grapplers, pneumatic impact wrenches).

- Protect monitoring and sampling instruments by bagging. Make openings in the bags for sample ports and sensors that must contact site materials.

- Wear disposable outer garments and use disposable equipment where appropriate.

- Cover equipment and tools with a coating that can be removed during decontamination.

- Encase the source of contaminants with plastic sheets or overpack containers.

SOPs should maximize worker protection. For example, proper procedures for dressing prior to entering the exclusion zone minimize the potential for contaminants to bypass the protective clothing and escape decontamination. In general, all fasteners should be used (i.e., zippers fully closed, all buttons buttoned, all snaps snapped). Gloves and boots should be tucked under the cuffs of the arms and legs of outer clothing, and hoods (if not attached) should be outside of the collar. This prevents contaminants from running inside the gloves, boots, and jackets or suit.

TYPES OF CONTAMINATION

Contamination of PPE can be categorized as *surface contamination* or *permeation* (the diffusion of molecules from one substance into another). Surface contaminants are usually easier to detect and remove than contaminants that have permeated the PPE. Pollutants that permeate a material and are not removed by decontamination may be gradually released to the surface, where they may cause an unexpected exposure.

Five major factors affect the extent of permeation:

- *Contact time.* The longer a contaminant is in contact with an object, the greater the probability and extent of permeation. For this reason, minimizing contact time is one of the most important objectives of a decontamination program.

- *Concentration.* Molecules flow from areas of high concentration to areas of low concentration. As concentrations of wastes increase, the potential for permeation of PPC increases.

- *Temperature.* An increase in temperature generally increases the permeation rate of contaminants.

- *Size of contaminant molecules and pore space.* Permeation increases as the contaminant molecule becomes smaller and the pore space of the material to be permeated increases.

- *Physical state of wastes.* As a rule, gases and low-viscosity liquids tend to permeate more readily than high-viscosity liquids or solids.

Any openings in PPE, such as cuts, punctures, inadequately protected zippers, or open areas between two garments (e.g., between cuffs and gloves), may result in direct worker contact with waste materials. Prior to each use, PPE should be checked to ensure that no such openings expose workers to wastes. Similarly, any injuries to the skin surface, such as cuts or scratches, may enhance the potential for chemicals or infectious agents to penetrate into the body. Particular care should be taken to protect these areas.

DECONTAMINATION METHODS

All personnel, clothing, equipment, and samples leaving the contaminated area of a site (generally referred to as the exclusion zone) must be decontaminated to remove any harmful chemicals or infectious organisms that may have adhered. Decontamination methods either (1) physically remove contaminants by one of several processes, (2) inactivate contaminants by chemical detoxification or disinfection/sterilization, or (3) remove contaminants by a combina-

TABLE 8.1 Decontamination Methods

Removal

Contaminant removal
    Water rinse using pressurized or gravity flow
    Chemical leaching and extraction
    Evaporation/vaporization; should be used only with care to prevent possible inhalation hazards
    Pressurized air jets
    Scrubbing/scraping; commonly done using brushes, scrapers, or sponges and water-compatible solvent cleaning solutions
    Steam jets; commonly used with solvent cleaning solutions
Removal of contaminated surfaces
    Disposal of deeply permeated materials (e.g., clothing, floor mats, seats)
    Disposal of protective coverings/coatings

Inactivation

Chemical detoxification
    Halogen stripping
    Neutralization
    Oxidation/reduction
    Thermal degradation
Disinfection/sterilization
    Chemical disinfection
    Dry heat sterilization
    Gas/vapor sterilization
    Irradiation
    Steam sterilization

tion of both physical and chemical means. Various types of decontamination methods are listed in Table 8.1.

Because of the potential hazards associated with options such as high-pressure washes, some of the more aggressive decontamination procedures are not suitable for decontamination of clothing, hand-held equipment, and personnel. In addition, PPE and the hazards associated with the method of decontamination should be considered when selecting a particular decontamination procedure.

PHYSICAL REMOVAL

In many cases, gross contamination can be removed by physical means involving dislodging/displacement, rinsing, wiping, and evaporation. Contaminants that can be removed by physical means can be categorized as follows:

- *Loose contaminants.* Dusts and vapors that cling to equipment and workers or become trapped in small openings, such as the weave of fabrics, can be removed with water or a liquid rinse. Removal of electrostatically attached materials can be facilitated by coating the clothing or equipment with antistatic solutions. These are available commercially as wash additives or antistatic sprays.

- *Adhering contaminants.* Some contaminants adhere by forces other than electrostatic attraction. Adhesive qualities vary greatly with the specific contaminants and the temperature. For example, contaminants such as glue, cement, resin, and mud have much greater adhesive properties than elemental mercury and consequently are difficult to remove by physical means. Physical removal methods for gross contaminants include scraping, brushing, and wiping. Removal of adhesive contaminants can be enhanced through methods such as solidifying, freezing (e.g., using dry ice or ice water), adsorption or absorption (e.g., with powdered lime or kitty litter), or melting (changing high-viscosity solids to lower-viscosity liquids).

- *Volatile liquids.* Volatile liquid contaminants can be removed from protective clothing or equipment by evaporation followed by a water rinse. Evaporation of volatile liquids can be enhanced by using steam jets. With any evaporation or vaporization process, care must be taken to prevent worker inhalation of the vaporized chemicals.

CHEMICAL REMOVAL

Physical removal of gross contamination should be followed by a wash/rinse process using cleaning solutions. These cleaning solutions normally work by one or more of the following methods:

- *Solubilization.* Chemical removal of surface contaminants via solubilization (i.e., the process of dissolving contaminants with a solvent) is successful only if the contaminants have a greater affinity for the solvent than for the surface that is being cleaned. Solvents must be chemically compatible with the contaminants and the equipment being cleaned. This is particularly important when decontaminating PPC made of organic materials that could be damaged or dissolved by organic solvents. In addition, care must be taken in selecting, using, and disposing of any organic solvents that may be flammable or potentially toxic. Organic solvents include alcohols, ethers, ketones, aromatics, straight-chain alkanes, and common petroleum products. Halogenated solvents, such as carbon tetrachloride, are incompatible with PPE and are toxic. They should be used only for decontamination in cases where other cleaning agents will not remove the contaminant. Table 8.2 provides a general guide to the solubility of several contaminant categories in four types of solvents: water, dilute acids, dilute bases, and organic solvents.

- *Surfactants.* Surfactants augment physical cleaning methods by reducing adhesion between contaminants and the surface being cleaned and by preventing redeposit of the contaminants. Household detergents are among the most common surfactants. Some deter-

TABLE 8.2 General Guide to Solubility of Contaminants in Four Solvent Types

Solvent	Contaminant
Water	Low-chain hydrocarbons Inorganic compounds Salts Some organic acids and other polar compounds
Dilute acids	Basic (caustic) compounds Amines Hydrazines
Dilute bases—detergent, soap	Acidic compounds Phenols Thiols Some nitro and sulfonic compounds
Organic solvents—alcohols, ethers, ketones, aromatics, straight-chain alkanes (e.g., hexane), common petroleum products (e.g., fuel oil, kerosene)	Nonpolar compounds (e.g., some organic compounds) (*Warning:* Some organic solvents can solubilize the material from which the protective clothing is made.)

gents can be used with organic solvents to improve the dissolving and dispersal of contaminants into the solvent.

- *Solidification.* Solidifying liquid or gel contaminants can enhance their physical removal. The mechanisms of solidification are (1) moisture removal through the use of absorbents such as ground clay or powdered lime, (2) chemical reactions via polymerization catalysts and chemical reagents, and (3) freezing using ice water.

- *Rinsing.* Rinsing removes contaminants through dilution, physical attraction, and solubilization. Multiple rinses with clean solutions remove more contaminants than a single rinse with the same volume of solution. Continuous rinsing will remove even more contaminants than multiple rinses.

- *Disinfection/sterilization.* Chemical disinfectants are the most practical means of disinfecting/sterilizing infectious agents because standard sterilization techniques are impractical for large equipment and for personal protective clothing and equipment.

Many factors, such as cost, availability, and ease of implementation, influence the selection of a decontamination method. From a health and safety standpoint, two key questions must be addressed:

- Is the decontamination method effective for the specific substances present?
- Does the method itself pose any health or safety hazards?

EFFECTIVENESS TESTING

Decontamination methods vary in their effectiveness for removing different substances. The effectiveness of any decontamination method should be assessed at the beginning of a program and periodically throughout its life. If contaminated materials are not being removed or are penetrating PPC, the decontamination program must be revised. The following methods may be useful in assessing the effectiveness of decontamination.

VISUAL TESTING

There is no reliable test to determine immediately how effective decontamination is. In some cases, effectiveness can be estimated by visual observation.

- *Natural light.* Discoloration, stains, corrosive effects, visible dirt, or alterations in clothing fabric may indicate that contaminants have not been removed. However, not all contaminants leave visible traces; many can permeate clothing and are not easily observed.
- *Ultraviolet (UV) light.* Certain contaminants, such as polycyclic aromatic hydrocarbons, which are common in many refined oils and solvent wastes, fluoresce and can be visually detected when exposed to UV light. Ultraviolet light can be used to observe contamination of skin, clothing, and equipment. However, certain areas of the skin may fluoresce naturally, thereby reducing the reliability of the test. In addition, use of UV light can increase the risk of skin cancer and eye damage. Therefore, a qualified health care professional must assess the benefits and risks associated with UV light prior to its use at a waste site.

SWIPE SAMPLING

Swipe testing provides information on the effectiveness of decontamination. In this procedure, a cloth or paper patch (swipe) is wiped over the surface of the potentially contaminated object and then analyzed in a laboratory. Both the inner and outer surfaces of PPC should be tested. Skin also may be tested using swipe samples.

RINSE SOLUTION ANALYSIS

Another way to test the effectiveness of decontamination procedures is to analyze for contamination left in the cleaning solutions. Field test kits or pH paper allow these analyses to be performed in real time. Elevated levels of contaminants in the final rinse solution may suggest that additional cleaning and rinsing are needed.

PERMEATION TESTING

Permeation testing for the presence of chemical contaminants requires that pieces of the protective garments be sent to a laboratory for analysis.

SWAB SAMPLING

To assess the effectiveness of disinfection, swab samples can be taken from contaminated surfaces and sent to a testing laboratory for analysis. Concentrations of active infectious organisms also can be measured in the spent solutions to determine whether sufficient levels of active disinfectants were used. The solutions should be tested periodically to ensure that the concentration of disinfectant is satisfactory. If tests indicate that contaminants are not being removed from the surface or are permeating clothing, consider whether more extensive decontamination methods are needed or protective measures should be changed (e.g., use less permeable materials or disposable outer garments). Use pH paper by dipping it into the wash and/or rinse solution or wiping it across any moist surface that may be contaminated. The results are immediate and can be compared to the pH scale on the container.

HEALTH AND SAFETY HAZARDS

While decontamination is performed to protect health and safety, it can pose hazards under certain circumstances. Decontamination methods may

- be incompatible with the hazardous substances being removed (i.e., a method may react with contaminants to produce an explosion, heat, or toxic products)
- be incompatible with the clothing or equipment being decontaminated (e.g., some organic solvents will solubilize the fabrics from which protective clothing is made)
- pose a direct health hazard to workers (e.g, vapors from chemical decontamination solutions may be hazardous if inhaled, or they may be flammable)

The chemical and physical compatibility of the decontamination solutions and other materials must be determined before they are used. Any method that permeates, degrades, damages, or otherwise impairs the safe functioning of the PPE is incompatible with it and should not be used. If a method does pose a direct health hazard, measures must be taken (e.g., wear a respirator if using a solution that produces hazardous vapors) to protect both decontamination personnel and the workers being decontaminated.

LEVELS OF DECONTAMINATION

Several site-specific factors determine the levels and types of decontamination procedures required at a site. Some of these factors follow:

- the chemical, physical, and toxicological properties of the wastes
- the pathogenicity of infectious wastes
- the amount, location, and containment of contaminants
- the potential for and location of exposure based on assigned worker duties, activities, and functions

- the potential for wastes to permeate, degrade, or penetrate materials used for protective clothing and equipment, vehicles, tools, and buildings and other structures
- the proximity of incompatible wastes
- the reasons for leaving or removing equipment from the site
- the methods available for protecting workers during decontamination
- the impact of the decontamination process and compounds on worker safety and health

DECONTAMINATION FACILITY DESIGN

At a hazardous waste site, decontamination facilities should be located in the CRZ—that is, in the area between the contamination zone (often called the exclusion zone) and the clean zone (often called the support zone).

Decontamination procedures must provide an organized process by which levels of contamination are reduced. The decontamination process should consist of a series of procedures performed in a specific sequence. For example, outer, more heavily contaminated items (such as boots and gloves) should be decontaminated and removed first, followed by inner, less contaminated items (such as jackets and pants). Each procedure should be performed at a separate station to prevent cross-contamination. The sequence of stations is called the *decontamination line*. Separate decontamination lines may be required if incompatible wastes are being handled at different work locations in the exclusion zone.

Stations should be separated physically to prevent cross-contamination and should be arranged in order of decreasing contamination, preferably in a straight line. Separate flow patterns and stations should be provided to isolate workers from different contamination zones containing incompatible wastes. Entry and exit points should be conspicuously marked and the entry to the CRZ from the exclusion zone should be separate from the entry to the exclusion zone from the CRZ. Dressing stations for entry to the CRZ should be separate from redressing areas for exit from the CRZ. Personnel who wish to enter clean areas of the decontamination facility, such as locker rooms, should be completely decontaminated.

The doffing station for respiratory protective equipment should always occur after the garments are removed to maximize respiratory protection while decontaminating.

For further information on decontamination procedures for more typical levels of personal protection, see Appendix I and Chapter 6, Personal Protective Equipment.

DECONTAMINATION EQUIPMENT SELECTION

Table 8.3 lists recommended equipment for decontamination of personnel and protective clothing and equipment. In selecting decontamination equipment, consider whether the equipment itself can be decontaminated for reuse or can be

TABLE 8.3 Recommended Equipment for Decontamination of Personnel and Personal Protective Clothing and Equipment

Drop cloths of plastic or other suitable material for placement of heavily contaminated equipment, monitoring equipment, tools, and outer protective clothing

Collection containers, such as drums or suitably lined trash cans, for storing disposable clothing and heavily contaminated protective clothing or equipment that must be discarded

Lined box with absorbents for wiping or rinsing off gross contaminants and liquid contaminants

Benches so workers can sit down to remove boot covers

Large galvanized tubs, stock tanks, or children's wading pools to hold wash and rinse solutions; should be at least large enough for a worker to place a booted foot in and have either no drain or a drain connected to a collection tank or appropriate treatment system

Wash solutions selected to wash off and reduce the hazards associated with the contaminants

Rinse solutions selected to remove contaminants and contaminated wash solutions

Long-handled, soft-bristled brushes to help wash and rinse off contaminants

Paper or cloth towels for drying protective clothing and equipment

Lockers and cabinets for storage of decontaminated clothing and equipment

Metal or plastic cans or drums for contaminated wash and rinse solutions

Plastic sheeting, sealed pads with drains, or other appropriate methods for containing and collecting contaminated wash and rinse solutions spilled during decontamination

Shower facilities for full body wash or, at a minimum, personal wash sinks (with drains connected to a collection tank or appropriate treatment system)

Soap or wash solution, washcloths, and towels

Lockers and closets for clean clothing and personal item storage

easily disposed of. Table 8.4 lists recommended equipment for decontamination of heavy equipment and vehicles. Note that other types of equipment not listed in Tables 8.3 and 8.4 may be appropriate in certain situations.

DISPOSAL METHODS

All equipment used for decontamination must be disposed of properly. Buckets, brushes, clothing, tools, and other contaminated equipment should be collected, placed in appropriate containers, and labeled. Also, all spent cleaning solutions should be collected and disposed of properly. Disposable PPC and clothing that cannot be adequately decontaminated should be placed in appropriate containers for disposal.

TABLE 8.4 Recommended Equipment for Decontamination of Heavy Equipment and Vehicles

Pads for collection of contaminated wash and rinse solutions (with drains or pumps connected to storage tanks or appropriate treatment systems)

Long-handled brushes for general exterior cleaning

Wash solutions selected to remove and reduce the hazards associated with the contamination

Rinse solutions selected to remove contaminants and contaminated wash solutions

Pressurized sprayers for washing and rinsing, particularly hard-to-reach areas

Curtains, enclosures, or spray booths to contain splashes from pressurized sprays

Long-handled brushes, rods, and shovels for dislodging contaminants and contaminated soil caught in tires and the underside of vehicles and equipment

Containers for the removal of contaminants and contaminated soil caught in tires and the underside of vehicles and equipment

Wash and rinse buckets for use in the decontamination of operator areas inside vehicles and equipment

Brooms and brushes for cleaning operator areas inside vehicles and equipment

Containers for storage and disposal of contaminated wash and rinse solutions, damaged or heavily contaminated parts, and equipment to be discarded

All decontamination wastes should be evaluated and characterized to determine appropriate disposal options in accordance with the EPA's hazardous waste regulations in 40 CFR 261, 265, 264, and 278 and any applicable state and local regulations.

PERSONNEL PROTECTION

Decontamination workers who initially come in contact with personnel and equipment leaving the exclusion zone require more protection from contaminants than decontamination workers assigned to the last station in the decontamination line. In some cases, decontamination personnel should wear the same levels of PPE as workers in the exclusion zone. In other cases, decontamination personnel may be sufficiently protected by wearing protection that is one level lower (e.g., wearing Level C protection while decontaminating workers who are wearing Level B protection).

The level of protection required varies with the type of decontamination equipment used. For example, workers using a steam jet may need a different type of respiratory protection than other decontamination personnel because of the high moisture levels produced by steam jets. In some situations, the cleaning solutions used and the wastes removed during decontamination may generate harmful vapors. Appropriate equipment and clothing for protecting decontamination personnel should be selected by a qualified health and safety expert.

All decontamination workers are in a contaminated area and must decontaminate themselves before entering the clean support zone. The extent of their decontamination should be determined by the types of contaminants they may have contacted and the type of work they performed.

EMERGENCY DECONTAMINATION

In addition to routine decontamination procedures, emergency decontamination procedures must be established. In an emergency, the primary concern is to prevent the loss of life or severe injury to site personnel. If immediate medical treatment is required to save a life, decontamination should be delayed until the victim is stabilized. If decontamination can be performed without interfering with essential lifesaving techniques or first aid, or if a worker has been contaminated with an extremely toxic or corrosive material that could cause severe injury or loss of life, decontamination must be performed immediately. If an emergency due to a heat-related illness develops, the victim's PPC should be removed as soon as possible to reduce heat stress.

During an emergency, provisions must be made to notify and protect medical personnel and arrange for the collection and disposal of contaminated clothing and equipment.

BIBLIOGRAPHY

Esposito, M.P., J.L. McArdle, A.H. Crone, J.S. Greber, R. Clark, S. Brown, J.B. Hallowell, A. Langham, and C.D. McCandlish. *Guide for Decontaminating Buildings, Structures, and Equipment at Superfund Sites.* U.S. Environmental Protection Agency (EPA) Publication No. 600/2–85–028. Cincinnati: EPA, March 1985.

Lippitt, J.M., T.G. Prothero, and L.P. Wallace. "Contamination Reduction/Removal Methods." In *Protecting Personnel at Hazardous Waste Sites,* S.P. Levine and W.F. Martin. eds., Boston: Butterworth-Heinemann, 1985.

Rosen, M.J. *Surfactants and Interfacial Phenomena.* New York: John Wiley & Sons, 1978.

9

SITE EMERGENCIES

*T*he nature of work at hazardous waste sites makes emergencies a continual possibility, no matter how infrequently they may occur. Emergencies happen quickly and unexpectedly and require an immediate response. At a hazardous waste site, an emergency may be as limited as a worker experiencing heat stress or as vast as an explosion that spreads toxic fumes throughout a community. Any hazard on-site can precipitate an emergency: chemicals, biological agents, radiation, or physical hazards may act alone or in concert to create explosions, fires, spills, toxic atmospheres, or other dangerous and harmful situations. Table 9.1 lists common causes of site emergencies.

Site emergencies are characterized by their potential for complexity: uncontrolled toxic chemicals may be numerous and unidentified, and their effects may be synergistic. Hazards may potentiate one another—for example, a flam-mable spill feeding a fire. Rescue personnel attempting to remove injured workers may themselves become victims. This variability means that advance planning, including anticipation of different emergency scenarios and thorough preparation for contingencies, is essential to protect worker and community health and safety.

This chapter outlines important factors to be considered when planning for and responding to emergencies. It defines the nature of site emergencies, lists the types of emergencies that may occur, and outlines an emergency contingency plan and its components, which include personnel roles, lines of authority, training, communication systems, site mapping, site security and control, refuges, evacuation routes, decontamination, a medical program, step-by-step emergency response procedures, documentation, and reporting to outside agencies. Backup information is detailed in other chapters.

PLANNING

When an emergency occurs, decisive action is required. Rapidly made choices may have far-reaching, long-term consequences. Delays of minutes can create life-threatening situations. Personnel must be ready to rescue or respond immediately; equipment must be on hand and in good working order. To handle emergencies effectively, planning is essential. For this purpose, a contingency plan must be developed. Under Superfund regulations (40 CFR 300.41, 42, 43), contingency (emergency) response plans are required at sites where Superfund work is being conducted.

A *contingency plan* is a written document that sets forth comprehensive policies and procedures for emergency response. It should incorporate the following items:

- personnel
    roles
    lines of authority
    training
    communication
- site
    safe distances

TABLE 9.1 Causes of Emergencies at Hazardous Waste Sites

Worker Related

Minor accidents (slips, trips, falls)
Chemical exposure
Medical problems (heat stress, heatstroke, cold exposure, aggravation of preexisting conditions)
Personal protective equipment failure (air source failure, tearing or permeation of protective clothing, face piece fogging)
Physical injury (from hot or flying objects, loose clothing entangling in machinery, serious falls)
Electrocution (exposure, burns, shock, possible death)

Waste Related

Fire
Explosion
Leak
Release of toxic vapors
Incompatible reaction
Collapse of containers
Deflagration

mapping
security and control
refuges
evacuation routes and procedures
decontamination stations

- emergency medical treatment and first aid
- PPE and emergency equipment
- emergency alerting and response procedures
- documentation procedures
- emergency recognition and prevention
- reporting
- critique of response and follow-up

Overall, a contingency plan should

- be designed as a discrete section of the site safety plan
- be compatible and integrated with the pollution response, disaster, fire, and emergency plans of local, state, and federal agencies
- be rehearsed regularly using drills and mock situations
- be reviewed periodically in response to new or changing site conditions or information

Situation-specific contingency plans should be outlined at the time of the emergency before an investigation or response. No one should attempt an emergency response or rescue until backup personnel and evacuation routes are identified.

PERSONNEL

This category includes not only on-site and off-site personnel with specific emergency response roles but also others who may be on-site, such as contractors, other agency representatives, and visitors. Emergency personnel and their responsibilities are covered in detail in Chapter 3 as part of the overall organizational structure. This information is summarized in Table 9.2.

Emergency personnel may be deployed in a variety of ways. Depending on the nature and scope of the emergency, the size of the site, and the number of personnel, the emergency response cadre might include individuals, small or large teams, or a number of interacting teams. Although deployment is determined on a site-by-site basis, general guidelines and recommendations are given here.

ON-SITE PERSONNEL

The contingency plan should identify all individuals and teams who will participate in an emergency response and define their roles. All personnel, whether directly involved in an emergency response or not, should know their responsibilities in an emergency. They also must know the names of those in authority and the extent of that authority.

TABLE 9.2 Personnel Involved in Emergency Response

Site safety officer
Has authority to stop work if any operation threatens worker or public health or safety
Knows emergency procedures, evacuation routes, and the telephone numbers of the ambulance, medical facility, poison control center, fire department, and police department
Notifies local police emergency officers
Provides emergency medical care on-site

Command post supervisor
Notifies emergency support personnel by telephone or radio in case rescue operations are required
Assists the site safety officer in a rescue if necessary

Emergency response team
Stands by, partially dressed in protective gear, near the exclusion zone and rescues any workers whose health or safety are endangered
Includes, in the event of a chemical release, the National Response Center (NRC), which can contact regional response teams, emergency response teams, or strike teams of the National Strike Force, if necessary
May include state emergency response personnel (varies among states)

Personnel decontamination station operators
Perform emergency decontamination

24-hour medical team
Includes ambulance personnel, personnel at local clinics or hospitals, and physicians
Transports and treats victims

Communication personnel
Include local emergency service networks, which provide communication links for mutual aid
Include civil defense organizations and local radio and television stations, which provide information to the public during an emergency

Environmental scientists
Predict the immediate and future movement of released hazardous substances through the geologic and hydrologic environment
Assess the effects of this movement on groundwater and surface water quality
Predict the exposure levels of people and the ecosystem to the materials

Fire fighters
Respond to fires that occur at a site
Rescue victims

Meteorologists
Determine the probable movement of released toxic gases
Estimate the expected concentration of gases in the community and the expected duration of exposure

Public safety personnel
Include the county sheriff, industrial security forces, the National Guard, and the police
Control site access, crowds, and traffic

Public evacuation personnel
  Include civil defense organizations, which plan evacuations
  Include the National Guard and other military, the Red Cross, the Salvation Army, and municipal transportation systems, which mobilize transit equipment and assist in evacuations

Leader In an emergency situation, one person must be able to assume total control and decision-making responsibility on-site. The leader must

- be identified in the emergency response plan (for example, the project team leader, SSO, or field team leader)
- be backed up by a specified alternate or alternates
- have authority to resolve all disputes about health and safety requirements and precautions
- be authorized to seek and purchase supplies as necessary
- have authority over everyone entering the site, including contractors, field investigation teams (FITs), technical assistance teams (TATs), fire departments, and police
- have the clear support of management

Teams Although individuals (for example, the SSO) may perform certain tasks in emergencies, in most cases teams provide greater efficiency and safety. Teams composed of on-site personnel may be created for specific emergency purposes, such as decontamination, rescue, and entry. Other options include the following:

- Emergency response teams can be used during a particularly dangerous operation or at large sites with multiple work parties in the exclusion zone. Their sole function is to remain near hazardous work areas, partially dressed in protective gear, ready for full suiting and immediate rescue of any endangered worker. This is an expensive option but may be justified in some situations.
- Emergency support personnel can be used at any size site. Each work team is provided with a special backup member who handles general safety and emergency support. This person
    is alert for hazards
    enforces SOPs
    monitors the air
    has veto authority over any proposed action
    inspects emergency and protective equipment
    is certified in first aid and cardiopulmonary resuscitation (CPR) procedures
    is familiar with site accident history, wastes to be handled, and changes in equipment or work plans
    knows the work crew and its capabilities and can recognize personality and attitude changes that may be caused by exposure to wastes
    in an emergency, stays out of the hazardous area but remains close enough to communicate with workers

summons aid and gives information to the person in charge
wears PPE and is capable of rescuing an unconscious worker in full gear

- Hazard- and worker-related emergency teams are two groups with discrete response capabilities—one for hazardous situations, the other for worker injuries. These are two separate categories calling for different procedures, equipment, and training (e.g., containment and fire fighting for hazards versus rescue and first aid for victims), although there is some overlap and many emergencies combine the two.

OFF-SITE PERSONNEL

These may include individual experts, such as meteorologists or toxicologists (see Table 9.2 for details), and representatives or groups from local, state, and federal organizations offering rescue, response, or support (see Table 9.3 for a list of typical organizations). As part of advance planning, site personnel should

- make arrangements with individual experts to provide guidance as needed
- make arrangements with the appropriate agencies (e.g., local fire department, state environmental agency, EPA regional office) for support
- alert these authorities to the types of emergencies that may arise
- determine their estimated response time and resources
- identify backup facilities
- provide training and information about hazards on-site and special procedures for handling them
- establish a contact person and means of notification at each agency

LINES OF AUTHORITY

The organizational structure should show a clear chain of command. All workers should know their own position and authority, but the chain of command must be flexible enough to handle multiple emergencies—for instance, a rescue and a spill response or two rescues with a fire and spill response.

TRAINING

Since an immediate, informed response is essential in an emergency, all site personnel and others entering the site (visitors, contractors, off-site emergency response groups, and other agency representatives) must have some level of emergency training. Any training program should

- relate directly to potential site-specific situations
- be brief and repeated often

TABLE 9.3 Examples of Agencies and Groups Involved in Emergencies

Agency or Group	Rescue[a]	Response[b]	Support[c]
Federal			
Army Corps of Engineers			X
Bureau of Explosives			X
Coast Guard[d]	X	X	X
Department of Defense			X
Department of Justice			X
Department of Transportation (DOT)			X
Environmental Protection Agency (EPA)[d]	X	X	X
Federal Aviation Administration (FAA)			X
Federal Emergency Management Agency (FEMA)			X
National Institute of Occupational Safety and Health (NIOSH)			X
Occupational Safety and Health Administration (OSHA)			X
State			
Civil defense			X
Environmental Protection Agency (EPA)[d]	X	X	X
Office of the attorney general			X
Department of health			X
Local			
Ambulance and rescue services	X	X	X
Cleanup contractor	X	X	X
Disposal companies	X	X	
Fire department	X	X	X
Field investigation teams (FITs)			X
Hospital			X
Police			X
Red Cross			X
Salvation Army			X
Technical assistance teams (TATs)	X	X	X
Transporters			X
Utility companies (electric, gas, water, phone)			X

[a]Rescue = extricating and/or providing on-the-spot emergency treatment to victims.
[b]Response = controlling and stabilizing hazardous conditions.
[c]Support = providing technical assistance, equipment, and/or resources.
[d]This agency provides an on-scene coordinator (OSC), depending on jurisdiction.

- be realistic and practical

- provide an opportunity for special skills to be practiced regularly

- feature frequent drills (e.g., site-specific mock rescue operations)

- ensure that training records are maintained in a training logbook

Everyone entering the site must be made aware of the hazards and the actions that may trigger them. Everyone also must know what to do in case of an emergency.

Viditors should be briefed on basic emergency procedures, such as decontamination, emergency signals, and evacuation routes.

Personnel without defined emergency response roles (e.g., contractors or federal agency representatives) must receive a level of training that includes, at a minimum, the following points:

- hazard recognition

- site SOPs

- signaling an emergency (the alarm used, how to summon help, what information to give, and whom to give it to)

- evacuation routes and refuges
- the person or station to report to when an alarm sounds

For more details on training, see Chapter 11.

EMERGENCY RECOGNITION AND PREVENTION

On a day-to-day basis, individual personnel should be alert for indicators of potentially hazardous situations and for symptoms in themselves and others that warn of hazardous conditions and exposures. Rapid recognition of dangerous situations can avert an emergency. Before passing out daily work assignments, hold regular tailgate safety meetings. Discuss the following points:

- tasks to be performed
- time constraints (e.g., length of rest breaks and time for air tank changes)
- hazards that may be encountered (including their effects), how to recognize or monitor them, concentration limits, and other danger signals
- emergency procedures

After daily work assignments are completed, hold a debriefing session to review the work accomplished and the problems observed.

COMMUNICATION

In an emergency, crucial messages must be conveyed quickly and accurately. Site staff must be able to communicate information such as the location of injured personnel, orders to evacuate the site, and notice of blocked evacuation routes, even through noise and confusion. Outside support sources must be reached, help obtained, and measures for public notification ensured, if necessary. To do this, a separate set of internal emergency signals should be developed and rehearsed daily. External communication systems and procedures should be clear and accessible to all workers.

INTERNAL COMMUNICATION

Internal emergency communication systems are used to alert workers to danger, convey safety information, and maintain site control. Any effective system or combination of systems may be used. Radios or field telephones often are used when work teams are far from the command post. Alarms or short, clear messages can be conveyed by audible signals (e.g., bullhorns, megaphones, sirens, bells, or whistles) or visual signals (e.g., colored flags, flares, lights, or hand and body movements). Every system must have a backup. For example, hand signals may be used as a backup if radio communications fail. All internal systems should be

- clearly understood by all personnel
- checked and practiced daily
- intrinsically safe (spark-free)

A special set of emergency signals also should be set up. These should be

- different from the ordinary signals
- brief and exact
- limited in number so that they are easily remembered

Examples include signals for stop, evacuate, help, and all clear. Any set of signals may be used to convey these messages as long as all personnel understand their meaning. See Table 9.4 for examples.

When designing and practicing communication systems, remember the following points:

- Background noise on-site will interfere with talking and listening.
- Wearing PPE will impede hearing and limit vision (e.g., the ability to recognize hand and body signals).
- Inexperienced radio operators may need practice in speaking clearly.

EXTERNAL COMMUNICATION

Off-site sources must be contacted to get assistance or to inform officials about hazardous conditions that may affect public or environmental safety. The telephone is the most common mode of off-site communication. Phone hookups are considered a necessity at all but the most remote sites.

- All personnel must be familiar with the protocol (phone number or emergency code and contact person) for contacting public emergency aid teams such as fire departments, ambulance units, and hospitals.

TABLE 9.4 Internal Emergency Communication Signals

Devices	Example Signals
Radio—citizens band or FM	Established code words
Audible signal—bull horn, siren, or whistle,	One short blast: attention getter One long blast: evacuate area by nearest emergency exit Two short blasts: localized problem (not dangerous to workers) Two long blasts: all clear
Visual signals—hand signals or whole body	Hand gripping throat: our of air/can't breathe Hands on top of head: need assistance
movements	Thumbs up: OK/I'm all right/I understand Thumbs down: no/negative Grip partner's wrist or both hands around partner's waist: leave area immediately

- If there is no site telephone system, all personnel must know the location of the nearest public telephone. A supply of change must be available near the phone.

SITE MAPPING

A site map should be drawn as part of the planning that can reduce or minimize emergencies. It serves as a graphic record of the locations and types of hazards, a reference source, and a method of documentation. This map can be a duplicate of the one developed for the health and safety plan (see Chapter 3), but it should focus on areas where emergencies may develop. Use pins and colored flags to mark changes in personnel deployment, hazardous areas, and equipment locations. The map should highlight the following:

- hazardous areas, especially potential IDLH conditions
- site terrain—topography, buildings, barriers
- evacuation routes
- site accessibility by land, sea, and air
- work crew locations
- changes (e.g., work activities, vandalism, or accidents)
- off-site populations or environments at potential risk

When using the map, consider both the accuracy of the data and the potential for overestimating or underestimating a hazard.

Use the map for planning and training. Develop potential emergency scenarios, focusing on alternative strategies.

When an emergency occurs, pinpoint the problem areas and add pertinent information, such as weather and wind conditions, temperature, and weather forecast. Use the map to design the emergency plan. For example, you could use it to

- define zones
- determine evacuation routes
- identify emergency first aid, decontamination, and command post stations

Even if the emergency develops so fast that the map cannot be used for on-the-spot planning, familiarity with it will aid in making informed decisions.

SAFE DISTANCES AND REFUGES

SAFE DISTANCES

No single recommendation can be given for safe distances because of the wide variety of hazardous substances and releases found at sites. For example, a small chlorine leak may call for an isolation distance of only 250 feet, while a large leak may require an evacuation area of 2½ miles or more.

Safe distances can be determined only at the time of an emergency, based on a combination of site- and incident-specific factors. However, planning for potential emergency scenarios will help familiarize personnel with points to consider. To establish safe distances, account for the following:

- toxicological properties of the substance
- physical state of the substance
- quantity released
- rate of release
- method of release
- vapor pressure of the substance
- vapor density of the substance relative to air
- wind speed and direction
- atmospheric stability
- altitude of release
- air temperature and temperature change with altitude
- local topography (e.g., barriers may enhance or retard a cloud or plume and attenuate a blast)

PUBLIC EVACUATION

If an incident may threaten the health or safety of the surrounding community, the public will need to be informed and possibly evacuate the area. Site management should plan for this in coordination with the appropriate local, state, and federal groups, such as the civil defense, county sheriff, local radio and television stations, municipal transportation systems, National Guard, and police.

SAFETY STATIONS

On-site refuges, also called safety stations, can be set up for localized emergencies that do not require site evacuation. These refuges should be used only for emergency response to workers' brief, essential needs, such as short rest breaks, strategy meetings, or mild cases of muscle strain and heat stress. Locate the refuge in a relatively safe but not necessarily clean area—for example, along the upwind fence line or in specially cleared places such as the periphery of an exclusion zone.* Do not use the refuge for clean activities such as eating, drinking, or air changes. The following items typically are located in a refuge area:

- a clean (if possible, shaded) sitting/resting area
- water
- a wind indicator
- a communication system with the command post

---

*In an emergency, as in daily work activities, the site is divided into three areas: exclusion (contaminated) zone, CRZ, and support (clean) zone.

- first aid supplies (e.g., eyewash, stretchers, and blankets)
- special monitoring devices (e.g., extra detector tubes and personal monitors)
- bolt cutters
- fire extinguishers
- hand tools

Other refuges can be set up in the support zone or, in the case of sitewide evacuations, off-site at the safe exit destination. These refuges provide for emergency needs, such as first aid for injured personnel; clean, dry clothing and wash water for chemical exposure victims; and communications with the command post. In a sitewide evacuation, they can be used to house evacuation exit equipment, reducing security problems. Stock these refuges with items such as the following:

- decontamination supplies
- oxygen and air
- water
- special testing equipment (e.g., pH and cyanide paper)
- first aid supplies
- a communication system

Following are some general rules for designating safe distances and refuges:

- Far is better than near.
- For lighter-than-air or hot gases or vapors, low is better than high.
- For heavier-than-air or cold gases or vapors, high is better than low.
- Upwind is better than downwind.
- Massive shielding is better than flimsy shielding.

SITE SECURITY AND CONTROL

In an emergency, the leader must know who is on-site and must be able to control entry of personnel into the hazardous areas to prevent additional injury and exposure. Only necessary rescue and response personnel should be allowed into the exclusion zone.

One control technique is a checkpoint or series of checkpoints through which all personnel entering or exiting the site must pass—for example, a support zone checkpoint and an exclusion zone checkpoint. Identification or authorization must be presented to a checkpoint control manager, who records each person's

- name (and affiliation if off-site personnel)
- status (in or out)

- time of entry
- anticipated exit time
- zones or areas to be entered
- team or buddy
- task being performed
- location of task
- PPE worn and air time left
- rescue and response equipment used

The emergency area checkpoint control manager should inform the supervisor if a person remains in the emergency area beyond his anticipated exit time.

In an emergency, it is vital for the leader and rescue personnel to determine rapidly where workers are located and who may be injured. Each site must have a *passive locator system*—that is, a written record of the location of all personnel on-site at any time. Any such system should be

- graphic (such as a drawing with a written key)
- roughly drawn to scale, with the scale and visible landmarks included
- kept current
- easy to locate
- stored outside the exclusion zone

A good passive locator system is a site map with flags or colored pins identifying each worker.

Active locator systems also can be used. These are worn or carried by individual personnel and are activated by actions such as flipping a switch, a decrease in air supply, or a fall. They have the advantage of locating individuals precisely.

EVACUATION ROUTES AND PROCEDURES

A severe emergency, such as a fire or an explosion, may cut workers off from the normal exit near the command post. Therefore, alternate routes for evacuating victims and endangered personnel must be established in advance, marked, and kept clear. Routes should be directed (1) from the exclusion zone through an upwind CRZ to the support zone and (2) from the support zone to an off-site location in case conditions necessitate a general site evacuation. The following guidelines will help in establishing safe evacuation routes:

- Place the evacuation routes in the predominantly upwind direction of the exclusion zone. At a very large site or one with many obstacles, some exits may be placed along the downwind fence line, normally an undesirable location. If this is done, workers must know that they are not out until they reach the designated safety area.

- Run the evacuation routes through the CRZ. Even if there is not enough time to process the evacuees through decontamination procedures, there needs to be a mechanism for accounting for all personnel.

- Consider the accessibility of potential routes. Take into account obstructions such as locked gates, trenches, pits, tanks, drums, and other barriers, as well as the extra time and equipment needed to maneuver around or through them.

- Develop two or more routes that lead to safe areas and that are separate or remote from each other. Multiple routes are necessary in case one is blocked by a fire, spill, or vapor cloud. These routes must not overlap because if a common point were obstructed by a fire or another emergency, all intersecting routes would be blocked.

- Mark routes daily as being safe or unsafe according to wind direction and other factors.

- Mark evacuation routes with materials such as barricade tape, flagging, or traffic cones. Equally important, mark areas that do not offer a safe escape or that should not be used in an emergency. These might include low ground, which can fill with gases or vapors, or routes blocked by natural barriers such as cliffs or streams.

- Consider the mobility constraints of personnel wearing protective clothing and equipment. They will have difficulty crossing even small streams and going up and down banks.

  Place ladders or scaffolding across any cut or excavation that is more than 3 feet deep. For long cuts, place ladders at least every 25 feet. For sure footing, place plywood or planks on top of ladders.

  Provide ladders for rapid descent from areas or structures elevated more than 3 feet.

  Use only ladders capable of supporting a 250-pound load.

  Secure ladders and scaffolding to prevent slipping.

  Place standard cleated ramps (chicken board) across ditches and other similar obstacles. Add a railing and toe boards if the board is narrow or steeply sloped.

  Check the toe and body clearance of ladders to make sure that personnel wearing PPC and SCBAs can use them.

  Check the clearance of access ports, such as crawl spaces, hatches, manholes, and tunnels to make sure that personnel wearing a protective ensemble can get through. In any case, access ports should be at least 36 inches in diameter if possible. (Standard tank manways are smaller.)

- Make escape routes known to everyone who goes on-site.

DECONTAMINATION

When planning for decontamination in medical emergencies, develop procedures for the following:

- decontaminating the victim

- protecting medical personnel

- disposing of contaminated protective equipment and wash solutions

See Chapter 8, Decontamination, for details on decontamination techniques and procedures.

EQUIPMENT

In an emergency, equipment will be necessary to rescue and treat victims, protect response personnel, and mitigate hazardous conditions on-site (e.g., to contain chemicals or fight fires). Some regular equipment can be used in emergencies. Because of its high cost, most heavy equipment (e.g., bulldozers, drum movers, and pumps) is used for both regular work assignments and emergencies. Make sure that all equipment is in working order, fueled, and available when an emergency occurs. Provide safe and unobstructed access for all fire-fighting and emergency equipment at all times. Consider adopting the following work procedures:

- Refuel all heavy equipment when there is still one-half to one-quarter tank of fuel left.

- Require all equipment repairs to take place at the time the problem is discovered.

- Separate two similar pieces of equipment (e.g., two front-end loaders or a bulldozer and a front-end loader). Park them in different spots on-site and do not use them at the same time in a hazardous area unless it is absolutely necessary. This will minimize the possibility of both pieces of equipment being damaged in the same explosion or fire.

  PPE also should be maintained:

- Refill all empty SCBA tanks and prepare them for emergencies immediately after normal use.

- Stock higher levels of PPE than required for anticipated hazards (e.g., a Level C site should have Level A and B equipment available for emergencies).

Basic equipment that should be available at any site is listed in Table 9.5. Special equipment should be obtained depending on the specific types of emergencies that may occur at a site and the capabilities of backup off-site personnel. For example, if the nearest fire department is small and carries only one bucket of foaming solution because of its high cost and short shelf-life, a site may need to stock a large quantity of foam.

When determining the type and quantity of special equipment, consider the following:

- the types of emergencies that may arise (For each emergency, consider a probable and worst-case scenario.)

TABLE 9.5 On-Site Equipment and Supplies for Emergency Response

Personal Protection	Medical	Hazard Mitigation
ESCBA[a] or SCBA,[b] which can be brought to the victim to replace or supplement his SCBA Personal protective equipment and clothing specialized for known site hazards	Air splints and equipment Antiseptics Blankets Decontamination solutions appropriate for on-site chemical hazards Emergency eyewash Emergency showers or wash stations Ice Reference books containing basic first aid procedures and information on treatment of specific chemical injuries Resuscitator Safety harness Stretchers Water in portable containers Wire basket litter (Stokes litter), which can be used to carry a victim in bad weather and on difficult terrain, allows easy decontamination of the victim and is itself easy to decontaminate	Fire fighting supplies Spill-containment absorbents and oil booms Special hazardous use tools, such as remote pneumatic impact wrenches and sparkless brass wrenches and picks Containers to hold contaminated material

[a]Escape self-contained breathing apparatus
[b]Self-contained breathing apparatus

- the types of hazards to which site personnel may be exposed and the appropriate containment, mitigative, and protective measures
- the capabilities and estimated response times of off-site emergency personnel
- the number of site personnel who could be victims during an emergency
- the probable number of personnel available for response

MEDICAL TREATMENT AND FIRST AID

Toxic exposures and hazardous situations that cause injuries and illnesses vary from site to site. Medical treatment may range from bandaging minor cuts and abrasions to lifesaving techniques. In many cases, essential medical help may not be immediately available. For this reason, it is vital to train on-site emergency personnel in on-the-spot treatment techniques, to establish and maintain telephone contact with medical experts (e.g., toxicologists), and to develop liaisons with local hospitals and ambulance services. Guidelines for establishing an emergency medical program are detailed in Chapter 10. When designing this program, consider these essential points:

- Train an abundance of personnel in emergency treatment such as first aid and CPR. Training should be thorough, frequent, and geared to site-specific hazards.
- Establish liaisons with local medical personnel—for example, a 24-hour on-call physician, medical specialists, a local hospital, an ambulance service, and a poison control center. Inform and educate these personnel about site-specific hazards so that they can provide optimum help if an emergency occurs. Develop procedures for contacting them and familiarize all on-site emergency personnel with these procedures.
- Set up on-site emergency first aid stations. Make sure they are well supplied and restocked immediately after each emergency.

EMERGENCY RESPONSE PROCEDURES

Response operations usually follow a sequence that starts with notifying personnel of trouble and continues through the preparation of equipment and personnel for the next emergency.

NOTIFICATION

Alert personnel to the emergency. Sound a site alarm to

- notify personnel
- stop work activities if necessary
- lower background noise to speed communication
- begin emergency procedures

Notify on-site emergency response personnel about the emergency and include essential information:

- what happened
- where it happened
- to whom it happened
- when it happened
- how it happened
- the extent of damage
- what aid is needed

ASSESSMENT

Evaluate available information about the incident and emergency response capabilities. Determine, to the extent possible, the following:

- what happened
    - type of incident
    - cause of incident
    - extent of hazardous material release and transport
    - extent of damage to structures, equipment, and terrain
- casualties
    - victims—number, location, and condition
    - treatment required
    - missing personnel
- what could happen
    - types of chemicals on-site
    - potential for fire, explosion, or release of hazardous substances
    - location of all personnel on-site relative to hazardous areas
    - potential for danger to off-site population or environment
- what can be done
    - equipment and personnel resources needed for victim rescue and hazard mitigation
    - number of uninjured personnel available for response
    - resources available on-site
    - resources available from outside groups and agencies
    - time for outside resources to reach the site
    - hazards involved in rescue and response

RESCUE AND RESPONSE

Based on the available information, decide what type of action to take and proceed with the necessary steps, some of which may be done concurrently:

- Enforce the buddy system. Allow no one to enter an exclusion zone or hazardous area without a partner. At all times, personnel in the exclusion zone should be in line-of-sight or communications contact with the command post supervisor or her designee.
- Survey casualties.
    - Locate all victims and assess their condition.
    - Determine the resources needed for stabilization and transport.
- Assess existing and potential hazards to site personnel and off-site population. Determine
    - whether and how to respond
    - the need for evacuation of on-site personnel and the off-site population
    - the resources needed for evacuation and response
- Allocate resources. Allocate on-site personnel and equipment to rescue and response operations.
- Request aid. Contact the required off-site personnel or facilities, such as the ambulance service, fire department, and police department.
- Bring the hazardous situation under complete or temporary control. Try to prevent the spread of the emergency.
- Remove or assist victims from the area.
- Decontaminate uninjured personnel in the CRZ. If the emergency makes this area unsafe, establish a new decontamination area at an appropriate distance. Decontaminate victims before or after stabilization as their medical condition indicates.
- Stabilize the situation.
    - Administer any medical procedures that are necessary before the victims can be moved.
    - Stabilize or permanently fix the hazardous condition (e.g., empty filled runoff dikes).
    - Attend to what caused the emergency and anything (e.g., drums or tanks) damaged or endangered by the emergency.
- Take measures to minimize chemical contamination of the transport vehicle and ambulance and hospital personnel. Adequately protected rescuers should decontaminate the victims before transport. If this is not possible, cover the victims with sheets. Before transport, determine the level of protection necessary for transport personnel. Provide them with disposable coveralls, disposable gloves, and supplied air, as necessary, for their protection. If appropriate, have response personnel accompany victims to the medical facility to advise on decontamination.
- Evacuate personnel and monitor the situation.
    - Move on-site personnel to a safe distance upwind of the incident.
    - Monitor the incident for significant changes. The hazards may diminish, permitting personnel to reenter the site, or increase and require public evacuation.
    - Inform public safety personnel when there is a potential or actual need to evacuate the off-site population. Do not attempt large-scale public evacuation.

This is the responsibility of government authorities (see Table 9.3).

FOLLOW-UP

Before normal site activities are resumed, personnel must be fully prepared and equipped to handle another emergency.

- Restock all equipment and supplies. Replace or repair damaged equipment. Clean and refuel equipment for future use.
- Review and revise all aspects of the contingency plan according to new site conditions and lessons learned from the emergency response. When reviewing the information, consider questions such as the following:

  What caused the emergency?

  Was it preventable? If so, how?

  Were inadequate or incorrect orders given or actions taken? Were these the result of bad judgment, wrong or insufficient information, or poor procedures? Can procedures or training be improved?

  How does the incident affect the site profile? How are other site cleanup activities affected?

  How is community safety affected?

  Who is liable for damage payments?

DOCUMENTATION

As soon as conditions return to normal, investigate the incident, putting all findings in writing. This is important in all cases but especially so when the incident has resulted in personal injury, on-site property damage, or damage to the surrounding environment. Documentation may be used as evidence in future legal action, for assessment of liability by insurance companies, and for review by government agencies. Methods of documenting can include a written transcript taken from tape recordings made during the emergency or a bound field book (not a loose-leaf notebook) with notes. The document must have the following qualities:

- *Accuracy.* All information must be recorded objectively.
- *Authenticity.* A chain-of-custody procedure should be used. Each person making an entry must date and sign the document. Keep the number of documenters to a minimum (to avoid confusion and because they may have to give testimony at hearings or in court). Nothing should be erased. If details change or revisions are needed, the person making the notation should mark a horizontal line through the old material and initial the change.
- *Completeness.* At a minimum, include the following: chronological history of the incident

TABLE 9.6 Examples of Agencies That Require Reports After Emergencies

Agency	Emergency
U.S. Coast Guard	Pollution of navigable waterway
U.S. Department of Transportation (DOT)	Highway accident involving a chemical spill or discharge
U.S. Environmental Protection Agency (EPA)	Chemical spill or discharge
Insurance carriers	As required
Occupational Safety and Health Administration (OSHA)	Injury or Illness
State agencies	As required

facts about the incident and when they became available

title and names of personnel and composition of teams

decisions made and by whom; orders given (to whom, by whom, and when); and actions taken (who did what, when, where, and how)

types of samples and test results; air-monitoring results

possible exposures of site personnel

history of all injuries or illnesses during or as a result of the emergency

After an emergency, various agencies and groups will require reports for their own documentation purposes, to support requests for payment, and to provide evidence in case of legal action. See Table 9.6 for a list of agencies that may require reports.

BIBLIOGRAPHY

LeFevre, M.J. *First Aid Manual for Chemical Accidents.* 2d ed. New York: Van Nostrand Reinhold, 1989.

National Fire Protection Association. *Fire Protection Guide on Hazardous Materials.* 7th ed. Boston, MA: National Fire Protection Association, 1978.

Robinson, J.S., ed. *Hazardous Chemicals Spills Cleanup.* Parkridge, N.J.: Noyes Data Corporation, 1980.

U.S. Department of Transportation (DOT). *Hazardous Materials Emergency Response Guide Book.* DOT Publication No. 5800.5. Washington, D.C.: DOT, 1990.

U.S. Coast Guard Department of Transportation (DOT). *Chemical Hazards Response Information (CHRIS).* Washington, D.C.: DOT, 1978.

Zajic, J.E., and W.A. Himmelman. *Highly Hazardous Materials Spills and Emergency Planning.* New York: Marcel-Dekker, 1978.

10

MEDICAL MONITORING
PROGRAM

*W*orkers handling hazardous wastes may be exposed to toxic chemicals, biological hazards, radiation, or physical dangers. They may develop heat stress while wearing PPE or working under temperature extremes, or they may face life-threatening emergencies such as explosions and fires. All these factors can interact to create psychological and physiological stress. Therefore, a medical program is essential to assess and monitor workers' health and fitness both prior to employment and during the course of work, to provide emergency and other treatment as needed, and to keep accurate records for future reference. Information from a site medical monitoring program also may be used to conduct future epidemiological studies; to adjudicate claims and determine benefits; to provide evidence in litigation; and to report workers' medical conditions to federal, state, and local agencies, as required by law.

This chapter presents general guidelines for a medical program for personnel at hazardous waste sites. In addition, it supplies a table of common toxins found at uncontrolled waste sites with recommended medical monitoring procedures and a sample occupational medical history form.

The recommendations in this chapter assume that workers will have adequate protection from exposures through administrative and engineering controls and appropriate PPE and decontamination procedures. Medical monitoring should be used only as a backup to other controls.

Each site should develop its own medical monitoring program based on its specific needs, location, and potential exposures. The program should be designed by the SSO in conjunction with an experienced occupational physician or other qualified occupational health consultant. The screening and examination protocols described here provide general outlines of a sample medical monitoring program.

The ideal director of a site medical program is a board-certified occupational medicine physician or a doctor who has had extensive experience managing occupational health services. If an occupational physician is not available, consider using one of the following:

- a capable primary care physician with an interest in the program

- a group of local physicians to give the examinations, with review and/or overall direction by an occupational medicine consultant

A site medical program should provide the following components:

- preemployment screening

- periodic monitoring

- termination examination

- emergency treatment

- episodic treatment

- record keeping

PROGRAM REVIEW AND EVALUATION

An effective medical monitoring program depends on active worker involvement. See Table 10.1 for an outline of a recommended medical program. In addition, management must have a firm commitment to worker health and safety and must express this not only by medical monitoring and treatment but also through management directives and informal encouragement to maintain good health through

exercise (such as aerobics), proper diet, and avoidance of alcohol and drug abuse. In particular, management should

- require prospective employees to provide a complete and detailed occupational and medical history

- require workers to report any exposures, regardless of degree

- require workers to bring any unusual physical or psycho-

TABLE 10.1 Recommended Medical Program

Component	Recommended	Optional
Preassignment screening	Medical history Occupational history Physical examination Determination of fitness to work wearing protective equipment	Baseline monitoring for specific exposures Other routine baseline tests Freezing preemployment serum specimen for later testing
Periodic monitoring	Yearly medical occupational history and physical examination; testing based on examination results and exposures More frequent testing based on exposures to specific hazards	Yearly testing with routine medical tests
Emergency treatment	Provide emergency first aid on-site Develop liaison with local hospital and medical specialists Arrange for transport of victims Transfer medical records; give details of incident and medical history to next care provider	
Nonemergency treatment	Develop mechanism for evaluation of possible site-related illness and episodic health care	
Program record keeping and review	Maintain and provide access to medical records in accordance with OSHA regulations Report and record occupational injuries and illnesses Review site safety plan regularly to determine whether additional testing is needed Review program periodically; focus on current site hazards, exposures, and industrial hygiene standards Post yearly summary of occupational illnesses and injuries	

logical conditions to the doctor's attention (Employee training must emphasize that vague disturbances or apparently minor complaints, such as skin irritation or headaches, may be precursors of a low-level exposure response or other illness.)

• promote positive attitudes toward health maintenance and discourage inappropriate worker biases, such as the idea that acknowledging and seeking treatment for health problems is weak or cowardly

MONITORING

A hazardous waste site medical monitoring program is designed for fieldworkers who may be exposed to toxic substances on a regular basis. Designing such a program is difficult because information about exposures and risks for these workers is limited. The design of tests and frequency of monitoring will be dependent on the potential hazardous exposures on each work site.

Specific testing protocols should be established in coordination with the physician being consulted for the medical monitoring services. Table 10.2 provides examples of tests frequently performed by occupational physicians in establishing baseline medical profiles.

Standard occupational medical tests were developed in factories and other enclosed industrial environments. They were based on the presence of specific identifiable toxic chemicals and the possibility of a significant degree of exposure. Some of these tests may not be totally appropriate for hazardous waste sites, since available data suggest that site workers have low-level exposure to many chemicals concurrently, as well as brief high-level exposure to some chemicals.

The recommendations in this chapter are based on known health risks for hazardous waste workers, a review of available data on their exposures, and an assessment of several established medical monitoring programs. Because conditions and hazards vary considerably at each site, only general guidelines are given.

A medical monitoring program for every employee is highly desirable. When developing test protocols for a specific site, bear in mind the following points:

TABLE 10.2 Tests Frequently Performed by Occupational Physicians

Function	Test	Example
Liver		
General	Blood tests	Total protein, albumin, globulin, total bilirubin (direct bilirubin if total is elevated)
Obstruction	Blood test	Alkaline phosphatase
Cell injury	Blood tests	Gamma glutamyl transpeptidase (GGTP), lactic dehydrogenase (LDH), serum glutamic-oxaloacetic transaminase (SGOT), serum glutamic-pyruvic transaminase (SGPT)
Kidney	Blood tests	Blood urea nitrogen (BUN), creatinine, uric acid
Multiple systems and organs	Urinalysis	Including color; appearance; specific gravity; pH; qualitative glucose, protein, bile, and acetone; occult blood; microscopic examination of centrifuged sediment
Blood-forming function	Blood tests	Complete blood count (CBC) with differential and platelet evaluation, including white cell count (WBC), red cell count (RBC), hemoglobin (HGB), hematocrit or packed cell volume (HCT), desired erythrocyte indices; reticulocyte count may be appropriate if there is a likelihood of exposure to hemolytic chemicals

- Determine monitoring needs on a case-by-case basis, taking into account the worker's medical and occupational history, plus current and potential exposures on-site.
- Consider the routine job tasks of each worker. For instance, a heavy equipment operator exposed to significant noise levels would require a different monitoring protocol from a field sample collector with minimal noise exposure. An administrator may need only a preemployment screening for ability to wear PPE, if this is an occasional requirement rather than a more comprehensive program.
- Consider that most testing recommendations, even those for specific toxic substances, have not been critically evaluated for efficacy.
- Take into account that toxicity can vary, not only with the amount and duration of exposure but also with individual factors such as age, sex, weight, stress, diet, medications taken, and off-site exposures (e.g., hobbies such as painting or furniture refinishing).

While it is often impossible to identify every toxic substance that exists at each hazardous waste site, certain groups of toxins are more likely to be present than others. These include the following:

- aromatic hydrocarbon solvents
- asbestos
- dioxin
- halogenated aliphatic hydrocarbon solvents
- heavy metals
- herbicides
- organochlorine insecticides
- organophosphate and carbamate insecticides
- PCBs

Table 10.3 lists these groups, along with representative compounds, uses, health effects, and available medical monitoring procedures.

PREEMPLOYMENT SCREENING

Preemployment screening has two major functions: (1) determination of physical fitness, including the ability to work while wearing PPE, and (2) provision of baseline data for future exposures or injuries.

DETERMINATION OF PHYSICAL FITNESS

Workers at hazardous waste sites are often required to perform strenuous tasks (e.g., moving 55-gallon drums) and wear PPE, such as respirators and protective suits, that may cause heat stress and other problems. To ensure that prospective employees are able to meet work requirements, the preemployment screening should focus on the following areas:

Medical History

- Make sure the worker fills out the occupational medical history questionnaire. (See Appendix F for a sample

TABLE 10.3 Common Toxins Found at Hazardous Waste Sites: Health Effects and Medical Monitoring

Chemical Group	Compound	Uses	Target Organs/ Systems	Potential Health Effects	Medical Monitoring
Aromatic hydro-carbon solvents	Benzene Ethyl benzene Toluene Xylene	Commercial solvents and intermediates for synthesis in the chemical and pharmaceutical industries	Blood Bone marrow Central nervous system (CNS) Eyes Respiratory system Skin Liver Kidneys	*All cause* CNS depression: decreased alertness, headache, sleepiness, loss of consciousness Defatting dermatitis *Benzene* suppresses bone-marrow function, causing blood changes. Chronic exposure can cause leukemia. *Note:* Because other aromatic hydrocarbons may be contaminated with benzene during distillation, benzene-related health effects should be considered when exposure to any of these agents is suspected.	Laboratory tests Complete blood count (CBC) Reticulocyte count Platelet count Measurement of kidney and liver function
Asbestos	Chrysotile asbestos Crocidolite asbestos Amosite asbestos Tremolite Actinolite Anthophyllite	A variety of industrial uses including: Building Construction Cement work Insulation Fireproofing Pipes and ducts for water, air, and chemicals Automobile brake pads and linings	Lungs Gastrointestinal system	Chronic effects Lung cancer Mesothelioma Asbestosis Gastrointestinal malignancies Asbestos exposure coupled with cigarette smoking has been shown to have a synergistic effect in the development of lung cancer and asbestosis.	History and physical examination should focus on the lungs, heart, and gastrointestinal system. Laboratory tests include a stool guaiac evaluation as a check for possible hidden gastrointestinal malignancy. A high-quality chest X ray and pulmonary function test may help to identify long-term changes (manifest in 10 to 30 years) associated with asbestos diseases; however, early identification of low-dose exposure is unlikely.

Dioxin (see Herbicides)

Halogenated aliphatic	Carbon tetrachloride Chloroform Ethyl bromide Ethyl chloride Ethylene dibromide Ethylene dichloride Methyl chloride Methyl chloroform Methylene chloride Tetrachloroethane Tetrachloroethylene (perchloroethylene) Trichloroethylene Vinyl chloride	Commercial solvents and intermediates in organic synthesis	CNS Kidneys Liver Skin	*All* cause: CNS depression: decreased alertness, headaches, sleepiness, loss of consciousness Kidney: decreased urine flow, swelling (especially around eyes), anemia Liver: fatigue, malaise, dark urine, liver enlargement, jaundice *Vinyl chloride* is a known carcinogen; several others in this group are potential carcinogens.	Occupational/general/medical history emphasizing prior exposure to these and other toxic agents Medical history of liver, kidney, CNS, and skin diseases Laboratory testing for liver and kidney function
Heavy metals	Arsenic Beryllium Cadmium Chromium Lead Mercury	Wide variety of industrial and commercial uses	Multiple organs and systems Blood Cardiopulmonary Gastrointestinal Kidney Liver Lung CNS Skin	*All* are toxic to the kidneys. *Each heavy metal* has its own characteristic symptom cluster. For example, *lead* causes decreased mental ability, weakness (especially hands), numbness of hands and feet, headache, abdominal cramps, diarrhea, abdominal pain resembling ulcer symptoms, and anemia. Lead also can affect the blood-forming mechanism, kidneys, and heart. *Long-term effects* also vary. Lead toxicity can cause permanent brain damage. Cadmium can cause chronic lung disease. Chromium, beryllium, and cadmium have been implicated as human carcinogens.	History taking and physical exam should search for symptom clusters associated with specific metal exposure (e.g., for lead, look for neurological deficit, anemia, gastrointestinal symptoms). Laboratory tests Quantitative and qualitative measurements of metallic content in blood, urine, and tissues (e.g., blood lead level; urine screen for arsenic, mercury, chromium, cadmium) CBC Reticulocyte count Measurement of kidney and liver function Chest-X ray or pulmonary function testing to identify changes associated with chromium, beryllium, or cadmium exposure

TABLE 10.3 (Continued)

Chemical Group	Compound	Uses	Target Organs/ Systems	Potential Health Effects	Medical Monitoring
Herbicides	Chlorophenoxy compounds 2,4-dichlorophen-oxyacetic acid (2,4-D) 2,4,5 trichloro-phenoxyacetic acid (2,4,5-T) Dioxin (tetrachloro-dibenzo-p-dioxin, TCDD), which occurs as a trace contaminant in these compounds, poses the most serious health risk.	Vegetation control	Kidney Liver CNS Skin	*Dioxin* causes chloracne and may aggravate preexisting liver and kidney diseases. Its cancer-causing potential is currently being investigated. *Chlorophenoxy* compounds cause liver and kidney abnormalities in laboratory animals. In humans, exposure can cause weakness or numbness of the arms and legs and may result in long-term nerve damage.	History and physical exam should focus on the skin and nerve systems. Laboratory tests Measurement of liver and kidney function Urinalysis
Organochlorine insecticides	*Chlorinated ethane*—DDT *Cyclodienes* Aldrin Chlordane Dieldrin Endrin *Chlorocyclohexane:* Lindane	Pest control	Kidney Liver CNS	*All* cause acute symptoms of apprehension, irritability, dizziness, disturbed equilibrium, tremor, and convulsions. *Cyclodienes* may cause convulsions without any other initial symptoms. *Chlorocyclohexanes* can cause anemia. *Cyclodienes and chlorocyclohexanes* cause liver toxicity and can cause permanent kidney damage, including tumors. They also cause cancer in laboratory animals.	History and physical exam should focus on the nerve system. Laboratory tests Measurement of kidney and liver function CBC for exposure to chlorocyclohexanes

Substance	Uses	Target organs	Health effects	Medical monitoring

Organophosphate and carbamate insecticides	Organophosphates
	Chlorfenvinfos
	  (benzyl alcohol)
	Diazinon
	Dichlorovos
	Dimethoate
	Trichlorfon
	Malathion
	Methylparathion
	Parathion
	Carbamates
	Aldicarb
	Baygon
	Zectran
		Pest control	CNS
			Liver
			Kidney
				All cause a chain of internal reactions leading to neuromuscular blockage. Depending on the extent of poisoning, acute symptoms range from headaches, fatigue, dizziness, increased salivation and crying, profuse sweating, nausea, vomiting, cramps, and diarrhea to tightness in the chest, muscle twitching, and slowing of the heartbeat. Severe cases may result in rapid onset of unconsciousness and seizures. A delayed effect may be weakness and numbness in the feet and hands. Long-term permanent nerve damage is possible.	Physical exam should focus on the nerve system.
					
					Laboratory tests
					Plasma/RBC
					  cholinesterase levels
					  for recent exposure
					Measurement of kidney
					  and liver function for
					  delayed neurotoxicity
					  and other effects

Polychlorinated biphenyls (PCBs)		Wide variety of industrial uses	Liver CNS Respiratory system Skin	*All* cause various skin ailments, including chloracne; may cause liver toxicity; carcinogenic to animals; cancer-causing potential in humans still being investigated.	Physical exam should focus on the skin and liver. Laboratory tests Serum PCB levels Measurement of liver function Triglycerides Cholesterol

document.) Review the questionnaire before seeing the worker. In the examining room, discuss the questionnaire with the worker, paying special attention to prior occupational exposures to chemical and physical hazards.

- Cover past illnesses and chronic diseases, particularly asthma, lung diseases, and cardiovascular disease.
- Review symptoms, especially shortness of breath or labored breathing on exertion, other chronic respiratory symptoms, chest pain, high blood pressure, and heat intolerance.
- Identify individuals who are vulnerable to particular substances (e.g., someone with a history of severe asthmatic reaction to a specific chemical).

Physical Examination

- Conduct a comprehensive physical exam of all body organs, focusing on the pulmonary, cardiovascular, and musculoskeletal systems.
- Note conditions that could increase susceptibility to heatstroke, such as obesity and lack of physical exercise.
- Note conditions that could interfere with respirator use, such as missing or arthritic fingers, facial scars, poor eyesight, or perforated eardrums.

Ability to Work Wearing PPE

- Disqualify individuals clearly unable to perform based on their medical history and physical exam (e.g., those with severe lung or heart disease).
- Provide additional testing for ability to wear PPE (e.g., chest X ray, pulmonary function testing, electrocardiogram) where necessary.
- Base the determination on the individual worker's profile: medical history and physical exam, age, previous exposures, and testing.
- If wearing a respirator is a job requirement, make a written assessment of the worker's capacity to perform while wearing one.*

BASELINE DATA FOR FUTURE EXPOSURES

Preemployment screening can be used to establish baseline data to determine whether subsequent exposures have had adverse effects on the worker. For this purpose, medical screening tests and biological monitoring tests can be used.

---

*The OSHA respirator standard (29 CFR 1910, Part 134) states that no employee be assigned to a task that requires the use of a respirator unless it has been determined that the person is physically able to perform under such conditions.

At present, there are no specific guidelines for prescribing these tests. Consider these general recommendations:

- Develop a battery of tests based on the worker's past occupational and medical history and an assessment of significant potential exposures.
- Use standard established testing for specific toxins in situations where workers may receive significant exposures to these agents. For example, long-term cleanup of a PCB waste facility can be monitored with preemployment and periodic serum PCB testing. Standard procedures are available for determining levels of other substances such as lead, cadmium, arsenic, and organophosphate pesticides.
- Where applicable, draw preemployment blood specimens and freeze serum for later testing. PCBs and some pesticides are examples of agents amenable to such monitoring.

SAMPLE PREEMPLOYMENT EXAMINATION

The following items are recommended for a preemployment examination:

Occupational and Medical History
Do a complete medical history emphasizing these systems: nerve, skin, lung, blood-forming, cardiovascular, gastrointestinal, ear, nose, reproductive, and genitourinary.

Physical Examination
Include at least the following points:

- height, weight, temperature, pulse, respiration, and blood pressure
- head, nose, and throat
- eyes—include vision tests that measure refraction, depth perception and color vision. These tests should be administered by a qualified technician or physician. Vision quality is essential to safety, to read instruments and labels accurately, to avoid physical hazards, and to respond to color-coded labels and signals.
- ears—include audiometric tests performed at 500, 1000, 2000, 3000, 4000, and 6000 hertz (Hz) pure tone in an approved booth. Tests should be administered by a qualified technician and results read by a certified audiologist. The integrity of the eardrum must be established. Persons with perforated eardrums who work around hazardous substances risk airborne chemicals entering the body unless they wear PPE that covers the head. If the eardrum is perforated, the examining physician should consult the appropriate site personnel.
- chest (heart and lungs)
- peripheral vascular system
- abdomen and rectum (including hernia exam)

- spine and rest of the musculoskeletal system
- genitourinary system
- skin
- nerves

WORKER'S ABILITY TO PERFORM WHILE WEARING PPE

To determine a worker's ability to perform while wearing PPE, additional tests may be necessary. These might include the following:

Chest X Ray A 14 × 17-inch posterior/anterior view is recommended. Lateral or oblique views are required only if indicated. Services should be provided by a certified radiology technician, with interpretation by a certified radiologist. Where available, chest X rays taken in the last 12-month period should be obtained and used for review. Chest X rays should not be repeated more than once a year, unless otherwise determined by the examining physician.

Electrocardiogram (EKG) At least one standard 12-lead resting EKG is recommended. Interpretation should be done by an internist or cardiologist. A stress test may be administered at the discretion of the examining physician.

Pulmonary Function Testing The following abbreviations are often used in occupational safety and health publications: FEF = forced expiratory flow; MMEFR = maximal expiratory flow rate; MVV = maximal voluntary ventilation; FRC = functional residual capacity; RV = residual volume; TLC = total lung capacity.

Pulmonary function testing should include forced expiratory volume in 1 second (FEV_1) and forced vital capacity (FVC), with interpretation and comparison to normal predicted values for age, height, race, and sex. Other factors such as FEF, MMEFR, MVV, FRC, RV, and TLC may be included. A permanent record of flow curves should be placed in the worker's medical chart.

The following items are optional:

- specific baseline monitoring to establish data relating to a particular toxin when there is a likelihood of potential on-site exposure to that toxin
- freezing a serum specimen for later testing in cases where there are known toxic agents amenable to such testing
- other routine baseline profiles depending on the worker's history and physical exam; these may include tests for liver, kidney, lung, or blood-forming functions (See Table 10.2 for examples of tests frequently performed by occupational physicians.)

General tests offered in clinical batteries should probably be omitted in baseline testing. While they can serve a nonoccupational preventive medicine function, they may provide little or no useful information for worker placement. In addition, they present an increased risk of false-positive results. Examples include tests for calcium, carbon dioxide content, cholesterol, chloride, glucose, inorganic phosphorus, potassium, sodium, and triglycerides.

All analyses should be performed by a laboratory demonstrating satisfactory performance in an established interlaboratory testing program at least equivalent to the Centers for Disease Control (CDC) program.

MEDICAL MONITORING EXAMINATION

PERIODIC SCREENING

Periodic medical monitoring examinations should be developed and used in conjunction with preemployment screening examinations. An ongoing comparison of sequential medical reports with the baseline data is essential in determining biological trends that may mark early signs of chronic adverse health effects.

The frequency and content of monitoring will vary depending on the nature of the work and exposures. Generally, medical monitoring examinations have been recommended at least yearly. More frequent monitoring may be necessary depending on the extent of potential or actual exposures, the type of chemicals involved, the duration of the work assignment, and the individual worker's profile. For example, workers participating in the cleanup of a PCB-contaminated building were initially examined monthly for serum PCB levels. Review of the data from the first few months revealed no appreciable evidence of PCB exposure. The frequency of PCB testing was then reduced.

Periodic screening should include the following elements:

- Interval medical history focusing on changes in health status, illnesses, and possible work-related symptoms. The examining physician should have information about the worker's interval exposure history, including exposure monitoring done at the job site, supplemented by a worker-reported exposure history and general information on possible exposures at previous sites.
- Physical examination
- Additional medical testing depending on available exposure information, medical history, and examination results. Testing should be specific for the possible medical effects of the worker's exposure. Multiple testing for a large range of potential exposures is not useful and can be counterproductive, since it may produce false-positive results because of chance or other factors.

   Pulmonary function tests should be administered if the individual has been exposed to irritating or toxic substances or if the individual has breathing difficulties, especially when wearing a respirator.

   Audiometric tests are recommended annually for per-

sonnel subject to high noise exposures (an 8-hour, time-weighted average of 85dBA[b] or more, where dBA[b] = decibels on the A-weighted scale), those required to wear hearing protection, or as otherwise indicated.
- Vision tests are recommended annually to check for vision degradation.
- Blood and urine tests are recommended for workers subject to exposure to heavy metals and toxic substances.

SAMPLE MEDICAL MONITORING EXAMINATION

The basic medical monitoring examination is the same as the preemployment screening, modified according to current conditions, such as changes in the worker's symptoms, site hazards, or exposures. See "Sample Preemployment Examination" earlier in this chapter for details.

TERMINATION EXAMINATION

At the end of employment at a hazardous waste site, all personnel should have a medical examination as described in the previous two sections. If the last examination (prescreening or monitoring) has been done within 6 months, a termination exam is not medically indicated.

EMERGENCY TREATMENT

Provisions for emergency treatment and acute nonemergency treatment should be made at each site. Preplanning is vital.

When developing plans, procedures, and equipment lists, consider the range of actual and potential hazards specific to the site: chemical, physical (such as heat and/or cold stress, falls, and trips), and biological (animal bites and plant poisoning, as well as hazardous biological wastes). Take into account that not only site workers but also contractors, visitors, and other agency personnel (particularly fire fighters) may require emergency treatment.

Following are recommended guidelines for establishing an emergency treatment program:

- Train a team of site personnel in emergency first aid. This should include a Red Cross or equivalent certified course in CPR and first aid training that emphasizes treatment for explosion and burn injuries, heat stress, and acute chemical toxicity. (See Table 10.4 for symptoms of exposure and heat stress that indicate potential medical emergencies.)

- Train personnel in emergency decontamination procedures in coordination with the emergency response plan. (See Chapter 9 for details.)

- Predesignate roles and responsibilities to be assumed by personnel in an emergency.

- Establish an on-site emergency first aid station capable of providing (1) stabilization for patients requiring off-

TABLE 10.4 Signs and Symptoms of Chemical Exposure and Heat Stress That Indicate Potential Medical Emergencies

Type of Hazard	Signs and Symptoms	
Chemical hazard	Behavioral changes Breathing difficulties Changes in complexion or skin color Coordination difficulties Coughing Dizziness Drooling Diarrhea Fatigue Irritability	Irritation of eyes, nose, respiratory tract, skin, or throat Light-headedness Nausea Sneezing Sweating Tearing Tightness in the chest
Heat exhaustion	Clammy skin Confusion Dizziness Fainting Fatigue	Light-headedness Nausea Rapid pulse Slurred speech Profuse sweating
Heatstroke	Confusion Convulsions Hot skin, high temperature (yet may feel chilled)	Incoherent speech Staggering gait Unconsciousness Sweating stops

site treatment and (2) general first aid (e.g., minor cuts, sprains, and abrasions).
- Locate the station in the clean area adjacent to the decontamination area to facilitate emergency decontamination.
- Provide a standard first aid kit or equivalent supplies, plus additional items such as emergency/deluge showers, stretchers, portable water, ice, emergency eyewash, decontamination solutions, stretchers, and fire-extinguishing blankets.
- Restock supplies and equipment immediately after each use and check them regularly.

- Arrange for a physician who can be paged on a 24-hour basis.

- Set up an on-call team of medical specialists for emergency consultations (e.g., toxicologist, dermatologist, hematologist, allergist, ophthalmologist, cardiologist, neurologist).

- Establish a protocol for monitoring heat stress.
- Make plans in advance for emergency transportation to and treatment at a nearby medical facility.

    Educate local emergency transport and hospital personnel about possible on-site medical problems: types of hazards and their consequences, potential for exposure, and scope and function of the site medical program.

    Assist the hospital in developing procedures for site-related emergencies. This will help to protect hospital personnel and patients and to minimize delays due to concerns about hospital safety or contamination.

    For specific illnesses or injuries, provide details of the incident and the worker's past medical history to the appropriate hospital staff. This is especially important when specific medical treatment is required (e.g., for exposure to cyanide or organophosphate pesticides).

Depending on the site's location and potential hazards, it may be important to identify additional medical facilities capable of sophisticated response to chemical or other exposures.

- Post conspicuously (with duplicates near the telephones) the names, phone numbers, addresses, and procedures for contacting the following facilities:
    on-call physician
    medical specialists
    ambulance
    medical facility
    emergency, fire, and police services
    poison control hot line
- Provide maps and directions.
- Make sure that all personnel know the way to the nearest medical facility.
- Designate a radio for emergency use.
- Review these procedures daily with all site personnel at safety meetings before beginning the work shift.

MEDICAL RECORDS

Proper record keeping is essential at hazardous waste sites because of the nature of the work and risks. Employees may work at a large number of geographically separate sites during their careers, and long-term adverse exposure effects may not manifest for many years. Workers should be informed about their exposures. Medical testing is important to help them take appropriate precautions. Other and subsequent medical care providers also should be informed about workers' exposures.

OSHA regulations mandate that, unless a specific occupational safety and health standard provides a different time period, the employer must

- maintain and preserve medical records on exposed workers for 30 years after they leave employment
- make available to workers, their authorized representatives, and OSHA inspection staff the results of medical testing and full medical records
- maintain records of occupational injuries and illnesses and provide a yearly summary report to OSHA (Form 200)

In addition, the Privacy Act of 1974 requires that workers for whom medical records will be kept read and sign a statement that explains the authority for collecting information, rules of confidentiality, and the disclosure of information. This statement becomes part of the worker's records. The employee may make written requests for release of this information. Such requests also become a permanent part of the record.

PROGRAM REVIEW

Regular maintenance and review of medical records and test results aid medical personnel, site officers, and parent company and/or agency managers in assessing the effectiveness of the health and safety program. The SSO, medical consultant, and/or management representative should

- evaluate each accident or illness promptly to determine the cause and make necessary changes in health and safety procedures
- periodically evaluate the efficacy of specific medical testing in the context of potential site exposures
- add or delete medical tests as suggested by current industrial hygiene and environmental data
- review potential exposures and safety plans at new sites to determine whether additional testing is required
- review emergency treatment procedures used

Arrangements should be made for episodic, nonemergency medical care for hazardous waste workers. In conjunction with the medical monitoring program, off-site medical care should ensure that any potential job-related symptoms or illnesses are evaluated in the context of the worker's exposure. Off-site medical personnel also should investigate and treat non–job-related illnesses that may put the worker at risk because of task requirements (e.g., a bad cold or flu that might interfere with respirator use). Arrange for treating physicians to have access to the worker's medical records. Keep a copy of the worker's medical records at the site (with provisions for security and confidentiality) and, when appropriate, at a nearby hospital.

BIBLIOGRAPHY

Chase, K. "Medical Surveillance of Clean-up Workers at the Binghamton State Office Building." Paper presented at the Expert Panel Meeting, Binghamton, New York, 1982.
Dinman, B.D. "Medical Aspects of the Occupational Envi-

ronment.'' In *The Industrial Environment: Its Evaluation and Control,* ed. National Institute for Occupational Safety and Health, 1973.

International Technical Information Institute. *Toxic and Hazardous Industrial Chemicals Safety Manual.* International Technical Information Institute, August 1980.

Melius, J.M. ''Medical Surveillance for Hazardous Waste Workers.'' In *Protecting Personnel at Hazardous Waste Sites,* S.P. Levine and W.F. Martin, eds. Boston: Butterworth-Heinemann, 1985.

Melius, J.M., and W.E. Halperin. ''Medical Screening of Workers at Hazardous Waste Disposal Sites.'' In *Hazardous Waste Disposal: Assessing the Problem,* J. Highland, ed. Ann Arbor, Mich.: Ann Arbor Science Publishers, 1982.

National Institute for Occupational Safety and Health. *Occupational Safety and Health Guidance Manual for Hazardous Waste Site Activities.* U.S. Department of Health and Human Services Publication No. 85–115. Washington, D.C.: GPO, October 1985.

National Institute for Occupational Safety and Health (NIOSH) and Occupational Safety and Health Administration (OSHA). *Pocket Guide to Chemical Hazards.* U.S. Department of Health and Human Services (DHHS) Publication No. 90–117. Washington, D.C.: DHHS, June 1990.

11

TRAINING

Anyone who enters a hazardous waste site must be able to recognize and understand the potential health and safety hazards associated with the cleanup of the site. Personnel working on the site must be thoroughly familiar with work practices and procedures contained in the site safety plan (see Chapter 3, Planning and Organization). Site workers must be trained to work safely wherever there is a reasonable possibility of employee exposure to safety or health hazards.

The training program objectives for hazardous waste site activities include the following:

- to ensure that workers are aware of the potential hazards they may encounter

- to provide the knowledge and skills necessary to perform the work with minimal risk to worker health and safety

- to ensure that workers are aware of the use and limitations of safety equipment

- to ensure that workers can safely avoid or escape from hazardous situations that may occur

The minimum content of the training program may be found in 29 CFR 1910. Workers may not participate in or supervise field activities until they have been trained to a level required by their job function and responsibility.

TRAINING PROGRAMS

OSHA regulation 29 CFR 1910.120 identifies the hazardous waste worker and the type of training as follows:

(e)*Training*—(1) *General* (i) All employees working on site (such as but not limited to equipment operators, general laborers and others) exposed to hazardous substances, health hazards, or safety hazards and their supervisors and management responsible for the site shall receive training meeting the requirements of this paragraph before they are permitted to engage in hazardous waste operations that could expose them to hazardous substances, safety, or health hazards, and they shall receive review training as specified in this paragraph.

(ii) Employees shall not be permitted to participate in or supervise field activities until they have been trained to a level required by their job function and responsibility.

(2) *Elements to be covered.* The training shall thoroughly cover the following:
(i) Names of personnel and alternates responsible for site safety and health;
(ii) Safety, health and other hazards present on the site;
(iii) Use of personal protective equipment;
(iv) Work practices by which the employee can minimize risks from hazards;

(v) Safe use of engineering controls and equipment on the site;
(vi) Medical surveillance requirements, including recognition of symptoms and signs which might indicate overexposure to hazards; and

(3) *Initial training.* (i) General site workers (such as equipment operators, general laborers and supervisory personnel) engaged in hazardous substance removal or other activities which expose or potentially expose workers to hazardous substances and health hazards shall receive a minimum of 40 hours of instruction off the site, and a minimum of three days actual field experience under the direct supervision of a trained, experienced supervisor.

(ii) Workers on site only occasionally for a specific limited task (such as, but not limited to, ground water monitoring, land surveying, or geo-physical surveying) and who are unlikely to be exposed over permissible exposure limits and published exposure limits shall receive a minimum of 24 hours of instruction off the site, and the minimum of one day actual field experience under the direct supervision of a trained, experienced supervisor.

(iii) Workers regularly on site who work in areas which have been monitored and fully characterized indicating that exposures are under permissible exposure limits and published exposure limits where respirators are not necessary, and the characterization indicates that there are no health hazards or the possibility of an emergency developing, shall receive a minimum of 24 hours of instruction off the site and the minimum of one day actual field experience under the direct supervision of a trained, experienced supervisor.

(iv) Workers with 24 hours of training who are covered by paragraphs (e)(3)(ii) and (e)(3)(iii) of this section, and who become general site workers or who are required to wear respirators, shall have the additional 16 hours and two days of training necessary to total the training specified in paragraph (e)(3)(i).

(4) *Management and supervisor training.* On-site management and supervisors directly responsible for, or who supervise employees engaged in hazardous waste operations shall receive 40 hours initial training, and three days of supervised field experience (the training may be reduced to 24 hours and one day if the only area of their responsibility is employees covered by paragraphs (e)(3)(ii) and (e)(3)(iii) and at least eight additional hours of specialized training at the time of job assignment on such topics as, but not limited to, the employer's safety and health program and the associated employee training program, personal protective equipment program, spill containment program, and health hazard monitoring procedure and techniques.

(5) *Qualifications for trainers.* Trainers shall be qualified to instruct employees about the subject matter that is being presented in training. Such trainers shall have satisfactorily completed a training program for teaching the subjects they are expected to teach, or they shall have the academic credentials and instructional experience necessary for teaching the subjects. Instructors shall demonstrate competent instructional skills and knowledge of the applicable subject matter.

(6) *Training certification.* Employees and supervisors that have received and successfully completed the training and field experience specified in paragraphs (e)(1) through (e)(4) of this section shall be certified by their instructor or the head instructor and trained supervisor as having successfully completed the necessary training. A written certificate shall be given to each person so certified. Any person who

has not been so certified or who does not meet the requirements of paragraph (e)(9) of this section shall be prohibited from engaging in hazardous waste operations.

(7) *Emergency response.* Employees who are engaged in responding to hazardous emergency situations at hazardous waste clean-up sites that may expose them to hazardous substances shall be trained in how to respond to such expected emergencies.

(8) *Refresher training.* Employees specified in paragraph (e)(1) of this section, and managers and supervisors specified in paragraph (e)(4) of this section, shall receive eight hours of refresher training annually on the items specified in paragraph (e)(2) and/or (e)(4) of this section, any critique of incidents that have occurred in the past year that can serve as training examples of related work, and other relevant topics.

(9) *Equivalent training.* Employers who can show by documentation or certification that an employee's work experience and/or training has resulted in training equivalent to that training required in paragraphs (e)(1) through (e)(4) of this section shall not be required to provide the initial training requirements of those paragraphs to such employees. However, certified employees or employees with equivalent training new to a site shall receive appropriate, site specific training before site entry and have appropriate supervised field experience at the new site. Equivalent training includes any academic training or the training that existing employees might have already received from actual hazardous waste site work experience.

CONTENT OF TRAINING PROGRAM

The training program must contain fundamental information such as effects and risks of safety and health hazards, as well as site-specific information such as the names of site personnel in charge. Table 11.1 lists the course content proposed by OSHA for workers at hazardous waste cleanup projects and RCRA treatment storage and disposal (TSD) facilities.

Table 11.1 Proposed Content of Training Course[a]

		40-hr	24-hr	16-hr	8-hr[b]
1	Overview of the applicable paragraphs of 29 CFR 1910.120 and the elements of an employer's effective occupational safety and health program.	X	X		
2	Effect of chemical exposures to hazardous substances (i.e., toxicity, carcinogens, irritants, sensitizers, etc.).	X	X	X	
3	Effects of biological and radiological exposures.	X	X		
4	Fire and explosion hazards (i.e., flammable and combustible liquids, reactive materials).	X	X	X	
5	General safety hazards, including electrical	X	X	X	

		40-hr	24-hr	16-hr	8-hr[b]
	hazards, powered equipment hazards, walking-working surface hazards and those hazards associated with hot and cold temperature extremes.				
6	Confined space, tank and vault hazards and entry procedures.	X	X	X	
7	Names of personnel and alternates, where appropriate, responsible for site safety and health at the site.	X		X	
8	Specific safety, health and other hazards that are to be addressed at a site and in the site safety and health plan.	X	X		
9	Use of personal protective equipment and the implementation of the personal protective equipment program.	X	X		
10	Work practices that will minimize employee risk from site hazards.	X	X		
11	Safe use of engineering controls and equipment and any new relevant technology or procedure.	X	X		
12	Content of the medical surveillance program and requirements, including the recognition of signs and symptoms of overexposure to hazardous substances.	X	X		
13	The contents of an effective site safety and health plan.	X	X		
14	Use of monitoring equipment with ''hands-on'' experience and the implementation of the employee and site monitoring program.	X	X	X	
15	Implementation and use of the informational program.	X		X	
16	Drum and container handling procedures and the elements of a spill containment program.	X	X	X	
17	Selection and use of material handling equipment.	X		X	
18	Methods for assessment of risk and handling of radioactive wastes.	X		X	
19	Methods for handling shock-sensitive wastes.	X		X	
20	Laboratory waste pack handling procedures.	X		X	
21	Container sampling procedures and safeguards.	X		X	
22	Safe preparation procedures for shipping and transport of containers.	X		X	
23	Decontamination program and procedures.	X	X	X	
24	Emergency response plan and procedures including first-aid.	X	X		
25	Safe site illumination levels.	X		X	
26	Site sanitation procedures and equipment for employee needs.	X		X	
27	Review of the applicable appendices to 29 CFR 1910.120.	X	X	X	

(continued)

Table 11.1 (Continued)

		40-hr	24-hr	16-hr	8-hr[b]
28	Overview and explanation of OSHA's hazard communication standard (29 CFR 1910.1200).	X	X	X	
29	Sources of reference, additional information and efficient use of relevant manuals and hazard coding systems.	X	X	X	
30	Principles of toxicology and biological monitoring.	X	X		
31	Rights and responsibilities of employees and employers under OSHA and CERCLA.	X	X		
32	"Hands-on" field exercises and demonstrations.	X	X		
33	Review of employer's training program and personnel responsible for that program.		X		
34	Final examination.	X	X	X	
35	Management of hazardous wastes and their disposal.				X
36	Federal, state and local agencies to be contacted in the event of a release of hazardous substances.				X
37	Management of emergency procedures in the event of a release of hazardous substances.				X

[a] Source: From OSHA Hazardous Waste Training 29 CFR 1910.120.
[b] Eight hour course for managers

ORGANIZATIONS OFFERING TRAINING PROGRAMS

A wide variety of hazardous waste training programs have been developed by other government and industrial organizations in the United States. The following is a partial list of such organizations. Many of these programs are available to the public for a registration fee.

Ecological Safety Services, Inc.
2901 Wilcrest drive
Suite 200
Houston, TX 77042
(713) 780–2316

Hygiene Safety and Training, Inc.
P.O. Box 837
Kittanning, PA 16201
(412) 543–6680

J.J. Keller and Associates, Inc.
Breezewood Lane
Neenah, WI 54956
(414) 722–2848
(DOT hazardous materials training only)

J.T. Baker Chemical Company
222 Red School Lane
Phillipsburg, NJ 08865
(908) 859–2151

TYPES OF TRAINING

GENERAL SITE WORKERS

General site workers, including equipment operators, general laborers, technicians, and other supervised personnel, should have training that provides an overview of the site, specific hazards and their risks, hazard recognition, and how to properly use the engineered controls and other means of controlling the site's hazards and risks. General site workers should receive close supervision from a trained, experienced supervisor at least during the first 24 hours following training. Some employees require additional follow-up training to develop good work practices on new tasks. Daily safety reviews just prior to commencing site work for the shift is a good way to give refresher training, make sure that everyone understands the tasks for the day, and inform workers of any new conditions on the site.

A few general site workers who may occasionally supervise others or must deal with special hazards should receive additional training in the following areas:

- site surveillance
- management of hazardous wastes and their disposal
- use and decontamination of fully encapsulating protective clothing and equipment
- federal, state, and local agencies to be contacted in the event of a release of hazardous substances

- management of emergency procedures in the event of a release of hazardous substances

ON-SITE MANAGEMENT AND SUPERVISORS

On-site management and supervisors, such as team leaders, who are responsible for directing others should receive the same training as the general site workers for whom they are responsible. They also need additional training to enhance their ability to provide guidance and make informed decisions. This training should include supervisory skills, planning and management of site cleanup operations, and techniques to communicate with the press and community.

HEALTH AND SAFETY STAFF

Those with specific responsibilities for health and safety guidance on-site should be familiar with the training provided to general site workers and their supervisors and should receive advanced training in hazardous substance health and safety sampling, monitoring, surveillance, evaluation, and control procedures.

ON-SITE EMERGENCY PERSONNEL

Those who have emergency roles in addition to their ordinary duties must have a thorough grounding in emergency response. Training should be directly related to their specific roles and should include subjects such as the following:

- emergency chain of command
- communication methods and signals
- how to call for help
- emergency equipment and its use
- emergency evacuation while wearing PPE
- removing injured personnel from enclosed spaces
- off-site support and how to use it

These personnel should obtain certification in first aid and CPR and practice treatment techniques regularly, with an emphasis on (1) recognizing and treating chemical and physical injuries and (2) recognizing and treating heat and cold stress.

OFF-SITE EMERGENCY PERSONNEL

Off-site emergency personnel include, for example, local fire fighters and ambulance crews, who often provide front-line response and run the risk of acute hazard exposure equal to that of any on-site worker. These personnel must be trained to recognize and deal effectively with on-site hazards. Lack of training may lead to their inadvertently worsening an emergency by improper actions (e.g., spraying water on a water-reactive chemical and causing an explosion). Inadequate knowledge of the on-site emergency chain of command may cause confusion and delays. Site manage-

ment should, at a minimum, supplement off-site personnel emergency training with the following information:

- site-specific hazards
- appropriate response techniques
- site emergency procedures
- decontamination procedures

VISITORS

Visitors to the site, including elected and appointed officials, reporters, and senior-level management, should receive a safety briefing. These visitors should not be permitted in the exclusion zone (see Chapter 7) unless they have been trained, fit-tested, and medically approved for respirator use. An observation tower in the clean zone reduces the need for visitors to enter the contaminated area.

RECORD OF TRAINING

A record of training should be maintained to confirm that every person assigned to a task has had adequate training for that task and that every employee's training is up to date. It is very important to document that the training is effective. Performance measurements prior to site entry are good personnel management and protection against future liability.

BIBLIOGRAPHY

"Accreditation of Training Programs for Hazardous Waste Operations." Federal Register 55, no. 18 (26 January 1990).

Fournier, S. *Hazardous Waste: Training Manual for Supervisors.* Business Legal Reports, 1985.

J.J. Keller & Associates, Inc. *Hazard Communication Guide.* Neenah, WI: J.J. Keller & Associates, Inc., 1987.

Payne, J.L., and C.B. Strong. "Taking Technology Off the Shelf: Texas A & M's Hazardous Material Control Program." Paper presented at the Conference on Hazardous Material Spills, Louisville, Ky., May 1980.

Payne, J.L., and C.B. Strong. "Training." In *Protecting Personnel at Hazardous Waste Sites,* S.P. Levine and W.F. Martin, eds. Boston: Butterworth-Heinemann, 1985.

Shaye, M.K. *Hazardous Waste Workers Health & Safety Training Requirements & 29 CFR 1910.120.* Detroit: Spill Control Association, 1988.

Swiss, J.J., W.S. Davis, and R.G. Simmons. "On-Scene Response Training Program." Paper presented at the Conference on Hazardous Material Spills, Milwaukee, April 1982.

U.S. Environmental Protection Agency. *Training Course Catalogue—EPA.* GPO Publication No. 91072969. Cincinnati: EPA, June 1990.

alc	alcohol	inhal	inhalation	pg	picogram
amorph	amorphous	insol	insoluble	pk	peak concentration
anhyd	anhydrous	intox	intoxication	pmol	picomole
aq	aqueous	ipr	intraperitoneal	PNS	peripheral nervous
atm	atmosphere	irr	irritant, irritating,		system effects
autoign temp	autoignition temperature		irritation	ppb	parts per billion (v/V)
bp	boiling point	IR	infrared	pph	parts per hundred (v/V)
BPR	blood pressure effects	IRR	irritant effects		(percent)
b range	boiling range		(systemic)	ppm	parts per million (v/V)
bz	benzene	itr	intratracheal	ppt	parts per trillion (v/V)
C	Centigrade/Celsius	iv	intravenous	PROP	properties
carc(s)	carcinogen(s)	kg	kilogram	psi	pounds per square inch
CARC	carcinogenic effects	L	liter	PSY	psychotropic effects
cc	cubic centimeter	liq	liquid	PUL	pulmonary system
CL	ceiling concentration	M	meter		effects
compd(s)	compound(s)	M	minute(s)	rbt	rabbit
conc	concentration,	m	meta	refr	refractive
	concentrated	m^3	cubic meter	resp	respiratory
contg	containing	mem	membrane	rhomb	rhombic
corr	corrosive	min	minimum	S, sec	second(s)
cryst	crystal(s), crystalline	μg, ug	microgram	scu	subcutaneous
CUM	cumulative effects	μmol, umol	micromole	SEV	severe irritation effects
CVS	cardiovascular effects	mg	milligram	SKN	systemic skin effects
d	density	mg/m^3	milligrams per cubic	slt	slight
D	day		meter	sltly	slightly
dBA	decibel	mg/L	milligrams per liter	sol	soluble
decomp	decomposition	misc	miscible	soln	solution
deliq	deliquescent	ml	milliliter	solv(s)	solvent(s)
dil	dilute	MLD	mild irritation effects	spont	spontaneous(ly)
eth	ether	mm	millimeter	subl	sublimes
exper	experimental (animal)	MMI	mucous membrane	susp	suspected
expl	explosive		effects	SYS	systemic effects
expos	exposure	mo	month	t_{re}	rectal temperature
eye	administration into eye	mod	moderately	ta	ambient air temperature
	(irritant)	MOD	moderate irritation	ta adj	adjusted air temperature
EYE	systemic eye effects		effects	tech	technical
F	Fahrenheit	mol	mole	temp	temperature
fbr	fibroblasts	mp	melting point	TER	teratogenic effects
flamm	flammable	mR	milliroentgen	TFX	toxic effects
flash p	flash point	mR/hr	milliroentgen per hour	tox	toxic, toxicity
fp	freezing point	MSK	musculoskeletal effects	uel	upper explosive limits
GIT	gastrointestinal tract	μ, u	micron	unk	unknown
	effects	mumem	mucous membrane	UNS	toxic effects unspecified
g/L	grams per liter	MUT	mutagen		in source
glac	glacial	mw	molecular weight	UV	ultraviolet
GLN	glandular effects	N	nitrogen	vap d	vapor density
gran	granular, granules	NEO	neoplastic effects	vap press	vapor pressure
hr	hour	nonflamm	nonflammable	visc	viscosity
hexag	hexagonal	NTP	National Toxicology	v/V	volume per volume
hmn	human		Program	W	week(s)
H_2	hydrogen	o-	ortho	Y	year(s)
htd	heated	ocu	ocular	>	greater than
htg	heating	p-	para	<	less than
ims	intramuscular	par	parenteral	<=	equal to or less than
incomp	incompatible	petr eth	petroleum ether	=>	equal to or greater than

ACGIH	American Conference of Governmental Industrial Hygienists	IDLH	immediately dangerous to life or health
ALR	air-line respirator	IUPAC	International Union for Pure and Applied Chemistry
ANSI	American National Standards Institute	LEL	lower explosive limit
APR	air-purifying respirators	LFL	lower flammable limit
CA	carcinogen	MMEFR	maximal expiratory flow rate
CAA	Clean Air Act	MSDS	Material Safety Data Sheet
CAS	Chemical Abstracts Service	MSHA	Mine Safety and Health Administration
CC	closed cup	MVV	maximal voluntary ventilation
CDC	Center for Disease Control	NCRP	North Carolina Research Park
CELDS	Computer-Aided Environmental Legislative Data System	NEPA	National Environmental Protection Agency
CEQ	Council on Environmental Quality	NFPA	National Fire Protection Agency
CERCLA	Comprehensive Environmental Response, Compensation, and Liability Act (also called Superfund)	NIOSH	National Institute for Occupational Safety and Health
CFR	Code of Federal Regulations	OC	open cup
CGI	combustible gas indicator	OSHA	Occupational Safety and Health Administration
CNS	central nervous system	OVA	organic vapor analyzer
COC	Cleveland Open Cup	PCB	polychlorinated biphenyl
CPC	chemical protective clothing	PDS	personnel decontamination station
CPR	cardiopulmonary resuscitation	PEL	permissible exposure limit or published exposure limit
CRC	contamination reduction corridor	PID	photoionization detector
CRZ	contamination reduction zone	PNA	polynuclear aromatic hydrocarbons
CWA	Clean Water Act	PPC	personal protective clothing
DOT	U.S. Department of Transportation	PPE	personal protective equipment
EKG	electrocardiogram	PVC	polyvinyl chloride
EPA	U.S. Environmental Protection Agency	RCRA	Resource Conservation and Recovery Act
ESCBA	escape-only self-contained breathing apparatus	REL	recommended exposure limits
ETA	equivocal tumorigenic agent	RV	residual volume
FEF	forced expiratory flow	SARA	Superfund Amendments and Reauthorization Act
FEMA	Federal Emergency Management Administration	SCBA	self-contained breathing apparatus
FEV_1	forced expiratory volume in one second	SMAC 23	Sequential Multiple Analyzer Computer
FID	flame ionization detector	SOP	standard operating procedure
FIFRA	Federal Insecticide, Fungicide and Rodenticide Act	SSO	site safety officer
FIT	field investigation team	STEL	short-term exposure limit
FRC	functional residual capacity	TAT	technical assistance team
FVC	forced vital capacity	TCC	taglibue closed cup
GC	gas chromatography	TD	toxic dose
GI	gastrointestinal	THR	toxic and hazard review
GM	Geiger-Müller	TLC	total lung capacity
HAZMAT	Hazardous Materials Response Team	TLV	threshold limit value
HAZWOPER	Hazardous Waste Operations and Emergency Response	TLV-C	threshold limit value-ceiling
		TSCA	Toxic Substances Control Act
HNU	name of company that manufactures a type of photoionizer used to detect organic gases and vapors	TSD	transportation, storage and delivery
		TWA	time-weighted average
		UEL	upper explosive limit
		UFL	upper flammable limit
HR	hazard rating	ULC	underwriters laboratory classification
HW	hazardous waste	USCG	U.S. Coast Guard
IARO	International Agency for Research on Cancer		

CHEMICAL FORMULAS

Ag_2O	silver oxide	H_2	hydrogen gas	N_4PCP	sodium pentachlorophenate
Al	aluminum	$HCHO$	formaldehyde		
$AlCl_3$	aluminum chloride	HCl	hydrochloric acid	NF_3	nitrogen fluoride
BF_3	boron trifluoride	HF	hydrofluoric acid	NH_3	ammonia
B_2O_3	boron oxide	HgF_2	mercuric fluoride	$NH_4{}^+$	ammonium radical
BO_x	boron oxides	HI	hydriodic acid	NH_4NO_3	ammonium nitrate
Br_2	bromine gas	HNO_3	nitric acid	NH_4OH	ammonium hydroxide
BrF_3	bromine trifluoride	H_2O	water	N_2O_4	nitrogen oxide
$CaCl_2$	calcium chloride	H_2O_2 or		NO_x	nitrogen oxides
$Ca(CN)_2$	calcium cyanide	$HOOH$	hydrogen peroxide	$NOCl$	nitrosyl chloride
CaO_x	calcium oxides	$HOAc$	acetic acid	O_2	oxygen gas
$Ca(OCl)_2$	calcium oxychloride	$HOCl$	hypochlorous acid	O_3	ozone
CCl_4	carbon tetrachloride	H_2S	hydrogen sulfide	OF_2	oxygen fluoride
CdO	cadmium oxide	H_2SO_4	sulfuric acid	OsO_4	osmium tetraoxide
$Cd(OH)_2$	cadmium hydroxide	H_2SO_3	sulfurous acid	PCl_3	phosphorus trichloride
C_6H_6	benzene	$H_2S_2O_3$	thiosulfuric acid	P_2O_3	phosphorus trioxide
$CHCl_3$	chloroform	IF_7	iodine heptafluoride	P_2O_5	phosphorus pentoxide
CH_3OH	methanol	$KClO_3$	potassium chlorate	PO_x	phosphorus oxides
Cl_2	chlorine gas	K_2CrO_4	potassium chromate	Rb_2C_2	rubidium carbide
ClF_3	chlorine trifluoride	KHC	potassium carbide	SCl_2	sulfur chloride
ClO_2	chlorine oxide	KOH	potassium hydroxide	SiO_2	silica
CN	cyanide	LiH	lithium hydride	SO_2	sulfur dioxide
CO	carbon monoxide	$LiOH$	lithium hydroxide	SO_x	sulfur oxides
CO_2	carbon dioxide	$Mg(C_2H_5)_2$	magnesium ethyl	2,3,7,8-	
$COCl_2$	phosgene	MgO	magnesia	$TCDD$	dioxin
CoO_x	cobolt oxides	Na_2C_2	sodium carbide	TeO	tellurium oxide
CrO_3	chromium trioxide	$NaClO_3$	sodium perchlorate	$Tl(NO_3)_3$	thallium nitrate
Cr_2O_3	chromium oxide	NaK	sodium-potassium alloy	Tl_2O	thallous oxide
CS_2	carbon bisulfide	NaN_3	sodium nitride	VO_x	vanadium oxides
Cs_2O	cesium oxide	$NaNO_3$	sodium nitrate	$ZnCl_2$	zinc chloride
$CuFeS_2$	copper iron sulfide	Na_2O	sodium oxide	$ZnCrO_4$	zinc chromate
$EtOH$	ethanol	Na_2O_2	sodium peroxide	$ZnCr_2O_7$	zinc dichromate
F_2	fluorine gas	$NaOBr$	sodium oxybromide	ZnO	zinc oxide
Fe_2O_3	iron oxide	$NaOCl$	sodium oxychloride		
F_2O_2	fluorine oxide	$NaOH$	sodium hydroxide		

acetic acid	$HOAc$	calcium oxychloride	$Ca(OCl)_2$	fluorine gas	F_2
aluminum	Al	carbon bisulfide	CS_2	fluorine oxide	F_2O_2
aluminum chloride	$AlCl_3$	carbon dioxide	CO_2	formaldehyde	$HCHO$
ammonia	NH_3	carbon monoxide	CO	hydriodic acid	HI
ammonium radical	$NH_4{}^+$	carbon tetrachloride	CCl_4	hydrochloric acid	HCl
ammonium hydroxide	NH_4OH	cesium oxide	Cs_2O	hydrofluoric acid	HF
ammonium nitrate	NH_4NO_3	chlorine gas	Cl_2	hydrogen gas	H_2
benzene	C_6H_6	chlorine oxide	ClO_2	hydrogen peroxide	H_2O_2 or
boron oxide	B_2O_3	chlorine trifluoride	ClF_3		$HOOH$
boron oxides	BO_x	chloroform	$CHCl_3$	hydrogen sulfide	H_2S
boron trifluoride	BF_3	chromium oxide	Cr_2O_3	hypochlorous acid	$HOCl$
bromine gas	Br_2	chromium trioxide	CrO_3	iodine heptafluoride	IF_7
bromine trifluoride	BrF_3	cobalt oxides	CoO_x	iron oxide	Fe_2O_3
cadmium hydroxide	$Cd(OH)_2$	copper iron sulfide	$CuFeSe_2$	lithium hydride	LiH
cadmium oxide	CdO	cyanide	CN	lithium hydroxide	$LiOH$
calcium chloride	$CaCl_2$	dioxin	2,3,7,8-	magnesia	MgO
calcium cyanide	$Ca(CN)_2$		$TCDD$	magnesium ethyl	$Mg(C_2H_5)_2$
calcium oxides	CaO_x	ethanol	$EtOH$	mercuric fluoride	HgF_2

methanol	CH_3OH	potassium chlorate	$KClO_3$	sulfur chloride	SCl_2
nitric acid	HNO_3	potassium chromate	K_2CrO_4	sulfur dioxide	SO_2
nitrogen fluoride	NF_3	potassium hydroxide	KOH	sulfur oxides	SO_x
nitrogen oxide	N_2O_4	rubidium carbide	Rb_2C_2	sulfuric acid	H_2SO_4
nitrogen oxides	NO_x	silica	SiO_2	sulfurous acid	H_2SO_3
nitrosyl chloride	$NOCl$	silver oxide	Ag_2O	tellurium oxide	TeO_x
osmium tetraoxide	OsO_4	sodium carbide	Na_2C_2	thallium nitrite	$Tl(NO_3)_3$
oxygen fluoride	OF_2	sodium hydroxide	$NaOH$	thallous oxide	Tl_2O
oxygen gas	O_2	sodium nitrate	$NaNO_3$	thiosulfuric acid	$H_2S_2O_3$
ozone	O_3	sodium nitride	NaN_3	vanadium oxides	VO_x
phosgene	$COCl_2$	sodium oxide	Na_2O	zinc chloride	$ZnCl_2$
phosphorus oxides	PO_x	sodium oxybromide	$NaOBr$	zinc chromate	$ZnCrO_4$
phosphorus pentoxide	P_2O_5	sodium oxychloride	$NaOCl$	zinc dichromate	$ZnCr_2O_7$
phosphorus trichloride	PCl_3	sodium perchlorate	$NaClO_3$	zinc oxide	ZnO
phosphorus trioxide	P_2O_3	sodium peroxide	Na_2O_2		
potassium carbide	KHC	sodium-potassium alloy	NaK		

acute exposure Exposure to a substance in a short time span and generally at high concentrations.

alpha radiation A type of ionizing radiation consisting of alpha particles, which are two protons and two neutrons bound together, with an electrical charge of +2. An alpha particle is equivalent to a helium nucleus.

autoreactive A compound that is reactive under normal conditions without initiation by heat or other compounds or change in conditions.

breakthrough time The elapsed time between initial contact of the hazardous chemical with the outside surface of protective clothing material and the time at which the chemical can be detected at the inside surface of the material by means of the chosen analytic technique.

buddy system A system of organizing employees into work groups in such a manner that each employee is designated to be observed by at least one other employee in the group. The purpose of the buddy system is to provide rapid assistance to employees in the event of an emergency.

bulk container A cargo container, such as that attached to a tank truck or tank car, used for transporting substances in large quantities.

bulking The mixing together of chemicals in large quantities for transport.

bung A cap or screw used to cover the small opening in the top of a metal drum or barrel.

canister A purifying device for an air-purifying respirator that is held in a harness attached to the body or attached to the chin part of a face piece, is connected to the face piece by a breathing tube, and removes particulates or specific chemical gases or vapors from the ambient air as it is inhaled through the canister.

carboy A bottle or rectangular container for holding liquids with a capacity of approximately 5 to 15 gallons; made of glass, plastic, or metal and often cushioned in a protective container.

cartridge A purifying device for an air-purifying respirator that attaches directly to the face piece and removes particulates or specific chemical gases or vapors from the ambient air as it is inhaled through the cartridge.

chronic exposure Exposure to a substance over a long period of time, usually at low doses.

clean-up operation An operation where hazardous substances are removed, contained, incinerated, neutralized, stabilized, or in any other manner processed or handled with the ultimate goal of making the site safer for people or the environment.

closed-circuit SCBA A type of self-contained breathing apparatus (SCBA) that recycles exhaled air by removing carbon dioxide and replenishing oxygen. Also called a *rebreather SCBA*.

colorimetric tube An instrument for the chemical analysis of liquids by comparison of the color of the given liquid with standard colors.

CLO A unit of measure for CPC thermal heating values. Based on heat transfer rates through clothing at room temperature.

combustible Capable of burning.

contamination control line The boundary between the support zone and the contamination reduction zone.

contamination reduction corridor (CRC) The part of the contamination reduction zone where the personnel decontamination stations are located.

contamination reduction zone (CRZ) The area on a site where decontamination takes place, preventing cross-contamination from contaminated areas to clean areas.

continuous-flow repirator A respiratory protection device that maintains a constant flow of air into the face piece at all times. Airflow is independent of user respiration.

crazing The formation of minute cracks (as in the lens of a face piece).

cross-contamination The transfer of a chemical contaminant from one person, piece of equipment, or area to another that was previously not contaminated with that substance.

decontamination The removal of hazardous substances from employees and their equipment to the extent necessary to preclude the occurrence of foreseeable adverse health effects.

decontamination line A specific sequence of decontamination stations within the contamination reduction zone for decontaminating personnel or equipment.

degradation A chemical reaction between chemical and structural materials (in, for example, protective clothing or equipment) that results in damage to the structural material.

demand respirator A respiratory protection device that supplies air or oxygen to the user in response to negative pressure created by inhalation.

dermal Pertaining to skin.

disinfection The application of a chemical that kills bacteria.

dosimeter An instrument for measuring doses of radioactivity or other chemical exposures based on collection media.

dress-out area A section of the support zone where personnel suit up for entry into the exclusion zone.

emergency response A response effort by employees from outside the immediate release area or by other designated responders (e.g., mutual-aid groups or local fire departments) to a situation that results, or is likely to result, in an uncontrolled release of a hazardous substance.

escape-only SCBA (ESCBA) A type of self-contained breathing apparatus (SCBA) that is approved for escape purposes only. It does not carry the safety features necessary for longer work periods.

etiologic agent A microorganism that may cause human disease.

exclusion zone The contaminated area of a site.

explosive A chemical that is capable of burning or bursting suddenly and violently.

facility Any site, area, building, structure, installation, equipment, pipe or pipeline (including any pipe into a sewer or publicly owned treatment works), well, pit, pond, lagoon, impoundment, ditch, storage container, motor vehicle, rolling stock, or aircraft where a hazardous substance has been deposited, stored, disposed of, or placed. Does not include any consumer product in consumer use or any waterborne vessel.

filter A purifying device for an air-purifying respirator that removes particulates and/or metal fumes from the ambient air as it is inhaled.

flammable Capable of being easily ignited or burning with extreme rapidity.

flammable gas Any compressed gas meeting the requirements for lower flammability limit, flammability limit range, flame projection, or flame propagation criteria as specified in 49 CFR 173.300(b).

flammable liquid Any liquid having a flash point below 100°F as determined by tests listed in 49 CFR 173.115(d). A *pyrophoric liquid* ignites spontaneously in dry or moist air at or above 130°F.

flammable solid Any solid material, other than an explosive, that can be ignited readily and when ignited burns so vigorously and persistently as to create a serious transportation hazard (49 CFR 173.150).

flash point The minimum temperature at which a liquid gives off enough vapors to form an ignitable mixture with the air near the surface of the liquid.

gamma radiation A type of ionizing radiation consisting of high-energy, short-wavelength electromagnetic radiation.

grappler An implement used to hold and manipulate objects from a distance.

hazardous materials response team (HAZMAT) An organized group of employees, designated by the employer, expected to handle and control actual or potential leaks or spills of hazardous substances requiring possible close approach to the substance.

hazardous substance Any substance designated by the following regulations: Sections 101(14) and 101(33) of CERCLA; 49 CFR 172.101.

hazardous waste A waste or combination of wastes as defined in 40 CFR 261.3 or 49 CRF 171.6.

hazardous waste operation Any operation conducted within the scope of 40 CFR 261.3 or 49 CFR 171.6.

health hazard A chemical, mixture of chemicals, or pathogen for which there is statistically significant evidence based on at least one scientific study that acute or chronic health effects may occur in exposed employees.

hot line The outer boundary of a site's exclusion zone.

immediately dangerous to life or health (IDLH) The maximum concentration from which one could escape within 30 minutes without any escape-impairing symptoms or any irreversible health effects.

incompatible Incapable of being combined without a dangerous effect (e.g., descriptive of two or more substances that produce an unfavorable chemical reaction if they come in contact).

injection The introduction of chemicals into the body through puncture wounds.

ionizing radiation High-energy radiation that causes irradiated substances to form ions, which are electrically charged particles.

LC$_{50}$ Abbreviation for the median lethal concentration of a substance that will kill 50% of the animals exposed to that concentration.

LD$_{50}$ Abbreviation for the median lethal dose of a substance that will kill 50% of the animals exposed to that dose.

manifest A list of cargo.

open-circuit SCBA A self-contained breathing apparatus (SCBA) in which the user exhales air directly into the atmosphere.

overpack 1. An oversized drum into which a leaking drum can be placed and sealed. 2. To overpack such a drum.

oxygen deficiency The concentration of oxygen by volume below which atmosphere-supplying respiratory equipment must be provided. It exists in atmospheres where the percentage of oxygen by volume is less than 19.5%.

palletize To place on a pallet or to transport or store by means of a pallet.

particulate Formed of separate, small, solid pieces.

penetration The chemical penetration of protective clothing through openings such as seams, buttonholes, zippers, or breathing air ports.

percutaneous Effected or performed through the skin.

permeation Seepage and sorption of a chemical through a material (e.g., the material making up protective clothing or equipment).

permissible exposure limit (PEL) The exposure, inhalation, or dermal permissible exposure limit specified in 29 CFR 1910, G and Z.

postemergency response That portion of an emergency response performed after the immediate threat of a release has been stabilized or eliminated and cleanup of the site has begun.

pressure-demand respirator A respiratory protection device that supplies air to the user and maintains a slight positive pressure in the face piece at all times. It supplies additional air in response to the negative pressure created by inhalation.

protection factor The ratio of the ambient concentration of an airborne substance to the concentration of the substance inside the respirator at the breathing zone of the wearer. The protection factor is a measure of the degree of protection the respirator offers.

published exposure limit (PEL) The recommended exposure limits published in *Recommendations of Occupational Health Standards* (NIOSH 1986).

qualified person A person with specific training, knowledge, and experience in the area for which he or she has the responsibility and the authority to control.

radiation Energy in the form of electromagnetic waves.

reagent A substance used in a chemical reaction to detect, measure, examine, or produce other substances.

redress area A section of the exclusion zone where decontaminated personnel put on clothing for use in the support zone.

self-contained breathing apparatus (SCBA) A respiratory protection device that supplies clean air to the user from a compressed air source carried by the user.

site safety supervisor (SSO) The individual located on a hazardous waste site who is responsible to the employer and has the authority and knowledge necessary to implement the site safety and health plan and verify compliance with applicable safety and health requirements.

small-quantity generator A generator of hazardous wastes that in any calendar month generates no more than 2205 pounds (1000 kilograms) of hazardous wastes.

sorbent material A substance that takes up other materials either by absorption or adsorption.

staging area An area in which items are arranged in some order.

standard operating procedure (SOP) Established or prescribed tactical or administrative method to be followed routinely for the performance of a designated operation or in a designated situation.

Superfund A common name for the Comprehensive Environmental Response, Compensation and Liability Act (CERCLA) of 1980.

supplied-air respirator A respiratory protection device that supplies air to the user from a source that is not worn by the user but is connected to the user by a hose. Also called an *airline respirator*.

support zone The uncontaminated area of a site where workers will not be exposed to hazardous conditions.

surfactant A decontamination agent that reduces adhesion forces between contaminants and the surfaces being cleaned.

swab A piece of cotton or gauze on the end of a slender stick used for obtaining a piece of tissue or secretion for bacteriologic examination.

swipe A patch of cloth or paper that is wiped over a surface and analyzed for the presence of a substance.

threshold The intensity or concentration below which a stimulus or substance produces a specified effect.

uncontrolled hazardous waste site An area where an accumulation of hazardous waste creates a threat to the health and safety of individuals, the environment, or both.

Appendix E
SAMPLE SITE SAFETY PLAN

This appendix provides a generic plan based on a plan developed by the U.S. Coast Guard for responding to hazardous chemical releases.* This generic plan can be adapted for designing a site safety plan for hazardous waste site cleanup operations. It is not all-inclusive and should be used only as a guide, not a standard.

---

*U.S. Coast Guard, *Policy Guidance for Response to Hazardous Chemical Releases*, USCG Pollution Response COMDTINST–M16465.30.

A. SITE DESCRIPTION

Date _____ Location _____

Hazards _____

Area affected _____

Surrounding population _____

Topography _____

Weather conditions _____

Additional information _____

B. ENTRY OBJECTIVES - The objective of the initial entry to the contaminated area is to _____ (describes actions, tasks to be accomplished; i.e., identify contaminated soil; monitor conditions, etc.)

C. ONSITE ORGANIZATION AND COORDINATION - The following personnel are designated to carry out the stated job functions on site. (Note: One person may carry out more than one job function.)

PROJECT TEAM LEADER _____

SCIENTIFIC ADVISOR _____

SITE SAFETY OFFICER _____

PUBLIC INFORMATION OFFICER _____

SECURITY OFFICER _____

RECORDKEEPER _____

FINANCIAL OFFICER _____

FIELD TEAM LEADER _____

FIELD TEAM MEMBERS _____

FEDERAL AGENCY REPS (i.e., EPA, NIOSH) _____

STATE AGENCY REPS _____

LOCAL AGENCY REPS _____

CONTRACTOR(S) _____

D. ONSITE CONTROL

All personnel arriving or departing the site should log in and out with the Recordkeeper. All activities on site must be cleared through the Project Team Leader.

_____ (Name of individual or agency) has been designated to coordinate access control and security on site. A safe perimeter has been established at _____ (distance or description of controlled area)

No unauthorized person should be within this area.

The onsite Command Post and staging area have been established at _____

The prevailing wind conditions are _____. This location is upwind from the Exclusion Zone.

Control boundaries have been established, and the Exclusion Zone (the contaminated area), hotline, Contamination Reduction Zone, and Support Zone (clean area) have been identified and designated as follows: _____ (describe boundaries and/or attach map of controlled area)

These boundaries are identified by: _____ (marking of zones, i.e., red boundary tape - hotline; traffic cones - Support Zone; etc.)

E. HAZARD EVALUATION

The following substance(s) are known or suspected to be on site. The primary hazards of each are identified.

Substances Involved (chemical name)	Concentrations (If Known)	Primary Hazards (e.g., toxic on inhalation)

The following additional hazards are expected on site: (i.e., slippery ground, uneven terrain, etc.)

Hazardous substance information form(s) for the involved substance(s) have been completed and are attached.

F. PERSONAL PROTECTIVE EQUIPMENT

Based on evaluation of potential hazards, the following levels of personal protection have been designated for the applicable work areas or tasks:

Location	Job Function	Level of Protection
Exclusion Zone		A B C D Other
		A B C D Other
		A B C D Other
		A B C D Other
Contamination Reduction Zone		A B C D Other
		A B C D Other
		A B C D Other
		A B C D Other

Specific protective equipment for each level of protection is as follows:

Level A   Fully-encapsulating suit
          SCBA
          (disposable coveralls)

Level B   Splash gear (type)
          SCBA

Level C   Splash gear (type)
          Full-face canister resp.

Level D

Other

The following protective clothing materials are required for the involved substances:

Substance (chemical name)	Material (material name, e.g., Viton)

If air-purifying respirators are authorized, _____ (filtering medium) is the appropriate canister for use with the involved substances and concentrations. A competent individual has determined that all criteria for using this type of respiratory protection have been met.

NO CHANGES TO THE SPECIFIED LEVELS OF PROTECTION SHALL BE MADE WITHOUT THE APPROVAL OF THE SITE SAFETY OFFICER AND THE PROJECT TEAM LEADER.

G. ONSITE WORK PLANS

Work party(s) consisting of _____ persons will perform the following tasks:

(name)	(function)
Project Team Leader	
Work Party #1	
Work Party #2	
Rescue Team (required for entries to IDLH environments)	
Decontamination Team	

The work party(s) were briefed on the contents of this plan at _____.

H. COMMUNICATION PROCEDURES

Channel _____ has been designated as the radio frequency for personnel in the Exclusion Zone. All other onsite communications will use channel _____.

Personnel in the Exclusion Zone should remain in constant radio communication or within sight of the Project Team Leader. Any failure of radio communication requires an evaluation of whether personnel should leave the Exclusion Zone.

_____ (Horn blast, siren, etc.) is the emergency signal to indicate that all personnel should leave the Exclusion Zone. In addition, a loud hailer is available if required.

The following standard hand signals will be used in case of failure of radio communications:

Hand gripping throat ------------ Out of air, can't breathe
Grip partner's wrist or ---------- Leave area immediately
   both hands around waist
Hands on top of head ------------ Need assistance
Thumbs up ----------------------- OK, I am all right, I understand
Thumbs down --------------------- No, negative

Telephone communication to the Command Post should be established as soon as practicable. The phone number is _____.

I. DECONTAMINATION PROCEDURES

Personnel and equipment leaving the Exclusion Zone shall be thoroughly decontaminated. The standard level _____ decontamination protocol shall be used with the following decontamination stations: (1) _____

(2) _____   (3) _____   (4) _____   (5) _____
(6) _____   (7) _____   (8) _____   (9) _____
(10) _____ Other _____

Emergency decontamination will include the following stations:

_____
_____
_____

The following decontamination equipment is required:

_____
_____

_____ (Normally detergent and water) will be used as the decontamination solution.

J. SITE SAFETY AND HEALTH PLAN

1. _____ (name) is the designated Site Safety Officer and is directly responsible to the Project Team Leader for safety recommendations on site.

2. Emergency Medical Care

_____ (names of qualified personnel) are the qualified EMTs on site. _____ (medical facility names), at _____ (address), phone _____ is located _____ minutes from this location. _____ (name of person) was contacted at _____ (time) and briefed on the situation, the potential hazards, and the substances involved. A map of alternative routes to this facility is available at _____ (normally Command Post).

Local ambulance service is available from _____ at phone _____. Their response time is _____ minutes. Whenever possible, arrangements should be made for onsite standby.

First-aid equipment is available on site at the following locations:

First-aid kit   _____
Emergency eye wash   _____
Emergency shower   _____
(other)   _____

Emergency medical information for substances present:

Substance	Exposure Symptoms	First-Aid Instructions
_____	_____	_____
_____	_____	_____
_____	_____	_____

List of emergency phone numbers:

Agency/Facility	Phone #	Contact
Police	_____	_____
Fire	_____	_____
Hospital	_____	_____
Airport	_____	_____
Public Health Advisor	_____	_____

3. Environmental Monitoring

The following environmental monitoring instruments shall be used on site (cross out if not applicable) at the specified intervals.

Combustible Gas Indicator - continuous/hourly/daily/other
O_2 Monitor   - continuous/hourly/daily/other
Colorimetric Tubes   - continuous/hourly/daily/other
   (type) _____
HNU/OVA _____   - continuous/hourly/daily/other
Other _____   - continuous/hourly/daily/other
_____   - continuous/hourly/daily/other

4. Emergency Procedures (should be modified as required for incident)

The following standard emergency procedures will be used by onsite personnel. The Site Safety Officer shall be notified of any onsite emergencies and be responsible for ensuring that the appropriate procedures are followed.

Personnel Injury in the Exclusion Zone: Upon notification of an injury in the Exclusion Zone, the designated emergency signal _____ shall be sounded. All site personnel shall assemble at the decontamination line. The rescue team will enter the Exclusion Zone (if required) to remove the injured person to the hotline. The Site Safety Officer and Project Team Leader should evaluate the nature of the injury, and the affected person should be decontaminated to the extent possible prior to movement to the Support Zone. The onsite EMT shall initiate the appropriate first aid, and contact should be made for an ambulance and with the designated medical facility (if required). No persons shall reenter the Exclusion Zone until the cause of the injury or symptoms is determined.

Personnel Injury in the Support Zone: Upon notification of an injury in the Support Zone, the Project Team Leader and Site Safety Officer will assess the nature of the injury. If the cause of the injury or loss of the injured person does not affect the performance of site personnel, operations may continue, with the onsite EMT initiating the appropriate first aid and necessary follow-up as stated above. If the injury increases the risk to others, the designated emergency signal _____ shall be sounded and all site personnel shall move to the decontamination line for further instructions. Activities on site will stop until the added risk is removed or minimized.

Fire/Explosion: Upon notification of a fire or explosion on site, the designated emergency signal _____ shall be sounded and all site personnel assembled at the decontamination line. The fire department shall be alerted and all personnel moved to a safe distance from the involved area.

Personal Protective Equipment Failure: If any site worker experiences a failure or alteration of protective equipment that affects the protection factor, that person and his/her buddy shall immediately leave the Exclusion Zone. Reentry shall not be permitted until the equipment has been repaired or replaced.

Other Equipment Failure: If any other equipment on site fails to operate properly, the Project Team Leader and Site Safety Officer shall be notified and then determine the effect of this failure on continuing operations on site. If the failure affects the safety of personnel or prevents completion of the Work Plan tasks, all personnel shall leave the Exclusion Zone until the situation is evaluated and appropriate actions taken.

The following emergency escape routes are designated for use in those situations where egress from the Exclusion Zone cannot occur through the decontamination line: _____ (describe alternate routes to leave area in emergencies)

In all situations, when an onsite emergency results in evacuation of the Exclusion Zone, personnel shall not reenter until:

1. The conditions resulting in the emergency have been corrected.
2. The hazards have been reassessed.
3. The Site Safety Plan has been reviewed.
4. Site personnel have been briefed on any changes in the Site Safety Plan.

5. Personal Monitoring

The following personal monitoring will be in effect on site:

Personal exposure sampling: _____ (describe any personal sampling programs being carried out on site personnel. This would include use of sampling pumps, air monitors, etc.)

Medical monitoring: The expected air temperature will be _____ (_____ °F). If it is determined that heat stress monitoring is required (mandatory if over 70°F) the following procedures shall be followed: _____ (describe procedures in effect, i.e., monitoring body temperature, body weight, pulse rate)

_____
_____
_____

All site personnel have read the above plan and are familiar with its provisions.

Site Safety Oficer _____ (name) _____ (signature)
Project Team Leader _____
Other Site Personnel _____
_____
_____

MEDICAL OCCUPATIONAL HISTORY

Source: NIOSH Contract Reports, Hazardous Waste Project/Publication 1983–1985, Cincinnati, Ohio, Project Officer, William F. Martin.

CONFIDENTIAL

DATE OF VISIT: _____
AGE: _____

NAME: _____

EMPLOYER: _____

EMPLOYER'S ADDRESS: _____

EMPLOYER'S PHONE NUMBER: _____

JOB TITLE: _____

JOB DESCRIPTION: _____
_____
_____

SEX: __(F) __(M)   SOCIAL SECURITY NO. _____

YEARS OF SCHOOL COMPLETED: _____

NAME/ADDRESS/PHONE NUMBER OF PERSONAL PHYSICIAN:
NAME: _____
ADDRESS: _____
PHONE
NUMBER: _____

WHEN WERE YOU LAST EXAMINED BY HIM/HER? _____

WHEN WAS YOUR LAST CHEST X-RAY? _____ RESULTS: _____

- -

PERSONAL MEDICAL HISTORY

List significant medical illness, and all hospitalizations in the last 5 years:

ILLNESS OR CONDITION	HOSPITALIZATION?	APPROX. DATE OF HOSP.
a. _____	( )yes ( )no	_____
b. _____	( )yes ( )no	_____
c. _____	( )yes ( )no	_____

MEDICAL HISTORY BY ORGAN SYSTEMS

Have you ever been told by a doctor that you had any of the following conditions?

Cardiovascular

Heart murmur	( )yes	( )no :
Angina/chest pain	( )yes	( )no :
Heart attack	( )yes	( )no :
High blood pressure	( )yes	( )no :
Vascular disease in arms/legs	( )yes	( )no :
Other, specify		:
_____		:

Gastrointestinal

Peptic ulcer	( )yes	( )no :
Hiatus hernia	( )yes	( )no :
Hepatitis	( )yes	( )no :
Gall bladder disease	( )yes	( )no :
Liver disease/jaundice	( )yes	( )no :
Cirrhosis	( )yes	( )no :
Other, specify	( )yes	( )no :

Skin

Psoriasis	( )yes	( )no :
Eczema	( )yes	( )no :
Contact dermatitis	( )yes	( )no :
Other allergic skin reactions	( )yes	( )no :
Specify		:
_____		:

Genitourinary

Nephritis	( )yes	( )no
Kidney disease	( )yes	( )no
(indicate type)		
Urinary infection	( )yes	( )no
Kidney/urinary bladder stones	( )yes	( )no
Blood/protein in urine	( )yes	( )no :
Venereal disease	( )yes	( )no :
Other, specify	( )yes	( )no :
_____		:

Blood

Anemia	( )yes	( )no :
Problems with blood clotting/bleeding	( )yes	( )no :
Sickle cell	( )yes	( )no :
Other blood disorders	( )yes	( )no :
Specify		:
_____		:

Eye

Require glasses	( )yes	( )no :
Glaucoma	( )yes	( )no :
Cataracts	( )yes	( )no :
Optic neuritis	( )yes	( )no :
Eye infection(s)	( )yes	( )no :
Other, specify	( )yes	( )no :
_____		:

CONFIDENTIAL

FAMILY HISTORY

If any member of your family noted in the following table has had any of the stated conditions, please indicate by the appropriate code number.

CODE: Father   = 1   Brother/Sister = 4
        Mother   = 2   My children.   = 5
        Grandparent = 3

CODE NO.	CONDITION		CODE NO.	CONDITION
___	Allergy (asthma, eczema, hay fever) :		___	Hypertension
___	Bleeding disorder :		___	Kidney disease
___	Cancer or leukemia :		___	Migraine headaches
___	Cirrhosis :		___	Rheumatic heart disease
___	Congenital malformation :		___	Sickle cell disease
___	Diabetes :		___	Tuberculosis
___	Emphysema :		___	Other: please specify
___	Epilepsy (seizures) :		___	___

Is your father still living?   ( )yes ( )no
If "no", at what age did he die?   ___
What was the cause of death?   ___

Is your mother still living?   ( )yes ( )no
If "no", at what age did she die? ___
What was the cause of death?   ___

CONFIDENTIAL

MEDICAL HISTORY BY ORGAN SYSTEMS--continued

Pulmonary

Pneumonia	( )yes ( )no
Pleurisy	( )yes ( )no
Asthma	( )yes ( )no
Bronchitis	( )yes ( )no
Emphysema	( )yes ( )no
Bronchiectasis	( )yes ( )no
Tuberculosis	( )yes ( )no
Silicosis	( )yes ( )no
Asbestosis	( )yes ( )no
Other, specify	( )yes ( )no

Nervous System

Seizure disorder	( )yes ( )no
Stroke	( )yes ( )no
Peripheral neuritis	( )yes ( )no
Psychiatric illness	( )yes ( )no
Other nervous disorder	( )yes ( )no
Specify ___	

Ear, Nose and Throat

Chronic sinusitis	( )yes ( )no
Impaired hearing	( )yes ( )no
Ringing in the ears	( )yes ( )no
Easy nasal bleeding	( )yes ( )no
Nasal allergies	( )yes ( )no
Tonsillectomy	( )yes ( )no
Other, specify	( )yes ( )no

Musculoskeletal

Rheumatoid arthritis	( )yes ( )no
Other arthritis	( )yes ( )no
Back injury	( )yes ( )no
Degenerative disc disease	( )yes ( )no
Sciatica/disc herniation	( )yes ( )no
Bone lesions/ infections	( )yes ( )no
Other, specify	( )yes ( )no

General

Thyroid Disease/goiter	( )yes ( )no
Diabetes	( )yes ( )no
Gout	( )yes ( )no
Frequent night sweats/fever	( )yes ( )no
Hemorrhoids	( )yes ( )no
Hernia	( )yes ( )no
Cancer	( )yes ( )no
Specify type ___	
Specify type ___	
Dental/gum problems	( )yes ( )no
Specify ___	
Other, specify	( )yes ( )no

ALCOHOL USE

On the average, how much of each of the following do you drink per week?

Beer _____ cans
Wine _____ glasses
Whiskey/liquor _____ jiggers

MEDICATION

Please indicate any medications you are taking, including nonprescription medication (such as aspirin, laxatives, vitamins, etc.):

_____
_____

REPRODUCTIVE HISTORY

Have you or your spouse been unable
to have children?   ( )yes ( )no
If "yes", specify reason, if known: _____

Have you ever had any children born with a handicap or congenital
malformation? ( )yes ( )no
If "yes", specify: _____

ALLERGIES

Are you allergic to anything that you know of? ( )yes ( )no
If "yes", specify: _____

CONFIDENTIAL

GENERAL HEALTH

Have you been examined or treated by any doctor
within the past year?   ( )yes ( )no
If "yes", for what? _____

Have you lost more than five pounds within the
last 6 months?   ( )yes ( )no

Have you noticed any swelling or lumps in your
breast, neck, armpits, groin or elsewhere during
the past year?   ( )yes ( )no
If "yes", specify site _____

Have you experienced the following signs/symptoms within the past year:

Frequent headache/dizziness   ( )yes ( )no
Frequent bowel problems
(constipation or diarrhea)   ( )yes ( )no
Swelling of the lower
extremities or eyelids   ( )yes ( )no
Frequent shortness of breath,
cough or morning phlegm   ( )yes ( )no

Indicate what you believe your health status is now:

( )EXCELLENT   ( )GOOD   ( )FAIR   ( )POOR

CIGARETTE USE

Check the smoking history closest to your own:

( ) Never smoked regularly
( ) Used to smoke regularly
    How many years did you smoke? _____
    How many packs per day? _____
    How long ago did you stop? _____

Do you smoke now? ( )yes ( )no
If "yes", for how many years? _____
How many packs per day on the average? _____

Cigars/pipe:
ever smoke ( )yes ( )no

OCCUPATIONAL HISTORY

How long have you been in present job? _____ years _____ months

Indicate any job-related illness or injuries you have experienced since working in present job: _____

_____

Indicate any substance(s) that you work with that you consider hazardous:

_____

_____

Do you wear protective clothing on the job? _____

Have you had any problems wearing or using a respirator? _____

List all your previous jobs, beginning with the one you had immediately prior to your present job. Please indicate the dates of employment as well as any hazardous exposures:

JOB	DATES	EXPOSURES
1. _____	_____	_____
2. _____	_____	_____
3. _____	_____	_____
4. _____	_____	_____
5. _____	_____	_____
6. _____	_____	_____
7. _____	_____	_____

Do you have any hobbies (e.g., arts/crafts, gunning, furniture refinishing) or home construction/gardening activities that may expose you to any hazards? ( )yes ( )no
If "yes", specify activities and kind of materials used: _____

FOR PHYSICIAN'S USE ONLY

Physician's summary and elaboration of all pertinent data. (Physician shall comment on all positive answers. Physician may develop by interview any additional medical history he deems important, and record any significant findings here.)

Typed or printed name of physician _____
Signature _____ Date _____

HAZARDOUS SUBSTANCE DATA SHEET

Under "Name of Substance," list both the common name and the name approved by the International Union for Pure and Applied Chemistry (IUPAC). Enter both because it may be necessary to look up information under different names. If a compound has more than one common name, list all synonyms. Write the chemical formula after the name because some references index chemicals by their formulas.

Part I lists the physical and chemical properties of the compound. In the far right column labeled "Source," enter the reference from which the information was obtained. In this way, if the information is later found to be incorrect or conflicting, it may be corrected. It also makes it easier to refer back to a particular source if additional information is needed. The following properties are included in the data sheet:

- *Normal physical state.* Check the appropriate space for the physical state of the chemical at normal ambient temperatures (20°C to 25°C).
- *Molecular weight:* Express in grams per gram-mole (g/g-mole). Neutralization or any other chemical treatment requires an estimate of the number of moles of chemical present.
- *Density and specific gravity.* Only one is required. Density is expressed in grams per milliliter (g/ml); specific gravity is dimensionless. Indicate the temperature at which specific gravity is measured and circle the appropriate letter corresponding to degrees Fahrenheit (°F) or Celsius (°).
- *Solubility (water).* Expressed in parts per million (ppm) or milligrams per liter (mg/L) (1 ppm = 1 mg/L). Solubility is temperature dependent.
- *Solubility.* Enter any other material for which solubility data are needed. For instance, recovering a spilled material by solvent extraction may require solubility data for an organic compound.
- *Boiling point.* Expressed in °F or °C, it is the temperature at which the vapor pressure of the compound equals atmospheric pressure (760 millimeters of mercury at sea level). The boiling point is raised if any impurities are present.
- *Melting point.* Expressed in °F or °C, it is equivalent to the freezing point. The melting point is lowered if any impurities are present.
- *Vapor pressure.* Expressed in millimeters of mercury or atmospheres at a given temperature. The vapor pressure is strongly temperature dependent.
- *Vapor density.* A dimensionless quantity expressed relative to air.
- *Flash point.* Expressed in °F or °C. Indicate whether the figure is based on an open-cup or closed test.
- *Other.* Enter any miscellaneous data, such as biochemical oxygen demand, autoignition temperature, or odor threshold concentrations.

Part II is a compilation of five types of hazardous characteristics. In the far right column labeled "Source," enter the reference from which the information was obtained.

Section A lists toxicological hazards:

- *Inhalation.* Under "Concentrations," enter the threshold limit value (TLV) concentration (or any other pertinent value—PEL, IDLH, etc.). This is important for selecting levels of protection for workers who will be in the area.
- *Ingestion.* Enter the type of test (e.g., LD_{50}) and the toxicity level in milligrams per kilogram of body weight (mg/kg).
- *Skin/eye absorption and contact.* Determine from the references whether these hazards exist.
- *Carcinogenic, teratogenic, and mutagenic.* It is difficult to obtain concentration data on these hazards, since very little is known about the mechanisms that cause these effects.
- *Aquatic.* Expressed in ppm for a particular species.
- *Other.* Enter an IDLH concentration or any other pertinent miscellaneous information.

Section B contains fire hazard data:

- *Combustibility.* Applies to any compound that can be oxidized in air. Almost every organic compound is combustible.
- *Toxic by-products.* If the compound is combustible, enter yes, because all combustion processes yield some carbon monoxide. List the particular toxic by-products in the spaces below.
- *Flamability/explosiveness limits.* Expressed as a percentage by volume in air. Usually flammable limits and explosive limits are synonymous.

Section C contains reactivity data:

- *Reactivity hazard.* If the material is reactive, indicate the substances that are incompatible with the material.

Section D contains corrosivity data:

- *pH.* Some references give the pH of an aqueous solution at a given concentration. For example, the pH of a 0.5% solution of sodium hydroxide is 13. There is also space for listing the types of materials known to be corroded by the compound in question.
- *Neutralizing agent.* Some references list neutralizing materials that bring the pH of the affected area to neutral (pH = 7).

Section E contains radioactivity data:

- *Background.* List a background level. Background is usually on the order of 0.01 milliroentgens per hour (mR/hr).
- *Alpha, beta, gamma.* Exposure rates on some elements may be found in the *Radiological Health Handbook* (U.S. Department of Health and Human Service, N.D.).

Parts III, IV, and V describe the specific incident and recommend safety measures. Sometimes parts of these sections will be left blank because of a lack of accurate information. Enter available incident information as promptly as possible so that mitigation measures can be taken.

Part III describes the incident:

- *Quantity involved.* Usually express in barrels, gallons, or liters for a liquid and kilograms or pounds for a solid.
- *Release information.* Indicate whether the container is still leaking and the rate of discharge, if known.
- *Monitoring/sampling recommended.* Indicate what type of monitoring should be initiated to characterize an incident completely and whether on-site samples are necessary.

Part IV covers recommended protection:

- *Worker.* Decide on levels of protection for response personnel based on the chemical, physical, and toxicological properties of the materials in question. The OHM/TADS reference segment 108, ''Personal Safety Precautions,'' aids in this decision.
- *Public.* Based on the data in Parts I, II, and III and on the proximity of the incident to populated areas, make an initial public

hazard evaluation. The Oil and Hazardous Materials/Technical Assistance Data System (OHM/TADS) reference segment 111, ''Degree of Hazard to Public Health,'' is helpful in recommending action to protect public health. OHM/TADS is an EPA data base.

Part V, recommended site control, covers the establishment of the following:

- hot line
- decontamination line
- command post
- exclusion zone
- contamination reduction zone
- support zone

An example of the documentation needed to complete a hazardous substance data sheet also is shown, in this case for benzene. Nine sources of information were used in this example:

ACGIH *Documentation of the Threshold Limit Values (TLV)*
CHRIS, Volume 2
Condensed Chemical Dictionary by G. Hawley
Dangerous Properties of Industrial Materials by N.I. Sax
Department of Labor Industry Safety Data Sheets
DOT *1989 Emergency Response Guidebook*
The Merck Index
NIOSH/OSHA *Pocket Guide to Chemical Hazards*
OHM/TADS

SAMPLE HAZARDOUS SUBSTANCE DATA SHEET

NAME OF SUBSTANCE:

COMMON _____ CHEMICAL _____

I. PHYSICAL/CHEMICAL PROPERTIES   SOURCE

Normal physical state: Gas ____ Liquid ____ Solid ____
Molecular weight
Density   g/ml
Specific gravity   @   °F/°C
Solubility: water   @   °F/°C
Solubility: @   °F/°C
Boiling point   °F/°C
Melting point   °F/°C
Vapor pressure   mmHg @   °F/°C
Vapor density   @   °F/°C
Flash point   °F/°C
Other:

II. HAZARDOUS CHARACTERISTICS

A. TOXICOLOGICAL HAZARD   HAZARD   CONCENTRATIONS   SOURCE

Inhalation   Yes No
Ingestion   Yes No
Skin/eye absorption   Yes No
Skin/eye contact   Yes No
Carcinogenic   Yes No
Teratogenic   Yes No
Mutagenic   Yes No
Aquatic   Yes No
Other: Yes No

B. FIRE HAZARD   HAZARD   CONCENTRATIONS   SOURCE

Combustibility   Yes No
Toxic byproducts: Yes No

Flammability   Yes No
  LFL
  UFL

Explosiveness   Yes No
  LEL
  UEL

C. REACTIVITY HAZARD   Yes No   CONCENTRATIONS   SOURCE

SAMPLE HAZARDOUS SUBSTANCE DATA SHEET (Continued)

                          HAZARD   CONCENTRATIONS   SOURCE

D. CORROSIVITY HAZARD   Yes No
   ph
   Neutralizing agent:

E. RADIOACTIVE HAZARD   Yes No   EXPOSURE RATE   SOURCE
   Background
   Alpha particles
   Beta particles
   Gamma radiation

III. DESCRIPTION OF INCIDENT:
   Quantity involved
   Release information

   Monitoring/sampling recommended

IV. RECOMMENDED PROTECTION:
   Worker
   Public

V. RECOMMENDED SITE CONTROL:
   Hotline
   Decontamination line
   Command post location

SAMPLE COMPLETED HAZARDOUS SUBSTANCE DATA SHEET

NAME OF SUBSTANCE:
COMMON Benzol, Cyclohexatriene   CHEMICAL Benzene C_6H_6

I. PHYSICAL/CHEMICAL PROPERTIES

				SOURCE
Normal physical state:	Gas ___	Liquid ✓	Solid ___	
Molecular weight	78.11			CHRIS II
Density		g/ml		CHRIS II
Specific gravity	0.879 @ 20	°F/(°C)		CHRIS II
Solubility: water	820 ppm @ 25	°F/°C		OHM/TADS
Solubility:	@	°F/°C		
Boiling point	176	(°F)/°C		CHRIS II
Melting point	42	(°F)/°C		CHRIS II
Vapor pressure	100 mmHg @ 26.1	°F/(°C)		SAX
Vapor density	2.77 @ 20	°F/(°C)		OHM/TADS
Flash point (cc)	12	(°F)/°C		SAX
Other:				

II. HAZARDOUS CHARACTERISTICS

A. TOXICOLOGICAL HAZARD

	HAZARD	CONCENTRATIONS	SOURCE
Inhalation	(Yes) No	25 ppm TLV	CHRIS II
Ingestion	(Yes) No	0.1 H mg/kg (hum)	OHM/TADS
Skin/eye absorption	(Yes) No		SAX
Skin/eye contact	(Yes) No		SAX
Carcinogenic	(Yes) No		
Teratogenic	Yes No		
Mutagenic	Yes No		
Aquatic	Yes No		
Other: IDHL level	(Yes) No	2000 ppm	NIOSH/OSHA Guide

B. FIRE HAZARD

	HAZARD	CONCENTRATIONS	SOURCE
Combustibility	(Yes) No		
Toxic byproducts:	(Yes) No		
Flammability	(Yes) No		
LFL		1.3%	CHRIS II
UFL		7.9%	
Explosiveness	(Yes) No		
LEL		1.3%	CHRIS II
UEL		7.9%	

C. REACTIVITY HAZARD (Yes) No

CONCENTRATIONS	SOURCE
Oxidizing mat'ls, O_2, O_3 also Cl_2, ClO_3, perchlorates, peroxides, H_2SO_4	SAX, OHM/TADS

SAMPLE COMPLETED HAZARDOUS SUBSTANCE DATA SHEET (Continued)

	HAZARD	CONCENTRATIONS	SOURCE
D. CORROSIVITY HAZARD	Yes (No)		OHM/TADS
pH			
Neutralizing agent:			

	HAZARD	EXPOSURE RATE	SOURCE
E. RADIOACTIVE HAZARD	Yes (No)		CHRIS II
Background			
Alpha particles			
Beta particles			
Gamma radiation			

III. DESCRIPTION OF INCIDENT:

Quantity involved

Release information

Monitoring/sampling recommended

IV. RECOMMENDED PROTECTION:

Worker Wear self-contained (positive pressure if available) breathing apparatus and full protective clothing (DOT).

Public Isolate hazard area and deny entry to unnecessary people (DOT).

V. RECOMMENDED SITE CONTROL:

Hotline

Decontamination line

Command post location

Appendix H

CHEMICAL PROTECTIVE CLOTHING RECOMMENDATIONS

In this appendix, chemical protective clothing (CPC) recommendations for chemical classes are presented. The chemicals have been grouped into generic families (acids, amines, etc.), and general recommendations are made for each family that is represented by more than one chemical having CPC performance information for a given material. The recommendations are contained in three tables. This information is extracted from *Guidelines for the Selection of Chemical Protective Clothing* prepared by Arthur D. Little, Inc., for the U.S. Environmental Agency and Los Alamos National Laboratory, March 1983.

SCOPE AND LIMITATIONS

CHEMICALS

CPC recommendations have been developed for approximately 300 chemicals and 14 clothing materials. The chemicals are the liquids included in the Clean Water Act (CWA) Sections 311 and 307a, the Clean Air Act (CAA) Section 112, and the Resource Conservation and Recovery Act (RCRA) Sections P, U, F, and K. Also included are any other chemicals (principally liquids but including some gases) for which CPC vendors' recommendations or technical reports of permeation test results were available. Vendors' recommendations or permeation data were not available for all the liquids addressed in the aforementioned acts. Approximately 40% of the chemicals are included in OSHA Directive Subpart 2—Toxic and Hazardous Substances, 29 CFR 1910.1000, Tables Z–1 and Z–2. Note that except for the aqueous solutions, all liquids are single components; multicomponent organic solutions are not addressed.

CPC recommendations are provided for 14 materials for generic families of chemicals. Recommendations are not given for all materials in all classes. The criterion for being given a recommendation is that the class must contain more than one chemical with a CPC recommendation for the material of concern. In many cases, there was considerable variability among the recommendations for chemicals within a class; these are indicated by double asterisks (**) in Table H.3.

MATERIALS

The 14 principal materials from which CPC is fabricated are listed across the top of Table H.3. Where information on other materials was available, recommendations for these materials are in the rightmost column of the table. A general characterization of several of the physical properties of the materials is presented in Table H.1.

The 14 categories were reduced from the approximately 100 types and forms of clothing materials available and represent the materials of construction for well over 90% of the CPC considered in the *Guidelines for the Selection of Chemical Protective Clothing*. By grouping several types and forms of clothing into one category, it is likely that in some cases particularly good or particularly poor items have gone unnoted, since there can be significant differences in product quality among vendors. This is a compromise that was accepted and recognized in summary compilations. In general, however, a given material will exhibit the same performance relative to another material independent of whether the materials are free of films or coatings and independent of source. For example, if a butyl rubber glove is more resistant to a given chemical than a nitrile rubber glove, it is highly likely that butyl rubber gloves and clothing in both supported and unsupported form will be better barriers to that chemical than their nitrile counterparts. In other words, differences in performance between products of a given material will probably be small compared to performance differences between categories of materials. In using the tables, remember that their purpose is to provide a starting point for CPC selection. Selection based on the tables' recommendations does not guarantee protection, since in no way do they take into account key issues such as the application of the CPC or quality differences among CPC products.

PERFORMANCE INFORMATION

The information on which the recommendations are based is from three sources:

- CPC vendors' chemical resistance charts, which are often included in the product catalogs. The ratings in the charts of approximately 30 vendors (including the 5 largest manufacturers of CPC) were tabulated and reviewed by chemical and material classes. In total, more than 6000 individual ratings were included in the tabulation.

- CPC raw materials suppliers' chemical resistance charts.

- The technical literature that addresses chemical resistance and permeation testing of CPC materials and products. In all, more than 2000 individual test results (such as breakthrough time, permeation rate, and percent weight change) were tabulated.

The vendors use a variety of rating scales. Some have three grades, most have four grades, and a few have five or six grades. To compare ratings, a normalized four-grade system (A, B, C, D) was developed. Briefly, products with the highest rating in a four- or three-grade system or the highest two ratings in a six-grade system were given a normalized rating of A. A normalized rating of

153

TABLE H.1 Physical Characteristics of CPC Materials*

Material Designation by Material	Abrasion Resistance	Cut Resistance	Flexibility	Heat Resistance	Ozone Resistance	Puncture Resistance	Tear Resistance	Relative Cost
Butyl rubber (Butyl)	F	G	G	E	E	G	G	High
Natural rubber (Nat. Rub.)	E	E	E	F	P	E	E	Medium
Neoprene (Neop.)	E	E	G	G	E	G	G	High
Neoprene/styrene-butadiene rubber (Neop./SBR)	G	G	G	G	G	G	G	Medium
Neoprene/natural rubber (Neop./Nat. Rub.)	E	E	E	G	G	G	G	Medium
Nitrile rubber (Nitrile)	E	E	E	G	F	E	G	High
Nitrile rubber/polyvinyl chloride (Nitrile/PVC)	G	G	G	F	E	G	G	Medium
Polyethylene (PE)	F	F	G	F	F	P	F	Low
Chlorinated polyethylene (CPE)	E	G	G	G	E	G	G	Low
Polyurethane (PU)	E	G	E	G	G	G	G	High
Polyvinyl alcohol (PVA)	F	F	P	G	E	F	G	Medium
Polyvinyl chloride (PVC)	G	P	G	P	E	G	G	Low
Styrene-butadiene rubber (SBR)	E	G	G	G	F	F	F	Low
Viton	G	G	F	G	E	G	G	Very high

*Ratings are subject to variation depending on formulation, thickness, and whether the material is supported by fabric. E = excellent; G = good; F = fair; P = poor.

TABLE H.2 Description of Criteria for Recommendations

Character	Performance Data	Vendor Recommendations
RR	Breakthrough times greater than 1 hour reported by (normally) two or more testers	A or B ratings from three or more (apparently independent) vendors
R	None	Same as RR
rr	Some data suggesting breakthrough times of approximately 1 hour or more	A or B ratings from fewer than three vendors; no Cs or Ds* B or C ratings—with Bs predominating—from several vendors
r	None	Same as rr
NN	Breakthrough times less (usually significantly less) than 1 hour reported by (normally) two or more testers	C or D ratings from three or more (apparently independent) vendors
N	None	Same as NN
nn	Some data (usually high solubilities) suggesting breakthrough times of 1 hour are not likely	C or D ratings from fewer than three vendors B or C ratings—with Cs predominating—from several vendors
n	None	Same as nn

*Products of some materials (e.g., CPE and Vitron) are manufactured and rated by only one or two vendors.

B was given to the next highest vendor ranking, which was generally called "good" but in some three-grade systems was called "fair." A normalized rating of C was given to the third highest vendor ranking, except in the three-grade systems. Typically, vendors called this ranking "fair." Finally, all vendor rankings of "poor" and "not recommended" were given a normalized rating of D.

RECOMMENDATIONS

The recommendations in the tables resulted from a comprehensive analysis of all the available information. Briefly, a computerized data base of the information was developed. No attempt was made to validate any of the data before they were input. In a sense, there was a self-validation of the data, since the recommendation scale used in the tables takes into account the number of independent information sources that will in total either substantiate or throw into question individual performance claims. This is discussed in the next paragraph and becomes evident from review of Table H.2. The data base was organized such that any available information for a particular chemical and a particular clothing material could be retrieved in the form of a single printed report. The report was analyzed, and a recommendation was developed. No recommendation was made for a chemical-material pair for which there was no information.

There are eight grades of recommendations. Each is designed to represent a particular combination of performance, the number of sources substantiating that performance, and the consistency of the information. This is reflected by the number and size of the let-

ters that indicate the recommendation. The criteria and explanations for the recommendations are summarized in Table H.2. In all cases of inconsistencies between test results and manufacturers' recommendation information, the test results were more influential in forming the recommendation. All recommendations are conservative in that they reflect a cautious attitude toward CPC selection.

During the selection and eventual use of the CPC recommended in Table H.3, it is important to remember the following:

- The recommendations are based on the best information available. In some cases, this information is very limited. The recommendations are a guide, not a guarantee.

- The recommendations probably do not hold for extreme conditions (e.g., high and low temperatures, long-term contact, high abrasion), nor do they consider the problems associated with reuse.

- Certain products in each category may be better or poorer than the norm. Also, the quality of construction of even the better products can vary from batch to batch. In their present form, the recommendations do not address quality issues. The assessment of quality and uniformity of quality can best be gained through field experience.

- The double-letter recommendations are based primarily on breakthrough-time data. Permeation-rate data were given only secondary consideration.

The recommendations may be modified as additional performance information becomes available from the EPA.

TABLE H.3 CPC Recommendations by Chemical Class

	Butyl	Nat. Rub.	Neop.	Neop./SBR	Neop./Nat. Rub.	Nitrile	Nitrile/PVC	PE	CPE	PU	PVA	PVC	SBR	Viton	Other Materials
Acids, carboxylic—aliphatic and alicyclic															
Unsubstituted	R	**	RR	r	r	RR		rr	r	r	n	RR	**	r	NBR(r)
Polybasic	**	RR	RR			RR	r	r	r	r	n	RR		r	
Aldehydes															
Aliphatic	R	**	**			**	**	r	**		n	rr			NBR(r)
Aromatic and heterocyclic	r	**	**	n	**	N		r	**		rr	N		n	NBR(r)
Amides, carboxylic—aliphatic	rr		NN			NN	c	**		nn				nn	
Amines—aliphatic and alicyclic															
Primary	r	**	rr			RR	r	r	NN	r	rr	rr	**		NBR(r)
Secondary	*	n	n			**	**	**	nn	r	**	**	**	**	NBR(r)
Tertiary	r	rr	R	r		RR			rr	r	rr	RR		nr	
Esters, carboxylic															
Aliphatic	**	NN	NN			NN		**	**	r	RR	NN			NBR(*)
Acetates	r	**		n	**		c		NN			N	n		
Higher monobasic	**														
Aromatic—phthalates	rr	**	RR		r	RR					**	**		**	
Esthers—aliphatic	rr	nn	**		**	RR	c				RR	NN		**	
Halogen compounds															
Aliphatic															
Unsubstituted	nn	NN	NN		NN	NN		NN	NN		**	NN			
Substituted	**	**	**	n	**	NN	c	nn	nn	nn	**	NN	n	RR	(RR)
Aromatic															
Unsubstituted	n	N	N			**	c	r	NN		NN	N		n	
Substituted		rr	RR			rr				**	RR	rr			
Polynuclear	**	N	**		nn						NN	N		**	
Heterocyclic compounds															
Epoxy compounds	rr	n	**			**		**	**	nn	NN	n	n	c	
Furan derivatives	n	**	**			NN		nn	nn	n	**	NN	*	c	
Hydrazines	**	**	**			**						**		**	
Hydrocarbons															
Aliphatic and alicyclic	NN	NN	RR	n	NN	RR	c	nn	r	R	RR	NN	n	RR	NBR(r)
Aromatic	NN	NN	*	n	NN	*	c	nn	*	*	RR	NN	*	RR	NBR(r)

Hydroxyl compounds

	C1	C2	C3	C4	C5	C6	C7	C8	C9	C10	C11	C12	Designation*
Aliphatic and alicyclic													
Primary	RR	**	RR	**	RR	r	r	r	r	r	**	**	NBR(r)
Secondary	**	RR	RR		RR	r	r	r	r	r	RR	r	
Polyols	rr				RR	**	**	**	RR	**	**		
Aromatic	rr	**	RR	**	**	**	r	r	**	**	**	**	
Inorganic gases	r	n	R		n	r	rr	**	**	**			
Inorganic salts							rr						
Inorganic acids	**	**	**	n	**	**	**	nr	**	**	rr	rr	NBR(r), Saranex (rr)
Inorganic bases	R	RR	RR	r	RR	r	r	nr	r	RR	**	r	NBR(r)
NBR(r)													
Ketones, aliphatic	rr	NN	NN	n	NN	n	**	**	**	NN	r	NN	
Nitriles, aliphatic	**	r											
Nitro—unsubstituted	**	NN	**		NN	nn		nn	RR	nn		nn	
Organophosphorous compounds	rr		**		rr			**		**	**	**	
Peroxides	r	**	RR		RR			r		r		r	
Vinyl halides						**							

*Designation from Table H.1
**Classification from Table H.2

DECONTAMINATION PROCEDURES FOR THREE TYPICAL LEVELS OF PROTECTION

LEVEL A DECONTAMINATION: A WORST-CASE DECONTAMINATION PROTOCOL

EQUIPMENT WORN

This decontamination procedure is for workers wearing the following protective clothing and equipment:

- fully encapsulating suit with integral boots and gloves
- self-contained breathing apparatus (SCBA)
- hard hat (optional)
- chemical-resistant boots with steel toe and shank
- boot covers
- inner and outer gloves
- taped joints between gloves, boots, and suit

DECONTAMINATION PROCEDURES

Decontamination of this level of protection is performed at 19 separate stations.

Station 1: Segregated Equipment Drop Deposit equipment used on the site (tools, sampling devices and containers, monitoring instruments, radios, clipboards, etc.) on plastic drop cloths or in different containers with plastic liners. Each piece of equipment may be contaminated to a different degree; therefore, segregation at the drop reduces the potential for contamination. Equipment needed:

- containers of various sizes
- plastic liners
- plastic drop cloths

Station 2: Suit, Boot Covers, and Glove Wash Thoroughly wash and scrub fully encapsulating suit, outer boot covers, and gloves with a decontamination solution or detergent-water solution. Equipment needed:

- container (20 to 30 gallon)
- decontamination solution
- detergent-water solution
- two or three long-handled, soft-bristled scrub brushes

Station 3: Suit, Boot Covers, and Glove Rinse Rinse off the decontamination solution from Station 2 using copious amounts of water. Repeat as many times as necessary. Equipment needed:

Note: Additional information may be found in Chapter 6, Personal Protective Equipment.

- container (30 to 50 gallon)
- high-pressure spray unit and splash guard
- water
- two or three long-handled, soft-bristled scrub brushes

Station 4: Tape Removal Remove tape around boots and gloves and deposit it in a container with a plastic liner. Equipment needed:

- container (20 to 30 gallon)
- plastic liners

Station 5: Boot Cover Removal Remove boot covers and deposit them in a container with a plastic liner. Equipment needed:

- container (30 to 50 gallon)
- plastic liners
- bench or stool

Station 6: Outer Glove Removal Remove outer gloves and deposit them in a container with a plastic liner. Equipment needed:

- container (20 to 30 gallon)
- plastic liners

Station 7: Suit and Safety Boot Wash If design does not include Station 2, suits will be washed at this station. Thoroughly wash suit and boots. Scrub them with a long-handled, soft-bristled scrub brush and copious amounts of decontamination solution or detergent-water solution. Repeat as many times as necessary. Equipment needed:

- container (30 to 50 gallon)
- decontamination solution
- detergent-water solution
- two or three long-handled, soft-bristled scrub brushes

Station 8: Suit and Safety Boot Rinse If design does not include Station 3, suits will be rinsed at this station. Rinse off the decontamination or detergent-water solution using copious amounts of water. Repeat as many times as necessary. Equipment needed:

- container (30 to 50 gallon)
- high-pressure spray unit
- water
- two or three long-handled, soft-bristled scrub brushes

Station 9: Tank Change If a worker leaves the exclusion zone to change her air tank, this is the last step in the decontamination procedure. She exchanges the tank, dons new outer gloves and boots, and has the joints taped. She then returns to duty. Equipment needed:

- air tanks
- tape
- boot covers
- gloves

Station 10: Safety Boot Removal   Remove safety boots and deposit them in a container with a plastic liner. Equipment needed:

- container (30 to 50 gallon)
- plastic liners
- bench or stool
- bootjack

Station 11: Fully Encapsulating Suit and Hard Hat Removal With the assistance of a helper, remove fully encapsulating suit and hard hat. Hang the suit on a rack or lay it out on drop cloths. Equipment needed:

- rack
- drop cloths
- bench or stools

Station 12: SCBA Backpack Removal   While still wearing face piece, remove backpack and place it on the table. Disconnect hose from regulator valve and proceed to next station. Equipment needed: table.

Station 13: Inner Glove Wash   Wash with decontamination solution or detergent-water solution that will not harm skin. Repeat as many times as necessary. Equipment needed:

- basin or bucket
- decontamination solution
- detergent-water solution
- small table

Station 14: Inner Glove Rinse   Rinse with water. Repeat as many times as necessary. Equipment needed:

- water
- basin or bucket
- small table

Station 15: Face Piece Removal   Remove face piece. Deposit it in a container with a plastic liner. Avoid touching face with fingers. Equipment needed:

- container (30 to 50 gallon)
- plastic liners

Station 16: Inner Glove Removal   Remove inner gloves and deposit them in a container with a plastic liner. Equipment needed:

- container (20 to 30 gallon)
- plastic liners

Station 17: Inner Clothing Removal   Remove inner clothing. Place it in a container with a plastic liner. Do not wear inner clothing off the site, since small amounts of contaminants may have been transferred in removing fully encapsulating suit. Equipment needed:

- container (30 to 50 gallon)
- plastic liners

Station 18: Field Wash   Shower if highly toxic, skin-corrosive, or skin-absorbable materials are known or suspected to be present. Wash hands and face if shower is not available. Equipment needed:

- water
- soap
- small table
- basin or bucket
- field showers
- towels

Station 19: Redress   Put on clean clothes. A dressing trailer is needed in inclement weather. Equipment needed:

- table
- chairs
- lockers
- clothes

LEVEL A DECONTAMINATION (SITUATION 1) AND THREE MODIFICATIONS

The preceding description outlines each station included in a complete worst-case decontamination protocol. Different sites will present different hazard levels. Thus, at each individual site, this protocol must be modified accordingly. The following table illustrates the modifications that can be made in response to a variety of conditions.

Situation Number	Station Number																		
	1	2	3	4	5	6	7	8	9	10	11	12	13	14	15	16	17	18	19
1		x	x	x	x	x	x	x		x	x	x	x	x	x	x	x	x	x
2		x	x	x	x	x	x	x	x										
3		x				x	x		x	x	x				x	x	x	x	
4		x				x	x	x											

Situation 1 = The individual entering the CRZ is observed to be grossly contaminated, or extremely toxic substances are known or suspected to be present.

Situation 2 = Same as Situation 1 except that the individual needs a new air tank and will return to the exclusion zone.

Situation 3 = The individual entering the CRZ is expected to be minimally contaminated. Extremely toxic or skin-corrosive materials are not present. No outer gloves or boot covers are worn. Inner gloves are not contaminated.

Situation 4 = Same as Situation 3 except that the individual needs a new air tank and will return to the exclusion zone.

LEVEL B DECONTAMINATION
EQUIPMENT WORN

This decontamination procedure is for workers wearing the following protective clothing and equipment:

- one-piece, hooded chemical-resistant splash suit
- SCBA
- hard hat
- chemical-resistant boots with steel toe and shank
- boot covers
- inner and outer gloves
- taped joints between gloves, boots, and suit

DECONTAMINATION PROCEDURES

Stations 1 through 6 These stations are exactly the same for Level B decontamination as for Level A decontamination.

Station 7: Suit, SCBA, Boot, and Glove Wash If design does not include Station 2, wash suit at this station. Thoroughly wash suit, SCBA, boots, and gloves with a long-handled, soft-bristled scrub brush and copious amounts of decontamination solution or detergent-water solution. Wrap SCBA regulator (if belt-mounted type) with plastic to keep out water. Wash backpack assembly with sponges or cloth. Equipment needed:

- container (30 to 50 gallon)
- decontamination solution
- detergent-water solution
- two or three long-handled, soft-bristled scrub brushes
- small buckets
- sponges or cloths

Station 8: Suit, SCBA, Boot, and Glove Rinse If design does not include Station 3, rinse suit at this station. Rinse off the decontamination solution or detergent-water solution using copious amounts of water. Repeat as many times as necessary. Equipment needed:

- container (30 to 50 gallon)
- high-pressure spray unit and splash guard
- water
- small buckets
- two or three long-handled, soft-bristled scrub brushes
- sponges or cloths

Stations 9 and 10 These stations are exactly the same for Level B decontamination as for Level A decontamination.

Station 11: SCBA Backpack Removal While still wearing face piece, remove backpack and place it on a table. Disconnect hose from regulator valve and proceed to next station. Equipment needed: table.

Station 12: Splash Suit Removal With assistance, remove splash suit. Deposit it in a container with a plastic liner. Equipment needed:

- container (30 to 50 gallon)
- plastic liners
- bench or stool

Stations 13 through 19 These stations are exactly the same for Level B decontamination as for Level A decontamination.

LEVEL C DECONTAMINATION
EQUIPMENT WORN

The decontamination procedure outlined is for workers wearing the following protective clothing and equipment:

- one-piece, hooded chemical-resistant splash suit
- canister-equipped full face mask
- hard hat
- chemical-resistant boots with steel toe and shank
- boot covers
- inner and outer gloves
- taped joints between gloves, boots, and suit

DECONTAMINATION PROCEDURES

Stations 1 through 8 These stations are exactly the same for Level C decontamination as for Level B decontamination.

Station 9: Canister or Mask Change If the worker leaves the exclusion zone to change his canister or mask, this is the last step in the decontamination procedure. He exchanges his canister or mask, dons new outer glove and boot covers, and has joints taped. He then returns to duty. Equipment needed:

- canisters or masks
- tape
- boot covers
- gloves

Station 10 This station is exactly the same for Level C decontamination as for Level B decontamination.

Stations 11 through 18 These stations are exactly the same for Level C decontamination as Stations 12 through 19 for Level B decontamination.

BIBLIOGRAPHY

U.S. Coast Guard (USCG). *Policy Guidance for Response to Hazardous Chemical Releases.* USCG Pollution Response COMDTINST-M16465.30. Washington, D.C.: USCG.

U.S. Environmental Protection Agency. Office of Emergency and Remedial Response. Hazard Response Support Division. *Standard Operating Safety Guides.* Washington, D.C.: EPA, November 1984.

HEALTH AND SAFETY CHECKLIST

This checklist is provided to assist hazardous waste site supervisors, federal and state inspectors, and industry planners identify, evaluate, and control site hazards, ensure proper worker protection, and identify potential health and safety problems. It can be used for active disposal sites, abandoned sites undergoing remedial action, and emergency response operations.

The checklist is divided into three sections: (1) hazard recognition, (2) hazard evaluation, and (3) hazard control and worker protection. Individual checklists on specific subjects are grouped under the appropriate heading.

Although this checklist is intended primarily for individuals with training and experience in health- and safety-related fields, two footnotes have been added for less experienced users. These are designed to draw attention to potential problems, indicated by an asterisk (*) or an exclamation point (!) placed in a response box. The asterisk indicates a potentially serious health or safety problem that must be addressed but does not present an immediate hazard. The exclamation mark indicates a serious health or safety problem that requires immediate corrective action.

HAZARD RECOGNITION

Hazard recognition involves the identification of the potential for human exposure to dangerous chemical, physical, or biological agents. Data from both off-site reconnaissance and on-site surveys are used to establish the degree of hazard present at a site.

HAZARD EVALUATION

This section provides a guide to assessing the potential risk of exposure to chemical, physical, and biological hazards and unsafe conditions at hazardous waste sites. Safety hazards, which are often classified as physical hazards, are treated separately under the subsection "Evaluation of Safety Hazards."

HAZARD CONTROL AND WORKER PROTECTION

Hazard control and worker protection at waste sites entail providing trained, healthy workers with the proper protective equipment and ensuring that their environment is secure and monitored.

AUDIT OF INFORMATION SOURCES

Before conducting site investigations or environmental sampling, the investigator should audit the existing site information and sources of information. Such an information audit will help to ascertain the presence of hazardous materials and to determine the origin of many of the waste materials. It also will help to familiarize the investigator with the physical characteristics of the site. The checklist on the following pages is designed to promote a comprehensive review of information sources and to identify those data that are needed for an adequate assessment of waste site hazards. Proper consideration of existing information at a given site will avoid unnecessary and costly duplication of data collection efforts.

HEALTH AND SAFETY CHECKLIST

Sources of Information

Question	Response Yes	No

(1) Are any of the following site records available?
- Waste storage inventories and operating records
- Manifests or shipping papers
- Receipts, logbooks, or other business ledgers

(2) If available, have they been received for information on the presence of hazardous wastes materials?

(3) Have waste generators been contacted for additional information?

(4) Have available records from the following Federal, state, and local sources been reviewed?
- Pollution control agencies
- State Attorney General
- State Occupational Safety and Health (OSH) agencies
- State fire marshalls
- Workers' Compensation agencies
- County Commissioner's Office
- Local government officials
- Police departments

(5) Have the following local nongovernmental sources of information been investigated?
- Utility companies (to inquire about existing electric, telephone, or gas lines; sewers; water pipes, etc.)
- Local hospitals
- Local businessmen's groups
- News media
- Court records
- Neighbors

(6) Have interviews been conducted with former and current employees?

Type of Information

(7) Have past or present inquiries into the following information characterized the topographical and geological aspects of the site?
- Soil surveys
- USGS maps or other topographic maps
- USGS reports
- Geologic maps
- Site construction documents

- Historical and recent aerial photos
- Electromagnetic conductivity
- Ground-penetrating radar

(8) Has the above information been reviewed to determine:

- All possible contaminant flow routes?
- Location of any streams or ponds?
- Approximate depth of local aquifer and location?
- Type of surface material and bedrock?
- Depth to bedrock?

(9) Have data published by the National Oceanic and Atmospheric Administration (NOAA) been consulted to determine the following meter-ological conditions?

- Wind direction
- Precipitation
- Temperature profile

(10) Have aerial photographs (historical and recent) been compared for:

- Disappearance of natural depressions, quarries, or borrow pits?
- Variation in reforestation of disturbed areas?
- Mounting in disturbed areas or paved surfaces?
- Changes in vegetation around buildings?
- Modifications of grade?
- Any alteration of topographic features?
- Appearance and disappearance of traffic patterns?

INITIAL SITE INSPECTION

The twofold purpose of an initial site inspection is to verify and to supplement the information obtained from the audit. The inspection may generate additional data through a combination of observations and environmental sampling. This initial site inspection checklist will serve as a guide for considering key features of the waste site that might represent a hazardous or unsafe situation. (A discussion of environmental sampling requirements is presented in Section 1.3, Environmental Sampling.)

	Response	
Question	Yes	No

(11) Can the physical state (i.e., solid, liquid, gaseous) of some or all of the wastes be determined visually during a walk-through?

(12) How are the wastes contained?

- Paper or wood packages
- Metal barrels
- Plastic barrels
- Underground tanks
- Aboveground tanks
- Compressed gas cylinders
- Open pits
- Other

(13) In general, what is the condition of waste containers?

- Sound (undamaged)

- Visibly rusted or corroded (damaged but not leaking)

- Leaking

ENVIRONMENTAL SAMPLING

The environment at the hazardous waste site may be sampled many times during inspection, cleanup, or work activities. Sampling is also a necessary part of hazard recognition and evaluation. This checklist for environmental sampling is not to be considered definitive in any reasonable sense; rather, it is designed to prompt the user to consider major areas of concern.

Question	Response	
	Yes	No

(14) Have the following environmental media been sampled?

- Air
  - --On-site
  - --Off-site (around perimeter of site)

- Topsoil

- Deep soil (core sampling)

- Groundwater
  - --Upgradient of site
  - --Beneath site
  - --Downgradient of site

- Surface water

- Waste containments
  - --Surface impoundments/lagoons
  - --Tanks
  - --Drums
  - --Gas cylinders

(15) Has the environment been adequately sampled (or monitored) to identify sources of:

- Ionizing radiation, i.e., X-rays, radioisotopes, etc.?

- Biological or pathological hazards, i.e., hospital wastes?

- Disease-carrying vectors, i.e., rats, mice, insects, etc.?

- Concentrations of combustible or explosive gases?

(16) Have any of the following remote sensing or subsurface investigative methods been used to screen for buried wastes or contaminant plumes?

- Electromagnetic conductivity

- Magnetometry

- Metal detection

- Ground-penetrating radar

- Comparison of historic and recent aerial photos

EVALUATION OF CHEMICAL HAZARDS

The potential for serious exposure to toxic chemicals is a persistent problem associated with hazardous waste handling and disposal. An evaluation of this potential is necessary to avert or prevent serious health problems. This section of the checklist provides the user with only a general overview for evaluating chemical hazards. Because the assessment of chemical hazards is a serious and often complex problem, it is recommended that the National Institute for Occupational Safety and Health (NIOSH) Data Book on Chemical Hazards be used to supplement this checklist.

	Question	Response Yes	No
(17)	Has the ambient air within the contamination, decontamination, and clean zones been sampled for concentrations of toxic substances?		*
(18)	Have the measured concentrations exceeded the following limits:		
	– Levels "immediately dangerous to life and health" (IDLH)?	!	
	– Threshold limit values (TLVs)--both 8-hour time-weighted averages (TWAs) and short-term exposure levels (STELs)?	*	
	– OSHA Permissible Exposure Levels (PELs)?	*	
(19)	Has the ambient air within the areas planned for use as contamination, decontamination, and clean zones been sampled for concentrations of explosive substances?		*
(20)	Do the measured concentrations approach or exceed the lower explosive limits (LELs) of the substance(s)?	*	
(21)	Have any severe skin irritants been identified on the site?		
(22)	Have any contaminants that are highly toxic via dermal absorption been identified?		

EVALUATION OF PHYSICAL HAZARDS

Exposure to physical hazards is a potential problem at many hazardous waste sites. An evaluation of this potential is necessary to avert or prevent serious exposure to at least two harmful agents: ionizing radiation and extremes of heat and cold. This section of the checklist briefly reviews the potential hazard for such exposures. Further information on these and other physical agents can be obtained from the American Conference of Governmental Industrial Hygienists (ACGIH) document, "Threshold Limit Values for Chemical Substances and Physical Agents in the Work Environment." (Note: Safety hazards, which are often classified as physical hazards, are treated under a separate subsection entitled "Evaluation of Safety Hazards.")

	Question	Response Yes	No
(23)	Has the level of ionizing radiation been determined for the planned work zones?		*
	If the answer is "yes," see Question 24.		
(24)	Is the radiation level at or above 10.0 millirems per hour (mR/h)?	*	
	If the answer is "no," see Question 25.		
(25)	Is the radiation level significantly above (20X) background (normal background is 0.02 mR/h	!	
	If the answer is "yes," see Question 26.		
(26)	Has a complete detailed radiation survey been conducted?		*
(27)	Do site conditions predispose workers to heat or cold stress?	*	

EVALUATION OF BIOLOGICAL HAZARDS

The checklist for evaluating biological hazards is designed to assure that the user considers the potential for exposure to biological agents at hazardous waste sites. The questions primarily address the disease potential at a given time.

Question	Response Yes	No
(28) Did the information audit or initial on-site inspection reveal the presence of pathological wastes?		*
(29) Has the site been sampled for pathological agents?		*
(30) If pathological wastes have been identified on site, have the potential hazards associated with these wastes been investigated and documented?		*
(31) Have potential disease and/or contaminant-carrying vectors (rats, mice, bats, mosquitoes, etc.) been identified?		
(32) Have potential dermatitis agents (poison ivy, poison oak, poison suman, etc.) been identified?		
(33) If biological agents are present, have any steps been taken to eliminate or negate the potential hazards?		*

EVALUATION OF SAFETY HAZARDS

Like any other work environment, hazardous waste sites may present unsafe situations to the worker. Although this checklist cannot impart the depth of understanding required to evaluate the accident potential at these sites, it offers a few key safety considerations as a general overview of an often complex problem. The user is referred to the Occupational Safety and Health Administrations (OSHA) General Industry Safety and Health Standards for more specific and detailed information.

Questions	Response Yes	No
(34) Have the quantity and nature of reactive waste materials been established?		*
(35) Have incompatible materials been identified?		*
(36) Have highly corrosive wastes or other wastes that may cause severe dermal and mucous membrane irritation or allergenic reaction been identified?		*
(37) Have flash point tests been performed to determine what wastes are ignitable?		*
(38) Is the physical condition of the site safe for the performance of work activities (i.e., have deteriorated buildings, unguarded pits, buried wastes, or unstable structures been identified)?		*
(39) Have the overall condition and situation of containment vessels, piping, tanks, etc., been determined to be safe for the work activities required?		*
(40) Do the excavating and general labor activities at the site comply with OSHA's general industry safety standards (OSHA 29 CFR 1910)?		*

ADMINISTRATIVE CONTROLS

Administrative controls consist of those measures that assure that the workers are healthy and well-trained. The training is supported by a program to ensure the maintenance of the workers' health and safety. They checklist for assessing administrative controls is specifically designed to evaluate medical, training, safety, industrial hygiene, and emergency programs. The questions provide a general review of the basic requirements for each program.

Medical Program

Question	Response Yes	No
(41) Has a formal medical surveillance program been established?		
(42) Does the medical surveillance program include comprehensive preplacement examinations?		
(43) If so, does the preplacement examination include:		
– Work-up of medical history and physical examination?		
– Visual acuity test?		
– Hearing acuity test?		
– Pulmonary function test?		
– Chest X ray?		
– Complete urinalysis?		
– Complete blood analysis (count and chemistry)?		
– Baseline levels of toxic substances in body fluids?		
(44) Does the medical surveillance program include periodic physical examinations?		
(45) Are physical examinations performed at least annually?		

Medical Program (continued)

Question	Response Yes	No
(46) Does the medical surveillance program include a program of surveillance testing to determine worker exposures to specific toxic agents?		
(47) Does the program of surveillance testing for specific toxic agents comply with the applicable medical surveillance sections of OSHA 29 CFR 1910.1000-1046 (Subpart Z)?		
(48) Does the medical program certify employees for the use of respirators?		
(49) Is the medical program adequate to screen workers susceptible to heat stress?		
(50) Has a medical recordkeeping system been established?		
(51) Does the recordkeeping system for specific toxic agents comply with the applicable medical surveillance sections of OSHA 29 CFR 1910.1000-1046 (Subpart Z)?		
(52) Do the medical records include, at a minimum, the following:		
– Medical and work histories of each employee?		
– Results of all medical examinations?		
– Monitoring data on chemical and other toxic agents to which an employee might be exposed?		

Training Program

Question	Response Yes	No
(53) Do all personnel receive comprehensive safety and health training before commencing with their duties?		

(54) Does the training program include:

- First aid and cardiopulmonary resuscitation?
- Review of physical/chemical/toxicological properties of suspect hazardous materials?
- Hazard recognition (chemical, biological, and physical agents)?
- Principles of site safety?
- Standard operating procedures (e.g., use of "buddy system," decontamination procedures, zone delineation)?
- Use and maintenance of respiratory equipment protective clothing and other personal protective equipment?
- Simulated emergency situations and contingency plan implementation?
- Use of communications or alarm systems and protocols?
- Use and maintenance of fire-fighting and spill control equipment?

(55) Does the training program provide for the safety orientation of contract employees?

(56) Does the training include both classroom and on-the-job training?

(57) Is the program directed by a person(s) trained in hazardous waste management and/or site remedial actions?

(58) Are files kept of records documenting the training?

Safety and Industrial Hygiene Program

(59) Has a written safety and industrial hygiene program been established?

(60) Does the safety and industrial hygiene program discuss procedures regarding:

- Emergency equipment requirements, selection, and use (see questions 96 through 109)?
- Decontamination (see questions 79 through 89)?
- Personal work practices (see questions 141 through 148)?
- Procedural work practices (see questions 143 through 148)?

(61) Does the safety and industrial hygiene program discuss procedures regarding:

- Emergency equipment requirements, selection, and use (see questions 96 through 109)?
- Selection, use, and limitations of protective clothing (see questions 110 through 122)?
- Selection, use, maintenance, and limitations of respirators and eye protection (see questions 123 through 140)?
- The monitoring of hazards during normal operations?
- Periodic workplace or site inspections?
- Periodic employee safety performance reviews?
- Evaluation of the effectiveness of controls following implementation?
- Maintenance of accident records and follow-up investigation of accidents?

Emergency Program

(62) Is there a written emergency action/disaster control plan?

(63) Does the plan include provisions for:

  – Emergency response procedures for fires, spills, explosions, and personal injury?

  – Evacuation and notification procedures for incidents affecting both on-site workers and nearby populations?

  – Identifying individuals with specific health, safety, and emergency responsibilities (formation of emergency control team)?

  – Protocol for communications with public emergency aid teams (i.e., fire departments, paramedical units, and local hospitals)?

  – Provisions for the rapid dissemination of exposure information to the medical facility (i.e., chemical identity, magnitude of exposure, and health of those involved)?

(64) Have names and phone numbers of police, sheriff, fire department, hospitals, ambulance services, and other emergency response agencies/services been posted in a conspicuous locations?

(65) Has a program been developed to monitor meteorological conditions likely to promote off-site transport of emissions?

(66) Does each person have a written job description outlining:

  – Responsibilities and duties?

  – Specific authority and position in the chain of command?

  – Minimal performance requirements/

(67) Are "tailgate" meetings held before daily work assignments are begun?

(68) Are post-assignment debriefing sessions held? If so, see question 69.

(69) Do the meetings include discussions on:

  – Tasks to be performed?

  – How the work is to be done?

  – Possible time constraints (mandatory rest breaks, change air tank, etc.)?

(70) Is the topic of hazards (chemical, biological, physical, and safety) that may be encountered well covered; i.e., does it include:

  – Ways to recognize or monitor the hazards?

  – Effects of hazards?

  – Hazard clues or concentration limits that require work stoppage?

(71) Are Emergency Procedures reviewed?

(72) When unexpected events occur, are they investigated immediately as to their meaning and effect on present and future work activities?

(73) Before restarting work do all personnel participate in a discussion session following the unexpected occurrence?

ENGINEERING CONTROLS

Engineering controls at hazardous waste sites entail proper site security, zone delineation, and decontamination capabilities. Checklist questions evaluating site security and zone delineation require only a visual assessment of the site. Questions dealing with decontamination require knowledge of the hazardous agents involved and the proper decontamination procedures required. This questionnaire provides the primary health and safety considerations for establishing a site control plan.

Site Security and Zone Delineation

(74) Has the site been secured, including the restriction of general (public) access?

Question

Response
Yes No

(75) Is the site properly posted and do the signs convey an adequate warning?

(76) Have the contamination, decontamination, and clean zones been established?

(77) Have the zones been clearly delineated?

(78) Is the worker access to the zones controlled, e.g., log-in, log-out of site workers?

Decontamination

(79) Have decontamination station(s) been established between the contamination and clean zone?

(80) Are separate stations established for personnel and large equipment items?

(81) Is the layout of the station(s) designed to prevent cross-contamination between clean and contaminated areas?

(82) Have the proper cleaning and rinsing methods and solutions been selected for the waste contaminants and equipment involved?

(83) Are decontamination stations equipped with adequate supplies, equipment, clothing, and storage facilities?

(84) If infectious agents have been found on site, have disinfection procedures been implemented?

(85) Are procedures for evaluating the effectiveness of decontamination documented? Do they include:

- Emergency decontamination procedures?

- Detailed decontamination procedures for personnel and equipment?

(86) Are cleaning and rinsing solutions and expend-able equipment treated as hazardous waste and disposed of accordingly?

(87) Is all equipment maintained in safe operating condition?

(88) Is all equipment stored in secure, safe area?

(89) Upon completion of daily work activities, is all equipment serviced to a state of operational readiness?

Standard Equipment

(90) Are spare parts readily obtainable to make immediate repairs on equipment as needed?

(91) Is all equipment cleaned and decontaminated before storage?

(92) Are maintenance and repair logs kept for each piece of equipment?

(93) Are routine equipment inspections conducted regularly?

(94) Are all equipment operators trained on and familiar with their equipment?

(95) Is equipment (e.g., pumps, forklifts, compressors, earth moving equipment, drum handling devices) designed, constructed, or modified, as appropriate, to perform satis-factorily under the conditions posed by the hazardous waste handling tasks for which it is intended?

Emergency Equipment

(96) Are the necessary fire protection equipment and supplies available on site?

- Portable fire extinguishers for initial response to fires of all classes?

- Water at sufficient pressure and capacity?

- Sufficient length of water hose?

- Foam carts?

(97) Is the available fire protection equipment strategically located to promote rapid deployment, yet sufficiently isolated and protected from accidental damage?

(98) Are spill control and containment equipment and supplies available on site? *

(99) Are the available spill-control equipment and supplies strategically located to provide rapid deployment, yet sufficiently isolated from hazard areas?

(100) Are adequate first-aid supplies available on site? *

(101) Are first-aid stations equipped with:
- Standard first-aid kit? *
- Stretchers and blankets? *
- Self-contained breathing apparatus? *
- Portable oxygen supply? *
- Portable eye wash and shower? *

(102) Are first-aid stations located in the clean zone, but as close as possible to the workers?

(103) Is an internal communications or alarm system available?

(104) Do all employees have access to the communications or an alarm system?

(105) Is all emergency equipment tested and maintained on a regular basis?

(106) Has a secure area been established for remote opening of damaged drums or drums suspected of containing extremely hazardous, reactive, or explosive materials? *

(107) Is heavy equipment (e.g., bulldozers, fork-lift trucks, backhoes) properly equipped to protect operators from rollover, explosives, liquid splashes flying debris, etc.? *

(108) Do equipment and tools meet OSHA general industry standards (OSHA CFR 29 1910)? *

(109) Are spark-proof and explosion-proof equipment and tools (spark-proof wrenches and other tools, sampling equipment, drum moving equipment, explosion-proof motors, lighting, and ground-fault provisions for electrical wiring) used for on-site remedial/emergency activities? *

PERSONAL PROTECTIVE EQUIPMENT

Personal protective equipment is extremely important in the protection of workers' health and safety at hazardous waste sites. This checklist for evaluating proper selection and use of personal protective equipment has been designed specifically to address the areas of protective clothing, protective headwear, and respiratory and eye protection. Evaluation questions for protective clothing require knowledge concerning the type of contaminants being handled to ascertain whether the equipment will be effective and whether its selection represents the efficient use of resources. Evaluation questions for respiratory protection require knowledge concerning both the types and concentrations of airborne contaminants present in the work environment. Evaluation questions for protective headwear and eyewear require only a visual assessment of equipment usage.

During the initial site inspection and environmental sampling effort, the clothing, headwear, and respiratory and eye protection should be the most effective possible unless the data collected during the information audit indicate otherwise.

This checklist covers the basic health and safety considerations for selection of adequate personal protection equipment. The user is also urged to consult the NIOSH publication entitled "A Guide to Industrial Respiratory Protection, by J. Pritchard.

Protective Clothing

Question

Response
Yes No

(110) Does a high to moderate potential exist for any body part to come in contact with chemicals that are severe irritants or highly toxic via dermal absorption? (Potential situations are those where vapors may be generated or splashing may occur through work activities.)

If so, see next question; if not, see question 112.

(111) Are workers wearing fully encapsulated chemical protective suits of impermeable material, inner and outer chemical-resistant gloves, and outer chemical protective boots? (Note: Fully encapsulated denotes full coverage of neck, head, and face.)

(112) Does a moderate to low potential exist for severe dermal irritants or chemicals that are toxic via dermal absorption to come in contact with area of the body not usually protected by a fully encapsulated protective suit (e.g., back of neck, ears, wrists)?

If answer is "yes," see following question; if answer is "no," see question 114.

(113) Are workers wearing hooded single or two-piece chemical-resistant suits, inner and outer chemical-resistant gloves, and inner and outer chemical-protective boots?

(114) Does a potential exist for contact with moderate skin irritants or hazardous chemicals that exhibit low or no toxicity via dermal absorption?

If so, see the following question, if not, see question 116.

(115) Are workers wearing two-piece chemical-resistant suits or disposable coveralls with a chemical-protective apron, inner and outer gloves, and inner and outer boots?

(116) Does a potential exist for exposure to hazardous chemicals considered minor irritants and not toxic via dermal absorption?

If the answer is "yes," see the following question; if "no," review questions 110 through 115 for the required protection.

(117) Are workers wearing cotton or disposable synthetic coveralls made of breathable fabric, chemical-resistant gloves, and inner and outer boots?

(118) Does a potential exist for a flash fire in addition to chemical contacts?

If the answer is "yes," see the following question.

(119) Are workers wearing flame-retardent clothing that meets appropriate requirements for the chemical exposure potential expected?

(120) Does the clothing resist tearing the puncturing from physical hazards?

(121) Do weather conditions under which the clothing is used have a significant effect on the strength, durability or flexibility of the material?

Protective Headwear

(122) Are workers wearing approved hard hats?

Respiratory and Eye Protection

(123) Does the work atmosphere contain concentrations of contaminants "immediately dangerous to life and health" (IDLH) or above the lower explosive limit (LEL); is it oxygen deficient (less than 19.5% O$_2$)? (Note: The conditions most often occur in enclosed spaces, with inadequate ventilation).

(124) Do workers wear positive-pressure, closed-circuit, self-contained breathing apparatus (SCBA)?

(125) Does work atmosphere contain potentially high concentrations of unknown substances or at concentration levels less than IDLH but greater than 50 percent of the IDLH for the known substance(s)?

If the answer is "yes," see the following question; if "no," see question 127.

(126) As a minimum, coworkers wear pressure-demand, open-circuit, self-contained breathing apparatus (SCBA)?

(127) Does work atmosphere contain only known contaminants at concentrations less than 50 percent of the IDLH, but greater than each substances TLV?

If the answer is "yes," see the following question; if "no," see question 129.

(128) As a minimum, do workers wear full-face, powered, air-purifying, chemical-cartridge respirators with NIOSH-approved cartridges for specific contaminants and a high efficiency particulate filter, and are they equipped with emergency escape respirators? (Note: These air-purifying respirators should only be worn if the concentration of the known contaminant does not exceed the manufacturer's recommendation for the cartridge filter used. Also, known contaminants should have good warning properties.)

(129) Are all air contaminants known, and do concentrations of each fall at or below the ACGIH TLV?

If the answer is "yes," see the following question; if "no," review questions 123 through 129 for required protection.

(130) Are workers equipped with half-face, air-purifying, chemical-cartridge respirators with NIOSH-approved cartridges for specific contaminants and a high-efficiency particulate filter?

(131) Have all workers and their respirators been properly "fit-tested"?

(132) Is all respiratory equipment Government-approved?

(133) Is the air supply for the SCBA and air-supplied respirators at least Grade D breathing air (i.e., 19.5-23.5% O$_2$; no bad or nauseous odors)?

(134) Is at least one crew member certified by the manufacturer to perform minor repairs on the systems?

(135) When air lines are used, are they kept away from or protected from chemical contamination?

(136) Where self-contained breathing apparatus (air packs) is used:

– Have the air cylinders passed current hydrostatic tests?

– Do their low-pressure alarms leave adequate air for a person to reach the decontamination station and go through the necessary procedure?

– If the regulator should fail, can it be bypassed?

– Does the apparatus restrict movement and interfere with work tasks?

(137) Are chemicals present that can cause eye irritation?

(138) Do workers wear chemical safety goggles and/or other equipment that affords adequate eye protection? (Note: Only full-face respirators are allowed when contaminants that produce eye irritation are present.)

(139) If a splash potential exists that may bring the facial area in contact with irritant chemicals, do the workers wear full-face chemical safety shields?

WORK PRACTICES

Maintaining the health and safety of the workers ultimately rests on the workers themselves. Their awareness and commitment to established safety procedures, both in terms of personal and procedural activities, form the basis of any program designed to keep work-related risks to a minimum. The Work Practices Checklist addresses both personal and procedural activities. Questions concerning procedural activities require site-specific knowledge regarding safety and industrial hygiene programs.

Personal Work Practices

	Question	Response Yes	No
(140)	Do workers abstain from the following "forbidden" personal work practices:		
-	Eating, drinking, or smoking outside of designated areas for such activities?		*
-	Growing facial hair that interferes with respirator performance?		*
-	Coming in contact with potentially contaminated objects or media (sitting or leaning on drums, unnecessary transit of contaminated materials).		*
-	Removal of protective equipment while in a contaminated zone?		!
(141)	Do workers use personal protective equipment properly?		*

Procedural Work Practices

	Question	Response Yes	No
(142)	Do workers follow established decontamination procedures?		*
(143)	Do workers follow procedures to avoid physiological stress, e.g., fatigue, heat stress, dehydration?		*
(144)	Do workers adhere to the "buddy system"?		*
(145)	Are reusable protective clothing, respirators, and other personal protective equipment cleaned and maintained daily?	*	
(146)	Are personnel properly deployed and stationed within the decontamination and clean zones for operational and emergency assistance?	*	
(147)	Do workers comply with established protocol on zone access (i.e., signing log sheets, donning required protective equipment, decontamination, etc.)?	*	

CHEMICAL HAZARD DATA

NIOSH POCKET GUIDE TO CHEMICAL HAZARDS
DHHS (NIOSH) Publication No. 90–117

U.S. DEPARTMENT OF HEALTH AND HUMAN SERVICES
Public Health Service
Centers for Disease Control
National Institute for Occupational Safety and Health

June 1990

CONTENTS

I. INTRODUCTION

The *NIOSH Pocket Guide to Chemical Hazards* is intended as a source of general industrial hygiene information for workers, employers, and occupational health professionals. The *Pocket Guide* does not contain an analysis of all pertinent data; rather, it presents key information and data in abbreviated tabular form for 398 chemicals or substance groupings (e.g., cyanides, fluorides, manganese compounds, etc.) that are found in the work environment and that have existing Occupational Safety and Health Administration (OSHA) regulations. The industrial hygiene information found in the *Pocket Guide* should help users recognize and control occupational chemical hazards. The chemicals contained in this revision are substances previously listed in the OSHA Z-1, Z-2, and Z-3 Tables (many of which now have revised permissible exposure limits [PELs]). Additional substances that were included as part of the new OSHA Z-1-A Table on January 19, 1989, will be incorporated into the next edition of the *Pocket Guide*.

Background

In 1974, the National Institute for Occupational Safety and Health (NIOSH) (which is responsible for recommending health and safety standards) joined OSHA (whose jurisdictions include promulgation and enforcement activities) in developing a series of complete occupational health standards for substances with existing PELs. This joint effort was labeled the Standards Completion Program and involved the cooperative efforts of several contractors and personnel from various divisions within NIOSH and OSHA. The Standards Completion Program developed 380 substance-specific draft standards with supporting documentation that contained technical information and recommendations needed for the promulgation of new occupational health regulations. The *Pocket Guide* was developed to make the technical information in those draft standards more conveniently available to workers, employers, and occupational health professionals. The *Pocket Guide* is updated periodically to reflect new data regarding the toxicity of various substances and any changes in exposure standards or recommendations.

Data Collection and Application

The data collected for this revision were derived from a variety of sources, including NIOSH criteria documents and Current Intelligence Bulletins (CIBs), and recognized references in the fields of industrial hygiene, occupational medicine, toxicology, and analytical chemistry.

II. NIOSH RECOMMENDATIONS

Acting under the authority of the Occupational Safety and Health Act of 1970 (Public Law 91-596), NIOSH develops and periodically revises recommended exposure limits (RELs) for hazardous substances or conditions in the workplace. NIOSH also recommends appropriate preventive measures to reduce or eliminate the adverse health and safety effects of these hazards. To formulate these recommendations, NIOSH evaluates all known and available medical, biological, engineering, chemical, trade, and other information relevant to the hazard. These recommendations are then published and transmitted to OSHA and the Mine Safety and Health Administration (MSHA) for use in promulgating legal standards.

NIOSH recommendations are published in a variety of documents. Criteria documents recommend workplace exposure limits and appropriate preventive measures to reduce or eliminate adverse health effects and accidental injuries.

CIBs are issued to disseminate new scientific information about occupational hazards. A CIB may draw attention to a formerly unrecognized hazard, report new data on a known hazard, or present information on hazard control.

Special Hazard Reviews, Occupational Hazard Assessments, and Technical Guidelines support and complement the other standards development activities of the Institute. Their purpose is to assess the safety and health problems associated with a given agent or hazard (e.g., the potential for injury or for carcinogenic, mutagenic, or teratogenic effects) and to recommend appropriate control and surveillance methods. Although these documents are not intended to supplant the more comprehensive criteria documents, they are prepared to assist OSHA and MSHA in the formulation of regulations.

In addition to these publications, NIOSH periodically presents testimony before various Congressional committees and at OSHA and MSHA rulemaking hearings.

A complete list of occupational safety and health issues for which NIOSH has formal policies (e.g., recommendations for occupational exposure to chemical and physical hazards, engineering controls, work practices, safety considerations, etc.) is scheduled for publication in 1990.

III. HOW TO USE THIS POCKET GUIDE

The *Pocket Guide* has been designed to provide chemical-specific data to supplement general industrial hygiene knowledge. To maximize the amount of data provided in this limited space, abbreviations and codes have been used extensively. These abbreviations and codes, which have been designed to permit rapid comprehension by the regular user, are discussed for each column in the following subsections.

Chemical Name and Structure/Formula, CAS and RTECS Numbers, and DOT ID and Guide Numbers

Chemical Name and Structure/Formula. — The chemical name found in the OSHA General Industry Air Contaminants Standard (29 CFR* 1910.1000) is listed first. The chemical structure or formula is also provided; carbon-carbon double bonds (-C=C-) have been indicated where applicable. A page index for selected synonyms and trade names is included at the back of the *Pocket Guide* to help the user locate a specific substance.

CAS and RTECS Numbers. — Below the chemical structure/formula is the Chemical Abstracts Service (CAS) registry number. This number, in the format xxx-xx-x, is unique for each chemical and allows efficient searching on computerized data bases. A page index for all CAS registry numbers listed is included at the back of the *Pocket Guide* to help the user locate a specific substance.

Immediately below the CAS number is the *NIOSH Registry of Toxic Effects of Chemical Substances* (RTECS) number, in the format ABxxxxxxx. RTECS may be useful for obtaining additional toxicologic information on a specific substance.

DOT ID and Guide Number. — Under the RTECS number are the U.S. Department of Transportation (DOT) identification number and the corresponding guide number. Their format is xxxx xx. The Identification number (xxxx) indicates that the chemical is regulated by DOT. The Guide number (xx) refers to actions to be taken to stabilize an emergency situation; this information can be found in the DOT 1987 *Emergency Response Guidebook*, DOT P 5800.4 (Office of Hazardous Materials Transportation [DHM-51], Research and Special Programs Administration, U.S. Department of Transportation, 400 Seventh Street S.W., Washington, D.C. 20590).

Synonyms, Trade Names, and Conversion Factors

Common synonyms and trade names are listed alphabetically for each chemical. Factors for the conversion of ppm (parts per million parts of air) to mg/m^3 (milligrams per cubic meter of air) at 68°F and 1 atmosphere are listed for all liquids and gases.

Exposure Limits

The NIOSH RELs are listed first in this column. Unless noted otherwise, RELs are time-weighted average (TWA) con-

---

Code of Federal Regulations.

centrations for up to a 10-hour workday during a 40-hour work-week. A short-term exposure limit (STEL) is designated by "ST" preceding the value; unless noted otherwise, the STEL is a 15-minute TWA exposure that should not be exceeded at any time during a workday. A ceiling REL is designated by "C" preceding the value; unless noted otherwise, the ceiling value should not be exceeded at any time. Any substance that NIOSH considers to be an occupational carcinogen is designated by the notation "Ca" (see Appendix A, which contains a brief discussion of occupational carcinogens).

The OSHA permissible exposure limits (PELs), as found in Tables Z-1-A or Z-2 of the OSHA General Industry Air Contaminants Standard (29 CFR 1910.1000), are listed next. Unless noted otherwise, PELs are TWA concentrations that must not be exceeded during any 8-hour work shift of a 40-hour work-week. A STEL is designated by "ST" preceding the value and is measured over a 15-minute period unless noted otherwise. OSHA ceiling concentrations (designated by "C" preceding the value) must not be exceeded during any part of the workday; if instantaneous monitoring is not feasible, the ceiling must be assessed as a 15-minute TWA exposure. In addition, beryllium, ethylene dibromide, and methylene chloride (from Table Z-2) have PEL ceiling values that must not be exceeded except for a maximum peak over a specified period. Appendix B contains a brief discussion of carcinogens regulated by OSHA.

The American Conference of Governmental Industrial Hygienists (ACGIH) Threshold Limit Values (TLVs) are also listed when they are more restrictive than the OSHA PEL or NIOSH REL. Unless noted otherwise, TLVs are 8-hour TWA concentrations. Ceiling concentrations (which should not be exceeded at any time) are designated by "C". A STEL is designated by "ST" and is measured over a 15-minute period. Where applicable, the ACGIH A1 (confirmed human carcinogen) and A2 (suspected human carcinogen) designations for individual chemicals are included. The ACGIH TLVs for 1989-1990 were used for this printing of the *Pocket Guide*.

Concentrations are given in ppm, mg/m³, or fibers per cubic centimeter (fibers/cm³). The "[skin]" designation indicates the potential for dermal absorption; skin exposure should be prevented as necessary through the use of good work practices and gloves, coveralls, goggles, and other appropriate equipment. Appendix C contains more detailed discussions of the specific exposure limits for asbestos, the various chromium compounds, coal tar pitch volatiles, and cotton dust. Appendix D contains a brief discussion of substances with no established RELs.

IDLH

Immediately dangerous to life or health concentrations (IDLHs) represent the maximum concentration from which, in the event of respirator failure, one could escape within 30 minutes without a respirator and without experiencing any escape-impairing (e.g., severe eye irritation) or irreversible health effects. These values were determined during the Standards Completion Program only for the purpose of respirator selection. The notation "Ca" appears in this column for all substances that NIOSH considers to be potential human carcinogens (NIOSH recommends the most protective respirators for exposure to these substances). However, the IDLHs that were originally determined in the Standards Completion Program are shown in brackets following the "Ca" designation (note: carcinogenic effects were not considered in the Standards Completion Program). "Unknown" indicates that because of a lack of data, no IDLH could be determined. "LEL" indicates that the IDLH is based on the lower explosive limit. "N.E." indicates that no evidence could be found for the existence of an IDLH.

Physical Description

This column provides a brief description of the appearance and odor of each substance. Notations are made as to whether a substance can be shipped as a liquefied compressed gas or whether it has major use as a pesticide.

Chemical and Physical Properties

The following abbreviations are used for the chemical and physical properties given for each substance. "NA" indicates that a property is not applicable, and a question mark (?) indicates that it is unknown.

MW	. . . .	Molecular weight
BP	. . . .	Boiling point at 1 atmosphere, °F
Sol	. . . .	Solubility in water at 68°F*, % by weight (i.e., g/100 ml)
Fl.P	. . . .	Flash point, closed cup (unless annotated "(oc)" for open cup), °F
IP**	. . . .	Ionization potential, eV (electron volts)
VP	. . . .	Vapor pressure at 68°F*, mm Hg; "approx" indicates approximately
MLT	. . . .	Melting point for solids, °F
FRZ	. . . .	Freezing point for liquids and gases, °F
UEL	. . . .	Upper explosive limit in air, % by volume (at room temperature*)
LEL	. . . .	Lower explosive limit in air, % by volume (at room temperature*)
Sp.Gr	. . . .	Specific gravity at 68°F* referenced to water at 39.2°F(4°C)

*If noted after a specific entry, these properties may be reported at other temperatures.
**Ionization potentials are given as a guideline for the selection of photo-ionization detector lamps used in direct-reading instruments.

When possible, the flammability of a substance was determined and listed below the specific gravity. The following OSHA criteria (29 CFR 1910.106) were used to classify flammable or combustible liquids:

Class IA flammable liquid	Fl.P below 73°F and BP below 100°F.
Class IB flammable liquid	Fl.P below 73°F and BP at or above 100°F.
Class IC flammable liquid	Fl.P at or above 73°F and below 100°F.
Class II combustible liquid	Fl.P at or above 100°F and below 140°F.
Class IIIA combustible liquid	Fl.P at or above 140°F and below 200°F.
Class IIIB combustible liquid	Fl.P at or above 200°F.

Incompatibilities and Reactivities

This column lists important hazardous incompatibilities or reactivities of each substance.

Measurement Method

This column provides a brief, key word description of the suggested sampling and analysis method. Each description comprises four components: (1) collection method, (2) sample work-up, (3) analytical method, and (4) method number. The method number is from the 2nd or 3rd edition (i.e., II or III) of the *NIOSH Manual of Analytical Methods* (DHHS [NIOSH] Publication No. 84-100). If the 2nd edition is cited, this fact is noted along with the appropriate volume number [e.g., II(4)]. In a number of instances, the table cites the *1985 OSHA Analytical Methods Manual* and applicable method number (e.g., OSHA [#21]). "None available" indicates that no method has been developed by NIOSH or OSHA. The codes listed are explained in Table 1. Table 2 presents ordering information for the measurement methods cited.

Personal Protection and Sanitation

This column presents a summary of recommended practices for each toxic substance. These recommendations supplement general work practices (e.g., no eating, drinking, or smoking where chemicals are used). Table 3 explains the codes used. Each category is described as follows:

CLOTHING: Describes conditions that require chemical protective clothing.

GOGGLES: Describes conditions that require eye protection.

WASH: Tells when workers should clean the spilled chemical from the body (in addition to normal washing before eating, smoking, etc.).

CHANGE: Tells when routine changes of clothing are advisable.

REMOVE: Advises workers when to remove clothing that has been accidentally contaminated.

PROVIDE: Recommends the availability of eyewash and quick drench equipment.

Recommendations for Respirator Selection

This column provides a condensed table of allowable respirator use. Each line lists in ppm or mg/m^3 a maximum concentration for use (commonly referred to as MUC, maximum use concentration) or condition (i.e., emergency or planned entry into unknown concentrations or IDLH conditions [designated by the symbol §], or escape from suddenly occurring respiratory hazards) followed by a series of codes representing classes of respirators. Individual respirator codes are separated by diagonal marks (/). For each MUC, the table lists only those respirators that possess the minimum required protection factor and other use restrictions based on the *NIOSH Respirator Decision Logic* (DHHS [NIOSH] Publication No. 87-108). All recommended respirators of a given class can be used at any concentration equal to or less than the MUC listed for that class.

The recommendations for respirator selection for each substance are based on the most restrictive occupational exposure limit found among the NIOSH REL, the OSHA PEL, and the ACGIH TLV.

In the *Pocket Guide,* single-use respirators refer only to disposable respirators that have been tested under the provisions of 30 CFR 11.140-5 and that have an established protection factor of 5. To determine whether certain single-use respirators can be used for other applications, including those that require a higher protection factor, consult the *NIOSH Certified Equipment List* (DHHS [NIOSH] Publication No. 89-105). If a higher protection factor is required, the wearer must first be properly fitted with a respirator using either a quantitative fit test or the "improved" qualitative fit test, as accepted in the OSHA lead standard (29 CFR 1910.1025).

All respirators selected must be approved by NIOSH and MSHA under the provisions of 30 CFR 11. Since an air-purifying respirator must be specifically approved by NIOSH and MSHA for use with pesticides, pesticides are not identified as such in the respirator selection tables.

In addition, a complete respiratory protection program must be implemented and must fulfill all requirements of 29 CFR 1910.134. At a minimum, a respiratory protection program should include regular training, fit-testing, periodic environmental monitoring, maintenance, inspection, cleaning, and storage. Selection of a specific respirator within a given class of recommended respirators depends on the particular situation; this choice should be made only by a knowledgeable person. *REMEMBER:* air-purifying respirators will not protect users against oxygen-deficient atmospheres, and they are not to be used in IDLH conditions. The only respirators recommended for fire fighting are self-contained breathing apparatuses that have full facepieces and are operated in a pressure-demand or other positive-pressure modes. Additional information on the selection and use of respirators can be found in the *NIOSH Respirator Decision Logic* and the *NIOSH Guide to Industrial Respiratory Protection* (DHHS [NIOSH] Publication No. 87-116).

Codes for the various categories of respirators are defined in Table 4. In addition, the assigned protection factor (APF) is indicated for each respirator class.

Route of Health Hazard

This column lists the toxicologically important routes of entry for each substance, abbreviated as follows:

Inh . . . . Inhalation
Abs . . . . Skin absorption
Ing . . . . Ingestion
Con . . . . Skin and/or eye contact

Symptoms

This column lists the potential symptoms of exposure. Their abbreviations are defined in Table 5.

First Aid

This column lists first aid procedures for eye and skin contact, inhalation, and ingestion of the toxic substance. Abbreviations are defined in Table 6.

Target Organs

This column lists the organs that are affected by exposure to each substance. Abbreviations are defined in Table 5.

Table 1. — Codes for measurement methods

Code	Method/reagent	Code	Method/reagent
Collection method:*		Sample work-up:	
Ambersorb	Ambersorb® XE-347 tube		
Bag	Gas collection bag	CCl_4	Carbon tetrachloride
Bub	Bubbler	CFC-113	1,1,2-Trichloro-1,2,2-trifluoroethane
Carbo-B	Carbosieve® B tube		
Char	Charcoal tube	$CHCl_3$	Chloroform
Char (low-Ni)	Charcoal tube (low nickel content)	CH_2Cl_2	Methylene chloride
		CH_3CN	Acetonitrile
Char (pet)	Charcoal tube (petroleum-based)	CH_3COOH	Acetic acid
Chrom	Chromosorb tube	CS_2	Carbon disulfide
Dry tube	Drying tube	DMF	Dimethylformamide
Filter	Particulate filter	$FeCl_3$	Ferric chloride
Florisil	Florisil® tube	HCl	Hydrochloric acid
G-chrom P	Gas-chrom P® tube	HCOOH	Formic acid
Hydrar	Hydrar® sorbent tube	HNO_3	Nitric acid
Imp	Impinger	H_2O_2	Hydrogen peroxide
Mol-sieve	Molecular sieve tube	HPLC	High-pressure liquid chromatography
Porapak	Porapak® tube		

(Continued)

Table 1. — Codes for measurement methods (Continued)

Code	Method/reagent	Code	Method/reagent
Si gel	Silica gel tube	H_2SO_4	Sulfuric acid
Tenax GC	Tenax® GC tube	KOH	Potassium hydroxide
T-Sorb	Thermosorb® tube	LTA	Low-temperature ashing
Vertical elut	Vertical elutriator	Na_2CO_3	Sodium carbonate
XAD	XAD® tube	$NaHCO_3$	Sodium bicarbonate
Sample work-up (continued):		GC/PID	GC with photoionization detection
		GC/TEA	GC with thermal energy analyzer detection
NaOH	Sodium hydroxide		
NH_4OH	Ammonium hydroxide	Grav	Gravimetric
Pho-acid	Phosphomolybdic acid	HGA	Flameless atomic absorption with a high-temperature graphite analyzer
Thermal desorp	Thermal desorption apparatus		
THF	Tetrahydrofuran		
Analytical method:		HPLC/FLD	High-pressure liquid chromatography with fluorescence detection
AA	Atomic absorption spectrometry		
ECA	Electrochemical analysis	HPLC/UVD	High-pressure liquid chromatography with ultraviolet detection
GC	Gas chromatography		
GC/AlkFID	GC with alkaline flame ionization detection	IC	Ion chromatography
		ICP	Inductively coupled plasma
GC/ECD	GC with electron capture detection	IR	Infrared spectrometry
		ISE	Ion-specific electrode
GC/EConD	GC with electrolytic conductivity detection	PCM	Phase contrast microscopy
		PES	Plasma emission spectroscopy
GC/FID	GC with flame ionization detection	PLR	Polarography
GC/FPD	GC with flame photometric detection for sulfur, nitrogen, or phosphorus	Titrate	Titration
		Vis	Visible spectrophotometry
		XRF	X-ray fluorescence spectrometry
GC/NPD	GC with nitrogen/phosphorus detection	XRD	X-ray diffraction spectrometry

In the chemical listing, an asterisk following an adsorbent tube code (e.g., Char, Si gel*, XAD-2*, etc.) indicates that a special coating must be added. The figure "2" in parentheses following a collection device (e.g., Filter(2), Imp(2), Char(2), etc.) indicates that two are used in series.

Table 2. — Ordering information for measurement methods

Manual	Publication No.	Ordering No.	Manual	Publication No.	Ordering No.
NIOSH Manual of Analytical Methods, 2nd edition:			Supplement to 3rd edition	NIOSH 85-117	PB-86-116-266
Vol. I	NIOSH 77-157-A	PB-274-845	2nd supplement to 3rd edition	NIOSH 87-117	PB-88-204-722
Vol. II	NIOSH 77-157-B	PB-276-624			
Vol. III	NIOSH 77-157-C	PB-276-838	3rd supplement to 3rd edition	NIOSH 89-127	PB-90-162-470
Vol. IV	NIOSH 78-175	PB-83-105-437			
Vol. V	NIOSH 79-141	PB-83-105-445	OSHA Analytical Methods Manual, 1985 edition	—	4540*, 4541 (fiche)*
Vol. VI	NIOSH 80-125	PB-82-157-728			
Vol. VII	NIOSH 82-100	PB-83-105-452			
NIOSH Manual of Analytical Methods, 3rd edition	NIOSH 84-100	PB-85-179-018			

*Denotes ordering number of the American Conference of Governmental Industrial Hygienists (ACGIH), 6500 Glenway Ave., Building D-7, Cincinnati, OH 45211 (513-661-7881). All other ordering numbers are for the National Technical Information Service (NTIS), Springfield, VA 22161 (703-487-4650).

Table 3. — Personal protection and sanitation codes

Code	Definition
CLOTHING	Wear appropriate equipment to prevent:
Any poss liq/Repeat vap	Any possibility of liquid contact and repeated or prolonged vapor contact with skin
Any poss	Any possibility of skin contact
Any poss wet	Any possibility that skin will become wet with contaminant
Any poss >x%/Repeat y%	Any possibility of skin contact with liquids containing >x% of contaminant and repeated or prolonged skin contact with liquids containing y%
Any poss pH <x/Repeat pH >x	Any possibility of skin contact with solutions having a pH <x and repeated or prolonged skin contact with solutions having a pH >x
Prevent skin freezing	Self explanatory
Prevent contam or freezing	Contamination or freezing of skin
Prevent wet or freezing	Wetting or freezing of skin
Repeat	Repeated or prolonged skin contact
Reason prob	Reasonable probability of skin contact
GOGGLES	Wear eye protection to prevent:
Any poss	Any possibility of eye contact
Any poss >x%/ Reason prob <y%	Any possibility of eye contact with liquids containing >x% of contaminant and reasonable probability of eye contact with liquid containing <y%
Any poss molt/ Reason prob liq-sol	Any possibility of eye contact with substance in a molten form and reasonable probability of eye contact with a liquid or solid containing the substance
Any poss (xxx)	Any possibility that substance xxx will contact the eyes
Reason prob	Reasonable probability of eye contact
Repeat	Repeated or prolonged eye contact
WASH	Workers should wash:
Daily	At the end of each work shift
Daily (reason prob)	At the end of each work shift when there is reasonable probability of contact with the contaminant
Immed contam	Immediately when skin becomes contaminated
Immed contam >x%/ Prompt contam y%	Immediately when skin is contaminated with liquids containing >x% contaminant and promptly when skin is contaminated with y%
Immed contam >x%/ Prompt wet y%	Immediately when skin is contaminated with liquids containing >x% contaminant and promptly when skin becomes wet with y%
Immed contam/daily	Immediately when skin becomes contaminated and at the end of each work shift
Immed contam pH <x/ Prompt wet pH >x	Immediately when skin becomes contaminated with liquid of pH <x and promptly when skin becomes wet with liquid of pH >x
Immed wet	Immediately when skin becomes wet
Immed wet/Prompt contam	Immediately when skin becomes wet and promptly when skin becomes contaminated
Immed wet/Prompt contam/daily	Immediately when skin becomes wet, promptly when skin becomes contaminated, and at the end of each work shift
Prompt contam	Promptly when skin becomes contaminated
Prompt contam/daily	Promptly when skin becomes contaminated and at the end of each work shift
Prompt wash soap	Promptly wash with soap when skin becomes contaminated
Prompt wet	Promptly when skin becomes wet
CHANGE	Work clothing should be changed daily:
After work if any poss contam	If there is any possibility that the clothing may be contaminated
After work if reason prob contam	If it is reasonably probable that the clothing may be contaminated
REMOVE	Remove clothing:
Immed contam	Immediately if it becomes contaminated
Immed wet	Immediately if it becomes wet
Immed wet (flamm)	Immediately if it becomes wet (to avoid flammability hazard)
Immed wet/immed non-imperv contam	Immediately if it becomes wet or if it is non-impervious clothing that becomes contaminated
Immed non-imperv contam	Immediately if it is non-impervious clothing that becomes contaminated
Immed non-imperv contam >x%/ Prompt non-imperv contam y%	Immediately if it is non-impervious clothing that becomes contaminated with liquids containing >x% of contaminant and promptly if clothing is contaminated with y%
Immed non-imperv wet/ prompt non-imperv contam	Immediately if it is non-impervious clothing that becomes wet and promptly if it is non-impervious clothing that becomes contaminated
Prompt non-imperv contam	Promptly if it is non-impervious clothing that becomes contaminated
Prompt non-imperv wet	Promptly if it is non-impervious clothing that becomes wet

(Continued)

Table 3. — Personal protection and sanitation codes (Continued)

Code	Definition
PROVIDE	The following equipment should be available:
Eyewash	Self-explanatory
Quick drench	Self-explanatory
Eyewash (xxx)	Eyewash if substance xxx is present
Eyewash >x%	Eyewash if liquids containing >x% of contaminant are present
OTHER CODES	
Liq	Liquid
Molt	Molten
N.R.	No recommendation applies in this category
Sol	Solid
Soln	Solution containing the contaminant
Vap	Vapor

Table 4. — Symbols, code components, and codes used for respirator selection

Item	Definition
Symbol:	
¥	At any detectable concentration
§	Emergency or planned entry into unknown concentrations or IDLH conditions
£	Substance causes eye irritation or damage; eye protection needed
*	Substancxe reported to cause eye irritation or damage; may require eye protection
^	If not present as a fume
¿	Only nonoxidizable sorbents are allowed (not charcoal)
†	End of service life indicator (ESLI) required
APF	Assigned protection factor
Code component:	
CCR	Chemical cartridge respirator
D	Dust respirator (if an independent code); or a dust filter
F	Full facepiece
Fu	Fume filter
GMF	Air-purifying, full-facepiece respirator (gas mask) with a chin-style, front- or back-mounted canister
M	Mist filter
PAPR	Powered, air-purifying respirator
SA	Supplied-air respirator
SCBA	Self-contained breathing apparatus
AG	Acid gas cartridge or canister
CF	Continuous flow mode
HiE	Air-purifying respirator with a high-efficiency particulate filter (if an independent code); or a high-efficiency particulate filter
HiEF	Air-purifying, full-facepiece respirator with a high-efficiency particulate filter
OV	Organic vapor cartridge or canister
PD,PP	Pressure-demand or other positive-pressure mode
S	Chemical cartridge or canister providing protection against the compound of concern
T	Tight-fitting facepiece
XS	Except single-use respirator
XSQ	Except single-use and quarter-mask respirator
Code:	
CCRFAGHiE (APF = 50)	Any chemical cartridge respirator with a full facepiece and acid gas cartridge(s) in combination with a high-efficiency particulate filter
CCRFOV (APF = 50)	Any chemical cartridge respirator with a full facepiece and organic vapor cartridge(s)
CCRFOVDMFu (APF = 50)	Any chemical cartridge respirator with a full facepiece and organic vapor cartridge(s) in combination with a dust, mist, and fume filter
CCRFOVHiE (APF = 50)	Any chemical cartridge respirator with a full facepiece and organic vapor cartridge(s) in combination with a high-efficiency particulate filter
CCRFS (APF = 50)	Any chemical cartridge respirator with a full facepiece and cartridge(s) providing protection against the compound of concern
CCRFSHiE (APF = 50)	Any chemical cartridge respirator with a full facepiece and cartridge(s) providing protection against the compound of concern and having a high-efficiency particulate filter
CCROV (APF = 10)	Any chemical cartridge respirator with organic vapor cartridge(s)
CCROVAG (APF = 10)	Any chemical cartridge respirator with organic vapor and acid gas cartridge(s)

(Continued)

Table 4. — Symbols, code components, and codes used for respirator selection (Continued)

Item	Definition
Code (continued):	
CCROVDM (APF = 10)	Any chemical cartridge respirator with organic vapor cartridge(s) in combination with a dust and mist filter
CCROVDMFu (APF = 10)	Any chemical cartridge respirator with organic vapor cartridge(s) in combination with a dust, mist, and fume filter
CROVHiE (APF = 10)	Any chemical cartridge respirator with organic vapor cartridge(s) in combination with a high-efficiency particulate filter
CCRS (APF = 10)	Any chemical cartridge respirator with cartridge(s) providing protection against the compound of concern
D (APF = 5)	Any dust respirator
DM (APF = 5)	Any dust and mist respirator
DMF (APF = 10)	Any dust and mist respirator with a full facepiece
DMFu (APF = 10)	Any dust, mist, and fume respirator
DMXSQ (APF = 10)	Any dust and mist respirator except single-use and quarter-mask respirators
DXSQ (APF = 10)	Any dust respirator except single-use and quarter-mask respirators
GMFAG (APF = 50)	Any air-purifying, full-facepiece respirator (gas mask) with a chin-style, front- or back-mounted acid gas canister
GMFAGHiE (APF = 50)	Any air-purifying, full-facepiece respirator (gas mask) with a chin-style, front- or back-mounted acid gas canister having a high-efficiency particulate filter
GMFOV (APF = 50)	Any air-purifying, full-facepiece respirator (gas mask) with a chin-style, front- or back-mounted organic vapor canister
GMFOVAG (APF = 50)	Any air-purifying, full-facepiece respirator (gas mask) with a chin-style, front- or back-mounted organic vapor and acid gas canister
GMFOVAGHiE (APF = 50)	Any air-purifying, full-facepiece respirator (gas mask) with a chin-style, front- or back-mounted organic vapor and acid gas canister having a high efficiency particulate filter
GMFOVDMFu (APF = 50)	Any air-purifying, full-facepiece respirator (gas mask) with a chin-style, front- or back-mounted organic vapor canister in combination with a dust, mist, and fume filter
GMFOVHiE (APF = 50)	Any air-purifying, full-facepiece respirator (gas mask) with a chin-style, front- or back-mounted organic vapor canister having a high-efficiency particulate filter
GMFS (APF = 50)	Any air-purifying, full-facepiece respirator (gas mask) with a chin-style, front- or back-mounted canister providing protection against the compound of concern
GMFSHiE (APF = 50)	Any air-purifying, full-facepiece respirator (gas mask) with a chin-style, front- or back-mounted canister providing protection against the compound of concern and having a high-efficiency particulate filter
HiE (APF = 10)	Any air-purifying respirator with a high-efficiency particulate filter
HiEF (APF = 50)	Any air-purifying, full-facepiece respirator with a high-efficiency particulate filter
PAPRAG (APF = 25)	Any powered, air-purifying respirator with acid gas cartridge(s)
PAPRAGHiE (APF = 25)	Any powered, air-purifying respirator with acid gas cartridge(s) in combination with a high-efficiency particulate filter
PAPRD (APF = 25)	Any powered, air-purifying respirator with a dust filter
PAPRDM (APF = 25)	Any powered, air-purifying respirator with a dust and mist filter
PAPRDMFu (APF = 25)	Any powered, air-purifying respirator with a dust, mist, and fume filter
PAPRHiE (APF = 25)	Any powered, air-purifying respirator with a high-efficiency particulate filter
PAPROV (APF = 25)	Any powered, air-purifying respirator with organic vapor cartridge(s)
PAPROVAG (APF = 25)	Any powered, air-purifying respirator with organic vapor and acid gas cartridge(s)
PAPROVDM (APF = 25)	Any powered, air-purifying respirator with organic vapor cartridge(s) in combination with a dust and mist filter
PAPROVDMFu (APF = 25)	Any powered, air-purifying respirator with organic vapor cartridge(s) in combination with a dust, mist, and fume filter
PAPRS (APF = 25)	Any powered, air-purifying respirator with cartridge(s) providing protection against the compound of concern
PAPRTHiE (APF = 50)	Any powered, air-purifying respirator with a tight-fitting facepiece and a high-efficiency particulate filter
PAPRTOV (APF = 50)	Any powered, air-purifying respirator with a tight-fitting facepiece and organic vapor cartridge(s)
PAPRTOVHiE (APF = 50)	Any powered, air-purifying respirator with a tight-fitting facepiece and organic vapor cartridge(s) in combination with a high-efficiency particulate filter
PAPRTS (APF = 50)	Any powered, air-purifying respirator with a tight-fitting facepiece and cartridge(s) providing protection against the compound of concern
SA (APF = 10)	Any supplied-air respirator
SA:CF (APF = 25)	Any supplied-air respirator operated in a continuous-flow mode
SAF (APF = 50)	Any supplied-air respirator with a full facepiece
SAF:PD,PP (APF = 2000)	Any supplied-air respirator that has a full facepiece and is operated in a pressure-demand or other positive-pressure mode
SAF:PD,PP:ASCBA (APF = 10,000)	Any supplied-air respirator that has a full facepiece and is operated in a pressure-demand or other positive-pressure mode in combination with an auxiliary self-contained breathing apparatus operated in pressure-demand or other positive-pressure mode
SA:PD,PP (APF = 1000)	Any supplied-air respirator operated in a pressure-demand or other positive-pressure mode

(Continued)

Table 4. — Symbols, code components, and codes used for respirator selection (Continued)

Item	Definition
Code (continued):	
SAT:CF (APF = 50)	Any supplied-air respirator that has a tight-fitting facepiece and is operated in a continuous-flow mode
SCBA (APF = 10)	Any self-contained breathing apparatus
SCBAE	Any appropriate escape-type, self-contained breathing apparatus
SCBAF (APF = 50)	Any self-contained breathing apparatus with a full facepiece
SCBAF:PD,PP (APF = 10,000)	Any self-contained breathing apparatus that has a full facepiece and is operated in a pressure-demand or other positive-pressure mode

Table 5. — Abbreviations for symptoms of exposure and target organs

Abbreviation	Symptom/organ	Abbreviation	Symptom/organ
abdom	Abdominal	inflamm	Inflammation
album	Albuminuria	inj	Injury
anes	Anesthesia	insom	Insomnia
anor	Anorexia	irreg	Irregular
anos	Anosmia	irrit	Irritation
appre	Apprehension	irrity	Irritability
arrhy	Arrhythmias	jaun	Jaundice
asphy	Asphyxia	kera	Keratitis
BP	Blood pressure	lac	Lacrimation
breath	Breathing	lar	Laryngeal
bron	Bronchitis	lass	Lassitude
broncopneu	Bronchopneumonia	leucyt	Leukocytosis
bronspas	Bronchospasm	leupen	Leukopenia
BUN	Blood urea nitrogen	li-head	Lightheadedness
[carc]	Carcinogen	low-wgt	Weight loss
card	Cardiac	mal	Malaise
chol	Cholinesterase	malnut	Malnutrition
cirr	Cirrhosis	monocy	Monocytosis
CNS	Central nervous system	muc memb	Mucous membrane
conf	Confusion	musc	Muscle
conj	Conjunctivitis	narco	Narcosis
constip	Constipation	nau	Nausea
constric	Constriction	nec	Necrosis
convuls	Convulsions	neph	Nephritis
CVS	Cardiovascular system	ner	Nervousness
cyan	Cyanosis	numb	Numbness
depres	Depressant/depression	opac	Opacity
derm	Dermatitis	palp	Palpitations
diarr	Diarrhea	para	Paralysis
dist	Disturbance	pares	Paresthesia
dizz	Dizziness	perf	Perforation
drow	Drowsiness	peri neur	Peripheral neuropathy
dysp	Dyspnea	periorb	Periorbital
emphy	Emphysema	phar	Pharyngeal
eosin	Eosinophilia	photo	Photophobia
epis	Epistaxis	pig	Pigmentation
equi	Equilibrium	pneu	Pneumonia
eryt	Erythema	pneuitis	Pneumonitis
euph	Euphoria	PNS	Peripheral nervous system
fail	Failure	polyneur	Polyneuropathy
fasc	Fasiculation	prot	Proteinuria
FEV	Forced expiratory volume	pulm	Pulmonary
fib	Fibrosis	RBC	Red blood cell
fibrl	Fibrillation	resp	Respiratory
ftg	Fatigue	retster	Retrosternal
func	Function	rhin	Rhinorrhea
GI	Gastrointestinal	salv	Salivation
gidd	Giddiness	sens	Sensitization
halu	Hallucinations	sez	Seizure
head	Headache	som	Somnolence
hemato	Hematopoietic	subs	Substernal
hemog	Hemoglobinuria	sweat	Sweating
hemorr	Hemorrhage	swell	Swelling
hypox	Hypoxemia	sys	System
incr	Increase(d)	tacar	Tachycardia
ict	Icterus		
inco	Incoordination		(Continued)

Table 5. — Abbreviations for symptoms of exposure and target organs (Continued)

Abbreviation	Symptom/organ	Abbreviation	Symptom/organ
tend	Tenderness	vis dist	Visual disturbance
trachbronc	Tracheobronchitis	vomit	Vomiting
venfib	Ventricular fibrillation	weak	Weakness
verti	Vertigo	wheez	Wheezing
vesic	Vesiculation		

Table 6. — Codes for first aid data

Code	Definition	Code	Definition
Eye:			chemical penetrates the clothing, immediately remove the clothing and flush the skin with water. If irritation persists after washing, get medical attention.
Irr immed	If this chemical contacts the eyes, immediately wash the eyes with large amounts of water, occasionally lifting the lower and upper lids. Get medical attention immediately. Contact lenses should not be worn when working with this chemical.		
		Soap flush prompt	If this chemical contacts the skin, promptly flush the contaminated skin with soap and water. If this chemical penetrates the clothing, promptly remove the clothing and flush the skin with water. If irritation persists after washing, get medical attention.
Irr immed (15 min)	If this chemical contacts the eyes, immediately wash the eyes with large amounts of water and continue flushing for 15 minutes, occasionally lifting the lower and upper lids. Get medical attention immediately. Contact lenses should not be worn when working with this chemical.		
		Soap prompt/molten flush immed	If this solid chemical or a liquid containing this chemical contacts the skin, promptly wash the contaminated skin with soap and water. If irritation persists after washing, get medical attention. If this molten chemical contacts the skin or nonimpervious clothing, immediately flush the affected area with large amounts of water to remove heat. Get medical attention immediately.
Irr prompt	If this chemical contacts the eyes, promptly wash the eyes with large amounts of water, occasionally lifting the lower and upper lids. Get medical attention if any discomfort continues. Contact lenses should not be worn when working with this chemical.		
Medical attention	Self-explanatory	Soap wash	If this chemical contacts the skin, wash the contaminated skin with soap and water.
Skin:			
Dust off solid; water flush	If this solid chemical contacts the skin, dust it off immediately and then flush the contaminated skin with water. If this chemical or liquids containing this chemical penetrate the clothing, promptly remove the clothing and flush the skin with water. Get medical attention immediately.	Soap wash immed	If this chemical contacts the skin, immediately wash the contaminated skin with soap and water. If this chemical penetrates the clothing, immediately remove the clothing, wash the skin with soap and water, and get medical attention promptly.
Medical attention for frostbite	If this chemical contacts the skin or mouth, stop the exposure immediately. If frostbite has occurred, get medical attention.	Soap wash prompt	If this chemical contacts the skin, promptly wash the contaminated skin with soap and water. If this chemical penetrates the clothing, promptly remove the clothing and wash the skin with soap and water. Get medical attention promptly.
Molten flush immed/ sol-liq soap wash prompt	If this molten chemical contacts the skin, immediately flush the skin with large amounts of water. Get medical attention immediately. If this chemical (or liquids containing this chemical) contacts the skin, promptly wash the contaminated skin with soap and water. If this chemical or liquids containing this chemical penetrate the clothing, immediately remove the clothing and wash the skin with soap and water. If irritation persists after washing, get medical attention.	Water flush	If this chemical contacts the skin, flush the contaminated skin with water. Where there is evidence of skin irritation, get medical attention.
		Water flush immed	If this chemical contacts the skin, immediately flush the contaminated skin with water. If this chemical penetrates the clothing, immediately remove the clothing and flush the skin with water. Get medical attention promptly.
Soap flush immed	If this chemical contacts the skin, immediately flush the contaminated skin with soap and water. If this		

(Continued)

Table 6. — Codes for first aid data (Continued)

Code	Definition	Code	Definition
Skin (continued):			
Water flush prompt	If this chemical contacts the skin, flush the contaminated skin with water promptly. If this chemical penetrates the clothing, immediately remove the clothing and flush the skin with water promptly. If irritation persists after washing, get medical attention.		person to fresh air at once. If breathing has stopped, perform mouth-to-mouth resuscitation. Keep the affected person warm and at rest. Get medical attention as soon as possible.
		Fresh air	If a person breathes large amounts of this chemical, move the exposed person to fresh air at once. Other measures are usually unnecessary.
Water wash immed	If this chemical contacts the skin, immediately wash the contaminated skin with water. If this chemical penetrates the clothing, immediately remove the clothing and wash the skin with water. If symptoms occur after washing, get medical attention immediately.		
		Fresh air; 100% O_2	If a person breathes large amounts of this chemical, move the exposed person to fresh air at once. If breathing has stopped, perform artificial respiration. When breathing is difficult, properly trained personnel may assist the affected person by administering 100% oxygen. Keep the affected person warm and at rest. Get medical attention as soon as possible.
Water wash prompt	If this chemical contacts the skin, promptly wash the contaminated skin with water. If this chemical penetrates the clothing, promptly remove the clothing and wash the skin with water. If irritation persists after washing, get medical attention.		
Breath:		Swallow:	
Resp support	If a person breathes large amounts of this chemical, move the exposed	Medical attention immed ..	If this chemical has been swallowed, get medical attention immediately.

IV. CHEMICAL LISTING

Chemical name, structure/formula, CAS and RTECS Nos., and DOT ID and guide Nos.	Synonyms, trade names, and conversion factors	Exposure limits (TWA unless noted otherwise)	IDLH	Physical description	Chemical and physical properties		Incompatibilities and reactivities	Measurement method (See Table 1)
					MW, BP, SOL FI.P, IP, Sp.Gr, flammability	VP, FRZ UEL, LEL		
Acetaldehyde CH₃CHO 75-07-0 AB1925000 1089 26	Acetic aldehyde, Ethanal, Ethyl aldehyde 1 ppm = 1.83 mg/m³	NIOSH Ca See Appendix A OSHA 100 ppm (180 mg/m³) ST 150 ppm (270 mg/m³)	Ca [10,000 ppm]	Colorless liquid or gas (above 69°F) with a pungent, fruity odor.	MW: 44.1 BP: 69°F Sol: Miscible Fl.P: -36°F IP: 10.22 eV Sp.Gr: 0.79 Class IA Flammable Liquid	VP: 740 mm FRZ: -190°F UEL: 60% LEL: 4.0%	Strong oxidizers, acids, bases, alcohols, ammonia & amines, phenols, ketones, HCN, H₂S [Note: Prolonged contact with air may cause formation of peroxides that may explode and burst containers; easily undergoes polymerization.]	XAD-2*; Toluene; GC/FID; III [#2538]
Acetic acid CH₃COOH 64-19-7 AF1225000 1842 29 (soln) 2790 60 (10-80% acid) 2789 29 (80% acid)	Acetic acid (aqueous), Ethanoic acid, Glacial acetic acid (pure compound), Methanecarboxylic acid [Note: Can be found in concentrations of 5-8% in vinegar.] 1 ppm = 2.50 mg/m³	NIOSH 10 ppm (25 mg/m³) ST 15 ppm (37 mg/m³) OSHA 10 ppm (25 mg/m³)	1000 ppm	Colorless liquid or crystals with a sour, vinegar-like odor. [Note: Pure compound is a solid below 62°F. Often used in an aqueous solution.]	MW: 60.1 BP: 244°F Sol: Miscible Fl.P: 102°F IP: 10.66 eV Sp.Gr: 1.05 Class II Combustible Liquid	VP: 11 mm FRZ: 62°F UEL(200°F): 19.9% LEL: 4.0%	Strong oxidizers (especially chromic acid, sodium peroxide & nitric acid), strong caustics [Note: Corrosive to metals.]	Char; HCOOH; GC/FID; III [#1603]
Acetic anhydride (CH₃CO)₂O 108-24-7 AK1925000 1715 39	Acetic acid anhydride, Acetic oxide, Acetyl oxide, Ethanoic anhydride 1 ppm = 4.24 mg/m³	NIOSH/OSHA C 5 ppm (20 mg/m³)	1000 ppm	Colorless liquid with a strong, pungent, vinegar-like odor.	MW: 102.1 BP: 282°F Sol: 12% Fl.P: 120°F IP: 10.00 eV Sp.Gr: 1.08 Class II Combustible Liquid	VP: 4 mm FRZ: -99°F UEL: 10.3% LEL: 2.7%	Water, alcohols, strong oxidizers (especially chromic acid), amines, strong caustics [Note: Corrosive to iron, steel & other metals. Reacts with water to form acetic acid.]	Bub; Reagent; Vis; III [#3506]
Acetone CH₃COCH₃ 67-64-1 AL3150000 1090 26	Dimethyl ketone, Ketone propane, 2-Propanone 1 ppm = 2.42 mg/m³	NIOSH 250 ppm (590 mg/m³) OSHA 750 ppm (1800 mg/m³) ST 1000 ppm (2400 mg/m³)	20,000 ppm	Colorless liquid with a fragrant, mint-like odor. [Note: Enforcement of the OSHA TWA for "doffers" in the cellulose acetate fiber industry was stayed on 9/5/89 until 9/1/90; further, the OSHA STEL does NOT apply to that industry.]	MW: 58.1 BP: 133°F Sol: Miscible Fl.P: 0°F IP: 9.69 eV Sp.Gr: 0.79 Class IB Flammable Liquid	VP: 180 mm FRZ: -140°F UEL: 13% LEL: 2.5%	Oxidizers, acids	Char; CS₂; GC/FID; III [#1300, Ketones I

Chemical name, structure/formula, CAS and RTECS Nos., and DOT ID and guide Nos.	Synonyms, trade names, and conversion factors	Exposure limits (TWA unless noted otherwise)	IDLH	Physical description	Chemical and physical properties		Incompatibilities and reactivities	Measurement method (See Table 1)
					MW, BP, SOL FI.P, IP, Sp.Gr, flammability	VP, FRZ UEL, LEL		
Acetonitrile CH₃CN 75-05-8 AL7700000 1648 28	Cyanomethane, Ethyl nitrile, Methyl cyanide 1 ppm = 1.71 mg/m³	NIOSH 20 ppm (34 mg/m³) OSHA 40 ppm (70 mg/m³) ST 60 ppm (105 mg/m³)	4000 ppm ACGIH [skin]	Colorless liquid with an aromatic odor.	MW: 41.1 BP: 179°F Sol: Miscible Fl.P(oc): 42°F IP: 12.20 eV Sp.Gr: 0.78 Class IB Flammable Liquid	VP: 73 mm FRZ: -49°F UEL: 16.0% LEL: 3.0%	Strong oxidizers	Char; Benzene; GC/FID; III [#1606]
2-Acetylaminofluorene C₁₅H₁₃NO 53-96-3 AB9450000	AAF, 2-AAF, 2-Acetaminofluorene, N-Acetyl-2-aminofluorene, FAA, 2-FAA, 2-Fluorenylacetamide	NIOSH Ca See Appendix A OSHA [1910.1014] See Appendix B	Ca	Tan, crystalline powder.	MW: 223.3 BP: ? Sol: Insoluble Fl.P: ? IP: ? Combustible Solid	VP: ? MLT: 381°F UEL: ? LEL: ?	None reported	None available
Acetylene tetrabromide CHBr₂CHBr₂ 79-27-6 KI8225000 2504 58	Symmetrical tetrabromo-ethane; TBE; Tetrabromoacetylene; Tetrabromoethane; 1,1,2,2-Tetrabromoethane 1 ppm = 14.37 mg/m³	NIOSH See Appendix D OSHA 1 ppm (14 mg/m³)	10 ppm	Pale-yellow liquid with a pungent odor. [Note: A solid below 32°F.]	MW: 345.7 BP: 474°F (Decomposes) Sol: 0.07% Fl.P: NA IP: ? Sp.Gr: 2.97 Noncombustible Liquid	VP: 0.02 mm FRZ: 32°F UEL: NA LEL: NA	Strong caustics; hot iron; reducing metals such as aluminum, magnesium & zinc	Si gel; THF; GC/FID; III [#2003]
Acrolein CH₂=CHCHO 107-02-8 AS1050000 1092 30 (inhibited)	Acraldehyde, Acrylaldehyde, Acrylic aldehyde, Allyl aldehyde, Propenal, 2-Propenal 1 ppm = 2.33 mg/m³	NIOSH/OSHA 0.1 ppm (0.25 mg/m³) ST 0.3 ppm (0.8 mg/m³)	5 ppm	Colorless or yellow liquid with a piercing, disagreeable odor.	MW: 56.1 BP: 127°F Sol: 40% Fl.P: -15°F IP: 10.13 eV Sp.Gr: 0.84 Class IB Flammable Liquid	VP: 210 mm FRZ: -126°F UEL: 31% LEL: 2.8%	Oxidizers, acids, alkalis, ammonia, amines [Note: Polymerizes readily unless inhibited—usually with hydroquinone.]	XAD-2*; Toluene; GC/NPD; III [#2501]

Personal protection and sanitation (See Table 3)		Recommendations for respirator selection — maximum concentration for use (MUC) (See Table 4)	Health hazards			
			Route	Symptoms (See Table 5)	First aid (See Table 6)	Target organs (See Table 5)
Clothing: Goggles: Wash: Change: Remove: Provide:	Repeat Any poss Prompt wet N.R. Immed wet (flamm) Eyewash	NIOSH ¥: SCBAF:PD,PP/SAF:PD,PP:ASCBA Escape: GMFOV/SCBAE	Inh Ing	Eye, nose, throat irrit; conj; cough; CNS depres; eye, skin burns; derm; delayed pulm edema; [carc]	Eye: Irr immed Skin: Water flush prompt Breath: Resp support Swallow: Medical attention immed	Resp sys, skin, kidneys
Clothing: Goggles: Wash: Change: Remove: Provide:	Any poss >50%/Repeat 10-49% Any poss Immed contam >50%/ Prompt 10-49% N.R. Immed non-imperv contam >50%/Prompt non- imperv contam 10-49% Eyewash (>5%)/Quick drench (>50%)	NIOSH/OSHA 250 ppm: SA:CFᵋ/PAPROVᵋ 500 ppm: CCRFOV/SCBAF/SAF/ GMFOV/PAPRTOVᵋ 1000 ppm: SAF:PD,PP §: SCBAF:PD,PP/SAF:PD,PP:ASCBA Escape: GMFOV/SCBAE	Inh	Conj, lac; irrit nose, throat; phar edema, chronic bron; burns eyes, skin; skin sens; dental erosion; black skin, hyperkeratosis	Eye: Irr immed Skin: Water flush immed Breath: Resp support Swallow: Medical attention immed	Resp sys, skin, eyes, teeth
Clothing: Goggles: Wash: Change: Remove: Provide:	Reason prob Any poss Immed contam N.R. Immed non-imperv contam Eyewash, quick drench	NIOSH/OSHA 125 ppm: SA:CFᵋ/PAPROVᵋ 250 ppm: CCRFOV/SCBAF/SAF/ GMFOV/PAPRTOVᵋ 1000 ppm: SAF:PD,PP §: SCBAF:PD,PP/SAF:PD,PP:ASCBA Escape: GMFOV/SCBAE	Inh Ing Con	Conj, lac, corneal edema, opac, photo; nasal, phar irrit; cough, dysp, bron; skin burns, vesic, sens derm	Eye: Irr immed Skin: Water flush immed Breath: Resp support Swallow: Medical attention immed	Resp sys, eyes, skin
Clothing: Goggles: Wash: Change: Remove:	Repeat Reason prob Prompt wet N.R. Immed wet (flamm)	NIOSH 1000 ppm: CCROV*/PAPROV*/SA*/ SCBA* 6250 ppm: SA:CF* 12,500 ppm: GMFOV/SCBAF/SAF 20,000 ppm: SAF:PD,PP §: SCBAF:PD,PP/SAF:PD,PP:ASCBA Escape: GMFOV/SCBAE	Inh Ing Con	Irrit eyes, nose, throat; head, dizz; derm	Eye: Irr immed Skin: Soap wash immed Breath: Resp support Swallow: Medical attention immed	Resp sys, skin

Personal protection and sanitation (See Table 3)		Recommendations for respirator selection — maximum concentration for use (MUC) (See Table 4)	Health hazards			
			Route	Symptoms (See Table 5)	First aid (See Table 6)	Target organs (See Table 5)
Clothing: Goggles: Wash: Change: Remove: Provide:	Repeat Reason prob Immed contam N.R. Immed wet (flamm) Quick drench	NIOSH 200 ppm: CCROV/SA/SCBA 500 ppm: SA:CF/PAPROV 1000 ppm: GMFOV/SCBAF/SAF/ CCRFOV 4000 ppm: SAF:PD,PP §: SCBAF:PD,PP/SAF:PD,PP:ASCBA Escape: GMFOV/SCBAE	Inh Abs Ing Con	Asphy; nau, vomit; chest pain; weak; stupor, convuls; eye irrit	Eye: Irr immed Skin: Water flush immed Breath: Resp support Swallow: Medical attention immed	Kidneys, liver, CVS, CNS, lungs, skin, eyes
Clothing: Goggles: Wash: Change: Remove: Provide:	Any poss Any poss Immed contam/daily After work if any poss contam Immed contam Eyewash, quick drench	NIOSH ¥: SCBAF:PD,PP/SAF:PD,PP:ASCBA Escape: HiEF/SCBAE	Inh Abs Ing Con	Reduced function of liver, kidneys, bladder, pancreas, [carc]	Eye: Irr immed Skin: Soap wash immed Breath: Resp support Swallow: Medical attention immed	Liver, bladder, kidneys, pancreas, skin, lungs
Clothing: Goggles: Wash: Change: Remove:	Repeat Reason prob Prompt contam N.R. Prompt non-imperv contam	OSHA 10 ppm: SA/SCBA §: SCBAF:PD,PP/SAF:PD,PP:ASCBA Escape: GMFOV/SCBAE	Inh Ing Con	Irrit eyes, nose; anor, nau; severe head; abdom pain; jaun; monocy	Eye: Irr immed Skin: Water flush prompt Breath: Resp support Swallow: Medical attention immed	Eyes, upper resp sys, liver
Clothing: Goggles: Wash: Change: Remove: Provide:	Any poss Any poss Immed contam N.R. Immed wet (flamm) Eyewash, quick drench	NIOSH/OSHA 2.5 ppm: SA:CF*/PAPROV* 5 ppm: CCRFOV/GMFOV/SCBAF/SAF §: SCBAF:PD,PP/SAF:PD,PP:ASCBA Escape: GMFOV/SCBAE	Inh Ing Con	Irrit eyes, skin, muc memb; abnormal pulm func; delayed pulm edema; chronic resp disease	Eye: Irr immed Skin: Water flush immed Breath: Resp support Swallow: Medical attention immed	Heart, eyes, skin, resp sys

Chemical name, structure/formula, CAS and RTECS Nos., and DOT ID and guide Nos.	Synonyms, trade names, and conversion factors	Exposure limits (TWA unless noted otherwise)	IDLH	Physical description	Chemical and physical properties		Incompatibilities and reactivities	Measurement method (See Table 1)
					MW, BP, SOL Fl.P, IP, Sp.Gr, flammability	VP, FRZ UEL, LEL		
Acrylamide $CH_2=CHCONH_2$ 79-06-1 AS3325000 2074 55	Acrylamide monomer, Acrylic amide, Propenamide, 2-Propenamide	NIOSH Ca See Appendix A 0.03 mg/m³ [skin] OSHA 0.03 mg/m³ [skin]	Ca [Unknown] ACGIH A2	White crystalline, odorless solid.	MW: 71.1 BP: 347-572°F (Decomposes) Sol(86°F): 216% Fl.P: 280°F IP: ? Sp.Gr(86°F): 1.12 Combustible Solid (may also be dissolved in flammable liquids).	VP: 0.007 mm MLT: 184°F UEL: ? LEL: ?	Strong oxidizers [Note: May polymerize violently upon melting.]	Filter/ Si gel; Methanol; GC/NPD; OSHA [#21]
Acrylonitrile $CH_2=CHCN$ 107-13-1 AT5250000 1093 30 (inhibited)	Acrylonitrile monomer, AN, Cyanoethylene, Propenenitrile, 2-Propenenitrile, VCN, Vinyl cyanide 1 ppm = 2.21 mg/m³	NIOSH Ca See Appendix A 1 ppm C 10 ppm [15-min] [skin] OSHA [1910.1045] 2 ppm C 10 ppm [15-min]	Ca [500 ppm] ACGIH A2	Colorless to pale-yellow liquid with an unpleasant odor. [Note: Odor can only be detected above the PEL.]	MW: 53.1 BP: 171°F Sol: 7% Fl.P: 30°F IP: 10.91 eV Sp.Gr: 0.81 Class IB Flammable Liquid	VP: 83 mm FRZ: -116°F UEL: 17% LEL: 3.0%	Strong oxidizers, acids & alkalis; bromine; amines [Note: Unless inhibited (usually with methylhydroquinone) may polymerize spontaneously or when heated or in presence of strong alkali. Attacks copper.]	Char; Acetone/ CS₂; GC/FID; III [#1604]
Aldrin $C_{12}H_8Cl_6$ 309-00-2 IO2100000 2761 55	1,2,3,4,10,10-Hexachloro-1,4,4a,5,8,8a-hexahydro-endo,exo-1,4:5,8-dimethanonaphthalene; HHDN; Octalene	NIOSH Ca See Appendix A 0.25 mg/m³ [skin] OSHA 0.25 mg/m³ [skin]	Ca [100 mg/m³]	Colorless to dark brown crystalline solid with a mild chemical odor. [Note: Formerly used as an insecticide.]	MW: 364.9 BP: Decomposes Sol: 0.003% Fl.P: NA IP: ? Sp.Gr: 1.60 Noncombustible Solid, but may be dissolved in flammable liquids.	VP: 0.00008 mm MLT: 219°F UEL: NA LEL: NA	Concentrated mineral acids, active metals, acid catalysts, acid oxidizing agents, phenol	Filter/Bub; Isooctane; GC/EConD; III [#5502]
Allyl alcohol $CH_2=CHCH_2OH$ 107-18-6 BA5075000 1098 28	AA, Allylic alcohol, 1-Propene-3-ol, Propenol, 2-Propenol, Vinyl carbinol 1 ppm = 2.42 mg/m³	NIOSH/OSHA 2 ppm (5 mg/m³) ST 4 ppm (10 mg/m³) [skin]	150 ppm	Colorless liquid with a pungent, mustard-like odor.	MW: 58.1 BP: 205°F Sol: Miscible Fl.P: 70°F IP: 9.63 eV Sp.Gr: 0.85 Class IB Flammable Liquid	VP: 17 mm FRZ: -200°F UEL: 18.0% LEL: 2.5%	Strong oxidizers, acids, carbon tetrachloride	Char; 2-Propanol/ CS₂; GC/FID; III [#1402, Alcohols III]

Chemical name, structure/formula, CAS and RTECS Nos., and DOT ID and guide Nos.	Synonyms, trade names, and conversion factors	Exposure limits (TWA unless noted otherwise)	IDLH	Physical description	Chemical and physical properties		Incompatibilities and reactivities	Measurement method (See Table 1)
					MW, BP, SOL Fl.P, IP, Sp.Gr, flammability	VP, FRZ UEL, LEL		
Allyl chloride $CH_2=CHCH_2Cl$ 107-05-1 UC7350000 1100 28	3-Chloropropene, 1-Chloro-2-propene, 3-Chloropropylene 1 ppm = 3.18 mg/m³	NIOSH/OSHA 1 ppm (3 mg/m³) ST 2 ppm (6 mg/m³)	300 ppm	Colorless, brown, yellow, or purple liquid with a pungent, unpleasant odor.	MW: 76.5 BP: 113°F Sol: 0.4% Fl.P: -25°F IP: 10.05 eV Sp.Gr: 0.94 Class IB Flammable Liquid	VP: 295 mm MLT: -210°F UEL: 11.1% LEL: 2.9%	Strong oxidizers, acids, amines, iron & aluminum chlorides, magnesium, zinc	Char; Benzene; GC/FID; III [#1000]
Allyl glycidyl ether $C_6H_{10}O_2$ 106-92-3 RR0875000 2219 29	AGE; 1-Allyloxy-2,3-epoxypropane; Glycidyl allyl ether; [(2-Propenyloxy)methyl] oxirane 1 ppm = 4.75 mg/m³	NIOSH 5 ppm (22 mg/m³) ST 10 ppm (44 mg/m³) [skin] OSHA 5 ppm (22 mg/m³) ST 10 ppm (44 mg/m³)	270 ppm	Colorless liquid with a pleasant odor.	MW: 114.2 BP: 309°F Sol: 14% Fl.P: 135°F IP: ? Sp.Gr: 0.97 Class II Combustible Liquid	VP: 2 mm FRZ: -148°F [forms glass] UEL: ? LEL: ?	Strong oxidizers	Tenax GC; Diethyl ether; GC/FID; II(4) [#S346]
4-Aminodiphenyl $C_6H_5C_6H_4NH_2$ 92-67-1 DU8925000	4-Aminobiphenyl, p-Aminobiphenyl, p-Aminodiphenyl, Biphenylamine, 4-Phenylaniline	NIOSH Ca See Appendix A OSHA [1910.1011] See Appendix B ACGIH A1 [skin]	Ca	Colorless crystals with a floral odor. [Note: Turns purple on contact with air.]	MW: 169.2 BP: 576°F Sol: Slight Fl.P: ? IP: ? Sp.Gr: 1.16 Combustible Solid, but must be preheated before ignition possible.	VP(227°F): 1 mm MLT: 127°F UEL: ? LEL: ?	None reported	Filter/ Si gel; 2-Propanol; GC/FID; II(4) [P&CAM #269]
2-Aminopyridine $NH_2C_5H_4N$ 504-29-0 US1575000 2671 55	alpha-Aminopyridine, alpha-Pyridylamine 1 ppm = 3.91 mg/m³	NIOSH/OSHA 0.5 ppm (2 mg/m³)	5 ppm	White powder, leaflets, or crystals with a characteristic odor.	MW: 94.1 BP: 411°F Sol: >100% Fl.P: 154°F IP: 8.00 eV Combustible Solid	VP: Low MLT: 137°F UEL: ? LEL: ?	Strong oxidizers	Tenax GC(2); Thermal desorp; GC/FID; II(4) [#S158]

Personal protection and sanitation (See Table 3)		Recommendations for respirator selection — maximum concentration for use (MUC) (See Table 4)	Health hazards			
			Route	Symptoms (See Table 5)	First aid (See Table 6)	Target organs (See Table 5)
Clothing: Goggles: Wash: Change: Remove: Provide:	Repeat Reason prob Immed contam After work if reason prob contam Immed non-imperv contam Quick drench	NIOSH ¥: SCBAF:PD,PP/SAF:PD,PP:ASCBA Escape: GMFOV/SCBAE	Inh Abs Ing Con	Ataxia, numb limbs, pares; musc weak; absent deep tendon reflex; hand sweat; ftg, lethargy; irrit eyes, skin; [carc]	Eye: Irr immed Skin: Water flush immed Breath: Resp support Swallow: Medical attention immed	CNS, PNS, skin, eyes
Clothing: Goggles: Wash: Change: Remove: Provide:	Repeat Reason prob Immed wet N.R. Immed wet (flamm) Quick drench	NIOSH ¥: SCBAF:PD,PP/SAF:PD,PP:ASCBA Escape: GMFOV/SCBAE	Inh Abs Ing Con	Asphy; irrit eyes; head; sneezing; nau, vomit; weak, li-head; skin vesic; scaling derm; [carc]	Eye: Irr immed Skin: Water wash immed Breath: Resp support Swallow: Medical attention immed	CVS, liver, kidneys, CNS, skin, brain tumor, lung and bowel cancer
Clothing: Goggles: Wash: Change: Remove: Provide:	Any poss Reason prob Immed contam After work if any poss contam Immed non-imperv contam Quick drench	NIOSH ¥: SCBAF:PD,PP/SAF:PD,PP:ASCBA Escape: GMFOVHiE/SCBAE	Inh Abs Ing Con	Head, dizz; nau, vomit, mal; myoclonic jerks of limbs; clonic, tonic convuls; coma; hema, azotemia; [carc]	Eye: Irr immed Skin: Soap wash immed Breath: Resp support Swallow: Medical attention immed	Cancer, CNS, liver, kidneys, skin
Clothing: Goggles: Wash: Change: Remove: Provide:	Any poss liq/Repeat vap >25 ppm Reason prob Immed contam N.R. Immed wet (flamm) Quick drench	NIOSH/OSHA 50 ppm: SA:CF*/PAPROV* 100 ppm: CCRFOV/GMFOV/SCBAF/SAF/PAPRTOV* 150 ppm: SAF:PD,PP §: SCBAF:PD,PP/SAF:PD,PP:ASCBA Escape: GMFOV/SCBAE	Inh Abs Ing	Eye irrit, tissue damage; irrit upper resp sys, skin; pulm edema	Eye: Irr immed Skin: Water flush immed Breath: Resp support Swallow: Medical attention immed	Eyes, skin, resp sys

Personal protection and sanitation (See Table 3)		Recommendations for respirator selection — maximum concentration for use (MUC) (See Table 4)	Health hazards			
			Route	Symptoms (See Table 5)	First aid (See Table 6)	Target organs (See Table 5)
Clothing: Goggles: Wash: Change: Remove: Provide:	Repeat Reason prob Immed contam N.R. Immed wet (flamm) Quick drench	NIOSH/OSHA 25 ppm: SA:CF* 50 ppm: SCBAF/SAF 300 ppm: SAF:PD,PP §: SCBAF:PD,PP/SAF:PD,PP:ASCBA Escape: GMFOV/SCBAE	Inh Abs Ing Con	Irrit eyes, nose, skin; pulm edema; in animals: liver, kidney damage	Eye: Irr immed Skin: Soap wash immed Breath: Resp support Swallow: Medical attention immed	Resp sys, skin, eyes, liver, kidneys
Clothing: Goggles: Wash: Change: Remove: Provide:	Reason prob Any poss Prompt contam N.R. Prompt non-imperv contam Eyewash	NIOSH/OSHA 50 ppm: SA/SCBA/CCROV 125 ppm: SA:CF*/PAPROV* 250 ppm: GMFOV/SCBAF/SAF/CCRFOV/SAT:CF 270 ppm: SAF:PD,PP §: SCBAF:PD,PP/SAF:PD,PP:ASCBA Escape: GMFOV/SCBAE	Inh Abs Ing Con	Derm; eye, nose irrit; pulm irrit, edema; narco	Eye: Irr immed Skin: Water flush prompt Breath: Resp support Swallow: Medical attention immed	Resp sys, skin
Clothing: Goggles: Wash: Change: Remove: Provide:	Any poss Any poss Immed contam/daily After work if any poss contam Immed contam Eyewash, quick drench	NIOSH ¥: SCBAF:PD,PP/SAF:PD,PP:ASCBA Escape: HiEF/SCBAE	Inh Abs Ing Con	Head, dizz; lethargy, dysp; ataxia, weak; methemoglobinemia; urinary burning; acute hemorrhagic cystitis; [carc]	Eye: Irr immed Skin: Soap wash immed Breath: Resp support Swallow: Medical attention immed	Bladder, skin
Clothing: Goggles: Wash: Change: Remove: Provide:	Reason prob Reason prob Immed contam After work if reason prob contam Immed non-imperv contam Quick drench	NIOSH/OSHA 5 ppm: SA*/SCBA* §: SCBAF:PD,PP/SAF:PD,PP:ASCBA Escape: GMFOVHiE/SCBAE	Inh Abs Ing Con	Head, dizz; excitement; nau; high BP; resp distress; weak; convuls; stupor	Eye: Irr immed Skin: Water flush immed Breath: Resp support Swallow: Medical attention immed	CNS, resp sys

Chemical name, structure/formula, CAS and RTECS Nos., and DOT ID and guide Nos.	Synonyms, trade names, and conversion factors	Exposure limits (TWA unless noted otherwise)	IDLH	Physical description	Chemical and physical properties		Incompatibilities and reactivities	Measurement method (See Table 1)
					MW, BP, SOL FI.P, IP, Sp.Gr, flammability	VP, FRZ UEL, LEL		
Ammonia NH₃ 7664-41-7 BO0875000 2672 60 (12-44% soln) 2073 15 (44% soln) 1005 15 (anhydrous)	Anhydrous ammonia, Aqua ammonia, Aqueous ammonia 1 ppm = 0.71 mg/m³	NIOSH 25 ppm (18 mg/m³) ST 35 ppm (27 mg/m³) OSHA ST 35 ppm (27 mg/m³)	500 ppm	Colorless gas with a pungent, suffo-cating odor. [Note: Often used in an aqueous solution. Easily liquefied under pressure.]	MW: 17.0 BP: -28°F Sol: 34% FI.P: NA (Gas) IP: 10.18 eV Combustible Gas, but difficult to burn.	VP: >1 atm FRZ: -108°F UEL: 28% LEL: 15%	Strong oxidizers, acids, halogens, salts of silver & zinc [Note: Corrosive to copper & galvanized surfaces.]	Passive sampler; none; IC; III [#6701]
Ammonium sulfamate NH₄SO₃NH₂ 7773-06-0 WO6125000 9089 31	Ammate® herbicide, Ammonium aminosulfonate, Monoammonium salt of sulfamic acid, Sulfamate	NIOSH/OSHA 10 mg/m³ (total) 5 mg/m³ (resp)	5000 mg/m³	Colorless to white crystalline, odorless solid. [herbicide]	MW: 114.1 BP: 320°F (Decomposes) Sol: 200% FI.P: NA IP: ? Noncombustible Solid	VP: Low MLT: 268°F UEL: NA LEL: NA	Acids, hot water [Note: Elevated temperatures cause a highly exothermic reaction with water.]	Filter; Water; IC; II(5) [#S348]
n-Amyl acetate CH₃COO[CH₂]₄CH₃ 628-63-7 AJ1925000 1104 26	Amyl acetic ester, Amyl acetic ether, 1-Pentanol acetate, Pentyl ester of acetic acid, Primary amyl acetate 1 ppm = 5.41 mg/m³	NIOSH/OSHA 100 ppm (525 mg/m³)	4000 ppm	Colorless liquid with a persistent banana-like odor.	MW: 130.2 BP: 301°F Sol: 0.2% FI.P: 77°F IP: ? Sp.Gr: 0.88 Class IC Flammable Liquid	VP(77°F): 5 mm FRZ: -95°F UEL: 7.5% LEL: 1.1%	Nitrates; strong oxidizers, alkalis & acids	Char; CS₂; GC/FID; III [#1450, Esters I]
sec-Amyl acetate CH₃COOCH(CH₃)C₃H₇ 626-38-0 AJ2100000 1104 26	1-Methylbutyl acetate, 2-Pentanol acetate, 2-Pentyl ester of acetic acid 1 ppm = 5.41 mg/m³	NIOSH/OSHA 125 ppm (650 mg/m³)	9000 ppm	Colorless liquid with a mild odor.	MW: 130.2 BP: 249°F Sol: Slight FI.P: 89°F IP: ? Sp.Gr: 0.87 Class IC Flammable Liquid	VP: 7 mm FRZ: -148°F UEL: ? LEL: ?	Nitrates; strong oxidizers, alkalis & acids	Char; CS₂; GC/FID; III [#1450, Esters I]

Chemical name, structure/formula, CAS and RTECS Nos., and DOT ID and guide Nos.	Synonyms, trade names, and conversion factors	Exposure limits (TWA unless noted otherwise)	IDLH	Physical description	Chemical and physical properties		Incompatibilities and reactivities	Measurement method (See Table 1)
					MW, BP, SOL FI.P, IP, Sp.Gr, flammability	VP, FRZ UEL, LEL		
Aniline and homologs C₆H₅NH₂ 62-53-3 BW6650000 1547 57	Aminobenzene, Aniline oil, Benzenamine, Phenylamine 1 ppm = 3.87 mg/m³	NIOSH Ca See Appendix A 2 ppm (8 mg/m³) [skin] OSHA 2 ppm (8 mg/m³) [skin]	Ca [100 ppm]	Colorless to brown oily liquid with an aromatic amine-like odor. [Note: A solid below 21°F.]	MW: 93.1 BP: 363°F Sol: 4% FI.P: 158°F IP: 7.70 eV Sp.Gr: 1.02 Class IIIA Combustible Liquid	VP: 0.6 mm FRZ: 21°F UEL: 11% LEL: 1.3%	Strong oxidizers, strong acids, toluene diisocyanate	Si gel; Ethanol; GC/FID; III [#2002, Aromatic Amines]
Anisidine (o-, p- isomers) NH₂C₆H₄OCH₃ o: 90-04-0 BZ5410000 p: 104-94-9 BZ5450000 2431 55	o: ortho-Aminoanisole, 2-Anisidine, o-Methoxyaniline p: para-Aminoanisole, 4-Anisidine, p-Methoxyaniline 1 ppm = 5.12 mg/m³	NIOSH Ca See Appendix A 0.5 mg/m³ [skin] OSHA 0.5 mg/m³ [skin]	Ca [50 mg/m³]	o: Red or yellow oily liquid with an amine-like odor. [Note: A solid below 41°F.] p: Yellow to brown crystalline solid with an amine-like odor.	MW: 123.2 BP: 437/475°F Sol: Insoluble/ Moderate FI.P: 244(oc)/ 86°F IP: 7.44 eV Sp.Gr: 1.10/1.07 o: Class IIIB Combustible Liquid p: Combustible Solid	VP: <0.1/ <0.1 mm FRZ/MLT: 41/135°F UEL: ?/? LEL: ?/?	Strong oxidizers	XAD-2; Methanol; HPLC/UVD; III [#2514]
Antimony and compounds (as Sb) Sb 7440-36-0 (Metal) CC4025000 (Metal) 2871 53 (Sb powder)	Metal: Antimony powder, Stibium Synonyms of other com-pounds vary depending upon the specific compound.	NIOSH/OSHA 0.5 mg/m³	80 mg/m³	Metal: Silver-white, lustrous, hard, brittle solid; scale-like crystals; or a dark gray, lustrous powder.	MW: 121.8 BP: 2975°F Sol: Insoluble FI.P: NA IP: NA Sp.Gr: 6.69 Metal: Noncombustible Solid in bulk form, but a moderate explosion hazard in the form of dust when exposed to flame.	VP: 0 mm (approx) MLT: 1166°F UEL: NA LEL: NA	Strong oxidizers, acids, halogenated acids [Note: Stibine is formed when antimony is exposed to nascent (freshly formed) hydrogen.]	Filter; Acid; AA; II(4) [P&CAM #261]
ANTU C₁₀H₇NHC(NH₂)S 86-88-4 YT9275000 1651 53	alpha-Naphthyl thio-carbamide, 1-Naphthyl thiourea, alpha-Naphthyl thiourea	NIOSH/OSHA 0.3 mg/m³	100 mg/m³	White crystalline or gray odorless powder. [rodenticide]	MW: 202.3 BP: Decomposes Sol: 0.06% FI.P: NA IP: ? Noncombustible Solid	VP: Low MLT: 388°F UEL: NA LEL: NA	Strong oxidizers, silver nitrate	Filter; Methanol; HPLC/UVD; II(5) [#S276]

Personal protection and sanitation (See Table 3)		Recommendations for respirator selection — maximum concentration for use (MUC) (See Table 4)	Health hazards				
			Route	Symptoms (See Table 5)	First aid (See Table 6)		Target organs (See Table 5)
Clothing: Goggles: Wash: Change: Remove: Provide:	Any poss >10%/Repeat <10% Any poss Immed contam >10%/ Prompt wet <10% N.R. Immed non-imperv >10%/ Prompt non-imperv contam <10% Eyewash 10%, quick drench >10%	NIOSH 250 ppm: CCRS*/SA*/SCBA* 500 ppm: PAPRS*/SCBAF/SAF/ GMFS/CCRFS/SA:CF* §: SCBAF:PD,PP/SAF:PD,PP:ASCBA Escape: GMFS/SCBAE	Inh Ing Con	Eye, nose, throat irrit; dysp, bronspas, chest pain; pulm edema; pink frothy sputum; skin burns, vesic	Eye: Skin: Breath: Swallow:	Irr immed Water flush immed Resp support Medical attention immed	Resp sys, eyes
Clothing: Goggles: Wash: Change: Remove:	N.R. N.R. N.R. N.R. N.R.	NIOSH/OSHA 50 mg/m³: DM 100 mg/m³: DMXSQ/SA 250 mg/m³: PAPRDM/SA:CF 500 mg/m³: HiEF/PAPRTHiE/ SAT:CF/SCBAF/SAF 5000 mg/m³: SA:PD,PP §: SCBAF:PD,PP/SAF:PD,PP:ASCBA Escape: HiEF/SCBAE	Inh Ing Con	Irrit eyes, nose, throat; cough, difficult breath	Eye: Skin: Breath: Swallow:	Irr immed Soap wash prompt Resp support Medical attention immed	Upper resp sys, eyes
Clothing: Goggles: Wash: Change: Remove:	Repeat Reason prob Prompt wet N.R. Immed wet (flamm)	NIOSH/OSHA 1000 ppm: PAPROV*/CCROV*/SA*/ SCBA* 2500 ppm: SA:CF* 4000 ppm: GMFOV/SCBAF/SAF §: SCBAF:PD,PP/SAF:PD,PP:ASCBA Escape: GMFOV/SCBAE	Inh Ing Con	Irrit eyes, nose; narco; derm	Eye: Skin: Breath: Swallow:	Irr immed Water flush prompt Resp support Medical attention immed	Eyes, skin, resp sys
Clothing: Goggles: Wash: Change: Remove:	Repeat Reason prob Prompt wet N.R. Immed wet (flamm)	NIOSH/OSHA 1000 ppm: PAPROV*/CCROV* 1250 ppm: SA*/SCBA* 3125 ppm: SA:CF* 6250 ppm: GMFOV/SCBAF/SAF 9000 ppm: SAF:PD,PP §: SCBAF:PD,PP/SAF:PD,PP:ASCBA Escape: GMFOV/SCBAE	Inh Ing Con	Irrit eyes, nose; narco; derm	Eye: Skin: Breath: Swallow:	Irr immed Water flush prompt Resp support Medical attention immed	Resp sys, eyes, skin

Personal protection and sanitation (See Table 3)		Recommendations for respirator selection — maximum concentration for use (MUC) (See Table 4)	Health hazards				
			Route	Symptoms (See Table 5)	First aid (See Table 6)		Target organs (See Table 5)
Clothing: Goggles: Wash: Change: Remove: Provide:	Reason prob Reason prob Immed contam N.R. Immed non-imperv contam Quick drench	NIOSH ¥: SCBAF:PD,PP/SAF:PD,PP:ASCBA Escape: GMFOV/SCBAE	Inh Abs Ing Con	Head; weak, dizz; cyan; ataxia; dysp on effort; tacar; eye irrit; [carc]	Eye: Skin: Breath: Swallow:	Irr immed Soap wash prompt Resp support Medical attention immed	Blood, CVS, liver, kidneys
Clothing: Goggles: Wash: Change: Remove: Provide:	Reason prob Reason prob Immed contam After work if reason prob contam Immed non-imperv contam Quick drench	NIOSH ¥: SCBAF:PD,PP/SAF:PD,PP:ASCBA Escape: HiEF/SCBAE	Inh Abs Ing Con	Head, dizz; cyan; RBC Heinz bodies; [carc]	Eye: Skin: Breath: Swallow:	Irr immed Soap wash immed Resp support Medical attention immed	Blood, kidneys, liver, CVS
Clothing: Goggles: Wash: Change: Remove:	Reason prob Reason prob Prompt contam After work if reason prob contam Prompt non-imperv contam	NIOSH/OSHA 5 mg/m³: DMXSQ^/SA/SCBA 12.5 mg/m³: PAPRDM^/SA:CF 25 mg/m³: HiEF/PAPRTHiE/ SAT:CF/SCBAF/SAF 80 mg/m³: SA:PD,PP §: SCBAF:PD,PP/SAF:PD,PP:ASCBA Escape: HiEF/SCBAE	Inh Con	Irrit nose, throat, mouth; cough; dizz; head; nau, vomit, diarr; stomach cramps; insom; anor; irrit skin; unable to smell properly; cardiac abnormalities in antimony trichloride exposures	Skin: Breath: Swallow:	Soap wash immed Resp support Medical attention immed	Resp sys, CVS, skin, eyes
Clothing: Goggles: Wash: Change: Remove:	N.R. N.R. N.R. After work if reason prob contam N.R.	NIOSH/OSHA 3 mg/m³: CCROVDMFu/SA/SCBA 7.5 mg/m³: PAPROVDMFu/SA:CF 15 mg/m³: CCRFOVHiE/PAPRTOVHiE/ SAT:CF/SCBAF/SAF/ GMFOVHiE 100 mg/m³: SA:PD,PP §: SCBAF:PD,PP/SAF:PD,PP:ASCBA Escape: GMFOVHiE/SCBAE	Inh Ing	After ingestion of large doses: vomit, dysp, cyan, coarse pulm rales; mild liver damage	Eye: Skin: Breath: Swallow:	Irr immed Soap wash prompt Resp support Medical attention immed	Resp sys

Chemical name, structure/formula, CAS and RTECS Nos., and DOT ID and guide Nos.	Synonyms, trade names, and conversion factors	Exposure limits (TWA unless noted otherwise)	IDLH	Physical description	Chemical and physical properties		Incompatibilities and reactivities	Measurement method (See Table 1)
					MW, BP, SOL Fl.P, IP, Sp.Gr, flammability	VP, FRZ UEL, LEL		
Arsenic (inorganic compounds as As) As 7740-38-2 (Metal) CG0525000 (Metal) 1558 53 (Metal)	Metal: Arsenia Synonyms of other compounds vary depending upon the specific compound. [Note: OSHA considers "Inorganic Arsenic" to mean copper acetoarsenite & all inorganic compounds containing arsenic except ARSINE.]	NIOSH Ca See Appendix A C 0.002 mg/m³ [15-min] OSHA [1910.1018] 0.010 mg/m³	Ca [100 mg/m³]	Metal: Silver-gray or tin-white, brittle, odorless solid.	MW: 74.9 BP: Sublimes Sol: Insoluble Fl.P: NA IP: NA Sp.Gr: 5.73 Metal: Noncombustible Solid in bulk form, but a slight explosion hazard in the form of dust when exposed to flame.	VP: 0 mm (approx) MLT: 1135EF (Sublimes) UEL: NA LEL: NA	Strong oxidizers, bromine azide [Note: Hydrogen gas can react with inorganic arsenic to form the highly toxic gas arsine.]	Filter; Acid; AA; III [#7900]
Arsine AsH₃ 7784-42-1 CG6475000 2188 18	Arsenic hydride, Arsenic trihydride, Arseniuretted hydrogen, Arsenious hydride, Hydrogen arsenide 1 ppm = 3.24 mg/m³	NIOSH Ca See Appendix A C 0.002 mg/m³ [15-min] OSHA 0.05 ppm (0.2 mg/m³)	Ca [6 ppm]	Colorless gas with a mild garlic-like odor.	MW: 78.0 BP: -81°F Sol: 20% Fl.P: NA (Gas) IP: 9.89 eV Flammable Gas	VP: >1 atm FRZ: -179°F UEL: ? LEL: ?	Strong oxidizers, chlorine, nitric acid [Note: Decomposes above 446°F. There is a high potential for the generation of arsine gas when inorganic arsenic is exposed to nascent (freshly formed) hydrogen.]	Char; HNO₃; HGA; III [#6001]
Asbestos Hydrated mineral silicates 1332-21-4 CI6475000 2212 31 (blue) 2590 31 (white)	Actinolite, Amosite (cummingtonite-grunerite), Anthophyllite, Chrysotile, Crocidolite, Tremolite	NIOSH Ca See Appendix A See Appendix C OSHA [1910.1001] OSHA [1910.1101] See Appendix C ACGIH See Appendix C	Ca	White or greenish (chrysotile), blue (crocidolite), or gray-green (amosite) fibrous, odorless solids.	MW: Varies BP: Decomposes Sol: Insoluble Fl.P: NA IP: NA Noncombustible Solid	VP: 0 mm (approx) MLT: 1112°F (Decomposes) UEL: NA LEL: NA	None reported	Filter; Acetone/ Triacetin; PCM; III [#7400, Fibers] [Note: Measurements may also be made with transmission electron microscopy using NIOSH #7402.]
Azinphos-methyl C₁₀H₁₂O₃PS₂N₃ 86-50-0 TE1925000 2783 55	Guthion®; 0,0-Dimethyl-S-4-oxo-1,2,3-benzotriazin-3(4H)-ylmethyl phosphorodithioate; Methyl azinphos [(CH₃O)₂P(S)SCH₂(N₃C₇H₄O)]	NIOSH/OSHA 0.2 mg/m³ [skin]	20 mg/m³	Colorless crystals or a brown, waxy solid. [insecticide]	MW: 317.3 BP: Decomposes Sol: 0.003% Fl.P: NA IP: ? Sp.Gr: 1.44 Noncombustible Solid	VP: Low MLT: 163°F UEL: NA LEL: NA	Strong oxidizers, acids	None available

Chemical name, structure/formula, CAS and RTECS Nos., and DOT ID and guide Nos.	Synonyms, trade names, and conversion factors	Exposure limits (TWA unless noted otherwise)	IDLH	Physical description	Chemical and physical properties		Incompatibilities and reactivities	Measurement method (See Table 1)
					MW, BP, SOL Fl.P, IP, Sp.Gr, flammability	VP, FRZ UEL, LEL		
Barium (soluble compounds as Ba) 1: Ba(NO₃)₂ 2: BaCl₂ 1: 10022-31-8/ CQ9625000 2: 10361-37-2/ CQ8750000 1446 42 (Barium nitrate)	1: Barium nitrate, Barium dinitrate 2: Barium chloride, Barium dichloride Synonyms of other soluble compounds vary depending upon the specific compound.	NIOSH/OSHA 0.5 mg/m³	1100 mg/m³	Barium nitrate & Barium chloride are white odorless solids.	MW: 261.4/ 208.3 BP: Decomposes /2840°F Sol: 9/38% Fl.P: NA/? IP: ?/? Sp.Gr: 3.24/3.86 Ba(NO₃)₂: Noncombustible Solid BaCl₂: Combustible Solid	VP: Low/Low MLT: 1098/ 1765°F UEL: NA/? LEL: NA/?	Acids, oxidizers [Note: Contact of barium nitrate with combustible material may cause fire.]	Filter; Water; AA; III [#7056]
Benzene C₆H₆ 71-43-2 CY1400000 1114 27	Benzol, Phenyl hydride 1 ppm = 3.25 mg/m³	NIOSH Ca See Appendix A 0.1 ppm ST 1 ppm OSHA [1910.1028] 1 ppm ST 5 ppm	Ca [3000 ppm] ACGIH A2	Colorless to light-yellow liquid with an aromatic odor. [Note: A solid below 42°F.]	MW: 78.1 BP: 176°F Sol: 0.07% Fl.P: 12°F IP: 9.24 eV Sp.Gr: 0.88 Class IB Flammable Liquid	VP: 75 mm FRZ: 42°F UEL: 7.9% LEL: 1.3%	Strong oxidizers, many fluorides & perchlorates, nitric acid	Char; CS₂; GC/FID; III [#1500, Hydrocarbons] [Note: Measurements may also be made with a portable GC using NIOSH #3700 (III).]
Benzidine NH₂C₆H₄C₆H₄NH₂ 92-87-5 DC9625000 1885 53	4,4'-Bianiline; 4,4'-Biphenyldiamine; 1,1'-Biphenyl-4,4'-diamine; 4,4'-Diaminobiphenyl; p-Diaminodiphenyl	NIOSH Ca See Appendix A OSHA [1910.1010] See Appendix B ACGIH A1 [skin]	Ca	Grayish-yellow, reddish-gray, or white crystalline powder.	MW: 184.3 BP: 752°F Sol(54°F): 0.04% Fl.P: ? IP: ? Sp.Gr: 1.25 Combustible Solid, but difficult to burn.	VP: Low MLT: 239°F UEL: ? LEL: ?	None reported	Filter/ Si gel; Reagent; HPLC/UVD; III [#5509]
Benzoyl peroxide (C₆H₅CO)₂O₂ 94-36-0 DM8575000 2085/2086/2087 49 2088/2089/2090 49	Benzoperoxide, Dibenzoyl peroxide	NIOSH/OSHA 5 mg/m³	7000 mg/m³	Colorless to white crystals or a granular powder with a faint benzaldehyde-like odor.	MW: 242.2 BP: ? Sol: <1% Fl.P: ? IP: ? Sp.Gr(77°F): 1.33 Combustible Solid (easily ignited and burns very rapidly).	VP: <1 mm MLT: 217°F UEL: ? LEL: ?	Combustible substances (wood, paper, etc.), acids, alkalis, alcohols, amines, ethers [Note: Containers may explode when heated. Extremely explosion-sensitive to shock, heat, and friction.]	Filter; Diethyl ether; HPLC/UVD; III [#5009]

Personal protection and sanitation (See Table 3)		Recommendations for respirator selection — maximum concentration for use (MUC) (See Table 4)	Health hazards				
			Route	Symptoms (See Table 5)	First aid (See Table 6)		Target organs (See Table 5)
Clothing:	Any poss	NIOSH	Inh	Ulceration of nasal septum,	Eye:	Irr immed (15 min)	Liver, kidneys,
Goggles:	Any poss	¥: SCBAF:PD,PP/SAF:PD,PP:ASCBA	Abs	derm, GI disturbances,	Skin:	Soap wash immed	skin, lungs,
Wash:	Immed contam/daily	Escape: GMFAGHiE/SCBAE	Con	peri neur, resp irrit,	Swallow:	Medical attention	lymphatic sys
Change:	After work if reason		Ing	hyperpig of skin, [carc]		immed	
	prob contam						
Remove:	Immed contam						
Provide:	Eyewash, quick drench						
Clothing:	N.R.	NIOSH	Inh	Head, malaise, weak,	Breath:	Resp support	Blood, kidneys,
Goggles:	N.R.	¥: SCBAF:PD,PP/SAF:PD,PP:ASCBA		dizz; dysp; abdom			liver
Wash:	N.R.	Escape: GMFS/SCBAE		and back pain; nau,			
Change:	N.R.			vomit; bronze skin; hema;			
Remove:	N.R.			jaun; peri neur; [carc]			
Clothing:	Any poss	NIOSH	Inh	Dysp, interstitial fib,	Eye:	Irr immed	Lungs
Goggles:	Reason prob	¥: SCBAF:PD,PP/SAF:PD,PP:ASCBA	Ing	restricted pulm function,			
Wash:	Daily	Escape: HiEF/SCBAE		finger clubbing, [carc]			
Change:	After work if reason						
	prob contam						
Clothing:	Reason prob	NIOSH/OSHA	Inh	Miosis; ache eyes; blurred	Eye:	Irr immed	Resp sys, CNS,
Goggles:	Reason prob	2 mg/m³: SA/SCBA/CCROVDMFu	Abs	vision; lac, rhin; head;	Skin:	Soap wash immed	CVS, blood chol
Wash:	Immed contam	5 mg/m³: PAPROVDMFu/SA:CF	Ing	tight chest, wheez, lar	Breath:	Resp support	
Change:	After work if any poss	10 mg/m³: CCRFOVHiE/PAPRTOVHiE/	Con	spas; salv; cyan; anor;	Swallow:	Medical attention	
	contam	SAT:CF/SCBAF/SAF/		nau, vomit, diarr; sweat;		immed	
Remove:	Immed non-imperv contam	GMFOVHiE		twitch, para, convuls; low			
Provide:	Quick drench	20 mg/m³: SA:PD,PP		BP, card irregularities			
		§: SCBAF:PD,PP/SAF:PD,PP:ASCBA					
		Escape: GMFOVHiE/SCBAE					

Personal protection and sanitation (See Table 3)		Recommendations for respirator selection — maximum concentration for use (MUC) (See Table 4)	Health hazards				
			Route	Symptoms (See Table 5)	First aid (See Table 6)		Target organs (See Table 5)
Recommendations vary depending upon the specific compound.		NIOSH/OSHA	Inh	Upper resp irrit;	Eye:	Irr immed	Heart, CNS,
		5 mg/m³: DMXSQ/SA/SCBA	Ing	gastroenteritis; musc	Skin:	Water flush immed	skin, resp sys,
		12.5 mg/m³: PAPRDM/SA:CF	Con	spasm; slow pulse,	Breath:	Resp support	eyes
		25 mg/m³: HiEF/PAPRTHiE/SAT:CF/		extrasystoles; hypokalemia;	Swallow:	Medical attention	
		SCBAF/SAF		irrit eyes, skin; skin		immed	
		250 mg/m³: SAF:PD,PP		burns			
		§: SCBAF:PD,PP/SAF:PD,PP:ASCBA					
		Escape: HiEF/SCBAE					
Clothing:	Repeat	NIOSH	Inh	Irrit eyes, nose, resp	Eye:	Irr immed	Blood, CNS,
Goggles:	Reason prob	¥: SCBAF:PD,PP/SAF:PD,PP:ASCBA	Abs	sys; gidd; head, nau,	Skin:	Soap wash prompt	skin, bone
Wash:	Prompt wash soap	Escape: GMFOV/SCBAE	Ing	staggered gait; ftg, anor,	Breath:	Resp support	marrow, eyes,
Change:	N.R.		Con	lass; derm; bone marrow	Swallow:	Medical attention	resp sys
Remove:	Immed wet (flamm)			depres; [carc]		immed	
Clothing:	Any poss	NIOSH	Inh	Hema; secondary anemia	Eye:	Irr immed (15 min)	Bladder,
Goggles:	Any poss	¥: SCBAF:PD,PP/SAF:PD,PP:ASCBA	Abs	from hemolysis; acute	Skin:	Soap wash immed	kidneys, liver,
Wash:	Immed contam/daily	Escape: HiEF/SCBAE	Ing	cystitis; acute liver	Breath:	Resp support	skin, blood
Change:	After work if any poss		Con	disorders; derm;	Swallow:	Medical attention	
	contam			painful and irregular		immed	
Remove:	Immed contam			urination; [carc]			
Provide:	Eyewash, quick drench						
Clothing:	Repeat	NIOSH/OSHA	Inh	Irrit skin, eyes, muc	Eye:	Irr immed	Skin, resp sys,
Goggles:	Reason prob	50 mg/m³: DMXSQ*/SA*/SCBA*	Ing	memb; sens derm	Skin:	Soap wash prompt	eyes
Wash:	Prompt contam	125 mg/m³: PAPRDM*/SA:CF*	Con		Swallow:	Medical attention	
Change:	After work if reason	250 mg/m³: HiEF/SCBAF/SAF/				immed	
	prob contam	PAPRTHiE*					
Remove:	Prompt non-imperv	7000 mg/m³: SAF:PD,PP					
	contam	§: SCBAF:PD,PP/SAF:PD,PP:ASCBA					
		Escape: HiEF/SCBAE					

Chemical name, structure/formula, CAS and RTECS Nos., and DOT ID and guide Nos.	Synonyms, trade names, and conversion factors	Exposure limits (TWA unless noted otherwise)	IDLH	Physical description	Chemical and physical properties		Incompatibilities and reactivities	Measurement method (See Table 1)
					MW, BP, SOL Fl.P, IP, Sp.Gr, flammability	VP, FRZ UEL, LEL		
Benzyl chloride $C_6H_5CH_2Cl$ 100-44-7 XS8925000 1738 59	Chloromethylbenzene, alpha-Chlorotoluene 1 ppm = 5.26 mg/m³	NIOSH C 1 ppm (5 mg/m³) [15-min] OSHA 1 ppm (5 mg/m³)	10 ppm	Colorless to slightly yellow liquid with a pungent, aromatic odor.	MW: 126.6 BP: 354°F Sol: 0.05% Fl.P: 153°F IP: ? Sp.Gr: 1.10 Class IIIA Combustible Liquid	VP(72°F): 1 mm FRZ: -38°F UEL: ? LEL: 1.1%	Oxidizers, acids, copper, aluminum, magnesium, iron, zinc, tin [Note: Can polymerize when in contact with all common metals except nickel & lead. Hydrolyzes in H_2O to benzyl alcohol.]	Char; CS₂; GC/FID; III [#1003, Halogenated Hydrocarbons]
Beryllium and compounds (as Be) Be 7440-41-7 DS1750000 1566 53 (cmpds) 1567 32 (powder)	Metal: None Synonyms of other compounds vary depending upon the specific compound.	NIOSH Ca See Appendix A Not to exceed 0.0005 mg/m³ OSHA 0.002 mg/m³ C 0.005 mg/m³ 0.025 mg/m³ [30-min maximum peak]	Ca [10 mg/m³] ACGIH A2	Metal: A hard, brittle, gray-white solid.	MW: 9.0 BP: 4532°F Sol: Insoluble Fl.P: NA IP: NA Sp.Gr: 1.85 Metal: Noncombustible Solid in bulk form, but a slight explosion hazard in the form of a powder or dust.	VP: 0 mm (approx) MLT: 2349°F UEL: NA LEL: NA	Acids, caustics, chlorinated hydrocarbons, oxidizers, molten lithium	Filter; Acid; HGA; III [#7102]
Boron oxide B_2O_3 1303-86-2 ED7900000	Boric anhydride, Boric oxide, Boron trioxide	NIOSH/OSHA 10 mg/m³	N.E.	Colorless, semi-transparent lumps or hard, white, odorless crystals.	MW: 69.6 BP: 3380°F Sol: 3% Fl.P: NA IP: NA Sp.Gr: 2.46 Noncombustible Solid	VP: 0 mm (approx) MLT: 842°F UEL: NA LEL: NA	Water [Note: Reacts slowly with water to form boric acid.]	Filter; none; Grav; III [#0500, Nuisance Dust (total)]
Boron trifluoride BF_3 7637-07-2 ED2275000 1008 15	Boron fluoride, Trifluoroborane 1 ppm = 2.82 mg/m³	NIOSH/OSHA C 1 ppm (3 mg/m³)	100 ppm	Colorless gas with a pungent, suffocating odor. [Note: Forms dense white fumes in moist air.]	MW: 67.8 BP: -148°F Sol: 106% (in cold H_2O) Fl.P: NA IP: 15.50 eV Nonflammable Gas	VP: >1 atm FRZ: -196°F UEL: NA LEL: NA	Alkali metals, calcium oxide [Note: Hydrolyzes in moist air or hot water to form boric acid, hydrogen fluoride, and fluoboric acid.]	None available

Chemical name, structure/formula, CAS and RTECS Nos., and DOT ID and guide Nos.	Synonyms, trade names, and conversion factors	Exposure limits (TWA unless noted otherwise)	IDLH	Physical description	Chemical and physical properties		Incompatibilities and reactivities	Measurement method (See Table 1)
					MW, BP, SOL Fl.P, IP, Sp.Gr, flammability	VP, FRZ UEL, LEL		
Bromine Br_2 7726-95-6 EF9100000 1744 59	Molecular bromine 1 ppm = 6.64 mg/m³	NIOSH/OSHA 0.1 ppm (0.7 mg/m³) ST 0.3 ppm (2 mg/m³)	10 ppm	Dark reddish-brown fuming liquid with suffocating, irritating fumes.	MW: 159.8 BP: 139°F Sol: 4% Fl.P: NA IP: 10.55 eV Sp.Gr: 3.12 Noncombustible Liquid, but accelerates the burning of combustibles.	VP: 172 mm FRZ: 19°F UEL: NA LEL: NA	Combustible organics (sawdust, wood, cotton, straw, etc.), aluminum, readily oxidizable materials, ammonia, hydrogen, acetylene, phosphorus, potassium, sodium [Note: Corrodes iron, steel, stainless steel & copper.]	Bub; none; IC; OSHA [#ID108]
Bromoform $CHBr_3$ 75-25-2 PB5600000 2515 58	Methyl tribromide, Tribromomethane 1 ppm = 10.51 mg/m³	NIOSH/OSHA 0.5 ppm (5 mg/m³) [skin]	Unknown	Colorless to yellow liquid with a chloroform-like odor. [Note: A solid below 47°F.]	MW: 252.8 BP: 301°F Sol: 0.1% Fl.P: NA IP: 10.48 eV Sp.Gr: 2.89 Noncombustible Liquid	VP: 5 mm FRZ: 47°F UEL: NA LEL: NA	Lithium, sodium, potassium, calcium, aluminum, zinc, magnesium, strong caustics, acetone [Note: Gradually decomposes, acquiring yellow color; air & light accelerate decomposition.]	Char; CS₂; GC/FID; III [#1003, Halogenated Hydrocarbons]
1,3-Butadiene $CH_2=CHCH=CH_2$ 106-99-0 EI9275000 1010 17 (inhibited)	Biethylene, Bivinyl, Butadiene, Divinyl, Erythrene, Vinylethylene 1 ppm = 2.25 mg/m³	NIOSH Ca See Appendix A Reduce exposure to lowest feasible concentration. OSHA 1000 ppm (2200 mg/m³)	Ca [20,000 ppm] [LEL] ACGIH A2, 10 ppm (22 mg/m³)	Colorless gas with a mild aromatic or gasoline-like odor. [Note: A liquid below 24°F.]	MW: 54.1 BP: 24°F Sol: Insoluble Fl.P: NA (Gas) <0°F (Liquid) IP: 9.07 eV Sp.Gr: 0.65(Liquid at 24°F) Flammable Gas Class IA Flammable Liquid	VP: >1 atm FRZ: -164°F UEL: 12.0% LEL: 2.0%	Phenol, chlorine dioxide, copper, crotonaldehyde [Note: May contain inhibitors (such as tri-butylcatechol) to prevent self-polymerization. May form explosive peroxides upon exposure to air.]	Char(2); CH₂Cl₂; GC/FID; III [#1024]
2-Butanone $CH_3COCH_2CH_3$ 78-93-3 EL6475000 1193/1232 26	Ethyl methyl ketone, MEK, Methyl acetone, Methyl ethyl ketone 1 ppm = 3.00 mg/m³	NIOSH/OSHA 200 ppm (590 mg/m³) ST 300 ppm (885 mg/m³)	3000 ppm	Colorless liquid with a moderately sharp, fragrant, mint- or acetone-like odor.	MW: 72.1 BP: 175°F Sol: 28% Fl.P: 16°F IP: 9.54 eV Sp.Gr: 0.81 Class IB Flammable Liquid	VP: 71 mm FRZ: -123°F UEL(200°F): 11.4% LEL(200°F): 1.4%	Strong oxidizers, amines, ammonia, inorganic acids, caustics, copper, isocyanates, pyridines	Ambersorb; CS₂; GC/FID; III [#2500]

Personal protection and sanitation (See Table 3)		Recommendations for respirator selection — maximum concentration for use (MUC) (See Table 4)	Health hazards				
			Route	Symptoms (See Table 5)	First aid (See Table 6)		Target organs (See Table 5)
Clothing: Goggles: Wash: Change: Remove: Provide:	Reason prob Any poss Immed contam N.R. Immed non-imperv contam Eyewash, quick drench	NIOSH/OSHA 10 ppm: SA*/PAPROVAG*/ CCROVAG*/GMFOVAG/SCBA* §: SCBAF:PD,PP/SAF:PD,PP:ASCBA Escape: GMFOVAG/SCBAE	Inh Ing Con	Irrit eyes, nose; weak; irrity; head; skin eruption; pulm edema	Eye: Skin: Breath: Swallow:	Irr immed Soap wash immed Resp support Medical attention immed	Eyes, resp sys, skin
Clothing: Goggles: Wash: Change: Remove: Provide:	Repeat Any poss Daily After work if any poss contam Prompt non-imperv contam Eyewash	NIOSH ¥: SCBAF:PD,PP/SAF:PD,PP:ASCBA Escape: HiEF/SCBAE	Inh	Resp symptoms, weak, ftg, weight loss, [carc]	Eye:	Irr immed	Lungs, skin, eyes, muc memb
Clothing: Goggles: Wash: Change: Remove:	Repeat Reason prob Prompt contam N.R. Prompt non-imperv contam	NIOSH/OSHA 50 mg/m³: DM* 100 mg/m³: DMXSQ^*/SA*/SCBA* 250 mg/m³: PAPRDM^*/SA:CF* 500 mg/m³: SCBAF/SAF/HiEF/ PAPRTHiE* 5000 mg/m³: SAF:PD,PP §: SCBAF:PD,PP/SAF:PD,PP:ASCBA Escape: HiEF/SCBAE	Inh Ing Con	Nasal irrit, conj, erythema	Eye: Skin: Breath: Swallow:	Irr immed Water flush prompt Fresh air Medical attention immed	Skin, eyes
Clothing: Goggles: Wash: Change: Remove:	N.R. N.R. N.R. N.R. N.R.	NIOSH/OSHA 10 ppm: SA*/SCBA* 25 ppm: SA:CF* 50 ppm: SCBAF/SAF 100 ppm: SAF:PD,PP §: SCBAF:PD,PP/SAF:PD,PP:ASCBA Escape: GMFS/SCBAE	Inh Con	Nasal irrit, epis; burns eyes, skin; in animals: pneu; kidney damage; epis	Eye: Skin: Breath:	Irr immed Water flush immed Resp support	Resp sys, kidneys, eyes, skin

Personal protection and sanitation (See Table 3)		Recommendations for respirator selection — maximum concentration for use (MUC) (See Table 4)	Health hazards				
			Route	Symptoms (See Table 5)	First aid (See Table 6)		Target organs (See Table 5)
Clothing: Goggles: Wash: Change: Remove: Provide:	Any poss Any poss Immed contam N.R. Immed contam Eyewash, quick drench	NIOSH/OSHA 2.5 ppm: SA:CFᴱ/PAPRSᶜᴱ 5 ppm: CCRFSᶜ/GMFSᶜ/SCBAF/SAF PAPRTSᶜᴱ 10 ppm: SAF:PD,PP §: SCBAF:PD,PP/SAF:PD,PP:ASCBA Escape: GMFS/SCBAE	Inh Ing Con	Dizz, head; lac, epis; cough, feeling of oppression, pulm edema, pneu; abdom pain, diarr; measle-like eruptions; severe burns eyes, skin	Eye: Skin: Breath: Swallow:	Irr immed Soap wash immed Resp support Medical attention immed	Resp sys, eyes, CNS
Clothing: Goggles: Wash: Change: Remove:	Repeat Reason prob Prompt contam N.R. Prompt non-imperv contam	NIOSH/OSHA 12.5 ppm: SA:CFᴱ/PAPROVᴱ 25 ppm: CCRFOV/GMFOV/SCBAF/SAF/ PAPRTOVᴱ 1000 ppm: SAF:PD,PP §: SCBAF:PD,PP/SAF:PD,PP:ASCBA Escape: GMFOV/SCBAE	Inh Abs Ing	Irrit eyes, resp sys; CNS depression; liver damage	Eye: Skin: Breath: Swallow:	Irr immed Soap wash prompt Resp support Medical attention immed	Skin, liver, kidneys, resp sys, CNS
Clothing: Goggles: Wash: Change: Remove:	Prevent skin freezing Reason prob Immed wet N.R. Immed wet (flamm)	NIOSH ¥: SCBAF:PD,PP/SAF:PD,PP:ASCBA Escape: GMFS/SCBAE	Inh Con	Irrit eyes, nose, throat; drow, li-head; frostbite; [carc]	Eye: Skin: Breath:	Irr immed Water flush immed Resp support	Eyes, resp sys, CNS
Clothing: Goggles: Wash: Change: Remove: Provide:	Repeat Reason prob N.R. N.R. Prompt non-imperv contam Eyewash	NIOSH/OSHA 1000 ppm: PAPROVᴱ/CCRFOV 3000 ppm: GMFOV/SA:CFᴱ/SCBAF/SAF §: SCBAF:PD,PP/SAF:PD,PP:ASCBA Escape: GMFOV/SCBAE	Inh Con	Irrit eyes, nose; head; dizz; vomit	Eye: Skin: Breath: Swallow:	Irr immed Water wash immed Fresh air Medical attention immed	CNS, lungs

Chemical name, structure/formula, CAS and RTECS Nos., and DOT ID and guide Nos.	Synonyms, trade names, and conversion factors	Exposure limits (TWA unless noted otherwise)	IDLH	Physical description	Chemical and physical properties		Incompatibilities and reactivities	Measurement method (See Table 1)
					MW, BP, SOL FI.P, IP, Sp.Gr, flammability	VP, FRZ UEL, LEL		
2-Butoxyethanol C₄H₉OCH₂CH₂OH 111-76-2 KJ8575000 2369 26	Butyl cellosolve, Butyl oxitol, Dowanol EB, Ektasolve EB, Ethylene glycol monobutyl ether, Jeffersol EB 1 ppm = 4.91 mg/m³	NIOSH/OSHA 25 ppm (120 mg/m³) [skin]	700 ppm	Colorless mobile liquid with a mild ether-like odor.	MW: 118.2 BP: 339°F Sol: Miscible FI.P: 143°F IP: ? Sp.Gr: 0.90 Class IIIA Combustible Liquid	VP: 0.8 mm FRZ: -107°F UEL(275°F): 12.7% LEL(200°F): 1.1%	Strong oxidizers, strong caustics	Char; Methanol/ CH₂Cl₂; GC/FID; III [#1403, Alcohols IV]
n-Butyl acetate CH₃COO[CH₂]₃CH₃ 123-86-4 AF7350000 1123 26	Butyl acetate, n-Butyl ester of acetic acid, Butyl ethanoate 1 ppm = 4.83 mg/m³	NIOSH/OSHA 150 ppm (710 mg/m³) ST 200 ppm (950 mg/m³)	10,000 ppm	Colorless liquid with a fruity odor.	MW: 116.2 BP: 258°F Sol: 1% FI.P: 72°F IP: 10.00 eV Sp.Gr: 0.88 Class IC Flammable Liquid	VP(77°F): 15 mm FRZ: -107°F UEL: 7.6% LEL: 1.7%	Nitrates; strong oxidizers, alkalis & acids	Char; CS₂; GC/FID; III [#1450, Esters I]
sec-Butyl acetate CH₃COOCH(CH₃)- CH₂CH₃ 105-46-4 AF7380000 1123 26	sec-Butyl ester of acetic acid, 1-Methylpropyl acetate 1 ppm = 4.83 mg/m³	NIOSH/OSHA 200 ppm (950 mg/m³)	10,000 ppm	Colorless liquid with a pleasant, fruity odor.	MW: 116.2 BP: 234°F Sol: 0.8% FI.P: 62°F IP: 9.91 eV Sp.Gr: 0.86 Class IB Flammable Liquid	VP(77°F): 24 mm FRZ: -100°F UEL: 9.8% LEL: 1.7%	Nitrates; strong oxidizers, alkalis & acids	Char; CS₂; GC/FID; III [#1450, Esters I]
tert-Butyl acetate CH₃COOC(CH₃)₃ 540-88-5 AF7400000 1123 26	tert-Butyl ester of acetic acid 1 ppm = 4.83 mg/m³	NIOSH/OSHA 200 ppm (950 mg/m³)	10,000 ppm	Colorless liquid with a fruity odor.	MW: 116.2 BP: 208°F Sol: Insoluble FI.P: 62-72°F IP: ? Sp.Gr: 0.87 Class IB Flammable Liquid	VP: ? FRZ: ? UEL: ? LEL: ?	Nitrates; strong oxidizers, alkalis & acids	Char; CS₂; GC/FID; III [#1450, Esters I]

Chemical name, structure/formula, CAS and RTECS Nos., and DOT ID and guide Nos.	Synonyms, trade names, and conversion factors	Exposure limits (TWA unless noted otherwise)	IDLH	Physical description	Chemical and physical properties		Incompatibilities and reactivities	Measurement method (See Table 1)
					MW, BP, SOL FI.P, IP, Sp.Gr, flammability	VP, FRZ UEL, LEL		
n-Butyl alcohol CH₃CH₂CH₂CH₂OH 71-36-3 EO1400000 1120 26	1-Butanol, n-Butanol, Butyl alcohol, 1-Hydroxybutane, n-Propyl carbinol 1 ppm = 3.08 mg/m³	NIOSH/OSHA C 50 ppm (150 mg/m³) [skin]	8000 ppm	Colorless liquid with a strong, characteristic, mildly alcoholic odor.	MW: 74.1 BP: 243°F Sol: 9% FI.P: 99°F IP: 10.04 eV Sp.Gr: 0.81 Class IC Flammable Liquid	VP: 6 mm FRZ: -129°F UEL: 11.2% LEL: 1.4%	Strong oxidizers	Char; 2-Propanol/ CS₂; GC/FID; III [#1401, Alcohols II]
sec-Butyl alcohol CH₃CHOHCH₂CH₃ 78-92-2 EO1750000 1120 26	2-Butanol, Butylene hydrate, 2-Hydroxybutane, Methyl ethyl carbinol 1 ppm = 3.08 mg/m³	NIOSH 100 ppm (305 mg/m³) ST 150 ppm (455 mg/m³) OSHA 100 ppm (305 mg/m³)	10,000 ppm	Colorless liquid with a strong, pleasant odor.	MW: 74.1 BP: 211°F Sol: 16% FI.P: 75°F IP: 10.10 eV Sp.Gr: 0.81 Class IC Flammable Liquid	VP(86°F): 24 mm FRZ: -175°F UEL(212°F): 9.8% LEL(212°F): 1.7%	Strong oxidizers	Char; 2-Propanol/ CS₂; GC/FID; III [#1401, Alcohols II]
tert-Butyl alcohol (CH₃)₃COH 75-65-0 EO1925000 1120 26	2-Methyl-2-propanol, Trimethyl carbinol 1 ppm = 3.08 mg/m³	NIOSH/OSHA 100 ppm (300 mg/m³) ST 150 ppm (450 mg/m³)	8000 ppm	Colorless solid or liquid with a camphor-like odor. [Note: Pure compound is a liquid above 77°F. Often used in aqueous solutions.]	MW: 74.1 BP: 180°F Sol: Miscible FI.P: 52°F IP: 9.70 eV Sp.Gr: 0.79 Class IB Flammable Liquid	VP(77°F): 42 mm FRZ: 77°F UEL: 8.0% LEL: 2.4%	Strong mineral acids, strong hydrochloric acid, oxidizers	Char; 2-Butanol/ CS₂; GC/FID; III [#1400, Alcohols I]
Butylamine CH₃CH₂CH₂CH₂NH₂ 109-73-9 EO2975000 1125 68	1-Aminobutane, n-Butylamine 1 ppm = 3.04 mg/m³	NIOSH/OSHA C 5 ppm (15 mg/m³) [skin]	2000 ppm	Colorless liquid with a fishy, ammonia-like odor.	MW: 73.2 BP: 172°F Sol: Miscible FI.P: 10°F IP: 8.71 eV Sp.Gr: 0.74 Class IA Flammable Liquid	VP: 82 mm FRZ: -58°F UEL: 9.8% LEL: 1.7%	Strong oxidizers, strong acids [Note: May corrode some metals in presence of water.]	Si gel; Methanol; GC/FID; II(4) [#S138]

Personal protection and sanitation (See Table 3)	Recommendations for respirator selection — maximum concentration for use (MUC) (See Table 4)	Route	Symptoms (See Table 5)	First aid (See Table 6)	Target organs (See Table 5)
Clothing: Repeat Goggles: Reason prob Wash: Immed contam Change: N.R. Remove: Immed non-imperv contam Provide: Quick drench	NIOSH/OSHA 250 ppm: SA*/SCBA* 625 ppm: PAPROV*/SA:CF* 700 ppm: GMFOV/CCRFOV/SCBAF/SAF §: SCBAF:PD,PP/SAF:PD,PP:ASCBA Escape: GMFOV/SCBAE	Inh Abs Ing Con	Irrit eyes, nose, throat; hemolysis, hemog	Eye: Irr immed Skin: Soap wash prompt Breath: Resp support Swallow: Medical attention immed	Liver, kidneys, lymphoid sys, skin, blood, eyes, resp sys
Clothing: Repeat Goggles: Reason prob Wash: Prompt wet Change: N.R. Remove: Immed wet (flamm)	NIOSH/OSHA 1000 ppm: PAPROV*/CCROV* 1500 ppm: SA*/SCBA* 3750 ppm: SA:CF* 7500 ppm: GMFOV/SCBAF/SAF 10,000 ppm: SAF:PD,PP §: SCBAF:PD,PP/SAF:PD,PP:ASCBA Escape: GMFOV/SCBAE	Inh Ing Con	Head; drow; dryness and irrit eyes, upper resp sys, skin	Eye: Irr immed Skin: Water flush prompt Breath: Resp support Swallow: Medical attention immed	Eyes, skin, resp sys
Clothing: Repeat Goggles: Reason prob Wash: Prompt wet Change: N.R. Remove: Immed wet (flamm)	NIOSH/OSHA 1000 ppm: PAPROV£/CCROV 5000 ppm: SA:CF£ 10,000 ppm: GMFOV/SCBAF/SAF §: SCBAF:PD,PP/SAF:PD,PP:ASCBA Escape: GMFOV/SCBAE	Inh Ing Con	Irrit eyes; head; drow; dryness upper resp sys, skin	Eye: Irr immed Skin: Water flush prompt Breath: Resp support Swallow: Medical attention immed	Eyes, skin, resp sys
Clothing: Repeat Goggles: Reason prob Wash: Prompt wet Change: N.R. Remove: Immed wet (flamm)	NIOSH/OSHA 1000 ppm: PAPROV£/CCRFOV 5000 ppm: SA:CF£ 10,000 ppm: GMFOV/SCBAF/SAF §: SCBAF:PD,PP/SAF:PD,PP:ASCBA Escape: GMFOV/SCBAE	Inh Ing Con	Itch, inflamm eyes; irrit upper resp tract; head; narco; derm	Eye: Irr immed Skin: Water flush prompt Breath: Resp support Swallow: Medical attention immed	Resp sys, eyes, skin

Personal protection and sanitation (See Table 3)	Recommendations for respirator selection — maximum concentration for use (MUC) (See Table 4)	Route	Symptoms (See Table 5)	First aid (See Table 6)	Target organs (See Table 5)
Clothing: Repeat Goggles: Reason prob Wash: Prompt wet Change: N.R. Remove: Immed wet (flamm)	NIOSH/OSHA 1000 ppm: PAPROV£/CCRFOV 1250 ppm: SA:CF£ 2500 ppm: GMFOV/SCBAF/SAF 8000 ppm: SAF:PD,PP §: SCBAF:PD,PP/SAF:PD,PP:ASCBA Escape: GMFOV/SCBAE	Inh Abs Ing Con	Irrit eyes, nose, throat; head, verti, drow; corneal inflamm, blurred vision, lac, photo; dry cracked skin	Eye: Irr immed Skin: Water flush prompt Breath: Resp support Swallow: Medical attention immed	Skin, eyes, resp sys
Clothing: Repeat Goggles: Reason prob Wash: Prompt wet Change: N.R. Remove: Immed wet (flamm)	NIOSH/OSHA 1000 ppm: PAPROV*/CCRFOV/SA*/SCBA* 2500 ppm: SA:CF* 5000 ppm: GMFOV/SCBAF/SAF 10,000 ppm: SAF:PD,PP §: SCBAF:PD,PP/SAF:PD,PP:ASCBA Escape: GMFOV/SCBAE	Inh Ing Con	Eye irrit, narco	Eye: Irr immed Skin: Water flush prompt Breath: Resp support Swallow: Medical attention immed	Eyes, CNS
Clothing: Repeat Goggles: Reason prob Wash: Prompt wet Change: N.R. Remove: Immed wet (flamm)	NIOSH/OSHA 1000 ppm: PAPROV£/CCRFOV 2500 ppm: SA:CF£ 5000 ppm: GMFOV/SCBAF/SAF 8000 ppm: SAF:PD,PP §: SCBAF:PD,PP/SAF:PD,PP:ASCBA Escape: GMFOV/SCBAE	Inh Ing Con	Drow; irrit skin, eyes	Eye: Irr immed Skin: Water flush prompt Breath: Resp support Swallow: Medical attention immed	Eyes, skin
Clothing: Any poss Goggles: Any poss Wash: Immed contam Change: N.R. Remove: Immed wet (flamm) Provide: Eyewash, quick drench	NIOSH/OSHA 50 ppm: SA*/SCBA*/CCRS* 125 ppm: SA:CF*/PAPRS* 250 ppm: CCRFS/GMFS/SCBAF/SAF/PAPRTS* 2000 ppm: SAF:PD,PP §: SCBAF:PD,PP/SAF:PD,PP:ASCBA Escape: GMFS/SCBAE	Inh Abs Ing Con	Irrit eyes, nose, throat; head; skin flush, burns	Eye: Irr immed Skin: Water flush immed Breath: Resp support Swallow: Medical attention immed	Resp sys, skin, eyes

Chemical name, structure/formula, CAS and RTECS Nos., and DOT ID and guide Nos.	Synonyms, trade names, and conversion factors	Exposure limits (TWA unless noted otherwise)	IDLH	Physical description	Chemical and physical properties		Incompatibilities and reactivities	Measurement method (See Table 1)
					MW, BP, SOL Fl.P, IP, Sp.Gr, flammability	VP, FRZ UEL, LEL		
tert-Butyl chromate (as CrO₃) ((CH₃)₃CO)₂CrO₂ 1189-85-1 GB2900000	di-tert-Butyl ester of chromic acid	NIOSH Ca See Appendix A 0.001 mg/m³ OSHA C 0.1 mg/m³ [skin]	Ca [30 mg/m³]	Liquid. [Note: Solidifies at 32-23°F.]	MW: 230.3 BP: ? Sol: ? Fl.P: ? IP: ?	VP: ? FRZ: 32-23°F UEL: ? LEL: ?	Reducing agents, moisture	None available
	1 ppm = 9.57 mg/m³							
n-Butyl glycidyl ether C₇H₁₄O₂ 2426-08-6 TX4200000	BGE; 1,2-Epoxy-3-butoxypropane	NIOSH C 5.6 ppm (30 mg/m³) [15-min] OSHA 25 ppm (135 mg/m³)	3500 ppm	Colorless liquid with an irritating odor.	MW: 130.2 BP: 327°F Sol: 2% Fl.P: 130°F IP: ?	VP(77°F): 3 mm FRZ: ? UEL: ? LEL: ?	Strong oxidizers, strong caustics	Char; CS₂; GC/FID; II(2) [#S81]
	1 ppm = 5.41 mg/m³				Sp.Gr(77°F): 0.91 Class II Combustible Liquid			
Butyl mercaptan CH₃CH₂CH₂CH₂SH 109-79-5 EK6300000	Butanethiol, 1-Butanethiol, n-Butanethiol, 1-Mercaptobutane	NIOSH C 0.5 ppm (1.8 mg/m³) [15-min] OSHA 0.5 ppm (1.5 mg/m³)	2500 ppm	Colorless liquid with a strong garlic-, cabbage-, or skunk-like odor.	MW: 90.2 BP: 209°F Sol: 0.06% Fl.P: 35°F IP: 9.15 eV	VP: 35 mm FRZ: -176°F UEL: ? LEL: ?	Strong oxidizers (such as dry bleaches), acids	Chrom-104; Acetone; GC/FID; II(4) [#S350]
2347 27	1 ppm = 3.75 mg/m³				Sp.Gr: 0.83 Class IB Flammable Liquid			
p-tert-Butyltoluene (CH₃)₃CC₆H₄CH₃ 98-51-1 XS8400000	1-Methyl-4-tert-butylbenzene	NIOSH/OSHA 10 ppm (60 mg/m³) ST 20 ppm (120 mg/m³)	1000 ppm	Colorless liquid with a distinct aromatic odor, somewhat like gasoline.	MW: 148.3 BP: 379°F Sol: Insoluble Fl.P: 155°F IP: 8.28 eV	VP(77°F): 0.7 mm FRZ: -62°F UEL: ? LEL: ?	Oxidizers	Char; CS₂; GC/FID; III [#1501, Aromatic Hydrocarbons]
2667 27	1 ppm = 6.16 mg/m³				Sp.Gr: 0.86 Class IIIA Combustible Liquid			

Chemical name, structure/formula, CAS and RTECS Nos., and DOT ID and guide Nos.	Synonyms, trade names, and conversion factors	Exposure limits (TWA unless noted otherwise)	IDLH	Physical description	Chemical and physical properties		Incompatibilities and reactivities	Measurement method (See Table 1)
					MW, BP, SOL Fl.P, IP, Sp.Gr, flammability	VP, FRZ UEL, LEL		
Cadmium dust (as Cd) Cd 7440-43-9 EU9800000 2570 53	Metal: none Synonyms of other compounds vary depending upon the specific compound.	NIOSH Ca See Appendix A Reduce exposure to lowest feasible concentration. OSHA 0.2 mg/m³ C 0.6 mg/m³	Ca [50 mg/m³]	Metal: Silver-white, blue-tinged lustrous, odorless solid.	MW: 112.4 BP: 1409°F Sol: Insoluble Fl.P: NA IP: NA	VP: 0 mm (approx) MLT: 610°F UEL: NA LEL: NA	Strong oxidizers; elemental sulfur, selenium & tellurium	Filter; Acid; AA; III [#7048]
					Sp.Gr(77°F): 8.65 Metal: Noncombustible Solid in bulk form, but will burn in powder form.			
Cadmium fume (as Cd) CdO/Cd 1306-19-0 (CdO) EV1930000 (CdO)	CdO: Cadmium monoxide, Cadmium oxide fume Cd: None	NIOSH Ca See Appendix A Reduce exposure to lowest feasible concentration. OSHA 0.1 mg/m³ C 0.3 mg/m³	Ca [9 mg/m³]	Odorless, yellow-brown, finely divided particulate dispersed in air.	MW: 128.4 BP: Decomposes Sol: Insoluble Fl.P: NA IP: NA [Note: See Cadmium dust for properties of Cd.]	VP: 0 mm (approx) MLT: 2599°F UEL: NA LEL: NA	Not applicable	Filter; Acid; AA; III [#7048]
					Sp.Gr: 8.15 (crystalline form)/6.95 (amorphous form) Noncombustible Solid			
Calcium arsenate (as As) Ca₃(AsO₄)₂ 7778-44-1 CG0830000 1573 53	Calcium salt (2:3) of arsenic acid, Cucumber dust, Tricalcium arsenate, Tricalcium ortho-arsenate	NIOSH Ca See Appendix A C 0.002 mg/m³ [15-min] OSHA [1910.1018] 0.010 mg/m³	Ca [100 mg/m³]	Colorless to white, odorless solid. [insecticide/ herbicide]	MW: 398.1 BP: Decomposes Sol(77°F): 0.01% Fl.P: NA IP: NA	VP: 0 mm (approx) MLT: ? UEL: NA LEL: NA	None reported [Note: Produces toxic fumes of arsenic when heated to decomposition.]	Filter; Acid; AA; III [#7900, Arsenic]
					Sp.Gr: 3.62 Noncombustible Solid			
Calcium oxide CaO 1305-78-8 EW3100000 1910 60	Burned lime, Burnt lime, Lime, Pebble lime, Quick lime, Unslaked lime	NIOSH 2 mg/m³ OSHA 5 mg/m³	Unknown	White or gray, odorless lumps or granular powder.	MW: 56.1 BP: 5162°F Sol: Reacts Fl.P: NA IP: NA	VP: 0 mm (approx) MLT: 4662°F UEL: NA LEL: NA	Water (liberates heat), fluorine [Note: Reacts with water to form calcium hydroxide.]	Filter; Acid; AA; III [#7020]
					Sp.Gr: 3.34 Noncombustible Solid			

Personal protection and sanitation (See Table 3)		Recommendations for respirator selection — maximum concentration for use (MUC) (See Table 4)	Health hazards			
			Route	Symptoms (See Table 5)	First aid (See Table 6)	Target organs (See Table 5)
Clothing:	Any poss	NIOSH	Inh	Lung, sinus cancer;	Eye: Irr immed	Resp sys, skin,
Goggles:	Any poss	¥: SCBAF:PD,PP/SAF:PD,PP:ASCBA	Abs	[carc]	Skin: Soap wash immed	eyes, CNS
Wash:	Immed contam/daily	Escape: GMFOVHiE/SCBAE	Ing		Breath: Resp support	
Change:	N.R.		Con		Swallow: Medical attention immed	
Remove:	Immed non-imperv contam					
Provide:	Eyewash, quick drench					
Clothing:	Any poss	NIOSH	Inh	Irrit eyes, nose; skin	Eye: Irr immed	Eyes, skin, resp
Goggles:	Reason prob	56 ppm: SA*/SCBA*/CCROV*	Ing	irrit, sens; narco	Skin: Soap wash immed	sys, CNS
Wash:	Prompt contam	140 ppm: SA:CF*/PAPROV*	Con		Breath: Resp support	
Change:	N.R.	280 ppm: SCBAF/SAF/CCRFOV/ GMFOV/PAPRTOV*			Swallow: Medical attention immed	
Remove:	Prompt non-imperv contam	3500 ppm: SAF:PD,PP §: SCBAF:PD,PP/SAF:PD,PP:ASCBA Escape: GMFOV/SCBAE				
Clothing:	Repeat	NIOSH/OSHA	Inh	In animals: narco, inco,	Eye: Irr immed	Resp sys;
Goggles:	Reason prob	5 ppm: SA/SCBA/CCROV	Ing	weak; cyan; pulm irrit;	Skin: Soap wash prompt	in animals: CNS,
Wash:	Prompt wet	12.5 ppm: SA:CF/PAPROV	Con	liver, kidney damage	Breath: Resp support	liver, kidneys
Change:	N.R.	25 ppm: SCBAF/SAF/CCRFOV/ PAPRTOV/GMFOV			Swallow: Medical attention immed	
Remove:	Immed wet (flamm)	500 ppm: SA:PD:PP* 1000 ppm: SAF:PD,PP §: SCBAF:PD,PP/SAF:PD,PP:ASCBA Escape: GMFOV/SCBAE				
Clothing:	Repeat	NIOSH/OSHA	Inh	Irrit eyes, skin; dry nose,	Eye: Irr immed	CVS, CNS, skin,
Goggles:	Reason prob	250 ppm: SA:CFᴱ/PAPROVᴱ	Ing	throat; head; low BP,	Skin: Water flush prompt	bone marrow,
Wash:	Prompt wet	500 ppm: CCRFOV/GMFOV/SCBAF/SAF	Con	tacar, abnormal	Breath: Resp support	eyes, upper resp
Change:	N.R.	1000 ppm: SAF:PD,PP		cardiovascular stress;	Swallow: Medical attention immed	sys
Remove:	Prompt non-imperv wet	§: SCBAF:PD,PP/SAF:PD,PP:ASCBA Escape: GMFOV/SCBAE		CNS depres; hemato depres		

Personal protection and sanitation (See Table 3)		Recommendations for respirator selection — maximum concentration for use (MUC) (See Table 4)	Health hazards			
			Route	Symptoms (See Table 5)	First aid (See Table 6)	Target organs (See Table 5)
Clothing:	N.R.	NIOSH	Inh	Pulm edema, dysp,	Eye: Irr immed	Resp sys,
Goggles:	Any poss (CdCl₂)	¥: SCBAF:PD,PP/SAF:PD,PP:ASCBA	Ing	cough, chest tight, subs	Skin: Soap wash	kidneys,
Wash:	Daily (reason prob)	Escape: HiEF/SCBAE		pain; head; chills, musc	Breath: Resp support	prostate, blood
Change:	After work if any poss contam			aches; nau, vomit, diarr; anos, emphy, prot, mild	Swallow: Medical attention immed	
Remove:	N.R.			anemia; [carc]		
Provide:	Eyewash (CdCl₂)					
Clothing:	N.R.	NIOSH	Inh	Pulm edema, dysp,	Breath: Resp support	Resp sys,
Goggles:	N.R.	¥: SCBAF:PD,PP/SAF:PD,PP:ASCBA		cough, tight chest, subs		kidneys, blood
Wash:	N.R.	Escape: HiEF/SCBAE		pain; head; chills, musc		
Change:	N.R.			aches; nau, vomit, diarr;		
Remove:	N.R.			emphy, prot, anos, mild anemia; [carc]		
Clothing:	Repeat	NIOSH	Inh	Weak; GI dist; peri neur;	Eye: Irr immed	Eyes, resp sys,
Goggles:	Any poss	¥: SCBAF:PD,PP/SAF:PD,PP:ASCBA	Abs	skin hyperpig, palmar	Skin: Soap wash prompt	liver, skin,
Wash:	Prompt contam/daily	Escape: HiEF/SCBAE	Ing	planter hyperkeratoses;	Breath: Resp support	lymphatics, CNS
Change:	After work if any poss contam		Con	derm; [carc]; in animals: liver damage	Swallow: Medical attention immed	
Remove:	Prompt non-imperv contam					
Clothing:	Reason prob	NIOSH	Inh	Irrit eyes, upper resp	Eye: Irr immed	Resp sys, skin,
Goggles:	Any poss	10 mg/m³: DM	Ing	tract; ulcer, perf	Skin: Water flush immed	eyes
Wash:	Prompt contam/daily	20 mg/m³: DMXSQ/SA/SCBA	Con	nasal septum;	Breath: Resp support	
Change:	After work if reason prob contam	50 mg/m³: PAPRHiE/SA:CF 100 mg/m³: HiEF/SCBAF/SAF/PAPRTHiE		pneu; derm	Swallow: Medical attention immed	
Remove:	Prompt non-imperv contam	250 mg/m³: SA:PD,PP §: SCBAF:PD,PP/SAF:PD,PP:ASCBA				
Provide:	Eyewash, quick drench	Escape: HiEF/SCBAE				

Chemical name, structure/formula, CAS and RTECS Nos., and DOT ID and guide Nos.	Synonyms, trade names, and conversion factors	Exposure limits (TWA unless noted otherwise)	IDLH	Physical description	Chemical and physical properties		Incompatibilities and reactivities	Measurement method (See Table 1)
					MW, BP, SOL FI.P, IP, Sp.Gr, flammability	VP, FRZ UEL, LEL		
Camphor (synthetic) $C_{10}H_{16}O$ 76-22-2 EX1225000 2717 32	2-Camphonone, Gum camphor, Laurel camphor, Synthetic camphor	NIOSH/OSHA 2 mg/m³	200 mg/m³	Colorless or white crystals with a penetrating, aromatic odor.	MW: 152.3 BP: 399°F Sol: Insoluble FI.P: 150°F IP: 8.76 eV Sp.Gr: 0.99 Combustible Solid	VP: 0.2 mm MLT: 345°F UEL: 3.5% LEL: 0.6%	Oxidizers (especially chromic anhydride)	Char; Methanol/ CS₂; GC/FID; III [#1301, Ketones II]
Carbaryl (Sevin®) $C_{10}H_7OOCNHCH_3$ 63-25-2 FC5950000 2757 55	alpha-Naphthyl N-methyl-carbamate, 1-Naphthyl N-Methyl-carbamate	NIOSH/OSHA 5 mg/m³	600 mg/m³	White or gray odorless solid. [pesticide]	MW: 201.2 BP: Decomposes Sol: 0.01% FL.P: NA IP: ? Sp.Gr: 1.23 Noncombustible Solid, but may be dissolved in flammable liquids.	VP(77°F): 0.00004 mm MLT: 293°F UEL: NA LEL: NA	Strong oxidizers, strongly alkaline pesticides	Filter; Reagent; Vis; III [#5006]
Carbon black C 1333-86-4 FF5800000	Acetylene black, Channel black, Furnace black, Lamp black, Thermal black	NIOSH/OSHA 3.5 mg/m³ NIOSH Ca See Appendix A 0.1 mg/m³ [Carbon black in presence of polycyclic aromatic hydrocarbons]	N.E. Ca	Black, odorless solid.	MW: 12.0 BP: Sublimes Sol: Insoluble FI.P: ? IP: NA Sp.Gr: 1.8-2.1 Combustible Solid that may contain flammable hydrocarbons.	VP: 0 mm (approx) MLT: Sublimes UEL: ? LEL: ?	Strong oxidizers such as chlorates, bromates & nitrates	Filter; none; Grav; III [#5000]
Carbon dioxide CO_2 124-38-9 FF6400000 1013 12 1845 21 (dry ice) 2187 21 (liquid)	Carbonic acid gas, Dry ice 1 ppm = 1.83 mg/m³	NIOSH 5,000 ppm (9000 mg/m³) ST 30,000 ppm (54,000 mg/m³) OSHA 10,000 ppm (18,000 mg/m³) ST 30,000 ppm (54,000 mg/m³)	50,000 ppm	Colorless, odorless gas. [Note: Normal constituent of air (about 3000 ppm). Solid form is utilized as dry ice.]	MW: 44.0 BP: Sublimes Sol(77°F): 0.2% FI.P: NA IP: 13.77 eV Noncombustible Gas	VP: >1 atm MLT: -109°F (Sublimes) UEL: NA LEL: NA	Dusts of various metals, such as magnesium, zirconium, titanium, aluminum, chromium & manganese are ignitable and explosive when suspended in carbon dioxide.	Bag; none; GC/FID; II(3) [#S249]

Chemical name, structure/formula, CAS and RTECS Nos., and DOT ID and guide Nos.	Synonyms, trade names, and conversion factors	Exposure limits (TWA unless noted otherwise)	IDLH	Physical description	Chemical and physical properties		Incompatibilities and reactivities	Measurement method (See Table 1)
					MW, BP, SOL FI.P, IP, Sp.Gr, flammability	VP, FRZ UEL, LEL		
Carbon disulfide CS_2 75-15-0 FF6650000 1131 28	Carbon bisulfide 1 ppm = 3.16 mg/m³	NIOSH 1 ppm (3 mg/m³) ST 10 ppm (30 mg/m³) [skin] OSHA 4 ppm (12 mg/m³) ST 12 ppm (36 mg/m³) [skin]	500 ppm	Colorless to faint-yellow liquid with a sweet ether-like odor. [Note: Reagent grades are foul smelling.]	MW: 76.1 BP: 116°F Sol: 0.3% FI.P: -22°F IP: 10.08 eV Sp.Gr: 1.26 Class IB Flammable Liquid	VP: 297 mm FRZ: -169°F UEL: 50.0% LEL: 1.3%	Strong oxidizers; chemically-active metals such as sodium, potassium & zinc; azides; rust; halogens; amines [Note: Vapors may be ignited by contact with ordinary light bulb.]	Char/ Dry tube; Toluene; GC/FPD; III [#1600]
Carbon monoxide CO 630-08-0 FG3500000 1016 18	Carbon oxide, Flue gas, Monoxide 1 ppm = 1.16 mg/m³	NIOSH/OSHA 35 ppm (40 mg/m³) C 200 ppm (229 mg/m³) [Note: On 9/5/89 OSHA stayed the start-up date for the Ceiling limit for the following steel industry operations — blast furnaces, vessel blowing at basic oxygen furnaces, and sinter plants.]	1500 ppm	Colorless, odorless gas. [Note: Shipped as a nonliquefied compressed gas.]	MW: 28.0 BP: -313°F Sol: 2% FI.P: NA (Gas) IP: 14.01 eV Flammable Gas	VP: >1 atm MLT: -337°F UEL: 74% LEL: 12.5%	Strong oxidizers, bromine trifluoride, chlorine trifluoride, lithium	Bag; none; ECA; II(4) [#S340]
Carbon tetrachloride CCl_4 56-23-5 FG4900000 1846 55	Carbon chloride, Carbon tet, Freon® 10, Halon® 104, Tetrachloromethane 1 ppm = 6.39 mg/m³	NIOSH Ca See Appendix A ST 2 ppm (12.6 mg/m³) [60-min] OSHA 2 ppm (12.6 mg/m³)	Ca [300 ppm] ACGIH A2 [skin]	Colorless liquid with a characteristic ether-like odor.	MW: 153.8 BP: 170°F Sol: 0.05% FI.P: NA IP: 11.47 eV Sp.Gr: 1.59 Noncombustible Liquid	VP: 91 mm FRZ: -9°F UEL: NA LEL: NA	Chemically-active metals such as sodium, potassium & magnesium; fluorine; aluminum [Note: Forms highly toxic phosgene gas when exposed to flames or welding arcs.]	Char; CS₂; GC/FID; III [#1003, Halogenated Hydrocarbons]
Chlordane $C_{10}H_6Cl_8$ 57-74-9 PB9800000 2762 28	1,2,4,5,6,7,8,8-Octachloro-3a,4,7,7a-tetrahydro-4,7-methanoindane 1 ppm = 17.04 mg/m³	NIOSH Ca See Appendix A 0.5 mg/m³ [skin] OSHA 0.5 mg/m³ [skin]	Ca [500 mg/m³]	Amber-colored, viscous liquid with a pungent, chlorine-like odor. [insecticide]	MW: 409.8 BP: Decomposes Sol: Insoluble FI.P: NA IP: ? Sp.Gr(77°F): 1.56 Noncombustible Liquid, but may be utilized in flammable solutions.	VP: 0.00001 mm FRZ: ? UEL: NA LEL: NA	Strong oxidizers, alkaline reagents	Filter/ Chrom-102; Toluene; GC/ECD; III [#5510]

Personal protection and sanitation (See Table 3)		Recommendations for respirator selection — maximum concentration for use (MUC) (See Table 4)	Health hazards			
			Route	Symptoms (See Table 5)	First aid (See Table 6)	Target organs (See Table 5)
Clothing: Goggles: Wash: Change: Remove:	Repeat Reason prob Prompt contam After work if reason prob contam Prompt non-imperv contam	NIOSH/OSHA 50 mg/m³: SA:CF^E/PAPROVDM^E 100 mg/m³: CCRFOVHiE/GMFOVHiE/ PAPRTOVHiE^E/SCBAF/SAF 200 mg/m³: SAF:PD,PP §: SCBAF:PD,PP/SAF:PD,PP:ASCBA Escape: GMFOVHiE/SCBAE	Inh Ing Con	Irrit eyes, skin, muc memb; nau, vomit, diarr; head, dizz, excitement, irrational, epileptiform convuls	Eye: Skin: Breath: Swallow:	Irr immed Soap wash immed Resp support Medical attention immed	CNS, eye, skin, resp sys
Clothing: Goggles: Wash: Change: Remove:	Repeat Reason prob Prompt contam After work if reason prob contam Prompt non-imperv contam	NIOSH/OSHA 50 mg/m³: SA*/SCBA* 125 mg/m³: SA:CF* 250 mg/m³: SCBAF/SAF 600 mg/m³: SAF:PD,PP §: SCBAF:PD,PP/SAF:PD,PP:ASCBA Escape: GMFOVHiE/SCBAE	Inh Abs Ing Con	Miosis, blurred vision, tear; nasal discharge, salv; sweat; abdom cramps, nau, vomit, diarr; tremor; cyan; convuls; skin irrit	Eye: Skin: Breath: Swallow:	Irr immed Soap wash prompt Resp support Medical attention immed	Resp sys, CNS, CVS, skin
Clothing: Goggles: Wash: Change: Remove:	N.R. Repeat Daily N.R. N.R.	NIOSH/OSHA 17.5 mg/m³: DM 35 mg/m³: DMXSQ/SA/SCBA 87.5 mg/m³: PAPRDM/SA:CF 175 mg/m³: HiEF/PAPRTHiE/SCBAF/SAF 1750 mg/m³: SA:PD,PP §: SCBAF:PD,PP/SAF:PD,PP:ASCBA Escape: HiEF/SCBAE	Inh Con	None expected; in presence of polycyclic aromatic hydrocarbons: [carc] In presence of polycyclic aromatic hydrocarbons: NIOSH ¥: SCBAF:PD,PP/SAF:PD,PP:ASCBA Escape: HiEF/SCBAE	Eye:	Irr prompt	None known
Clothing: Goggles: Wash: Change: Remove:	Prevent skin freezing N.R. N.R. N.R. N.R.	NIOSH 50,000 ppm: SA/SCBA §: SCBAF:PD,PP/SAF:PD,PP:ASCBA Escape: SCBAE	Inh Con	Head, dizz, restless, pares; dysp; sweat, mal; inc heart rate, pulse pressure; elevated BP; coma; asphy; convuls at high concentrations; frostbite (dry ice)	Eye: Skin: Breath:	Medical attention Medical attention for frostbite Resp support	Lungs, skin, CVS

Personal protection and sanitation (See Table 3)		Recommendations for respirator selection — maximum concentration for use (MUC) (See Table 4)	Health hazards			
			Route	Symptoms (See Table 5)	First aid (See Table 6)	Target organs (See Table 5)
Clothing: Goggles: Wash: Change: Remove:	Reason prob Reason prob Prompt contam N.R. Immed wet (flamm)	NIOSH 10 ppm: CCROV/SA/SCBA 25 ppm: SA:CF/PAPROV 50 ppm: CCRFOV/PAPRTOV/GMFOV/ SCBAF/SAF 500 ppm: SA:PD,PP §: SCBAF:PD,PP/SAF:PD,PP:ASCBA Escape: GMFOV/SCBAE	Inh Abs Ing Con	Dizz, head, poor sleep, ftg, ner, anor, low-wgt; psychosis; polyneur; Parkinson-like syndrome; ocular changes; coronary heart disease; gastritis; kidney, liver damage; eye, skin burns; derm	Eye: Skin: Breath: Swallow:	Irr immed Soap wash immed Resp support Medical attention immed	CNS, PNS, CVS, eyes, kidneys, liver, skin
Clothing: Goggles: Wash: Change: Remove:	Prevent skin freezing Reason prob N.R. N.R. Immed wet (flamm)	NIOSH/OSHA 350 ppm: SA/SCBA 875 ppm: SA:CF 1500 ppm: SCBAF/SAF/GMFS† §: SCBAF:PD,PP/SAF:PD,PP:ASCBA Escape: GMFS†/SCBAE	Inh Con (liq)	Head, tachypnea, nau, weak, dizz, conf, halu; cyan; depres S-T segment of electrocardiogram, angina, syncope	Breath:	Resp support	CVS, lungs, blood, CNS
Clothing: Goggles: Wash: Change: Remove:	Repeat Reason prob Prompt wet N.R. Prompt non-imperv contam	NIOSH ¥: SCBAF:PD,PP/SAF:PD,PP:ASCBA Escape: GMFOV/SCBAE	Inh Abs Ing Con	CNS depres; nau, vomit; liver, kidney damage; skin irrit; [carc]	Eye: Skin: Breath: Swallow:	Irr immed Soap wash immed Resp support Medical attention immed	CNS, eyes, lungs, liver, kidneys, skin
Clothing: Goggles: Wash: Change: Remove: Provide:	Any poss Reason prob Immed contam After work if any poss contam Immed non-imperv contam Quick drench	OSHA ¥: SCBAF:PD,PP/SAF:PD,PP:ASCBA Escape: GMFOVHiE/SCBAE	Inh Abs Ing Con	Blurred vision; conf; ataxia, delirium, cough; abdom pain, nau, vomit, diarr; irrity, tremor, convuls; anuria; in animals: lung, liver, kidney damage	Eye: Skin: Breath: Swallow:	Irr immed Soap wash immed Resp support Medical attention immed	CNS, eyes, lungs, liver, kidneys, skin

Chemical name, structure/formula, CAS and RTECS Nos., and DOT ID and guide Nos.	Synonyms, trade names, and conversion factors	Exposure limits (TWA unless noted otherwise)	IDLH	Physical description	Chemical and physical properties		Incompatibilities and reactivities	Measurement method (See Table 1)
					MW, BP, SOL Fl.P, IP, Sp.Gr, flammability	VP, FRZ UEL, LEL		
Chlorinated camphene $C_{10}H_{10}Cl_8$ 8001-35-2 XW5250000 2761 55	Chlorocamphene, Octachlorocamphene, Polychlorocamphene, Toxaphene	NIOSH Ca See Appendix A [skin] OSHA 0.5 mg/m³ ST 1 mg/m³ [skin]	Ca [200 mg/m³]	Amber, waxy solid with a mild chlorine- and camphor-like odor. [insecticide]	MW: 413.8 BP: Decomposes Sol: 0.0003% Fl.P: 275°F IP: ? Sp.Gr(77°F): 1.65 A Solid that does not burn, or burns with difficulty.	VP(77°F): 0.4 mm MLT: 149 to 194°F UEL: ? LEL: ?	Strong oxidizers [Note: Slightly corrosive to metals under moist conditions.]	Filter; Petroleum ether; GC/ECD; II(2) [#S67]
Chlorinated diphenyl oxide $(C_6H_2Cl_3)_2O$ 55720-99-5 KO0875000	Hexachlorodiphenyl oxide, Hexachlorophenyl ether	NIOSH/OSHA 0.5 mg/m³ 1 ppm = 15.67 mg/m³	Unknown	Light yellow, very viscous, waxy liquid.	MW: 376.9 BP: 446-500°F Sol: 0.1% Fl.P: NA IP: ? Sp.Gr: 1.60 Noncombustible Liquid	VP: 0.00006 mm FRZ: ? UEL: NA LEL: NA	Strong oxidizers	Filter; Isooctane; GC/EConD; III [#5025, Chlorinated diphenyl ether]
Chlorine Cl_2 7782-50-5 FO2100000 1017 20	Molecular chlorine	NIOSH/OSHA 0.5 ppm (1.5 mg/m³) ST 1 ppm (3 mg/m³) 1 ppm = 2.95 mg/m³	30 ppm	Greenish-yellow gas with a pungent, irritating odor. [Note: Shipped in steel cylinders.]	MW: 70.9 BP: -29°F Sol: 0.7% Fl.P: NA IP: 11.48 eV Noncombustible Gas, but a strong oxidizer.	VP: >1 atm FRZ: -150°F UEL: NA LEL: NA	Reacts explosively or forms explosive compounds with many common substances such as acetylene, ether, turpentine, ammonia, fuel gas, hydrogen & finely divided metals.	Bub; none; ISE; OSHA [#ID101]
Chlorine dioxide ClO_2 10049-04-4 FO3000000 1 ppm = 2.81 mg/m³	Chlorine oxide, Chlorine peroxide	NIOSH/OSHA 0.1 ppm (0.3 mg/m³) ST 0.3 ppm (0.9 mg/m³)	10 ppm	Yellow to red gas or a red-brown liquid (below 52°F) with an unpleasant odor similar to chlorine and nitric acid.	MW: 67.5 BP: 52°F Sol(77°F): 0.3% Fl.P: NA (Gas) ? (Liquid) IP: 10.36 eV Sp.Gr: 1.6 (Liquid at 32°F) Combustible Gas/Liquid	VP: >1 atm FRZ: -74°F UEL: ? LEL: ?	Organic materials, heat, phosphorus, potassium hydroxide, sulfur [Note: Unstable in light. A powerful oxidizer.]	None available

Chemical name, structure/formula, CAS and RTECS Nos., and DOT ID and guide Nos.	Synonyms, trade names, and conversion factors	Exposure limits (TWA unless noted otherwise)	IDLH	Physical description	Chemical and physical properties		Incompatibilities and reactivities	Measurement method (See Table 1)
					MW, BP, SOL Fl.P, IP, Sp.Gr, flammability	VP, FRZ UEL, LEL		
Chlorine trifluoride ClF_3 7790-91-2 FO2800000 1749 44	Chlorine fluoride, Chlorotrifluoride	NIOSH/OSHA C 0.1 ppm (0.4 mg/m³) 1 ppm = 3.85 mg/m³	20 ppm	Colorless gas or a greenish-yellow liquid (below 53°F) with a somewhat sweet, suffocating odor.	MW: 92.5 BP: 53°F Sol: Reacts Fl.P: NA IP: 13.00 eV Sp.Gr: 1.77 (Liquid at 53°F) Noncombustible Liquid, but contact with organic materials may result in SPONTANEOUS ignition.	VP: >1 atm FRZ: -105°F UEL: NA LEL: NA	Oxidizers, water, acids, combustible materials, sand, glass [Note: Reacts with water to form chlorine and hydrofluoric acid.]	None available
Chloroacetaldehyde $ClCH_2CHO$ 107-20-0 AB2450000 2232 55	Chloroacetaldehyde (40% aqueous solution), 2-Chloroacetaldehyde, 2-Chloroethanal	NIOSH/OSHA C 1 ppm (3 mg/m³) 1 ppm = 3.26 mg/m³	100 ppm	Colorless liquid with an acrid, penetrating odor.	MW: 78.5 BP: 186°F Sol: Miscible Fl.P: 190°F (40% soln) IP: 10.61 eV Sp.Gr: 1.19 Class IIIA Combustible Liquid	VP: 100 mm FRZ: -3°F (40% soln) UEL: ? LEL: ?	Oxidizers, acids	Si gel; Methanol; GC/ECD; II(5) [#S11]
alpha-Chloroacetophenone $C_6H_5COCH_2Cl$ 532-27-4 AM6300000 1697 58	2-Chloroacetophenone, Chloromethyl phenyl ketone, Mace®, Phenacyl chloride, Phenyl chloromethyl ketone, Tear gas	NIOSH/OSHA 0.3 mg/m³ (0.05 ppm)	100 mg/m³	Colorless to gray crystalline solid with a sharp, irritating odor.	MW: 154.6 BP: 472°F Sol: Insoluble Fl.P: 244°F IP: 9.44 eV Sp.Gr(59°F): 1.32 Combustible Solid	VP: 0.01 mm MLT: 134°F UEL: ? LEL: ?	Water, steam [Note: Slowly corrodes metals.]	Tenax GC (2); Thermal desorp; GC/FID; II(5) [P&CAM #291]
Chlorobenzene C_6H_5Cl 108-90-7 CZ0175000 1134 27	Benzene chloride, Chlorobenzol, MCB, Monochlorobenzene, Phenyl chloride	NIOSH See Appendix D OSHA 75 ppm (350 mg/m³) 1 ppm = 4.68 mg/m³	2400 ppm	Colorless liquid with an almond-like odor.	MW: 112.6 BP: 270°F Sol: 0.05% Fl.P: 85°F IP: 9.07 eV Sp.Gr: 1.11 Class IC Flammable Liquid	VP(77°F): 12 mm FRZ: -50°F UEL: 9.6% LEL: 1.3%	Strong oxidizers	Char; CS_2; GC/FID; III [#1003, Halogenated Hydrocarbons]

Personal protection and sanitation (See Table 3)		Recommendations for respirator selection — maximum concentration for use (MUC) (See Table 4)	Health hazards			
			Route	Symptoms (See Table 5)	First aid (See Table 6)	Target organs (See Table 5)
Clothing: Goggles: Wash: Change: Remove: Provide:	Any poss liq/Repeat sol Reason prob Prompt contam After work if reason prob contam sol Prompt non-imperv contam Quick drench	NIOSH ¥: SCBAF:PD,PP/SAF:PD,PP:ASCBA Escape: GMFOVHiE/SCBAE	Inh Abs Ing Con	Nau, conf, agitation, tremors, convuls, unconscious; dry, red skin; [carc]	Eye: Irr immed Skin: Soap wash prompt Breath: Resp support Swallow: Medical attention immed	CNS, skin
Clothing: Goggles: Wash: Change: Remove:	Repeat Reason prob Prompt contam After work if reason prob contam Prompt non-imperv contam	NIOSH/OSHA 5 mg/m³: SA/SCBA §: SCBAF:PD,PP/SAF:PD,PP:ASCBA Escape: GMFOVAGHiE/SCBAE	Inh Ing Con	Acne-form derm, liver damage	Eye: Irr immed Skin: Soap wash prompt Breath: Resp support Swallow: Medical attention immed	Skin, liver
Clothing: Goggles: Wash: Change: Remove: Provide:	Any poss Any poss Immed contam N.R. Immed non-imperv contam Eyewash, quick drench	NIOSH/OSHA 5 ppm: CCRS*/SA*/SCBA* 12.5 ppm: SA:CF*/PAPRS* 25 ppm: SCBAF/SAF/GMFS/PAPRTS*/ CCRFS 30 ppm: SAF:PD,PP §: SCBAF:PD,PP/SAF:PD,PP:ASCBA Escape: GMFS/SCBAE	Inh Con	Burning of eyes, nose, mouth; lac, rhin; cough, choking, subs pain; nau, vomit; head, dizz; syncope; pulm edema; pneu; hypox; derm; eye, skin burns	Eye: Irr immed Skin: Water flush immed Breath: Resp support	Resp sys
Clothing: Goggles: Wash: Change: Remove: Provide:	Any poss Any poss Immed contam N.R. Immed contam Eyewash, quick drench	NIOSH/OSHA 1 ppm: CCRS 2.5 ppm: SA:CFᵋ/PAPRSᵋ 5 ppm: CCRFS/GMFSᵋ/SCBAF/SAF 10 ppm: SAF:PD,PP §: SCBAF:PD,PP/SAF:PD,PP:ASCBA Escape GMFSᵋ/SCBAE	Inh Ing Con	Irrit eyes, nose, throat; cough, wheez, bron, pulm edema; chronic bron	Eye: Irr immed Skin: Soap wash immed Breath: Resp support Swallow: Medical attention immed	Resp sys, eyes

Personal protection and sanitation (See Table 3)		Recommendations for respirator selection — maximum concentration for use (MUC) (See Table 4)	Health hazards			
			Route	Symptoms (See Table 5)	First aid (See Table 6)	Target organs (See Table 5)
Clothing: Goggles: Wash: Change: Remove: Provide:	Any poss Any poss Immed contam N.R. Immed contam Eyewash, quick drench	NIOSH/OSHA 2.5 ppm: SA:CFᵋ 5 ppm: SCBAF/SAF 20 ppm: SAF:PD,PP §: SCBAF:PD,PP/SAF:PD,PP:ASCBA Escape: GMFS/SCBAE	Inh Ing Con	Eye, skin burns; in animals: lac, corneal ulcer; pulm edema	Eye: Irr immed Skin: Water flush immed Breath: Resp support Swallow: Medical attention immed	Skin, eyes
Clothing: Goggles: Wash: Change: Remove: Provide:	Any poss >0.1%/Repeat <0.1% Any poss Immed contam N.R. Immed non-imperv contam Eyewash, quick drench	NIOSH/OSHA 10 ppm: CCROV*/SA*/SCBA* 25 ppm: SA:CF*/PAPROV* 50 ppm: CCRFOV/GMFOV/SCBAF/SAF 100 ppm: SAF:PD,PP §: SCBAF:PD,PP/SAF:PD,PP:ASCBA Escape: GMFOV/SCBAE	Inh Ing Con	Irrit skin, eyes, muc memb; skin burns; eye damage; pulm edema; skin, resp sys sens; narco, coma	Eye: Irr immed Skin: Water flush immed Breath: Resp support Swallow: Medical attention immed	Eyes, skin, resp sys
Clothing: Goggles: Wash: Change: Remove: Provide:	Any poss Any poss Immed contam After work if reason prob contam Immed non-imperv contam Eyewash	NIOSH/OSHA 3 mg/m³: CCROVDM 7.5 mg/m³: SA:CFᵋ/PAPROVDMᵋ 15 mg/m³: CCRFOVHiE/SCBAF/SAF 100 mg/m³: SAF:PD,PP §: SCBAF:PD,PP/SAF:PD,PP:ASCBA Escape: GMFOVHiE/SCBAE	Inh Ing Con	Irrit eyes, skin, resp sys; pulm edema	Eye: Irr immed Skin: Soap wash immed Breath: Resp support Swallow: Medical attention immed	Eyes, skin, resp sys
Clothing: Goggles: Wash: Change: Remove:	Repeat Reason prob Prompt wet N.R. Immed wet (flamm)	OSHA 1000 ppm: PAPROVᵋ/CCRFOV 1875 ppm: SA:CFᵋ 2400 ppm: GMFOV/SCBAF/SAF §: SCBAF:PD,PP/SAF:PD,PP:ASCBA Escape: GMFOV/SCBAE	Inh Ing Con	Irrit skin, eyes, nose; drow, inco; in animals: liver, lung, kidney damage	Eye: Irr immed Skin: Soap wash prompt Breath: Resp support Swallow: Medical attention immed	Resp sys, eyes, skin, CNS, liver

Chemical name, structure/formula, CAS and RTECS Nos., and DOT ID and guide Nos.	Synonyms, trade names, and conversion factors	Exposure limits (TWA unless noted otherwise)	IDLH	Physical description	Chemical and physical properties		Incompatibilities and reactivities	Measurement method (See Table 1)
					MW, BP, SOL Fl.P, IP, Sp.Gr, flammability	**VP, FRZ UEL, LEL**		
o-Chlorobenzylidene malononitrile ClC$_6$H$_4$CH=C(CN)$_2$ 2698-41-1 OO3675000	2-Chlorobenzalmalononitrile, CS, OCBM	NIOSH/OSHA C 0.05 ppm (0.4 mg/m³) [skin]	2 mg/m³	White crystalline solid with a pepper-like odor.	MW: 188.6 BP: 590-599°F Sol: Insoluble Fl.P: ? IP: ? Combustible Solid	VP: 1 mm MLT: 199-203°F UEL: ? LEL: ?	Strong oxidizers	Filter/ Tenax GC Reagent; HPLC/UVD; II(5) [P&CAM #304]
Chlorobromomethane CH$_2$BrCl 74-97-5 PA5250000 1887 58	Bromochloromethane, CB, CBM, Fluorocarbon 1011, Halon® 1011, Methyl chlorobromide 1 ppm = 5.38 mg/m³	NIOSH/OSHA 200 ppm (1050 mg/m³)	5000 ppm	Colorless to pale yellow liquid with a chloroform-like odor.	MW: 129.4 BP: 155°F Sol: Insoluble Fl.P: NA IP: 10.77 eV Sp.Gr: 1.93 Noncombustible Liquid	VP(77°F): 160 mm FRZ: -124°F UEL: NA LEL: NA	Chemically-active metals such as calcium, powdered aluminum, zinc & magnesium	Char; CS$_2$; GC/FID; III [#1003, Halogenated Hydrocarbons]
Chlorodiphenyl (42% chlorine) C$_6$H$_4$ClC$_6$H$_3$Cl$_2$ (approx) 53469-21-9 TQ1356000 2315 31	Aroclor® 1242, PCB, Polychlorinated biphenyl 1 ppm = 10.72 mg/m³	NIOSH Ca See Appendix A 0.001 mg/m³ OSHA 1 mg/m³ [skin]	Ca [10 mg/m³]	Colorless to light colored, viscous liquid with a mild hydrocarbon odor.	MW: 258(approx) BP: 617-691°F Sol: Insoluble Fl.P: ? IP: ? Sp.Gr(77°F): 1.39 Nonflammable Liquid, but exposure in a fire results in the formation of a black soot containing PCBs, polychlorinated dibenzofurans, and chlorinated dibenzo-p-dioxins.	VP: 0.001 mm FRZ: -2°F UEL: ? LEL: ?	Strong oxidizers	Filter/ Florisil; Hexane; GC/ECD; III [#5503, PCBs]
Chlorodiphenyl (54% chlorine) C$_6$H$_3$Cl$_2$C$_6$H$_2$Cl$_3$ (approx) 11097-69-1 TQ1360000 2315 31	Aroclor® 1254, PCB, Polychlorinated biphenyl 1 ppm = 13.55 mg/m³	NIOSH Ca See Appendix A 0.001 mg/m³ OSHA 0.5 mg/m³ [skin]	Ca [5 mg/m³]	Colorless to pale yellow, viscous liquid or solid (below 50°F) with a mild, hydrocarbon odor.	MW: 326(approx) BP: 689-734°F Sol: Insoluble Fl.P: ? IP: ? Sp.Gr(77°F): 1.38 Nonflammable Liquid, but exposure in a fire results in the formation of a black soot containing PCBs, polychlorinated dibenzofurans, and chlorinated dibenzo-p-dioxins.	VP: 0.00006 mm FRZ: 50°F UEL: ? LEL: ?	Strong oxidizers	Filter/ Florisil; Hexane; GC/ECD; III [#5503, PCBs]

Chemical name, structure/formula, CAS and RTECS Nos., and DOT ID and guide Nos.	Synonyms, trade names, and conversion factors	Exposure limits (TWA unless noted otherwise)	IDLH	Physical description	Chemical and physical properties		Incompatibilities and reactivities	Measurement method (See Table 1)
					MW, BP, SOL Fl.P, IP, Sp.Gr, flammability	**VP, FRZ UEL, LEL**		
Chloroform CHCl$_3$ 67-66-3 FS9100000 1888 55	Methane trichloride, Trichloromethane 1 ppm = 4.96 mg/m³	NIOSH Ca See Appendix A ST 2 ppm (9.78 mg/m³) [60-min] OSHA 2 ppm (9.78 mg/m³)	Ca [1000 ppm] ACGIH A2	Colorless liquid with a pleasant odor.	MW: 119.4 BP: 143°F Sol(77°F): 0.5% Fl.P: NA IP: 11.42 eV Sp.Gr: 1.48 Noncombustible Liquid	VP: 160 mm FRZ: -82°F UEL: NA LEL: NA	Strong caustics; chemically-active metals such as aluminum or magnesium powder, sodium & potassium; strong oxidizers [Note: When heated to decomposition, forms phosgene gas.]	Char; CS$_2$; GC/FID; III [#1003, Halogenated Hydrocarbons]
bis-Chloromethyl ether (CH$_2$Cl)$_2$O 542-88-1 KN1575000 2249 55	BCME, bis-CME, Chloromethyl ether, Dichloromethyl ether, Dichlorodimethyl ether, Oxybis(chloromethane) 1 ppm = 4.78 mg/m³	NIOSH Ca See Appendix A OSHA [1910.1008] See Appendix B ACGIH A1, 0.001 ppm (0.005 mg/m³)	Ca	Colorless liquid with a suffocating odor.	MW: 115.0 BP: 223°F Sol: Reacts Fl.P: <66°F IP: ? Sp.Gr: 1.32 Class IA Flammable Liquid	VP(72°F): 30 mm FRZ: -43°F UEL: ? LEL: ?	Acids, water [Note: Reacts with water to form hydrochloric acid and formaldehyde.]	Imp; Reagent; GC/ECD OSHA [#10]
Chloromethyl methyl ether CH$_3$OCH$_2$Cl 107-30-2 KN6650000 1239 28	Chlorodimethyl ether, Chloromethoxymethane, CMME, Dimethylchloroether, Methylchloromethyl ether 1 ppm = 3.35 mg/m³	NIOSH Ca See Appendix A OSHA [1910.1006] See Appendix B ACGIH A2	Ca	Colorless liquid with an irritating odor.	MW: 80.5 BP: 138°F Sol: Reacts Fl.P(oc): 32°F IP: 10.25 eV Sp.Gr: 1.06 Class IA Flammable Liquid	VP: ? FRZ: -154°F UEL: ? LEL: ?	Water [Note: Reacts with water to form hydrochloric acid and formaldehyde.]	Imp; Hexane; GC/ECD; II(1) [P&CAM #220]
1-Chloro-1-nitropropane CH$_3$CH$_2$CHClNO$_2$ 600-25-9 TX5075000 1 ppm = 5.14 mg/m³	Korax, Lanstan®	NIOSH/OSHA 2 ppm (10 mg/m³)	2000 ppm	Colorless liquid with an unpleasant odor. [fungicide]	MW: 123.6 BP: 289°F Sol: 0.5% Fl.P(oc): 144°F IP: ? Sp.Gr: 1.21 Class IIIA Combustible Liquid	VP(77°F): 6 mm FRZ: ? UEL: ? LEL: ?	Strong oxidizers, acids	Chrom-108; Ethyl Acetate; GC/FID; II(5) [#S211]

Personal protection and sanitation (See Table 3)		Recommendations for respirator selection — maximum concentration for use (MUC) (See Table 4)	Health hazards				
			Route	Symptoms (See Table 5)	First aid (See Table 6)		Target organs (See Table 5)
Clothing: Goggles: Wash: Change: Remove:	Repeat Reason prob Prompt contam/daily After work any poss contam Prompt non-imperv contam	NIOSH/OSHA 2 mg/m³: SA:CFᵋ/SCBAF/SAF §: SCBAF:PD,PP/SAF:PD,PP:ASCBA Escape: GMFOVHiE/SCBAE	Inh Abs Ing Con	Pain, burn eyes, lac, conj; eryt eyelids, blepharospasm; irrit throat, cough, chest constric; head; eryt, vesic skin	Eye: Skin: Breath: Swallow:	Irr immed Soap wash immed Resp support Medical attention immed	Resp sys, skin, eyes
Clothing: Goggles: Wash: Change: Remove:	Repeat Reason prob Prompt wet N.R. Prompt non-imperv wet	NIOSH/OSHA 1000 ppm: PAPROVᵋ/CCRFOV 5000 ppm: SA:CFᵋ/GMFOV/SCBAF/SAF §: SCBAF:PD,PP/SAF:PD,PP:ASCBA Escape: GMFOV/SCBAE	Inh Ing Con	Disorientation, dizz; irrit eyes, throat, skin; pulm edema	Eye: Skin: Breath: Swallow:	Irr immed Soap wash prompt Resp support Medical attention immed	Skin, liver, kidneys, resp sys, CNS
Clothing: Goggles: Wash: Change: Remove:	Any poss Reason prob Prompt contam N.R. Prompt contam non-imperv	NIOSH ¥: SCBAF:PD,PP/SAF:PD,PP:ASCBA Escape: GMFOVHiE/SCBAE	Inh Abs Ing Con	Irrit eyes; chloracne; liver damage; [carc]	Eye: Skin: Breath: Swallow:	Irr immed Soap wash immed Resp support Medical attention immed	Skin, eyes, liver
Clothing: Goggles: Wash: Change: Remove:	Any poss Reason prob Immed contam N.R. Prompt non-imperv wet	NIOSH ¥: SCBAF:PD,PP/SAF:PD,PP:ASCBA Escape: GMFOVHiE/SCBAE	Inh Abs Ing Con	Irrit eyes, skin; acne-form derm; [carc]; in animals: liver damage	Eye: Skin: Breath: Swallow:	Irr immed Soap wash immed Resp support Medical attention immed	Skin, eyes, liver

Personal protection and sanitation (See Table 3)		Recommendations for respirator selection — maximum concentration for use (MUC) (See Table 4)	Health hazards				
			Route	Symptoms (See Table 5)	First aid (See Table 6)		Target organs (See Table 5)
Clothing: Goggles: Wash: Change: Remove:	Reason prob Reason prob Prompt wet N.R. Prompt non-imperv contam	NIOSH ¥: SCBAF:PD,PP/SAF:PD,PP:ASCBA Escape: GMFOV/SCBAE	Inh Ing Con	Dizz, mental dullness, nau, disorientation; head, ftg; anes; hepatomegaly; irrit eyes, skin; [carc]	Eye: Skin: Breath: Swallow:	Irr immed Soap wash prompt Resp support Medical attention immed	Liver, kidneys, heart, eyes, skin
Clothing: Goggles: Wash: Change: Remove: Provide:	Any poss Any poss Immed contam/daily After work if any poss contam Immed contam Eyewash, quick drench	NIOSH ¥: SCBAF:PD,PP/SAF:PD,PP:ASCBA Escape: GMFOV/SCBAE	Inh Abs Con	Irrit eyes, skin, muc memb of resp system; pulm congestion, edema; corneal damage, nec; reduced pulm function, cough, dysp, wheez; blood-stained sputum, bronchial secretions; [carc]	Eye: Skin: Breath: Swallow:	Irr immed (15 min) Soap wash immed Resp support Medical attention immed	Lungs, eyes, skin
Clothing: Goggles: Wash: Change: Remove: Provide:	Any poss Any poss Immed contam/daily After work if any poss contam Immed contam Eyewash, quick drench	NIOSH ¥: SCBAF:PD,PP/SAF:PD,PP:ASCBA Escape: GMFOV/SCBAE	Inh Abs Con	Irrit eyes, skin, muc memb; pulm edema, pulm congestion, pneu; burns, nec; cough, wheez, pulm congestion; blood stained-sputum; low-wgt; loss; bronchial secretions; [carc]	Eye: Skin: Breath: Swallow:	Irr immed (15 min) Soap wash immed Resp support Medical attention immed	Resp sys, skin, eyes, muc memb
Clothing: Goggles: Wash: Change: Remove:	Repeat Reason prob Prompt wet N.R. Prompt non-imperv wet	NIOSH/OSHA 50 ppm: SA:CF*/PAPROV* 100 ppm: CCRFOV/GMFOV/SCBAF/SAF 2000 ppm: SAF:PD,PP §: SCBAF:PD,PP/SAF:PD,PP:ASCBA Escape: GMFOV/SCBAE	Inh Ing Con	In animals: irrit eyes; pulm edema; liver, kidney, heart damage	Eye: Skin: Breath: Swallow:	Irr immed Soap wash Resp support Medical attention immed	In animals: resp sys, liver, kidneys, CVS

Chemical name, structure/formula, CAS and RTECS Nos., and DOT ID and guide Nos.	Synonyms, trade names, and conversion factors	Exposure limits (TWA unless noted otherwise)	IDLH	Physical description	Chemical and physical properties		Incompatibilities and reactivities	Measurement method (See Table 1)
					MW, BP, SOL Fl.P, IP, Sp.Gr, flammability	VP, FRZ UEL, LEL		
Chloropicrin CCl₃NO₂ 76-06-2 PB6300000 1580 56 2929 52 (flam. mixture) 1583 56 (mixture)	Nitrochloroform, Nitrotrichloromethane, Trichloronitromethane 1 ppm = 6.83 mg/m³	NIOSH/OSHA 0.1 ppm (0.7 mg/m³)	4 ppm	Colorless to faint yellow, oily liquid with an intensely irritating odor. [pesticide]	MW: 164.4 BP: 234°F Sol: 0.2% Fl.P: NA IP: ? Sp.Gr: 1.66 Noncombustible Liquid	VP: 20 mm FRZ: -93°F UEL: NA LEL: NA	Strong oxidizers [Note: With strong initiation, the heated material under confinement will detonate.]	None available
beta-Chloroprene CH₂=CCICH=CH₂ 126-99-8 EI9625000 1991 30 (inhibited)	2-Chloro-1,3-butadiene; Chlorobutadiene; Chloroprene 1 ppm = 3.68 mg/m³	NIOSH Ca See Appendix A C 1 ppm (3.6 mg/m³) [15-min] OSHA 10 ppm (35 mg/m³) [skin]	Ca [400 ppm]	Colorless liquid with a pungent, ether-like odor.	MW: 88.5 BP: 139°F Sol: Slight Fl.P: -4°F IP: 8.79 eV Sp.Gr: 0.96 Class IA Flammable Liquid	VP: 188 mm FRZ: -153°F UEL: 20.0% LEL: 4.0%	Peroxides & other oxidizers [Note: Polymerizes at room temperature unless inhibited with antioxidants.]	Char; CS₂; GC/FID; III [#1002]
Chromic acid and chromates (as CrO₃) H₂CrO₄ (acid) 7738-94-5 (acid) GB2450000 (acid) 1755 60 (acid soln) 1463 42 (acid, solid)	Chromic acid: none Synonyms of chromates (i.e., chromium(VI) compounds) vary depending upon the specific compound.	NIOSH Ca See Appendix A 0.001 mg/m³ See Appendix C OSHA C 0.1 mg/m³ See Appendix C ACGIH See Appendix C	Ca [30 mg/m³]	Appearance and odor vary depending upon the specific compound.	Properties vary depending upon the specific compound. Chromic acid is noncombustible, but will accelerate the burning of combustible materials.		Chromic acid: Combustible, organic, or other readily oxidizable materials (paper, wood, sulfur, aluminum, plastics, etc.); corrosive to metals	Filter; Reagent; Vis; III [#7600, Hexavalent Chromium]
Chromium metal (as Cr) Cr 7440-47-3 GB4200000	Chrome, Chromium	NIOSH 0.5 mg/m³ See Appendix C OSHA 1 mg/m³ See Appendix C ACGIH See Appendix C	N.E.	Blue-white to steel-gray, lustrous, brittle, hard solid.	MW: 52.0 BP: 4788°F Sol: Insoluble Fl.P: NA IP: NA Sp.Gr: 7.14 Noncombustible Solid in bulk form, but finely divided dust burns rapidly if heated in a flame.	VP: 0 mm (approx) MLT: 3452°F UEL: NA LEL: NA	Strong oxidizers such as hydrogen peroxide, alkalis	Filter; Acid; AA; III [#7024]

Chemical name, structure/formula, CAS and RTECS Nos., and DOT ID and guide Nos.	Synonyms, trade names, and conversion factors	Exposure limits (TWA unless noted otherwise)	IDLH	Physical description	Chemical and physical properties		Incompatibilities and reactivities	Measurement method (See Table 1)
					MW, BP, SOL Fl.P, IP, Sp.Gr, flammability	VP, FRZ UEL, LEL		
Chromium(II) and (III) compounds (as Cr)	Synonyms vary depending upon the specific Cr(II) or Cr(III) compound.	NIOSH/OSHA 0.5 mg/m³ See Appendix C ACGIH See Appendix C	N.E.	Appearance and odor vary depending upon the specific compound.	Properties vary depending upon the specific compound.		Water	Filter; Acid; AA; III [#7024]
Coal tar pitch volatiles (benzene-soluble fraction) 65996-93-2 GF8655000	Synonyms vary depending upon the specific compound (e.g., benzo(a)pyrene, phenanthrene, acridine, chrysene, anthracene & pyrene). [Note: NIOSH considers coal tar, coal tar pitch, and creosote to be coal tar products.]	NIOSH Ca See Appendix A 0.1 mg/m³ (cyclohexane-extractable fraction) See Appendix C OSHA 0.2 mg/m³	Ca [700 mg/m³] ACGIH A1	Black or dark-brown amorphous residue.	Properties vary depending upon the specific compound. Combustible Solid		Strong oxidizers	Filter; Benzene; Grav; III [#5023]
Cobalt metal, dust, and fume (as Co) Co 7440-48-4 (Metal) GF8750000 (Metal)	Metal: none Synonyms of other compounds vary depending upon the specific compound.	NIOSH/OSHA 0.05 mg/m³	20 mg/m³	Metal: Odorless, silver-gray to black solid.	MW: 58.9 BP: 5612°F Sol: Insoluble Fl.P: NA IP: NA Sp.Gr: 8.92 (Metal) Metal: Noncombustible Solid in bulk form, but finely divided dust will burn at high temperatures.	VP: 0 mm (approx) MLT: 2719°F UEL: NA LEL: NA	Strong oxidizers, ammonium nitrate	Filter; Acid; AA; III [#7027]
Copper dusts and mists (as Cu) Cu 7440-50-8 (Metal) GL5325000 (Metal)	Metal: none Synonyms of other compounds vary depending upon the specific compound.	NIOSH/OSHA 1 mg/m³	N.E.	Metal: Reddish, lustrous, malleable, odorless solid.	MW: 63.5 BP: 4703°F Sol: Insoluble Fl.P: NA IP: NA Sp.Gr: 8.94 (Metal) Metal: Noncombustible Solid in bulk form, but powdered form may ignite.	VP: 0 mm (approx) MLT: 1981°F UEL: NA LEL: NA	Oxidizers, alkalis, sodium azide, acetylene	Filter; Acid; AA; III [#7029]

Personal protection and sanitation (See Table 3)		Recommendations for respirator selection — maximum concentration for use (MUC) (See Table 4)	Health hazards			
			Route	Symptoms (See Table 5)	First aid (See Table 6)	Target organs (See Table 5)
Clothing: Goggles: Wash: Change: Remove: Provide:	Any poss Any poss Immed contam N.R. Immed non-imperv contam Eyewash, quick drench	NIOSH/OSHA 2.5 ppm: SA:CF^E/PAPROV^E 4 ppm: SCBAF/SAF/CCRFOV/GMFOV §: SCBAF:PD,PP/SAF:PD,PP:ASCBA Escape: GMFOV/SCBAE	Inh Ing Con	Eye irrit, lac; cough, pulm edema; nau, vomit; skin irrit	Eye: Irr immed Skin: Soap wash immed Breath: Resp support Swallow: Medical attention immed	Resp sys, skin, eyes
Clothing: Goggles: Wash: Change: Remove:	Any poss Reason prob Prompt contam N.R. Immed wet (flamm)	NIOSH ¥: SCBAF:PD,PP/SAF:PD,PP:ASCBA Escape: GMFOV/SCBAE	Inh Abs Ing Con	Irrit eyes, resp sys; ner, irrity; derm; alopecia; [carc]	Eye: Irr immed Skin: Soap wash immed Breath: Resp support Swallow: Medical attention immed	Resp sys, skin, eyes
Clothing: Goggles: Wash: Change: Remove: Provide:	Any poss Any poss Immed contam After work if reason prob contam Immed non-imperv contam Eyewash, quick drench	NIOSH ¥: SCBAF:PD,PP/SAF:PD,PP:ASCBA Escape: HiEF/SCBAE	Inh Ing Con	Resp sys irrit, nasal septum perf; liver, kidney damage; leucyt, leupen, monocy, eosin; eye inj, conj; skin ulcer, sens derm; [carc]	Eye: Irr immed Skin: Soap flush immed Breath: Resp support Swallow: Medical attention immed	Blood, resp sys, liver, kidneys, eyes, skin
Clothing: Goggles: Wash: Change: Remove:	Repeat Reason prob Prompt contam N.R. Prompt non-imperv contam	NIOSH 2.5 mg/m³: DM* 5 mg/m³: DMXSQ*/SA*/SCBA* 12.5 mg/m³: PAPRDM*/SA:CF* 25 mg/m³: HiEF/PAPRTHiE*/SCBAF/SAF 500 mg/m³: SAF:PD,PP §: SCBAF:PD,PP/SAF:PD,PP:ASCBA Escape: HiEF/SCBAE	Inh Ing	Histologic fibrosis of lungs	Eye: Irr immed Skin: Soap wash Breath: Resp support Swallow: Medical attention immed	Resp sys

Personal protection and sanitation (See Table 3)		Recommendations for respirator selection — maximum concentration for use (MUC) (See Table 4)	Health hazards			
			Route	Symptoms (See Table 5)	First aid (See Table 6)	Target organs (See Table 5)
Clothing: Goggles: Wash: Change: Remove:	Repeat Any poss Prompt contam N.R. Prompt non-imperv contam	NIOSH/OSHA 2.5 mg/m³: DM* 5 mg/m³: DMXSQ*/SA*/SCBA* 12.5 mg/m³: PAPRDM*/SA:CF* 25 mg/m³: HiEF/PAPRTHiE*/SCBAF/SAF 250 mg/m³: SAF:PD,PP §: SCBAF:PD,PP/SAF:PD,PP:ASCBA Escape: HiEF/SCBAE	Ing Con	Sens derm	Eye: Irr immed Skin: Water flush prompt Breath: Resp support Swallow: Medical attention immed	Skin
Clothing: Goggles: Wash: Change: Remove:	Reason prob Reason prob Daily (reason prob) After work if reason prob contam N.R.	NIOSH ¥: SCBAF:PD,PP/SAF:PD,PP:ASCBA Escape: GMFOVHiE/SCBAE	Inh Con	Derm, bron, [carc]	Eye: Irr immed Skin: Soap wash Breath: Resp support	Resp sys, bladder, kidneys, skin
Clothing: Goggles: Wash: Change: Remove:	Repeat N.R. Prompt contam After work if reason prob contam Prompt non-imperv contam	NIOSH/OSHA 0.25 mg/m³: DM^ 0.5 mg/m³: DMXSQ*^/DMFu*/SA*/SCBA* 1.25 mg/m³: PAPRDM*^/SA:CF*/PAPRDMFu* 2.5 mg/m³: HiEF/SCBAF/SAF 20 mg/m³: SAF:PD,PP §: SCBAF:PD,PP/SAF:PD,PP:ASCBA Escape: HiEF/SCBAE	Inh Con	Cough, dysp, decrease pulm func; low-wgt; derm; diffuse nodular fibrosis; resp hypersensitivity	Eye: Irr immed Skin: Soap wash Breath: Resp support Swallow: Medical attention immed	Resp sys, skin
Clothing: Goggles: Wash: Change: Remove:	Repeat Reason prob Prompt contam After work if reason prob contam Prompt contam non-imperv	NIOSH/OSHA 5 mg/m³: DM* 10 mg/m³: DMXSQ*^/SA*/SCBA* 25 mg/m³: PAPRDM*/SA:CF* 50 mg/m³: HiEF/SCBAF/SAF/PAPRTHiE* 2000 mg/m³: SAF:PD,PP §: SCBAF:PD,PP/SAF:PD,PP:ASCBA Escape: HiEF/SCBAE	Inh Ing Con	Irrit nasal muc memb, pharynx; nasal perforation; eye irrit; metallic taste; derm; in animals: lung, liver, kidney damage; anemia	Eye: Irr immed Skin: Soap wash prompt Breath: Resp support Swallow: Medical attention immed	Resp sys, skin, liver, kidneys, inc risk with Wilson's disease

Chemical name, structure/formula, CAS and RTECS Nos., and DOT ID and guide Nos.	Synonyms, trade names, and conversion factors	Exposure limits (TWA unless noted otherwise)	IDLH	Physical description	Chemical and physical properties		Incompatibilities and reactivities	Measurement method (See Table 1)
					MW, BP, SOL Fl.P, IP, Sp.Gr, flammability	VP, FRZ UEL, LEL		
Copper fume (as Cu) CuO/Cu 7440-50-8 (Cu) 1317-38-0 (CuO) GL5350000 (Cu)	CuO: Black copper oxide, Copper monoxide, Copper(II) oxide, Cupric oxide Cu: none	NIOSH/OSHA 0.1 mg/m³	N.E.	Finely divided black particulate dispersed in air. [Note: Exposure may occur in copper & brass plants and during the welding of copper alloys.]	MW: 79.5 BP: Decomposes Sol: Insoluble Fl.P: NA IP: NA [Note: See Copper Dusts & Mists for properties of Cu.] Sp.Gr: 6.4 (CuO) CuO: Noncombustible Solid	VP: 0 mm (approx) MLT: 1879°F (Decomposes) UEL: NA LEL: NA	Acetylene	Filter; Acid; AA; III [#7029]
Cotton dust (raw) GN2275000	None	NIOSH 0.200 mg/m³ See Appendix C OSHA [Z-1-A & 1910.1043] 1 mg/m³ (waste processing during waste recycling [sorting, blending, cleaning & willowing] & garretting) 0.200 mg/m³ (textile yarn manufacturing) 0.750 mg/m³ (textile slashing & weaving) 0.500 mg/m³ (all other operations) See Appendix C	N.E.	Colorless, odorless solid.	MW: ? BP: Decomposes Sol: Insoluble Fl.P: ? IP: NA Combustible Solid	VP: 0 mm (approx) MLT: Decomposes UEL: ? LEL: ?	Strong oxidizers	Vertical elut; none; Grav; OSHA [1910.1043]
Crag® herbicide C₆H₃Cl₂OCH₂CH₂-OSO₃Na 136-78-7 KK4900000	Crag® herbicide No. 1; Sesone; Sodium-2-(2,4-dichloro-phenoxy) ethyl sulfate	NIOSH/OSHA 10 mg/m³ (total) 5 mg/m³ (resp)	5000 mg/m³	Colorless to white crystalline, odorless solid. [herbicide]	MW: 309.1 BP: Decomposes Sol(77°F): 26% Fl.P: NA IP: ? Sp.Gr: 1.70 Noncombustible Solid	VP: 0.1 mm MLT: 473°F (Decomposes) UEL: NA LEL: NA	Strong oxidizers, acids	Filter; Water; Vis; II(5) [#S356]
Cresol (all isomers) CH₃C₆H₄OH 1319-77-3 GO5950000 2076 55	ortho-, meta-, or para-Cresol; Cresylic acid; 2-, 3-, or 4-Methyl phenol 1 ppm = 4.50 mg/m³	NIOSH 2.3 ppm (10 mg/m³) OSHA 5 ppm (22 mg/m³) [skin]	250 ppm	Colorless, yellow, brown, or pinkish, oily liquids or solids with a sweet, tarry odor.	MW: 108.2 BP: 376-397°F Sol: 2% Fl.P: 178°F(o) 187°F(m) 187°F(p) IP: 8.93 eV(o) 8.98 eV(m) 8.97 eV(p) Sp.Gr(77°F): 1.03 Class IIIA Combustible Liquids	VP(100°F): 1 mm FRZ: 52-95°F UEL: ? LEL(300°F): 1.4%(o) 1.1%(m) 1.1%(p)	Strong oxidizers, acids	Si gel; Acetone; GC/FID; III [#2001]

Chemical name, structure/formula, CAS and RTECS Nos., and DOT ID and guide Nos.	Synonyms, trade names, and conversion factors	Exposure limits (TWA unless noted otherwise)	IDLH	Physical description	Chemical and physical properties		Incompatibilities and reactivities	Measurement method (See Table 1)
					MW, BP, SOL Fl.P, IP, Sp.Gr, flammability	VP, FRZ UEL, LEL		
Crotonaldehyde CH₃CH=CHCHO 123-73-9 (trans-isomer) GP9625000 (trans-isomer) 1143 28 (inhibited)	2-Butenal, beta-Methylacrolein, Propylene aldehyde 1 ppm = 2.91 mg/m³	NIOSH/OSHA 2 ppm (6 mg/m³)	400 ppm	Water-white liquid with a suffocating odor. [Note: Turns pale-yellow on contact with air.]	MW: 70.1 BP: 219°F Sol: 18% Fl.P: 45°F IP: 9.73 eV Sp.Gr: 0.87 Class IA Flammable Liquid	VP: 19 mm FRZ: -101°F UEL: 15.5% LEL: 2.1%	Caustics, ammonia, strong oxidizers, nitric acid, amines [Note: Polymerization may occur at elevated temperatures, such as in fire conditions.]	Bub; none; PLR; II(5) [P&CAM #285]
Cumene C₆H₅CH(CH₃)₂ 98-82-8 GR8575000 1918 28	Cumol, Isopropyl benzene, 2-Phenyl propane 1 ppm = 5.00 mg/m³	NIOSH/OSHA 50 ppm (245 mg/m³) [skin]	8000 ppm	Colorless liquid with a sharp, penetrating, aromatic odor.	MW: 120.2 BP: 306°F Sol: Insoluble Fl.P: 96°F IP: 8.75 eV Sp.Gr: 0.86 Class IC Flammable Liquid	VP(77°F): 5 mm FRZ: -141°F UEL: 6.5% LEL: 0.9%	Oxidizers	Char; CS₂; GC/FID; III [#1501, Aromatic Hydro-carbons]
Cyanides (as CN) 1: KCN 2: NaCN 1: 151-50-8/TS8750000 2: 143-33-9/VZ7530000 1680 55 (KCN soln) 1689 55 (NaCN)	1: Potassium cyanide, Potassium salt of hydrocyanic acid 2: Sodium cyanide, Sodium salt of hydrocyanic acid Synonyms of other compounds vary depending upon the specific compound.	NIOSH C 5 mg/m³ (4.7 ppm) [10-min] OSHA 5 mg/m³	50 mg/m³	KCN & NaCN are white granular or crystalline solids with a faint almond-like odor.	MW: 65.1/49.0 BP: ?/2725°F Sol(77°F): 72/58% Fl.P: NA/NA IP: NA/NA Sp.Gr: 1.55 (KCN)/1.60 (NaCN) Noncombustible Solids, but contact with acids releases highly flammable hydrogen cyanide.	VP: 0/0 mm (approx) MLT: 1173/1047°F UEL: NA/NA LEL: NA/NA	Strong oxidizers, such as acids, acid salts, chlorates & nitrates	Filter/Bub; KOH; ISE; III [#7904]
Cyclohexane C₆H₁₂ 110-82-7 GU6300000 1145 26	Benzene hexahydride, Hexahydrobenzene, Hexamethylene, Hexanaphthene 1 ppm = 3.50 mg/m³	NIOSH/OSHA 300 ppm (1050 mg/m³)	10,000 ppm	Colorless liquid with a sweet, chloroform-like odor. [Note: A solid below 44°F.]	MW: 84.2 BP: 177°F Sol: Insoluble Fl.P: 0°F IP: 9.88 eV Sp.Gr: 0.78 Class IA Flammable Liquid	VP(77°F): 98 mm FRZ: 44°F UEL: 8% LEL: 1.3%	Oxidizers	Char; CS₂; GC/FID; III [#1500, Hydro-carbons]

Personal protection and sanitation (See Table 3)		Recommendations for respirator selection — maximum concentration for use (MUC) (See Table 4)	Health hazards				
			Route	Symptoms (See Table 5)	First aid (See Table 6)		Target organs (See Table 5)
Clothing: Goggles: Wash: Change: Remove:	N.R. N.R. N.R. N.R. N.R.	NIOSH/OSHA 1 mg/m³: DMFu/SA/SCBA 2.5 mg/m³: PAPRDMFu/SA:CF 5 mg/m³: HiEF/PAPRTHiE/SCBAF/SAF/ SAT:CF 200 mg/m³: SAF:PD,PP §: SCBAF:PD,PP/SAF:PD,PP:ASCBA Escape: HiEF/SCBAE	Inh Con	Metal fume fever: chills, musc ache, nau, fever, dry throat, cough, weak, lass; irrit eyes, upper resp tract; metallic or sweet taste; discoloration skin, hair	Breath	Resp support	Resp sys, skin, eyes, inc risk with Wilson's disease
Clothing: Goggles: Wash: Change: Remove:	N.R. N.R. N.R. N.R. N.R.	NIOSH 1 mg/m³: D 2 mg/m³: DXSQ/SA/HiE/SCBA 5 mg/m³: PAPRD/SA:CF 10 mg/m³: HiEF/PAPRTHiE/SCBAF/SAF/ SAT:CF 20 mg/m³: SA:PD,PP §: SCBAF:PD,PP/SAF:PD,PP:ASCBA Escape: HiEF/SCBAE	Inh	Tight chest, cough, wheez, dysp; decrease FEV; bron; mal; cough, fever, chills, upper resp symptoms after initial exposure	Eye: Breath:	Irr immed Fresh air	Resp sys, CVS
Clothing: Goggles: Wash: Change: Remove:	Repeat Reason prob Prompt contam After work if reason prob contam Prompt non-imperv contam	NIOSH/OSHA 50 mg/m³: DM 100 mg/m³: DMXSQ/SA/SCBA 250 mg/m³: PAPRDM/SA:CF 500 mg/m³: HiEF/SCBAF/SAF/ PAPRTHiE*/SAT:CF* 5000 mg/m³: SAF:PD,PP §: SCBAF:PD,PP/SAF:PD,PP:ASCBA Escape: HiEF/SCBAE	Inh Ing Con	None known	Eye: Skin: Breath: Swallow:	Irr immed Water wash prompt Resp support Medical attention immed	None known
Clothing: Goggles: Wash: Change: Remove: Provide:	Any poss Any poss Immed contam After work if any poss contam Immed non-imperv contam Eyewash, quick drench	NIOSH 23 ppm: CCROVDM/SA/SCBA 57.5 ppm: SA:CF/PAPROVDM 115 ppm: CCRFOVHiE/SCBAF/SAF/ GMFOVHiE/PAPRTHiE*/ SAT:CF* 250 ppm: SAF:PD,PP §: SCBAF:PD,PP/SAF:PD,PP:ASCBA Escape: GMFOVHiE/SCBAE	Inh Abs Ing Con	CNS effects; conf, depres, resp fail; dysp, irreg rapid resp, weak pulse; skin, eye burns; derm; lung, liver, kidney damage	Eye: Skin: Breath: Swallow:	Irr immed Soap wash immed Resp support Medical attention immed	CNS, resp sys, liver, kidneys, skin, eyes

Personal protection and sanitation (See Table 3)		Recommendations for respirator selection — maximum concentration for use (MUC) (See Table 4)	Health hazards				
			Route	Symptoms (See Table 5)	First aid (See Table 6)		Target organs (See Table 5)
Clothing: Goggles: Wash: Change: Remove: Provide:	Reason prob Any poss Immed contam N.R. Immed wet (flamm) Eyewash, quick drench	NIOSH/OSHA 20 ppm: CCROV*/SA*/SCBA* 50 ppm: SA:CF*/PAPROV* 100 ppm: CCRFOV/GMFOV/SCBAF/SAF 400 ppm: SAF:PD,PP §: SCBAF:PD,PP/SAF:PD,PP:ASCBA Escape: GMFOV/SCBAE	Inh Ing Con	Irrit eyes, resp sys; in animals: dysp, pulm edema, irrit skin	Eye: Skin: Breath: Swallow:	Irr immed Water flush immed Resp support Medical attention immed	Resp sys, eyes, skin
Clothing: Goggles: Wash: Change: Remove:	Repeat Reason prob Prompt wet N.R. Prompt non-imperv wet	NIOSH/OSHA 500 ppm: CCROV*/SA*/SCBA* 1000 ppm: PAPROV*/CCRFOV 1250 ppm: SA:CF* 2500 ppm: GMFOV/SCBAF/SAF 8000 ppm: SAF:PD,PP §: SCBAF:PD,PP/SAF:PD,PP:ASCBA Escape: GMFOV/SCBAE	Inh Abs Ing Con	Irrit eyes, muc memb; head; derm; narco, coma	Eye: Skin: Breath: Swallow:	Irr immed Water flush prompt Resp support Medical attention immed	Eyes, upper resp sys, skin, CNS
Clothing: Goggles: Wash: Change: Remove: Provide:	Any poss Any poss Immed contam After work if any poss contam Immed non-imperv contam Eyewash, quick drench	NIOSH/OSHA 50 mg/m³: SA/SCBA §: SCBAF:PD,PP/SAF:PD,PP:ASCBA Escape: GMFSHiE/SCBAE	Inh Abs Ing Con	Asphy and death can occur; weak, head, conf; nau, vomit; incr rate resp; slow gasping resp; irrit eyes, skin	Eye: Skin: Breath: Swallow:	Irr immed Soap wash immed Resp support Medical attention immed	CVS, CNS, liver, kidneys, skin
Clothing: Goggles: Wash: Change: Remove:	Repeat Reason prob Prompt wet N.R. Immed wet (flamm)	NIOSH/OSHA 1000 ppm: PAPROVᵋ/CCRFOV 7500 ppm: SA:CFᵋ 10,000 ppm: GMFOV/SCBAF/SAF §: SCBAF:PD,PP/SAF:PD,PP:ASCBA Escape: GMFOV/SCBAE	Inh Ing Con	Irrit eyes, resp sys; drow; derm; narco, coma	Eye: Skin: Breath: Swallow:	Irr immed Water flush prompt Resp support Medical attention immed	Eyes, resp sys, skin, CNS

Chemical name, structure/formula, CAS and RTECS Nos., and DOT ID and guide Nos.	Synonyms, trade names, and conversion factors	Exposure limits (TWA unless noted otherwise)	IDLH	Physical description	Chemical and physical properties		Incompatibilities and reactivities	Measurement method (See Table 1)
					MW, BP, SOL Fl.P, IP, Sp.Gr, flammability	VP, FRZ UEL, LEL		
Cyclohexanol C₆H₁₁OH 108-93-0 GV7875000	Anol, Cyclohexyl alcohol, Hexahydrophenol, Hexalin, Hydralin, Hydroxycyclohexane	NIOSH/OSHA 50 ppm (200 mg/m³) [skin]	3500 ppm	Sticky solid or colorless to light-yellow liquid (above 77°F) with a camphor-like odor.	MW: 100.2 BP: 322°F Sol: 4% Fl.P: 154°F IP: 10.00 eV	VP: 1 mm MLT: 77°F UEL: ? LEL: ?	Strong oxidizers	Char; 2-Propanol/ CS₂; GC/FID; III [#1402, Alcohols III]
	1 ppm = 4.17 mg/m³				Sp.Gr: 0.96 Class IIIA Combustible Liquid			
Cyclohexanone C₆H₁₀O 108-94-1 GW1050000	Cyclohexyl ketone, Pimelic ketone	NIOSH/OSHA 25 ppm (100 mg/m³) [skin]	5000 ppm	Water-white to pale-yellow liquid with a peppermint- or acetone-like odor.	MW: 98.2 BP: 312°F Sol(50°F): 15% Fl.P: 146°F IP: 9.14 eV	VP(77°F): 5 mm FRZ: -49°F UEL: 9.4% LEL(212°F): 1.1%	Oxidizers, nitric acid	Char; CS₂; GC/FID; III [#1300, Ketones I]
1915 26	1 ppm = 4.08 mg/m³				Sp.Gr: 0.95 Class IIIA Combustible Liquid			
Cyclohexene C₆H₁₀ 110-83-8 GW2500000	Benzene tetrahydride, Tetrahydrobenzene	NIOSH/OSHA 300 ppm (1015 mg/m³)	10,000 ppm	Colorless liquid with a sweet odor.	MW: 82.2 BP: 181°F Sol: Insoluble Fl.P: 11°F IP: 8.95 eV	VP(100°F): 160 mm FRZ: -154°F UEL ? LEL ?	Strong oxidizers [Note: Forms explosive peroxides with oxygen upon storage.]	Char; CS₂; GC/FID; III [#1500, Hydro-carbons]
2256 29	1 ppm = 3.42 mg/m³				Sp.Gr: 0.81 Class IA Flammable Liquid			
Cyclopentadiene C₅H₆ 542-92-7 GY1000000	1,3-Cyclopentadiene	NIOSH/OSHA 75 ppm (200 mg/m³)	2000 ppm	Colorless liquid with an irritating, terpene-like odor.	MW: 66.1 BP: 107°F Sol: Insoluble Fl.P(oc): 77°F IP: 8.56 eV	VP: ? FRZ: -121°F UEL: ? LEL: ?	Strong oxidizers [Note: Polymerizes to dicyclopentadiene upon standing.]	Chrom-104*; Ethyl acetate; GC/FID; III [#2523]
	1 ppm = 2.75 mg/m³				Sp.Gr: 0.80 Class IB Flammable Liquid			

Chemical name, structure/formula, CAS and RTECS Nos., and DOT ID and guide Nos.	Synonyms, trade names, and conversion factors	Exposure limits (TWA unless noted otherwise)	IDLH	Physical description	Chemical and physical properties		Incompatibilities and reactivities	Measurement method (See Table 1)
					MW, BP, SOL Fl.P, IP, Sp.Gr, flammability	VP, FRZ UEL, LEL		
2,4-D Cl₂C₆H₃OCH₂COOH 94-75-7 AG6825000	2,4-Dichlorophenoxyacetic acid; Dichlorophenoxyacetic acid	NIOSH/OSHA 10 mg/m³	500 mg/m³	White to yellow crystalline, odorless powder. [herbicide]	MW: 221.0 BP: Decomposes Sol: 0.05% Fl.P: NA IP: ?	VP: Low MLT: 280°F UEL: NA LEL: NA	Strong oxidizers	Filter; Methanol; HPLC/UVD; III [#5001]
2765 55					Sp.Gr(86°F): 1.57 Noncombustible Solid, but may be dissolved in flammable liquids.			
DDT (C₆H₄Cl)₂CHCCl₃ 50-29-3 KJ3325000	1,1,1-Trichloro-2,2-bis(p-chlorophenyl)ethane; p,p'-DDT; Dichlorodiphenyltrichloro-ethane	NIOSH Ca See Appendix A 0.5 mg/m³ OSHA 1 mg/m³ [skin]	Ca [N.E.]	Colorless crystals or off-white powder with a slight aromatic odor. [pesticide]	MW: 354.5 BP: 230°F (Decomposes) Sol: Insoluble Fl.P: 162-171°F IP: ?	VP: Low MLT: 227°F UEL: ? LEL: ?	Strong oxidizers, alkalis	Filter; Isooctane; GC/ECD; II(3) [#S274]
2761 55					Sp.Gr: 0.99 Combustible Solid			
Decaborane B₁₀H₁₄ 17702-41-9 HD1400000	Boron hydride, Decaboron tetradecahydride	NIOSH/OSHA 0.3 mg/m³ (0.05 ppm) ST 0.9 mg/m³ (0.15 ppm) [skin]	100 mg/m³	Colorless to white crystalline solid with an intense, bitter, chocolate-like odor.	MW: 122.2 BP: 415°F Sol: Slight Fl.P: 176°F IP: 9.88 eV	VP(77°F): 0.05 mm MLT: 211°F UEL: ? LEL: ?	Oxidizers, water, halogenated materials (especially carbon tetrachloride) [Note: May sponta-neously ignite on exposure to air. Decomposes slowly in hot water.]	None available
1868 34					Sp.Gr(77°F): 0.94 Combustible Solid			
Demeton C₈H₁₉O₃PS₂ 8065-48-3 TF3150000	O-O-Diethyl-(O and S)-2-(ethylthio)ethyl phosphoro-thioate mixture; Systox®	NIOSH/OSHA 0.1 mg/m³ [skin]	20 mg/m³	Amber, oily liquid with a sulfur-like odor. [insecticide]	MW: 258.3 BP: Decomposes Sol: 0.01% Fl.P: 113°F IP: ?	VP: 0.0003 mm FRZ: <-13°F UEL: ? LEL: ?	Strong oxidizers, alkalis, water	Filter/ XAD-2; Toluene; GC/FPD; III [#5514]
	1 ppm = 10.74 mg/m³				Sp.Gr: 1.12 Class II Combustible Liquid			

Personal protection and sanitation (See Table 3)		Recommendations for respirator selection — maximum concentration for use (MUC) (See Table 4)	Health hazards			
			Route	Symptoms (See Table 5)	First aid (See Table 6)	Target organs (See Table 5)
Clothing:	Repeat	NIOSH/OSHA	Inh	Irrit eyes, nose, throat,	Eye: Irr immed	Eyes, resp sys,
Goggles:	Reason prob	500 ppm: CCROV*/SA*/SCBA*	Abs	skin; narco	Skin: Water wash prompt	skin
Wash:	Prompt contam	1000 ppm: PAPROV*/CCRFOV	Ing		Breath: Resp support	
Change:	N.R.	1250 ppm: SA:CF*	Con		Swallow: Medical attention	
Remove:	Prompt non-imperv contam	2500 ppm: GMFOV/SCBAF/SAF			immed	
		3500 ppm: SAF:PD,PP				
		§: SCBAF:PD,PP/SAF:PD,PP:ASCBA				
		Escape: GMFOV/SCBAE				
Clothing:	Repeat	NIOSH/OSHA	Inh	Irrit eyes, muc memb;	Eye: Irr immed	Resp sys, eyes,
Goggles:	Reason prob	625 ppm: SA:CFᵋ/PAPROVᵋ	Abs	head; narco, coma; derm	Skin: Water flush prompt	skin, CNS
Wash:	Prompt wet	1000 ppm: CCRFOV	Ing		Breath: Resp support	
Change:	N.R.	1250 ppm: GMFOV/SCBAF/SAF	Con		Swallow: Medical attention	
Remove:	Immed wet (flamm)	5000 ppm: SAF:PD,PP			immed	
		§: SCBAF:PD,PP/SAF:PD,PP:ASCBA				
		Escape: GMFOV/SCBAE				
Clothing:	Repeat	NIOSH/OSHA	Inh	Irrit eyes, resp sys,	Eye: Irr immed	Skin, eyes, resp
Goggles:	Reason prob	1000 ppm: PAPROVᵋ/CCRFOV	Ing	skin; drow	Skin: Soap wash prompt	sys
Wash:	Prompt wet	7500 ppm: SA:CFᵋ	Con		Breath: Resp support	
Change:	N.R.	10,000 ppm: GMFOV/SCBAF/SAF			Swallow: Medical attention	
Remove:	Immed wet (flamm)	§: SCBAF:PD,PP/SAF:PD,PP:ASCBA			immed	
		Escape: GMFOV/SCBAE				
Clothing:	Repeat	NIOSH/OSHA	Inh	Irrit eyes, nose	Eye: Irr immed	Eyes, resp sys
Goggles:	Reason prob	750 ppm: CCROV/SA/SCBA	Ing		Skin: Soap wash prompt	
Wash:	Prompt wet	1000 ppm: PAPROV/CCRFOV/PAPRTOV	Con		Breath: Resp support	
Change:	N.R.	1875 ppm: SA:CF			Swallow: Medical attention	
Remove:	Immed wet (flamm)	2000 ppm: GMFOV/SCBAF/SAF			immed	
		§: SCBAF:PD,PP/SAF:PD,PP:ASCBA				
		Escape: GMFOV/SCBAE				

Personal protection and sanitation (See Table 3)		Recommendations for respirator selection — maximum concentration for use (MUC) (See Table 4)	Health hazards			
			Route	Symptoms (See Table 5)	First aid (See Table 6)	Target organs (See Table 5)
Clothing:	Repeat	NIOSH/OSHA	Inh	Weak, stupor,	Eye: Irr immed	Skin, CNS
Goggles:	Reason prob	100 mg/m³: CCROVDMFu/SA/SCBA	Abs	hyporeflexia, musc twitch;	Skin: Soap wash prompt	
Wash:	Prompt contam	250 mg/m³: PAPROVDMFu/SA:CF	Ing	convuls; derm;	Breath: Resp support	
Change:	After work if reason prob contam	500 mg/m³: CCRFOVHiE/GMFOVHiE/	Con	in animals: liver, kidney	Swallow: Medical attention	
Remove:	Prompt non-imperv contam	PAPRTOVHiE/SCBAF/SAF/ SAT:CF		damage	immed	
		§: SCBAF:PD,PP/SAF:PD,PP:ASCBA				
		Escape: GMFOVHiE/SCBAE				
Clothing:	Repeat	NIOSH	Inh	Pares tongue, lips, face;	Eye: Irr immed	CNS, kidneys,
Goggles:	Reason prob	¥: SCBAF:PD,PP/SAF:PD,PP:ASCBA	Abs	tremor; appre, dizz, conf,	Skin: Soap wash prompt	liver, skin, PNS
Wash:	Prompt contam	Escape: GMFOVHiE/SCBAE	Ing	mal, head, ftg; convuls;	Breath: Resp support	
Change:	After work if reason prob contam		Con	paresis hands; vomit; irrit	Swallow: Medical attention	
Remove:	Prompt non-imperv contam			eyes, skin; [carc]	immed	
Clothing:	Reason prob	NIOSH/OSHA	Inh	Dizz, head, nau, li-head,	Eye: Irr immed	CNS
Goggles:	Reason prob	3 mg/m³: SA/SCBA	Abs	drow; inco, local musc	Skin: Soap wash immed	
Wash:	Immed contam/daily	7.5 mg/m³: SA:CF	Ing	spasm, tremor, convuls;	Breath: Resp support	
Change:	After work if reason prob contam	15 mg/m³: SCBAF/SAF/SAT:CF	Con	ftg;	Swallow: Medical attention	
Remove:	Immed non-imperv contam	100 mg/m³: SAF:PD,PP		in animals: dysp; weak;	immed	
Provide:	Eyewash, quick drench	§: SCBAF:PD,PP/SAF:PD,PP:ASCBA		liver, kidney damage		
		Escape: GMFOVHiE/SCBAE				
Clothing:	Any poss	NIOSH/OSHA	Inh	Miosis, ache eyes, rhin,	Eye: Irr immed	Resp sys, CVS,
Goggles:	Any poss	1 mg/m³: SA/SCBA	Abs	head; tight chest, wheez,	Skin: Soap wash immed	CNS, skin, eyes,
Wash:	Immed contam	2.5 mg/m³: SA:CF	Ing	lar spasm, salv, cyan; anor,	Breath: Resp support	blood chol
Change:	N.R.	5 mg/m³: SCBAF/SAF/SAT:CF	Con	nau, vomit, abdom	Swallow: Medical attention	
Remove:	Immed non-imperv contam	20 mg/m³: SA:PD,PP		cramp, diarr; local sweat;	immed	
Provide:	Eyewash, quick drench	§: SCBAF:PD,PP/SAF:PD,PP:ASCBA		musc fasc, weak, para; gidd,		
		Escape: GMFOVHiE/SCBAE		conf, ataxia; convuls, coma;		
				low BP; card irregularities;		
				irrit eyes, skin		

Chemical name, structure/formula, CAS and RTECS Nos., and DOT ID and guide Nos.	Synonyms, trade names, and conversion factors	Exposure limits (TWA unless noted otherwise)	IDLH	Physical description	Chemical and physical properties		Incompatibilities and reactivities	Measurement method (See Table 1)
					MW, BP, SOL FI.P, IP, Sp.Gr, flammability	VP, FRZ UEL, LEL		
Diacetone alcohol CH₃COCH₂C(CH₃)₂OH 123-42-2 SA9100000 1148 26	Diacetone, 4-Hydroxy-4-methyl-2-pentanone, 2-Methyl-2-pentanol-4-one 1 ppm = 4.83 mg/m³	NIOSH/OSHA 50 ppm (240 mg/m³)	2100 ppm	Colorless liquid with a faint, minty odor.	MW: 116.2 BP: 334°F Sol: Miscible FI.P: 125°F IP: ? Sp.Gr: 0.94 Class II Combustible Liquid	VP: 1 mm FRZ: -47°F UEL: 6.9% LEL: 1.8%	Strong oxidizers, strong alkalis	Char; 2-Propanol/ CS₂; GC/FID; III [#1402, Alcohols III]
Diazomethane CH₂N₂ 334-88-3 PA7000000 1 ppm = 1.75 mg/m³	Azimethylene, Azomethylene, Diazirine, Diazomethane	NIOSH/OSHA 0.2 ppm (0.4 mg/m³)	2 ppm	Yellow gas with a musty odor. [Note: Shipped as a liquefied compressed gas.]	MW: 42.1 BP: -9°F Sol: Reacts FI.P: NA (Gas) IP: 9.00 eV Flammable Gas [EXPLOSIVE!]	VP: >1 atm FRZ: -229°F UEL: ? LEL: ?	Alkali metals, water, drying agents such as calcium arsenate [Note: May explode violently on heating, exposure to sunlight, or contact with rough edges such as ground glass.]	XAD-2*; CS₂; GC/FID; III [#2515]
Diborane B₂H₆ 19287-45-7 HQ9275000 1911 18	Boroethane, Boron hydride, Diboron hexahydride 1 ppm = 1.15 mg/m³	NIOSH/OSHA 0.1 ppm (0.1 mg/m³)	40 ppm	Colorless gas with a repulsive, sweet odor. [Note: Usually shipped in pressurized cylinders.]	MW: 27.7 BP: -135°F Sol: Reacts FI.P: NA (Gas) IP: 11.38 eV Flammable Gas	VP: >1 atm FRZ: -265°F UEL: 88% LEL: 0.8%	Water, halogenated compounds, aluminum, lithium, oxidized surfaces, acids [Note: Will ignite spontaneously in moist air at room temperature. Reacts with water to form hydrogen & boric acid.]	Filter/ Char*; H₂O₂; PES; III [#6006]
1,2-Dibromo-3-chloropropane CH₂BrCHBrCH₂Cl 96-12-8 TX8750000 2872 58	1-Chloro-2,3-dibromopropane; DBCP; Dibromochloropropane 1 ppm = 9.83 mg/m³	NIOSH Ca See Appendix A OSHA [1910.1044] 0.001 ppm	Ca	Dense yellow or amber liquid with a pungent odor at high concentrations. [Note: A solid below 43°F.] [pesticide]	MW: 236.4 BP: 384°F Sol(70°F): 0.1% FI.P(oc): 170°F IP: ? Sp.Gr: 2.05 Class IIIA Combustible Liquid	VP: 0.8 mm FRZ: 43°F UEL: ? LEL: ?	Chemically-active metals such as aluminum, magnesium & tin alloys [Note: Corrosive to metals.]	None available

Chemical name, structure/formula, CAS and RTECS Nos., and DOT ID and guide Nos.	Synonyms, trade names, and conversion factors	Exposure limits (TWA unless noted otherwise)	IDLH	Physical description	Chemical and physical properties		Incompatibilities and reactivities	Measurement method (See Table 1)
					MW, BP, SOL FI.P, IP, Sp.Gr, flammability	VP, FRZ UEL, LEL		
Dibutyl phosphate (C₄H₉O)₂(OH)PO 107-66-4 TB9605000 1 ppm = 8.74 mg/m³	Dibutyl acid o-phosphate, di-n-Butyl hydrogen phosphate, Dibutyl phosphoric acid	NIOSH/OSHA 1 ppm (5 mg/m³) ST 2 ppm (10 mg/m³)	125 ppm	Pale amber, odorless liquid.	MW: 210.2 BP: 212°F (Decomposes) Sol: Insoluble FI.P: ? IP: ? Sp.Gr: 1.06 Combustible Liquid	VP: 1 mm (approx) FRZ: ? UEL: ? LEL: ?	Strong oxidizers	Filter; CH₃CN; GC/FPD; III [#5017]
Dibutylphthalate C₆H₄(COOC₄H₉)₂ 84-74-2 TI0875000 1 ppm = 11.57 mg/m³	DBP; Dibutyl 1,2-benzene-dicarboxylate; Di-n-butyl phthalate	NIOSH/OSHA 5 mg/m³	9300 mg/m³	Colorless to faint yellow, oily liquid with a slight, aromatic odor.	MW: 278.3 BP: 644°F Sol(77°F): 0.5% FI.P: 315°F IP: ? Sp.Gr: 1.05 Class IIIB Combustible Liquid	VP: <0.01 mm FRZ: -31°F UEL: ? LEL(456°F): 0.5%	Nitrates; strong oxidizers, alkalis & acids; liquid chlorine	Filter; CS₂; GC/FID; III [#5020]
o-Dichlorobenzene C₆H₄Cl₂ 95-50-1 CZ4500000 1591 58	o-DCB; 1,2-Dichlorobenzene; ortho-Dichlorobenzene; o-Dichlorobenzol 1 ppm = 6.11 mg/m³	NIOSH/OSHA C 50 ppm (300 mg/m³)	1000 ppm	Colorless to pale-yellow liquid with a pleasant, aromatic odor. [herbicide]	MW: 147.0 BP: 357°F Sol: 0.01% FI.P: 151°F IP: 9.06 eV Sp.Gr: 1.30 Class IIIA Combustible Liquid	VP: 1 mm FRZ: 1°F UEL: 9.2% LEL: 2.2%	Strong oxidizers, aluminum, acids, acid fumes, chlorides	Char; CS₂; GC/FID; III [#1003, Halogenated Hydrocarbons]
p-Dichlorobenzene C₆H₄Cl₂ 106-46-7 CZ4550000 1592 58	p-DCB; 1,4-Dichlorobenzene; para-Dichlorobenzene; Dichlorocide	NIOSH Ca See Appendix A OSHA 75 ppm (450 mg/m³) ST 110 ppm (675 mg/m³)	Ca [1000 ppm]	Colorless or white crystalline solid with a mothball-like odor. [insecticide]	MW: 147.0 BP: 345°F Sol: 0.008% FI.P: 150°F IP: 8.98 eV Sp.Gr: 1.25 Combustible Solid, but may take some effort to ignite.	VP(77°F): 0.4 mm MLT: 128°F UEL: ? LEL: ?	Oxidizers	Char; CS₂; GC/FID; III [#1003, Halogenated Hydrocarbons]

Personal protection and sanitation (See Table 3)		Recommendations for respirator selection — maximum concentration for use (MUC) (See Table 4)	Health hazards			
			Route	Symptoms (See Table 5)	First aid (See Table 6)	Target organs (See Table 5)
Clothing: Goggles: Wash: Change: Remove:	Repeat Reason prob Prompt wet N.R. Prompt non-imperv wet	NIOSH/OSHA 1000 ppm: CCRFOV/PAPROV℥ 1250 ppm: SA:CF℥ 2100 ppm: GMFOV/SCBAF/SAF §: SCBAF:PD,PP/SAF:PD,PP:ASCBA Escape: GMFOV/SCBAE	Inh Ing Con	Irrit eyes, nose, throat, skin; corneal tissue damage; narco	Eye: Skin: Breath: Swallow:	Irr immed Water flush prompt Resp support Medical attention immed	Eyes, skin, resp sys
Clothing: Goggles: Wash: Change: Remove: Provide:	Prevent contam or freezing Any poss Prompt contam N.R. Immed wet (flamm) Eyewash	NIOSH/OSHA 2 ppm: SA*/SCBA* §: SCBAF:PD,PP/SAF:PD,PP:ASCBA Escape: GMFOV/SCBAE	Inh Ing Con	Cough, short breath; head; flush skin, fever; chest pain, pulm edema, pneuitis; irrit eyes; asthma; ftg	Eye: Skin: Breath: Swallow:	Irr immed Water flush immed Resp support Medical attention immed	Resp sys, eyes, skin
Clothing: Goggles: Wash: Change: Remove:	N.R. N.R. N.R. N.R. N.R.	NIOSH/OSHA 1 ppm: SA/SCBA 2.5 ppm: SA:CF 5 ppm: SCBAF/SAF/SAT:CF 40 ppm: SA:PD,PP §: SCBAF:PD,PP/SAF:PD,PP:ASCBA Escape: GMFS/SCBAE	Inh Con	Tight chest, precordial pain, short breath, non-productive cough, nau; head, li-head, verti, chills, fever, ftg, weak, tremor, musc fasc; in animals: liver, kidney damage; pulm edema; hemorr	Eye: Breath:	Irr immed Resp support	Resp sys, CNS
Clothing: Goggles: Wash: Change: Remove:	Any poss Any poss Immed contam/daily After work if any poss contam Immed contam	NIOSH ¥: SCBAF:PD,PP/SAF:PD,PP:ASCBA Escape: GMFOVHiE/SCBAE	Inh Abs Con	Drow; nau, vomit; irrit eyes, nose, throat, skin; pulm edema, [carc]	Eye: Skin: Breath: Swallow:	Irr immed Soap wash immed Resp support Medical attention immed	CNS, skin, liver, kidneys, spleen, reproductive sys, digestive sys

Personal protection and sanitation (See Table 3)		Recommendations for respirator selection — maximum concentration for use (MUC) (See Table 4)	Health hazards			
			Route	Symptoms (See Table 5)	First aid (See Table 6)	Target organs (See Table 5)
Clothing: Goggles: Wash: Change: Remove: Provide:	Reason prob Reason prob Prompt contam N.R. Prompt non-imperv contam Quick drench	NIOSH/OSHA 10 ppm: SA/SCBA 25 ppm: SA:CF 50 ppm: SCBAF/SAF/SAT:CF 125 ppm: SA:PD,PP §: SCBAF:PD,PP/SAF:PD,PP:ASCBA Escape: GMFOVHiE/SCBAE	Inh Ing Con	Irrit resp sys, skin; head	Eye: Skin: Breath: Swallow:	Irr immed Soap wash prompt Resp support Medical attention immed	Resp sys, skin
Clothing: Goggles: Wash: Change: Remove:	N.R. Reason prob N.R. N.R. N.R.	NIOSH/OSHA 50 mg/m³: DMF 125 mg/m³: PAPRDM℥/SA:CF℥ 250 mg/m³: HiEF/SCBAF/SAF 9300 mg/m³: SAF:PD,PP §: SCBAF:PD,PP/SAF:PD,PP:ASCBA Escape: HiEF/SCBAE	Inh Ing Con	Irrit upper resp tract, stomach	Eye: Skin: Breath: Swallow:	Irr immed Wash regularly Resp support Medical attention immed	Resp sys, GI tract
Clothing: Goggles: Wash: Change: Remove:	Repeat Reason prob Prompt contam N.R. Prompt non-imperv contam	NIOSH/OSHA 1000 ppm: PAPROV℥/CCRFOV §: SCBAF:PD,PP/SAF:PD,PP:ASCBA Escape: GMFOV/SCBAE	Inh Abs Ing Con	Irrit nose, eyes; liver, kidney damage; skin blister	Eye: Skin: Breath: Swallow:	Irr immed Soap wash prompt Resp support Medical attention immed	Liver, kidneys, skin, eyes
Clothing: Goggles: Wash: Change: Remove:	Repeat Reason prob Daily (reason prob) N.R. N.R.	NIOSH ¥: SCBAF:PD,PP/SAF:PD,PP:ASCBA Escape: GMFOV/SCBAE	Inh Ing Con	Head; eye irrit, swell periorb; profuse rhinitis; anor, nau, vomit; low-wgt, jaun, cirr; [carc] in animals: liver, kidney damage	Eye: Skin: Breath: Swallow:	Irr immed Soap wash Resp support Medical attention immed	Liver, resp sys, eyes, kidneys, skin

Chemical name, structure/formula, CAS and RTECS Nos., and DOT ID and guide Nos.	Synonyms, trade names, and conversion factors	Exposure limits (TWA unless noted otherwise)	IDLH	Physical description	Chemical and physical properties		Incompatibilities and reactivities	Measurement method (See Table 1)
					MW, BP, SOL Fl.P, IP, Sp.Gr, flammability	VP, FRZ UEL, LEL		
3,3'-Dichlorobenzidine (and its salts) C₆H₃ClNH₂C₆H₃ClNH₂ 91-94-1 DD0525000	4,4'-Diamino-3,3'-dichloro-biphenyl; Dichlorobenzidine base; o,o'-Dichlorobenzidine; 3,3'-Dichlorobiphenyl-4,4'-diamine; 3,3'-Dichloro-4,4'-biphenyl-diamine; 3,3'-Dichloro-4,4'-diamino-biphenyl	NIOSH Ca See Appendix A OSHA [1910.1007] See Appendix B ACGIH A2 [skin]	Ca	Gray to purple crystalline solid.	MW: 253.1 BP: 788°F Sol: Almost Insoluble Fl.P: ? IP: ?	VP: ? MLT: 271°F UEL: ? LEL: ?	None reported	Filter/ Si gel; Reagent; HPLC/UVD; III [#5509]
Dichlorodifluoromethane CCl₂F₂ 75-71-8 PA8200000 1028 12	Difluorodichloromethane, Fluorocarbon 12, Freon® 12, Halon® 122, Propellant 12, Refrigerant 12 1 ppm = 5.03 mg/m³	NIOSH/OSHA 1000 ppm (4950 mg/m³)	50,000 ppm	Colorless gas with an ether-like odor at extremely high concentrations. [Note: Shipped as a liquefied compressed gas.]	MW: 120.9 BP: -22°F Sol(77°F): 0.03% Fl.P: NA IP: 11.75 eV Nonflammable Gas	VP: >1 atm FRZ: -252°F UEL: NA LEL: NA	Chemically-active metals such as sodium, potassium, calcium, powdered aluminum, zinc & magnesium	Char(2); CH₂Cl₂; GC/FID; III [#1018]
1,3-Dichloro-5,5-dimethylhydantoin C₅H₆Cl₂N₂O₂ 118-52-5 MU0700000	Dactin, DDH, Halane	NIOSH/OSHA 0.2 mg/m³ ST 0.4 mg/m³	Unknown	White powder with a chlorine-like odor.	MW: 197.0 BP: ? Sol: 0.2% Fl.P: 346°F IP: ? Sp.Gr: 1.5 Combustible Solid	VP: ? MLT: 270°F UEL: ? LEL: ?	Water, strong acids, easily oxidized materials such as ammonia salts & sulfides	None available
1,1-Dichloroethane CHCl₂CH₃ 75-34-3 KI0175000 2362 27	Asymmetrical dichloroethane; Ethylidene chloride; 1,1-Ethylidene dichloride 1 ppm = 4.12 mg/m³	NIOSH/OSHA 100 ppm (400 mg/m³)	4000 ppm	Colorless, oily liquid with a chloroform-like odor.	MW: 99.0 BP: 135°F Sol: 0.6% Fl.P(oc): 22°F IP: 11.06 eV Sp.Gr: 1.18 Class IB Flammable Liquid	VP(77°F): 230 mm FRZ: -143°F UEL: ? LEL: 5.6%	Strong oxidizers, strong caustics	Char; CS₂; GC/FID; III [#1003, Halogenated Hydrocarbons]

Chemical name, structure/formula, CAS and RTECS Nos., and DOT ID and guide Nos.	Synonyms, trade names, and conversion factors	Exposure limits (TWA unless noted otherwise)	IDLH	Physical description	Chemical and physical properties		Incompatibilities and reactivities	Measurement method (See Table 1)
					MW, BP, SOL Fl.P, IP, Sp.Gr, flammability	VP, FRZ UEL, LEL		
1,2-Dichloroethylene ClCH=CHCl 540-59-0 KV9360000 1150 29	Acetylene dichloride, cis-Acetylene dichloride, trans-Acetylene dichloride, sym-Dichloroethylene 1 ppm = 4.03 mg/m³	NIOSH/OSHA 200 ppm (790 mg/m³)	4000 ppm	Colorless liquid (usually a mixture of the cis & trans isomers) with a slightly acrid, chloroform-like odor.	MW: 97.0 BP: 118-140°F Sol: 0.4% Fl.P: 36°F IP: 9.65 eV Sp.Gr(77°F): 1.27 Class IB Flammable Liquid	VP: 180-264 mm FRZ: -57 to -115°F UEL: 12.8% LEL: 5.6%	Strong oxidizers, strong alkalis, potassium hydroxide, copper	Char; CS₂; GC/FID; III [#1003, Halogenated Hydrocarbons]
Dichloroethyl ether (ClCH₂CH₂)₂O 111-44-4 KN0875000 1916 57	bis(2-Chloroethyl)ether; 2,2'-Dichlorodiethyl ether 1 ppm = 5.94 mg/m³	NIOSH Ca See Appendix A 5 ppm (30 mg/m³) ST 10 ppm (60 mg/m³) [skin] OSHA 5 ppm (30 mg/m³) ST 10 ppm (60 mg/m³) [skin]	Ca [250 ppm]	Colorless liquid with a chlorinated solvent-like odor.	MW: 143.0 BP: 352°F Sol: 1% Fl.P: 131°F IP: ? Sp.Gr: 1.22 Class IIIA Combustible Liquid	VP: 0.7 mm FRZ: -58°F UEL: ? LEL: 2.7%	Strong oxidizers [Note: Decomposes in presence of moisture to form hydrochloric acid.]	Char; CS₂; GC/FID; III [#1004]
Dichloromonofluoromethane CHCl₂F 75-43-4 PA8400000 1029 12	Dichlorofluoromethane, Fluorodichloromethane, Freon® 21, Halon® 112, Refrigerant 21 1 ppm = 4.28 mg/m³	NIOSH/OSHA 10 ppm (40 mg/m³)	50,000 ppm	Colorless gas with a slight, ether-like odor. [Note: A liquid below 48°F. Shipped as a liquefied compressed gas.]	MW: 102.9 BP: 48°F Sol(86°F): 0.7% Fl.P: NA IP: 12.39 eV Nonflammable Gas	VP: >1 atm FRZ: -211°F UEL: NA LEL: NA	Chemically-active metals such as sodium, potassium, calcium, powdered aluminum, zinc & magnesium; acid; acid fumes	Char(2); CS₂; GC/FID; III [#2516]
1,1-Dichloro-1-nitro-ethane CH₃CCl₂NO₂ 594-72-9 KI1050000 2650 57	Dichloronitroethane 1 ppm = 5.98 mg/m³	NIOSH/OSHA 2 ppm (10 mg/m³)	150 ppm	Colorless liquid with an unpleasant odor. [fumigant]	MW: 143.9 BP: 255°F Sol: 0.3% Fl.P: 136°F IP: ? Sp.Gr: 1.43 Class II Combustible Liquid	VP: 15 mm FRZ: ? UEL: ? LEL: ?	Strong oxidizers [Note: Corrosive to iron in presence of moisture.]	Char(pet); CS₂; GC/FID; III [#1601]

Personal protection and sanitation (See Table 3)	Recommendations for respirator selection — maximum concentration for use (MUC) (See Table 4)	Route	Symptoms (See Table 5)	First aid (See Table 6)	Target organs (See Table 5)
Clothing: Any poss Goggles: Any poss Wash: Immed contam/daily Change: After work if any poss contam Remove: Immed contam Provide: Eyewash, quick drench	NIOSH ¥: SCBAF:PD,PP/SAF:PD,PP:ASCBA Escape: HiEF/SCBAE	Inh Abs Ing Con	Allergic skin reaction, sens, derm; head, dizz; caustic burns; frequent urination, dysuria; hema; GI upsets; upper resp infection; [carc]	Eye: Irr immed (15 min) Skin: Soap wash immed Breath: Resp support Swallow: Medical attention immed	Bladder, liver, lung, skin, GI tract
Clothing: Prevent wet or freezing Goggles: Reason prob Wash: N.R. Change: N.R. Remove: Immed wet (flamm)	NIOSH/OSHA 10,000 ppm: SA/SCBA 25,000 ppm: SA:CF 50,000 ppm: SCBAF/SAF §: SCBAF:PD,PP/SAF:PD,PP:ASCBA Escape: GMFOV/SCBAE	Inh Con	Dizz, tremors, unconsciousness, card arrhy, card arrest	Eye: Irr immed Skin: Water flush immed Breath: Resp support	CVS, PNS
Clothing: Repeat Goggles: Any poss Wash: Prompt contam Change: After work if reason prob contam Remove: Prompt non-imperv contam Provide: Eyewash	NIOSH/OSHA 2 mg/m³: SA/SCBA 5 mg/m³: SA:CF/SCBAF/SAF §: SCBAF:PD,PP/SAF:PD,PP:ASCBA Escape: GMFSHiE/SCBAE	Inh Ing Con	Irrit eyes, muc memb, resp sys	Eye: Irr immed Skin: Soap wash prompt Breath: Resp support Swallow: Medical attention immed	Resp sys, eyes
Clothing: Repeat Goggles: Reason prob Wash: Immed wet Change: N.R. Remove: Immed wet (flamm)	NIOSH/OSHA 1000 ppm: SA/SCBA 2500 ppm: SA:CF 4000 ppm: SCBAF/SAF §: SCBAF:PD,PP/SAF:PD,PP:ASCBA Escape: GMFOV/SCBAE	Inh Ing Con	CNS depres; skin irrit; liver, kidney damage	Eye: Irr immed Skin: Soap flush prompt Breath: Resp support Swallow: Medical attention immed	Skin, liver, kidneys

Personal protection and sanitation (See Table 3)	Recommendations for respirator selection — maximum concentration for use (MUC) (See Table 4)	Route	Symptoms (See Table 5)	First aid (See Table 6)	Target organs (See Table 5)
Clothing: Repeat Goggles: Reason prob Wash: Prompt wet Change: N.R. Remove: Immed wet (flamm)	NIOSH/OSHA 1000 ppm: PAPROVᵋ/CCRFOV 4000 ppm: SA:CFᵋ/GMFOV/SCBAF/SAF §: SCBAF:PD,PP/SAF:PD,PP:ASCBA Escape: GMFOV/SCBAE	Inh Ing Con	Irrit eyes, resp sys; CNS depres	Eye: Irr immed Skin: Soap wash prompt Breath: Resp support Swallow: Medical attention immed	Resp sys, eyes, CNS
Clothing: Reason prob Goggles: Reason prob Wash: Immed contam Change: N.R. Remove: Immed non-imperv contam Provide: Quick drench	NIOSH ¥: SCBAF:PD,PP/SAF:PD,PP:ASCBA Escape: GMFOV/SCBAE	Inh Abs Ing Con	Lac; irrit nose, throat, cough; nau, vomit; [carc]; in animals: pulm irrit, edema; liver damage	Eye: Irr immed Skin: Soap wash Breath: Resp support Swallow: Medical attention immed	Resp sys, skin, eyes
Clothing: Any poss wet Goggles: Reason prob Wash: N.R. Change: N.R. Remove: Immed wet	NIOSH/OSHA 100 ppm: SA/SCBA 250 ppm: SA:CF 500 ppm: SCBAF/SAF 10,000 ppm: SA:PD,PP 20,000 ppm: SAF:PD,PP §: SCBAF:PD,PP/SAF:PD,PP:ASCBA Escape: GMFOV/SCBAE	Inh Ing Con	Asphy, card arrhy, card arrest	Eye: Irr immed Skin: Water flush immed Breath: Resp support Swallow: Medical attention immed	Resp sys, CVS
Clothing: Repeat Goggles: Reason prob Wash: Prompt contam Change: N.R. Remove: Prompt non-imperv contam	NIOSH/OSHA 20 ppm: SA/SCBA 50 ppm: SA:CF 100 ppm: SCBAF/SAF 150 ppm: SAF:PD,PP §: SCBAF:PD,PP/SAF:PD,PP:ASCBA Escape: GMFOV/SCBAE	Inh Ing Con	In animals: irrit eyes, skin; liver, heart, kidney damage; pulm edema, hemorr	Eye: Irr immed Skin: Soap wash immed Breath: Resp support Swallow: Medical attention immed	Lungs

Chemical name, structure/formula, CAS and RTECS Nos., and DOT ID and guide Nos.	Synonyms, trade names, and conversion factors	Exposure limits (TWA unless noted otherwise)	IDLH	Physical description	Chemical and physical properties		Incompatibilities and reactivities	Measurement method (See Table 1)
					MW, BP, SOL Fl.P, IP, Sp.Gr, flammability	VP, FRZ UEL, LEL		
Dichlorotetrafluoro-ethane CClF₂CClF₂ 76-14-2 KI1101000 1958 12	1,2-Dichlorotetra-fluoroethane; Freon® 114; Halon® 242; Refrigerant 114 1 ppm = 7.10 mg/m³	NIOSH/OSHA 1000 ppm (7000 mg/m³)	50,000 ppm	Colorless gas with a faint, ether-like odor at high concentrations. [Note: A liquid below 38°F. Shipped as a liquefied compressed gas.]	MW: 170.9 BP: 38°F Sol: 0.01% Fl.P: NA IP: 12.20 eV Nonflammable Gas	VP: >1 atm FRZ: -137°F UEL: NA LEL: NA	Chemically-active metals such as sodium, potassium, calcium, powdered aluminum, zinc & magnesium; acids; acid fumes	Char(2); CH₂Cl₂; GC/FID; III [#1018]
Dichlorvos (CH₃O)₂P(O)OCH=CCl₂ 62-73-7 TC0350000 2783 55	DDVP; 2,2-Dichlorovinyl dimethyl phosphate 1 ppm = 9.19 mg/m³	NIOSH/OSHA 1 mg/m³ [skin]	200 mg/m³	Colorless to amber liquid with a mild, chemical odor. [Note: Insecticide that may be absorbed on a dry carrier.]	MW: 221.0 BP: Decomposes Sol: 0.01% Fl.P(oc): >175°F IP: ? Sp.Gr(77°F): 1.42 Class III Combustible Liquid	VP(86°F): 0.01 mm FRZ: ? UEL: ? LEL: ?	Strong acids, strong alkalis [Note: Corrosive to iron & mild steel.]	XAD-2; Toluene; GC/FPD; II(5) [P&CAM #295]
Dieldrin C₁₂H₈Cl₆O 60-57-1 IO1750000 2761 55	1,2,3,4,10,10-Hexachloro-6,7-epoxy-1,4,4a,5,6,7,8,8a-octahydro-1,4-endo-exo-5,8-dimethano-naphthalene; HEOD	NIOSH Ca See Appendix A 0.25 mg/m³ [skin] OSHA 0.25 mg/m³ [skin]	Ca [450 mg/m³]	Colorless to light tan crystals with a mild, chemical odor. [insecticide]	MW: 380.9 BP: Decomposes Sol: 0.02% Fl.P: NA IP: ? Sp.Gr: 1.75 Noncombustible Solid	VP: Low MLT: 349°F UEL: NA LEL: NA	Strong oxidizers, active metals such as sodium, strong acids, phenols	Filter; Isooctane; GC/ECD; II(3) [#S283]
Diethylamine (C₂H₅)₂NH 109-89-7 HZ8750000 1154 68	Diethamine; N,N-Diethylamine; N-Ethylethanamine 1 ppm = 3.04 mg/m³	NIOSH/OSHA 10 ppm (30 mg/m³) ST 25 ppm (75 mg/m³)	2000 ppm	Colorless liquid with a fishy, ammonia-like odor.	MW: 73.1 BP: 132°F Sol: Miscible Fl.P: -15°F IP: 8.01 eV Sp.Gr: 0.71 Class IB Flammable Liquid	VP: 192 mm FRZ: -58°F UEL: 10.1% LEL: 1.8%	Strong oxidizers, strong acids, cellulose nitrate	Si gel; H₂SO₄/ Methanol GC/FID; III [#2010]

Chemical name, structure/formula, CAS and RTECS Nos., and DOT ID and guide Nos.	Synonyms, trade names, and conversion factors	Exposure limits (TWA unless noted otherwise)	IDLH	Physical description	Chemical and physical properties		Incompatibilities and reactivities	Measurement method (See Table 1)
					MW, BP, SOL Fl.P, IP, Sp.Gr, flammability	VP, FRZ UEL, LEL		
2-Diethylaminoethanol (C₂H₅)₂NCH₂CH₂OH 100-37-8 KK5075000 2686 29	Diethylaminoethanol; 2-Diethylaminoethyl alcohol; N,N-Diethylethanolamine; Diethyl-(2-hydroxyethyl)-amine; 2-Hydroxytriethylamine 1 ppm = 4.87 mg/m³	NIOSH/OSHA 10 ppm (50 mg/m³) [skin]	500 ppm	Colorless liquid with a nauseating, ammonia-like odor.	MW: 117.2 BP: 325°F Sol: Miscible Fl.P: 126°F IP: ? Sp.Gr: 0.89 Class II Combustible Liquid	VP: 21 mm FRZ: -94°F UEL: ? LEL: ?	Strong oxidizers, strong acids	Si gel; Methanol; GC/FID; III [#2007, Amino-ethanol Compounds]
Difluorodibromomethane CBr₂F₂ 75-61-6 PA7525000 1941 58	Dibromodifluoromethane, Freon® 12B2, Halon® 1202 1 ppm = 8.72 mg/m³	NIOSH/OSHA 100 ppm (860 mg/m³)	2500 ppm	Colorless, heavy liquid or gas (above 76°F) with a characteristic odor.	MW: 209.8 BP: 76°F Sol: Insoluble Fl.P: NA IP: 11.07 eV Sp.Gr(59°F): 2.29 Nonflammable Liquid	VP: 620 mm FRZ: -231°F UEL: NA LEL: NA	Chemically-active metals such as sodium, potassium, calcium, powdered aluminum, zinc & magnesium	Char(2); 2-Propanol; GC/FID; III [#1012]
Diglycidyl ether C₆H₁₀O₃ 2238-07-5 KN2350000	Diallyl ether dioxide; DGE; 2-Epoxypropyl ether; bis (2,3-Epoxypropyl) ether; di(2,3-Epoxypropyl) ether 1 ppm = 5.41 mg/m³	NIOSH Ca See Appendix A 0.1 ppm (0.5 mg/m³) OSHA 0.1 ppm (0.5 mg/m³)	Ca [25 ppm]	Colorless liquid with a strong, irritating odor.	MW: 130.2 BP: 500°F Sol: ? Fl.P: 147°F IP: ? Sp.Gr(77°F): 1.26 Class IIIA Combustible Liquid	VP(77°F): 0.09 mm FRZ: ? UEL: ? LEL: ?	Strong oxidizers	None available
Diisobutyl ketone ((CH₃)₂CHCH₂)₂CO 108-83-8 MJ5775000 1157 26	DIBK; sym-Diisopropyl acetone; 2,6-Dimethyl-4-heptanone; Isovalerone; Valerone 1 ppm = 5.92 mg/m³	NIOSH/OSHA 25 ppm (150 mg/m³)	2000 ppm	Colorless liquid with a mild, sweet odor.	MW: 142.3 BP: 334°F Sol: 0.05% Fl.P: 120°F IP: ? Sp.Gr: 0.81 Class II Combustible Liquid	VP: 2 mm FRZ: -43°F UEL(200°F): 7.1% LEL(200°F): 0.8%	Strong oxidizers	Char; CS₂; GC/FID; III [#1300, Ketones I]

Personal protection and sanitation (See Table 3)	Recommendations for respirator selection — maximum concentration for use (MUC) (See Table 4)	Health hazards			
		Route	Symptoms (See Table 5)	First aid (See Table 6)	Target organs (See Table 5)
Clothing: Any poss wet Goggles: Reason prob Wash: N.R. Change: N.R. Remove: Immed wet	NIOSH/OSHA 10,000 ppm: SA/SCBA 25,000 ppm: SA:CF 50,000 ppm: SCBAF/SAF §: SCBAF:PD,PP/SAF:PD,PP:ASCBA Escape: GMFOV/SCBAE	Inh Ing Con	Irrit resp sys; asphy; card arrhy, card arrest	Eye: Irr immed Skin: Water flush immed Breath: Resp support Swallow: Medical attention immed	Resp sys, CVS
Clothing: Repeat Goggles: Reason prob Wash: Immed contam Change: N.R. Remove: Immed non-imperv contam	NIOSH/OSHA 10 mg/m^3: SA/SCBA 25 mg/m^3: SA:CF 50 mg/m^3: SCBAF/SAF/SAT:CF 200 mg/m^3: SA:PD,PP §: SCBAF:PD,PP/SAF:PD,PP:ASCBA Escape: GMFOVHiE/SCBAE	Inh Abs Ing Con	Miosis, ache eyes; rhin; head; tight chest, wheez, lar spasm, salv; cyan; anor, nau, vomit, diarr; sweat; musc fasc, para, gidd, ataxia; convuls; low BP, card irregularities; irrit skin, eyes	Eye: Irr immed Skin: Soap wash immed Breath: Resp support Swallow: Medical attention immed	Resp sys, CVS, CNS, eyes, skin, blood chol
Clothing: Any poss Goggles: Any poss Wash: Immed contam Change: After work if any poss contam Remove: Immed non-imperv contam Provide: Eyewash, quick drench	NIOSH ¥: SCBAF:PD,PP/SAF:PD,PP:ASCBA Escape: GMFOVHiE/SCBAE	Inh Abs Ing Con	Head, dizz; nau, vomit, mal, sweat; myoclonic limb jerks; clonic, tonic convuls; coma; [carc]; in animals: liver, kidney damage	Eye: Irr immed Skin: Soap wash immed Breath: Resp support Swallow: Medical attention immed	CNS, liver, kidneys, skin
Clothing: Any poss liq Goggles: Any poss >0.5%/ Reason prob 0.5% Wash: Immed contam liq Change: N.R. Remove: Immed wet (flamm) Provide: Eyewash (>0.5%)/ Quick drench (liq)	NIOSH/OSHA 250 ppm: SA:CF^ɛ/PAPRS^ɛ 500 ppm: GMFS/SCBAF/SAF/CCRFS 2000 ppm: SAF:PD,PP §: SCBAF:PD,PP/SAF:PD,PP:ASCBA Escape: GMFS/SCBAE	Inh Abs Ing Con	Eye, skin, resp sys irrit	Eye: Irr immed Skin: Water flush immed Breath: Resp support Swallow: Medical attention immed	Resp sys, skin, eyes

Personal protection and sanitation (See Table 3)	Recommendations for respirator selection — maximum concentration for use (MUC) (See Table 4)	Health hazards			
		Route	Symptoms (See Table 5)	First aid (See Table 6)	Target organs (See Table 5)
Clothing: Reason prob Goggles: Any poss >5%/Reason prob <5% Wash: Immed contam Change: N.R. Remove: Immed non-imperv contam Provide: Eyewash (>5%), quick drench	NIOSH/OSHA 100 ppm: CCROV*/SA*/SCBA* 250 ppm: SA:CF*/PAPROV* 500 ppm: CCRFOV/PAPRTOV*/GMFOV/ SCBAF/SAF §: SCBAF:PD,PP/SAF:PD,PP:ASCBA Escape: GMFOV/SCBAE	Inh Abs Ing Con	Nau, vomit; irrit resp sys, skin, eyes	Eye: Irr immed Skin: Water flush immed Breath: Resp support Swallow: Medical attention immed	Resp sys, skin, eyes
Clothing: Repeat Goggles: Reason prob Wash: N.R. Change: N.R. Remove: Prompt non-imperv wet	NIOSH/OSHA 1000 ppm: SA/SCBA 2500 ppm: SA:CF/SCBAF/SAF §: SCBAF:PD,PP/SAF:PD,PP:ASCBA Escape: GMFOV/SCBAE	Inh Ing Con	In animals: irrit resp sys; CNS symptoms; liver damage	Eye: Irr immed Skin: Water flush immed Breath: Resp support Swallow: Medical attention immed	Resp sys
Clothing: Any poss Goggles: Reason prob Wash: Immed contam/daily Change: After work if any poss contam Remove: Immed non-imperv contam Provide: Quick drench	NIOSH ¥: SCBAF:PD,PP/SAF:PD,PP:ASCBA Escape: GMFOV/SCBAE	Inh Ing Con	Irrit eyes, resp sys; skin burns; [carc]; in animals: hemato sys, lung, liver, kidney damage	Eye: Irr immed Skin: Soap wash immed Breath: Resp support Swallow: Medical attention immed	Skin, eyes, resp sys
Clothing: Repeat Goggles: N.R. Wash: Prompt wet Change: N.R. Remove: Prompt non-imperv contam	NIOSH/OSHA 625 ppm: SA:CF^ɛ/PAPROV^ɛ 1000 ppm: CCRFOV 1250 ppm: GMFOV/SCBAF/SAF 2000 ppm: SAF:PD,PP §: SCBAF:PD,PP/SAF:PD,PP:ASCBA Escape: GMFOV/SCBAE	Inh Ing Con	Irrit eyes, nose, throat; head, dizz; derm	Eye: Irr immed Skin: Soap wash prompt Breath: Resp support Swallow: Medical attention immed	Resp sys, skin, eyes

Chemical name, structure/formula, CAS and RTECS Nos., and DOT ID and guide Nos.	Synonyms, trade names, and conversion factors	Exposure limits (TWA unless noted otherwise)	IDLH	Physical description	Chemical and physical properties		Incompatibilities and reactivities	Measurement method (See Table 1)
					MW, BP, SOL Fl.P, IP, Sp.Gr, flammability	VP, FRZ UEL, LEL		
Diisopropylamine ((CH₃)₂CH)₂NH 108-18-9 IM4025000	N-(1-Methylethyl)-2-propanamine	NIOSH/OSHA 5 ppm (20 mg/m³) [skin]	1000 ppm	Colorless liquid with an ammonia- or fish-like odor.	MW: 101.2 BP: 183°F Sol: Slight Fl.P: 20°F IP: 7.73 eV	VP: 70 mm FRZ: -141°F UEL: 7.1% LEL: 1.1%	Strong oxidizers, strong acids	Imp; KOH; GC/FID; II(4) [#S141]
1158 68	1 ppm = 4.21 mg/m³				Sp.Gr: 0.72 Class IB Flammable Liquid			
Dimethyl acetamide CH₃CON(CH₃)₂ 127-19-5 AB7700000	N,N-Dimethyl acetamide; DMAC	NIOSH/OSHA 10 ppm (35 mg/m³) [skin]	400 ppm	Colorless liquid with a weak ammonia- or fish-like odor.	MW: 87.1 BP: 329°F Sol: Miscible Fl.P(oc): 158°F IP: 8.81 eV	VP: 2 mm FRZ: -4°F UEL(320°F): 11.5% LEL(212°F): 1.8%	Carbon tetrachloride, other halogenated compounds when in contact with iron	Si gel; GC/FID; III [#2004]
	1 ppm = 3.62 mg/m³				Sp.Gr: 0.94 Class IIIA Combustible Liquid			
Dimethylamine (CH₃)₂NH 124-40-3 IP8750000	Dimethylamine (anhydrous), N-Methylmethanamine	NIOSH/OSHA 10 ppm (18 mg/m³)	2000 ppm	Colorless gas with an ammonia- or fish-like odor. [Note: A liquid below 44°F. Shipped as a liquefied compressed gas.]	MW: 45.1 BP: 44°F Sol(140°F): 24% Fl.P: NA (Gas) 20°F(Liquid) IP: 8.24 eV	VP: >1 atm FRZ: -134°F UEL: 14.4% LEL: 2.8%	Strong oxidizers, chlorine, mercury, acraldehyde, fluoride, maleic anhydride	Si gel; H₂SO₄/ Methanol; GC/FID; III [#2010]
1032 19	1 ppm = 1.87 mg/m³				Sp.Gr: 0.67 (Liquid at 44°F) Class IA Flammable Liquid			
4-Dimethylamino-azobenzene C₆H₅NNC₆H₄N(CH₃)₂ 60-11-7 BX7350000	Butter yellow; DAB; p-Dimethylaminoazobenzene; N,N-Dimethyl-4-aminoazo-benzene; Methyl yellow	NIOSH Ca See Appendix A OSHA [1910.1015] See Appendix B	Ca	Yellow, leaf-shaped crystals.	MW: 225.3 BP: Sublimes Sol: Insoluble Fl.P: ? IP: ?	VP: ? MLT: 237°F UEL: ? LEL: ?	None reported	G-chrom P; 2-Propanol; GC/FID; II(4) [P&CAM #284]

Chemical name, structure/formula, CAS and RTECS Nos., and DOT ID and guide Nos.	Synonyms, trade names, and conversion factors	Exposure limits (TWA unless noted otherwise)	IDLH	Physical description	Chemical and physical properties		Incompatibilities and reactivities	Measurement method (See Table 1)
					MW, BP, SOL Fl.P, IP, Sp.Gr, flammability	VP, FRZ UEL, LEL		
Dimethylaniline C₆H₅N(CH₃)₂ 121-69-7 BX4725000	N,N-Dimethylaniline; N,N-Dimethylbenzeneamine; N,N-Dimethylphenylamine	NIOSH/OSHA 5 ppm (25 mg/m³) ST 10 ppm (50 mg/m³) [skin]	100 ppm	Pale yellow, oily liquid with an amine-like odor. [Note: A solid below 36°F.]	MW: 121.2 BP: 378°F Sol: 2% Fl.P: 142°F IP: 7.14 eV	VP(85°F): 1 mm FRZ: 36°F UEL: ? LEL: ?	Strong oxidizers, strong acids, benzoyl peroxide	Si gel; Ethanol; GC/FID; III [#2002, Aromatic Amines]
2253 57	1 ppm = 5.04 mg/m³				Sp.Gr: 0.96 Class IIIA Combustible Liquid			
Dimethyl-1,2-dibromo-2,2-dichlorethyl phosphate C₄H₇O₄PBr₂Cl₂ 300-76-5 TB9450000	Dibrom; 1,2-Dibromo-2,2-dichloro-ethyl dimethyl phosphate; Naled [(CH₃O)₂P(O)OC(Br)-HCBr(Cl₂)]	NIOSH/OSHA 3 mg/m³ [skin]	1800 mg/m³	Colorless to white solid or straw-colored liquid (above 80°F) with a slightly pungent odor.	MW: 380.8 BP: Decomposes Sol: Insoluble Fl.P: NA IP: ?	VP: 0.0002 mm MLT: 80°F UEL: NA LEL: NA	Strong oxidizers, acids [Note: Corrosive to metals.]	None available
2783 55	1 ppm = 15.83 mg/m³				Sp.Gr(77°F): 1.96 Noncombustible Solid			
Dimethylformamide HCON(CH₃)₂ 68-12-2 LQ2100000	Dimethyl formamide; N,N-Dimethylformamide; DMF	NIOSH/OSHA 10 ppm (30 mg/m³) [skin]	3500 ppm	Colorless to pale-yellow liquid with a faint, amine-like odor.	MW: 73.1 BP: 307°F Sol: Miscible Fl.P: 136°F IP: 9.12 eV	VP(77°F): 4 mm FRZ: -78°F UEL: 15.2% LEL(212°F): 2.2%	Carbon tetrachloride; other halogenated compounds when in contact with iron; strong oxidizers; alkyl aluminums; inorganic nitrates	Si gel; Methanol; GC/FID; III [#2004]
2265 26	1 ppm = 3.04 mg/m³				Sp.Gr: 0.95 Class IIIA Combustible Liquid			
1,1-Dimethylhydrazine (CH₃)₂NNH₂ 57-14-7 MV2450000	Dimazine, DMH, UDMH, Unsymmetrical dimethyl-hydrazine	NIOSH Ca See Appendix A C 0.06 ppm (0.15 mg/m³) [2-hr] OSHA 0.5 ppm (1 mg/m³) [skin]	Ca [50 ppm] ACGIH A2	Colorless liquid with an ammonia- or fish-like odor.	MW: 60.1 BP: 147°F Sol: Miscible Fl.P: 5°F IP: 8.05 eV	VP(77°F): 157 mm FRZ: -72°F UEL: 95% LEL: 2%	Oxidizers, halogens, metallic mercury, fuming nitric acid, hydrogen peroxide [Note: May ignite SPONTANEOUSLY in contact with oxidizers.]	Bub; Reagent; Vis; II(3) [#S143]
1163 28	1 ppm = 2.50 mg/m³				Sp.Gr: 0.79 Class IB Flammable Liquid			

Personal protection and sanitation (See Table 3)		Recommendations for respirator selection — maximum concentration for use (MUC) (See Table 4)	Health hazards				
			Route	Symptoms (See Table 5)	First aid (See Table 6)		Target organs (See Table 5)
Clothing:	Repeat	NIOSH/OSHA	Ing	Nau, vomit; head; eye	Eye:	Irr immed	Resp sys, skin,
Goggles:	Any poss (soln >5%)	125 ppm: SA:CF$^£$/PAPROV$^£$	Abs	irrit, vis dist; pulm	Skin:	Water wash immed	eyes
Wash:	Prompt wet	250 ppm: CCRFOV/GMFOV/SCBAF/SAF	Ing	irrit	Breath:	Resp support	
Change:	N.R.	1000 ppm: SAF:PD,PP	Con		Swallow:	Medical attention	
Remove:	Immed wet (flamm)	§: SCBAF:PD,PP/SAF:PD,PP:ASCBA				immed	
Provide:	Eyewash (<5%)	Escape: GMFOV/SCBAE					
Clothing:	Any poss	NIOSH/OSHA	Inh	Jaun, liver damage;	Eye:	Irr immed	Liver, skin
Goggles:	Reason prob	100 ppm: SA/SCBA	Abs	depres, lethargy, halu,	Skin:	Water flush immed	
Wash:	Immed contam	250 ppm: SA:CF	Ing	delusions; irrit skin	Swallow:	Medical attention	
Change:	N.R.	400 ppm: SCBAF/SAF	Con			immed	
Remove:	Immed non-imperv contam	§: SCBAF:PD,PP/SAF:PD,PP:ASCBA					
Provide:	Quick drench	Escape: GMFOV/SCBAE					
Clothing:	Any poss liq/Repeat soln	NIOSH/OSHA	Inh	Irrit nose, throat;	Eye:	Irr immed	Resp sys, skin,
		250 ppm: SA:CF$^£$	Ing	sneezing, cough, dysp;	Skin:	Water flush immed	eyes
Goggles:	Any poss	500 ppm: SCBAF/SAF	Con	pulm edema; conj; burns	Breath:	Resp support	
Wash:	Immed contam liq/ Prompt contam soln	2000 ppm: SAF:PD,PP		skin, muc memb; derm	Swallow:	Medical attention	
Change:	N.R.	§: SCBAF:PD,PP/SAF:PD,PP:ASCBA				immed	
Remove:	Immed wet (flamm)	Escape: GMFS/SCBAE					
Provide:	Liq: Eyewash, quick drench						
Clothing:	Any poss	NIOSH	Inh	Enlarged liver; hepatic	Eye:	Irr immed	Liver, skin,
Goggles:	Any poss	¥: SCBAF:PD,PP/SAF:PD,PP:ASCBA	Abs	and renal dysfunction;	Skin:	Soap wash immed	bladder
Wash:	Immed contam/daily	Escape: HiEF/SCBAE	Ing	contact derm; cough,	Breath:	Resp support	
Change:	After work if any poss contam		Con	wheez, difficulty breath; bloody sputum;	Swallow:	Medical attention immed	
Remove:	Immed contam			bronchial secretions;			
Provide:	Eyewash, quick drench			frequent urination, hema, dysuria; [carc]			

Personal protection and sanitation (See Table 3)		Recommendations for respirator selection — maximum concentration for use (MUC) (See Table 4)	Health hazards				
			Route	Symptoms (See Table 5)	First aid (See Table 6)		Target organs (See Table 5)
Clothing:	Reason prob	NIOSH/OSHA	Inh	Anoxia symptoms: cyan,	Eye:	Irr immed	Blood, kidneys,
Goggles:	Reason prob	50 ppm: SA/SCBA	Abs	weak, dizz, ataxia	Skin:	Soap wash immed	liver, CVS
Wash:	Immed contam	100 ppm: SA:CF/SCBAF/SAF	Ing		Breath:	Resp support	
Change:	N.R.	§: SCBAF:PD,PP/SAF:PD,PP:ASCBA	Con		Swallow:	Medical attention	
Remove:	Immed non-imperv contam	Escape: GMFOV/SCBAE				immed	
Provide:	Quick drench						
Clothing:	Reason prob	NIOSH/OSHA	Inh	Miosis, lac; head; tight	Eye:	Irr immed	Resp sys, CNS,
Goggles:	Any poss	30 mg/m^3: DMFu/SA/HiE/SCBA	Abs	chest, wheez, lar spasm;	Skin:	Soap wash immed	CVS, skin, eyes,
Wash:	Immed contam	75 mg/m^3: PAPRDMFu/SA:CF	Ing	salv; cyan, anor, nau,	Breath:	Resp support	blood chol
Change:	After work if reason prob contam	150 mg/m^3: HiEF/PAPRTHiE/SAT:CF/ SCBAF/SAF	Con	vomit, abdom cramp, diarr; weak, twitch, para; gidd,	Swallow:	Medical attention immed	
Remove:	Immed non-imperv contam	1800 mg/m^3: SA:PD,PP		ataxia, convuls; low BP;			
Provide:	Eyewash	§: SCBAF:PD,PP/SAF:PD,PP:ASCBA		card irregularities; irrit			
		Escape: HiEF/SCBAE		skin, eyes			
Clothing:	Repeat	NIOSH/OSHA	Inh	Nau, vomit, colic; liver	Eye:	Irr immed	Liver, kidneys,
Goggles:	Reason prob	100 ppm: SA*/SCBA*	Abs	damage, hepatomegaly;	Skin:	Water flush prompt	CVS, skin
Wash:	Prompt contam	250 ppm: SA:CF*	Ing	high BP; face flush; derm;	Breath:	Resp support	
Change:	N.R.	500 ppm: SCBAF/SAF/SAT:CF*	Con	in animals: kidney, heart	Swallow:	Medical attention	
Remove:	Prompt non-imperv contam	3500 ppm: SAF:PD,PP		damage		immed	
		§: SCBAF:PD,PP/SAF:PD,PP:ASCBA					
		Escape: GMFOV/SCBAE					
Clothing:	Any poss	NIOSH	Inh	Irrit eyes, skin; choking,	Eye:	Irr immed	CNS, liver, GI
Goggles:	Any poss	¥: SCBAF:PD,PP/SAF:PD,PP:ASCBA	Abs	chest pain, dysp; lethargy;	Skin:	Water flush immed	tract, blood,
Wash:	Immed contam	Escape: GMFS/SCBAE	Ing	nau; anoxia; convuls; liver	Breath:	Resp support	resp sys, eyes,
Change:	N.R.		Con	inj; [carc]	Swallow:	Medical attention	skin
Remove:	Immed wet (flamm)					immed	
Provide:	Eyewash, quick drench						

Chemical name, structure/formula, CAS and RTECS Nos., and DOT ID and guide Nos.	Synonyms, trade names, and conversion factors	Exposure limits (TWA unless noted otherwise)	IDLH	Physical description	Chemical and physical properties		Incompatibilities and reactivities	Measurement method (See Table 1)
					MW, BP, SOL Fl.P, IP, Sp.Gr, flammability	VP, FRZ UEL, LEL		
Dimethylphthalate C$_6$H$_4$(COOCH$_3$)$_2$ 131-11-3 TI1575000 1 ppm = 8.07 mg/m³	Dimethyl ester of 1,2-benzenedicarboxylic acid; DMP	NIOSH/OSHA 5 mg/m³	9300 mg/m³	Colorless, oily liquid with a slight, aromatic odor. [Note: A solid below 42°F.]	MW: 194.2 BP: 543°F Sol: 0.4% Fl.P: 295°F IP: 9.64 eV Sp.Gr: 1.19 Class IIIB Combustible Liquid	VP: 0.01 mm FRZ: 42°F UEL: ? LEL(358°F): 0.9%	Nitrates; strong oxidizers, alkalis & acids	None available
Dimethyl sulfate (CH$_3$)$_2$SO$_4$ 77-78-1 WS8225000 1595 57	Dimethyl ester of sulfuric acid, Dimethylsulfate, Methyl sulfate 1 ppm = 5.24 mg/m³	NIOSH Ca See Appendix A 0.1 ppm (0.5 mg/m³) [skin] OSHA 0.1 ppm (0.5 mg/m³) [skin]	Ca [10 ppm] ACGIH A2	Colorless, oily liquid with a faint, onion-like odor.	MW: 126.1 BP: 370°F (Decomposes) Sol(64°F): 3% Fl.P: 182°F IP: ? Sp.Gr: 1.33 Class IIIA Combustible Liquid	VP: 0.1 mm FRZ: -25°F UEL: ? LEL: ?	Strong oxidizers, ammonia solutions [Note: Decomposes in water to sulfuric acid; corrosive to metals.]	Porapak-P; Diethyl ether; GC/EConD; III; [#2524]
Dinitrobenzene (all isomers) C$_6$H$_4$(NO$_2$)$_2$ o: 528-29-0 CZ7450000 m: 99-65-0 CZ7350000 p: 100-25-4 CZ7525000 1597 56	o: ortho-Dinitrobenzene; 1,2-Dinitrobenzene m: meta-Dinitrobenzene; 1,3-Dinitrobenzene p: para-Nitrobenzene; 1,4-Dinitrobenzene	NIOSH/OSHA 1 mg/m³ [skin]	200 mg/m³	Pale white or yellow solids.	MW: 168.1 BP: 606/572/570°F Sol: 0.05/0.02/0.01% Fl.P: 302/302°F/? IP: 10.71/10.43/10.50 eV Sp.Gr(64°F): 1.57/1.58/1.63 Combustible Solids	VP: ?/?/? MLT: 244/192/343°F UEL: ?/?/? LEL: ?/?/?	Strong oxidizers, caustics, metals such as tin & zinc [Note: Prolonged exposure to fire and heat may result in an explosion due to SPONTANEOUS decomposition.]	Filter/Bub; Methanol; HPLC/UVD; II(4) [#S214]
Dinitro-o-cresol CH$_3$C$_6$H$_2$OH(NO$_2$)$_2$ 534-52-1 GO9625000 1598 53	4,6-Dinitro-o-cresol; 3,5-Dinitro-2-hydroxy-toluene; 4,6-Dinitro-2-methyl phenol; DN; DNOC	NIOSH/OSHA 0.2 mg/m³ [skin]	5 mg/m³	Yellow, odorless solid. [insecticide]	MW: 198.1 BP: 594°F Sol: 0.01% Fl.P: NA IP: ? Sp.Gr: 1.1 (estimated) Noncombustible Solid	VP: 0.00005 mm MLT: 190°F UEL: NA LEL: NA	Strong oxidizers	Filter/Bub; 2-Propanol; HPLC/UVD; II(5) [#S166]

Chemical name, structure/formula, CAS and RTECS Nos., and DOT ID and guide Nos.	Synonyms, trade names, and conversion factors	Exposure limits (TWA unless noted otherwise)	IDLH	Physical description	Chemical and physical properties		Incompatibilities and reactivities	Measurement method (See Table 1)
					MW, BP, SOL Fl.P, IP, Sp.Gr, flammability	VP, FRZ UEL, LEL		
Dinitrotoluene C$_6$H$_3$CH$_3$(NO$_2$)$_2$ 25321-14-6 XT1300000 1600 56 (liquid) 2038 56 (solid)	Dinitrotoluol, DNT, Methyldinitrobenzene	NIOSH Ca See Appendix A 1.5 mg/m³ [skin] OSHA 1.5 mg/m³ [skin]	Ca [200 mg/m³]	Orange-yellow crystalline solid with a character-istic odor. [Note: Often shipped molten.]	MW: 182.2 BP: 572°F Sol: Insoluble Fl.P: 404°F IP: ? Sp.Gr: 1.32 Combustible Solid, but difficult to ignite.	VP: 1 mm MLT: 158°F UEL: ? LEL: ?	Strong oxidizers, caustics, metals such as tin & zinc [Note: Commercial grades will decompose at 482°F, with self-sustaining decomposition at 536°F.]	None available
Di-sec octyl phthalate C$_{24}$H$_{38}$O$_4$ 117-81-7 TI0350000 1 ppm = 16.23 mg/m³	DEHP, DOP, bis-(2-Ethylhexyl) phthalate, di(2-Ethylhexyl) phthalate, Octyl phthalate [C$_6$H$_4$(COOCH$_2$CH-(C$_2$H$_5$)C$_4$H$_9$)$_2$]	NIOSH Ca See Appendix A 5 mg/m³ ST 10 mg/m³ OSHA 5 mg/m³ ST 10 mg/m³	Ca [Unknown]	Colorless, oily liquid with a slight odor.	MW: 390.5 BP: 727°F Sol: Insoluble Fl.P(oc): 420°F IP: ? Sp.Gr: 0.99 Class IIIB Combustible Liquid	VP: <0.01 mm FRZ: -58°F UEL: ? LEL(474°F): 0.3%	Nitrates; strong oxidizers, acids & alkalis	Filter; CS$_2$; GC/FID; III [#5020, di(2-Ethyl-hexyl) phthalate]
Dioxane C$_4$H$_8$O$_2$ 123-91-1 JG8225000 1165 26	Diethylene dioxide; Diethylene ether; Dioxan; p-Dioxane; 1,4-Dioxane 1 ppm = 3.66 mg/m³	NIOSH Ca See Appendix A C 1 ppm (3.6 mg/m³) [30-min] OSHA 25 ppm (90 mg/m³) [skin]	Ca [2000 ppm]	Colorless liquid or solid (below 53°F) with a mild ether-like odor.	MW: 88.1 BP: 214°F Sol: Miscible Fl.P: 55°F IP: 9.13 eV Sp.Gr: 1.03 Class IB Flammable Liquid	VP: 29 mm FRZ: 53°F UEL: 22% LEL: 2.0%	Strong oxidizers, decaborane, triethynyl aluminum	Char; CS$_2$; GC/FID; III [#1602]
Diphenyl C$_6$H$_5$C$_6$H$_5$ 92-52-4 DU8050000	Biphenyl, Phenyl benzene	NIOSH/OSHA 1 mg/m³ (0.2 ppm)	300 mg/m³	Colorless to pale-yellow solid with a pleasant, characteristic odor.	MW: 154.2 BP: 489°F Sol: Insoluble Fl.P: 235°F IP: 7.95 eV Sp.Gr: 1.04 Combustible Solid	VP(160°F): 1 mm MLT: 156°F UEL(311°F): 5.8% LEL(232°F): 0.6%	Oxidizers	Tenax GC; CCl$_4$; GC/FID; III [#2530, Biphenyl]

Personal protection and sanitation (See Table 3)		Recommendations for respirator selection — maximum concentration for use (MUC) (See Table 4)	Health hazards			
			Route	Symptoms (See Table 5)	First aid (See Table 6)	Target organs (See Table 5)
Clothing: Goggles: Wash: Change: Remove:	N.R. Reason prob N.R. N.R. N.R.	NIOSH/OSHA 50 mg/m³: DMF 125 mg/m³: PAPRDMᴱ/SA:CFᴱ 250 mg/m³: HiEF/SCBAF/SAF 9300 mg/m³: SAF:PD,PP §: SCBAF:PD,PP/SAF:PD,PP:ASCBA Escape: HiEF/SCBAE	Inh Ing Con	Irrit upper resp sys, stomach pain	Eye: Irr prompt Skin: Wash regularly Breath: Resp support Swallow: Medical attention immed	Resp sys, GI tract
Clothing: Goggles: Wash: Change: Remove: Provide:	Any poss Any poss Immed contam N.R. Immed non-imperv contam Eyewash, quick drench	NIOSH ¥: SCBAF:PD,PP/SAF:PD,PP:ASCBA Escape: GMFS/SCBAE	Inh Abs Ing Con	Irrit eyes, nose; head; gidd; conj irrit; photo; periorb edema; dysphonia, aphonia, dysphagia, productive cough; chest pain; dysp, cyan; vomit, diarr; dysuria; analgesia; fever; ict; album, hema; skin, eye burns; delirium; [carc]	Eye: Irr immed Skin: Water flush immed Breath: Resp support Swallow: Medical attention immed	Eyes, resp sys, liver, kidneys, CNS, skin
Clothing: Goggles: Wash: Change: Remove: Provide:	Reason prob Reason prob Immed wet/Prompt contam After work if reason prob contam Immed non-imperv wet/ Prompt non-imperv contam Quick drench	NIOSH/OSHA 5 mg/m³: DM 10 mg/m³: DMXSQ/SA/HiE/SCBA 25 mg/m³: PAPRDM/SA:CF 50 mg/m³: HiEF/PAPRTHiE/SCBAF/ SAF/SAT:CF 200 mg/m³: SA:PD,PP §: SCBAF:PD,PP/SAF:PD,PP:ASCBA Escape: HiEF/SCBAE	Inh Abs Ing Con	Anoxia, cyan; vis dist, central scotomas; bad taste, burning mouth, dry throat, thirst; yellowing hair; eyes, skin; anemia; liver damage	Eye: Irr immed Skin: Soap wash immed Breath: Resp support Swallow: Medical attention immed	Blood, liver, CVS, eyes, CNS
Clothing: Goggles: Wash: Change: Remove:	Reason prob Reason prob Prompt contam/daily After work if reason prob contam Prompt non-imperv contam	NIOSH/OSHA 2 mg/m³: DM 5 mg/m³: PAPRDMᴱ/SA:CFᴱ/HiEF/ SCBAF/SAF §: SCBAF:PD,PP/SAF:PD,PP:ASCBA Escape: HiEF/SCBAE	Inh Abs Ing Con	Sense of well being; head, fever, lass, profuse sweat, excess thirst, tacar, hyperpnea, cough, short breath, coma	Eye: Irr immed Skin: Soap wash immed Breath: Resp support Swallow: Medical attention immed	CVS, endocrine sys, eyes

Personal protection and sanitation (See Table 3)		Recommendations for respirator selection — maximum concentration for use (MUC) (See Table 4)	Health hazards			
			Route	Symptoms (See Table 5)	First aid (See Table 6)	Target organs (See Table 5)
Clothing: Goggles: Wash: Change: Remove: Provide:	Any poss molt/Repeat liq Any poss molt Immed wet/Prompt contam/daily After if reason prob contam Immed non-imperv wet/ Prompt non-imperv contam Quick drench	NIOSH ¥: SCBAF:PD,PP/SAF:PD,PP:ASCBA Escape: GMFOVHiE/SCBAE	Inh Abs Ing Con	Anoxia, cyan; anemia, jaun; [carc]	Eye: Irr immed Skin: Soap wash immed Breath: Resp support Swallow: Medical attention immed	Blood, liver, CVS
Clothing: Goggles: Wash: Change: Remove:	N.R. N.R. N.R. N.R. N.R.	NIOSH ¥: SCBAF:PD,PP/SAF:PD,PP:ASCBA Escape: HiEF/SCBAE	Inh Con Ing	Irrit eyes, muc memb; [carc]	Eye: Skin: (Not a dermal hazard) Breath: Resp support Swallow: Medical attention immed	Eyes, upper resp sys, GI tract
Clothing: Goggles: Wash: Change: Remove:	Repeat Reason prob Prompt contam N.R. Immed wet (flamm)	NIOSH ¥: SCBAF:PD,PP/SAF:PD,PP:ASCBA Escape: GMFOV/SCBAE	Inh Abs Ing Con	Drow, head; nau, vomit; irrit eyes, nose, throat; liver damage; kidney failure; skin irrit; [carc]	Eye: Irr immed Skin: Water wash prompt Breath: Resp support Swallow: Medical attention immed	Liver, kidney, skin, eyes
Clothing: Goggles: Wash: Change: Remove:	Repeat Any poss molt/Repeat otherwise Prompt contam After work if reason prob contam Prompt non-imperv contam	NIOSH/OSHA 10 mg/m³: CCROVDM/SA/SCBA 25 mg/m³: SA:CF/PAPROVDM* 50 mg/m³: CCRFOVHiE/SCBAF/SAF/ GMFOVHiE/PAPRTOVHiE* 300 mg/m³: SAF:PD,PP §: SCBAF:PD,PP/SAF:PD,PP:ASCBA Escape: GMFOVHiE/SCBAE	Inh Abs Ing Con	Irrit throat, eyes; head, nau, ftg, numb limbs; liver damage	Eye: Irr immed Skin: Water flush immed Breath: Resp support Swallow: Medical attention immed	Liver, skin, CNS, upper resp sys, eyes

Chemical name, structure/formula, CAS and RTECS Nos., and DOT ID and guide Nos.	Synonyms, trade names, and conversion factors	Exposure limits (TWA unless noted otherwise)	IDLH	Physical description	Chemical and physical properties		Incompatibilities and reactivities	Measurement method (See Table 1)
					MW, BP, SOL Fl.P, IP, Sp.Gr, flammability	VP, FRZ UEL, LEL		
Dipropylene glycol methyl ether CH$_3$OC$_3$H$_6$OC$_3$H$_6$OH 34590-94-8 JM1575000	Dipropylene glycol monomethyl ether, Dowanol® 50B	NIOSH/OSHA 100 ppm (600 mg/m³) ST 150 ppm (900 mg/m³) [skin]	Unknown	Colorless liquid with a mild, ether-like odor.	MW: 148.2 BP: 374°F Sol: Miscible Fl.P: 180°F IP: ?	VP(79°F): 0.4 mm FRZ: -112°F UEL: ? LEL: ?	Strong oxidizers	Char; CS$_2$; GC/FID; II(2) [#S69]
	1 ppm = 6.16 mg/m³				Sp.Gr: 0.95 Class IIIA Combustible Liquid			
Endrin C$_{12}$H$_8$Cl$_6$O 72-20-8 I01575000 2761 55	1,2,3,4,10,10-Hexachloro-6,7-epoxy-1,4,4a,5,6,7,8,8a-octahydro-1,4-endo-endo-5,8-dimethano-naphthalene; Hexadrin	NIOSH/OSHA 0.1 mg/m³ [skin]	2000 mg/m³	Colorless to tan crystalline solid with a mild, chemical odor. [insecticide]	MW: 380.9 BP: Decomposes Sol: Insoluble Fl.P: NA IP: ? Sp.Gr: 1.70 Noncombustible Solid, but may be dissolved in flammable liquids.	VP: Low MLT: 392°F (Decomposes) UEL: NA LEL: NA	Strong oxidizers, strong acids, parathion	Filter/ Chrom-102; Toluene; GC/ECD; III [#5519]
Epichlorohydrin C$_3$H$_5$OCl 106-89-8 TX4900000 2023 30	1-Chloro-2,3-epoxypropane; 2-Chloropropylene oxide; gamma-Chloropropylene oxide	NIOSH Ca See Appendix A Reduce exposure to lowest feasible concentration. OSHA 2 ppm (8 mg/m³) [skin] 1 ppm = 3.85 mg/m³	Ca [250 ppm]	Colorless liquid with a slightly irritating, chloroform-like odor.	MW: 92.5 BP: 242°F Sol: 7% Fl.P: 93°F IP: 10.60 eV Sp.Gr: 1.18 Class IC Flammable Liquid	VP: 13 mm FRZ: -54°F UEL: 21.0% LEL: 3.8%	Strong oxidizers, strong acids, certain salts, caustics, zinc, aluminum, water [Note: May polymerize in presence of strong acids and bases, particularly when hot.]	Char; CS$_2$; GC/FID; III [#1010]
EPN C$_2$H$_5$O(C$_6$H$_5$)P(S)-OC$_6$H$_4$NO$_2$ 2104-64-5 TB1925000	O-Ethyl O-P-nitrophenyl benzenephosphonothioate, O-Ethyl O-P-nitrophenyl benzenethiophosphonate	NIOSH/OSHA 0.5 mg/m³ [skin]	50 mg/m³	Yellow solid with an aromatic odor. [Note: A brown liquid above 97°F.] [pesticide]	MW: 323.3 BP: ? Sol: Insoluble Fl.P: NA IP: ? Sp.Gr(77°F): 1.27 Noncombustible Solid	VP(212°F): 0.0003 mm MLT: 97°F UEL: NA LEL: NA	Strong oxidizers	Filter; Isooctane; GC/FPD; III [#5012]

Chemical name, structure/formula, CAS and RTECS Nos., and DOT ID and guide Nos.	Synonyms, trade names, and conversion factors	Exposure limits (TWA unless noted otherwise)	IDLH	Physical description	Chemical and physical properties		Incompatibilities and reactivities	Measurement method (See Table 1)
					MW, BP, SOL Fl.P, IP, Sp.Gr, flammability	VP, FRZ UEL, LEL		
Ethanolamine NH$_2$CH$_2$CH$_2$OH 141-43-5 KJ5775000 2491 60	2-Aminoethanol, beta-Aminoethyl alcohol, Ethylolamine, 2-Hydroxyethylamine, Monoethanolamine	NIOSH/OSHA 3 ppm (8 mg/m³) ST 6 ppm (15 mg/m³) 1 ppm = 2.54 mg/m³	1000 ppm	Colorless, viscous liquid or solid (below 51°F) with an unpleasant, ammonia-like odor.	MW: 61.1 BP: 339°F Sol: Miscible Fl.P: 185°F IP: 8.96 eV Sp.Gr: 1.02 Class IIIA Combustible Liquid	VP: 0.4 mm FRZ: 51°F UEL: ? LEL: ?	Strong oxidizers, strong acids	Si gel; Methanol/ Water; GC/FID; III [#2007, Amino-ethanol Compounds]
2-Ethoxyethanol C$_2$H$_5$OCH$_2$CH$_2$OH 110-80-5 KK8050000 1171 26	Cellosolve®, EGEE, Ethylene glycol monoethyl ether	NIOSH Reduce exposure to lowest feasible concentration. OSHA 200 ppm (740 mg/m³) [skin] 1 ppm = 3.75 mg/m³	* [6000 ppm] *Not applicable because of the NIOSH REL.	Colorless liquid with a sweet, pleasant, ether-like odor.	MW: 90.1 BP: 275°F Sol: Miscible Fl.P: 110°F IP: ? Sp.Gr: 0.93 Class II Combustible Liquid	VP: 4 mm FRZ: -130°F UEL(200°F): 15.6% LEL(200°F): 1.7%	Strong oxidizers	Char/ CH$_2$Cl$_2$; Methanol; GC/FID; III [#1403, Alcohols IV]
2-Ethoxyethyl acetate CH$_3$COOCH$_2$CH$_2$OC$_2$H$_5$ 111-15-9 KK8225000 1172 26	Cellosolve® acetate, EGEEA, Ethylene glycol monoethyl ether acetate, Glycol monoethyl ether acetate	NIOSH Reduce exposure to lowest feasible concentration. OSHA 100 ppm (540 mg/m³) [skin] 1 ppm = 5.49 mg/m³	* [2500 ppm] *Not applicable because of the NIOSH REL.	Colorless liquid with a mild odor.	MW: 132.2 BP: 313°F Sol: 23% Fl.P: 124°F IP: ? Sp.Gr: 0.98 Class II Combustible Liquid	VP: 2 mm FRZ: -79°F UEL: ? LEL: 1.7%	Nitrates; strong oxidizers, alkalis & acids	Char; CS$_2$; GC/FID; III [#1450, Esters I]
Ethyl acetate CH$_3$COOC$_2$H$_5$ 141-78-6 AH5425000 1173 26	Acetic ester, Acetic ether, Ethyl ester of acetic acid, Ethyl ethanoate	NIOSH/OSHA 400 ppm (1400 mg/m³) 1 ppm = 3.66 mg/m³	10,000 ppm	Colorless liquid with an ether-like, fruity odor.	MW: 88.1 BP: 171°F Sol(77°F): 10% Fl.P: 24°F IP: 10.01 eV Sp.Gr: 0.90 Class IB Flammable Liquid	VP: 74 mm FRZ: -117°F UEL: 11.5% LEL: 2.0%	Nitrates; strong oxidizers, alkalis & acids	Char; CS$_2$; GC/FID; II(2) [#S49]

Personal protection and sanitation (See Table 3)		Recommendations for respirator selection — maximum concentration for use (MUC) (See Table 4)	Health hazards			
			Route	Symptoms (See Table 5)	First aid (See Table 6)	Target organs (See Table 5)
Clothing: Goggles: Wash: Change: Remove:	N.R. N.R. N.R. N.R. N.R.	NIOSH/OSHA 1000 ppm: SA/SCBA 2500 ppm: SA:CF 5000 ppm: SCBAF/SAF/SAT:CF §: SCBAF:PD,PP/SAF:PD,PP:ASCBA Escape: GMFOVHiE/SCBAE	Inh Abs Ing Con	Irrit eyes, nose; weak, li-head, head	Eye: Irr immed Skin: Water wash prompt Breath: Resp support Swallow: Medical attention immed	Resp sys, eyes
Clothing: Goggles: Wash: Change: Remove: Provide:	Any poss Any poss Immed contam After work if any poss contam Immed non-imperv contam Eyewash, quick drench	NIOSH/OSHA 1 mg/m³: CCROVDMFu/SA/SCBA 2.5 mg/m³: SA:CF/PAPROVDMFu 5 mg/m³: CCRFOVHiE/SCBAF/SAF/ GMFOVHiE/PAPRTOVHiE/ SAT:CF 100 mg/m³: SA:PD,PP 200 mg/m³: SAF:PD,PP §: SCBAF:PD,PP/SAF:PD,PP:ASCBA Escape: GMFOVHiE/SCBAE	Inh Abs Ing Con	Epileptiform convuls; stupor, head, dizz; abdom discomfort, nau, vomit; insom; aggressiveness, conf; lethargy, weak; anor; in animals: liver damage	Eye: Irr immed Skin: Soap wash immed Breath: Resp support Swallow: Medical attention immed	CNS, liver
Clothing: Goggles: Wash: Change: Remove: Provide:	Any poss Any poss Immed contam N.R. Immed non-imperv contam Eyewash, quick drench	NIOSH ¥: SCBAF:PD,PP/SAF:PD,PP:ASCBA Escape: GMFOVAG/SCBAE	Inh Abs Ing Con	Nau, vomit; abdom pain; resp distress, cough; cyan; irrit eyes, skin with deep pain; [carc]	Eye: Irr immed Skin: Soap wash immed Breath: Resp support Swallow: Medical attention immed	Resp sys, skin, kidneys
Clothing: Goggles: Wash: Change: Remove: Provide:	Any poss Any poss Immed contam After work if any poss contam Immed non-imperv contam Eyewash, quick drench	NIOSH/OSHA 5 mg/m³: SA/SCBA 12.5 mg/m³: SA:CF 25 mg/m³: SCBAF/SAF/SAT:CF 50 mg/m³: SA:PD,PP §: SCBAF:PD,PP/SAF:PD,PP:ASCBA Escape: GMFOVHiE/SCBAE	Inh Abs Ing Con	Miosis, lac, rhin; head; tight chest, wheez; lar spasm; salv; cyan; anor; nau, abdom cramp, diarr; para, convuls; low BP; card irregularities; irrit skin, eyes	Eye: Irr immed Skin: Soap wash immed Breath: Resp support Swallow: Medical attention immed	Resp sys, CVS, CNS, eyes, skin, blood chol

Personal protection and sanitation (See Table 3)		Recommendations for respirator selection — maximum concentration for use (MUC) (See Table 4)	Health hazards			
			Route	Symptoms (See Table 5)	First aid (See Table 6)	Target organs (See Table 5)
Clothing: Goggles: Wash: Change: Remove: Provide:	Repeat Any poss Prompt contam After work if reason prob contam Prompt non-imperv contam Eyewash	NIOSH/OSHA 30 ppm: SA*/SCBA*/CCRS* 75 ppm: SA:CF*/PAPRS* 150 ppm: CCRFS/GMFS/SCBAF/SAF 1000 ppm: SAF:PD,PP §: SCBAF:PD,PP/SAF:PD,PP:ASCBA Escape: GMFS/SCBAE	Inh Ing Con	Resp sys, skin, eye irrit; lethargy	Eye: Irr immed Skin: Water flush prompt Breath: Resp support Swallow: Medical attention immed	Skin, eyes, resp sys
Clothing: Goggles: Wash: Change: Remove:	Repeat Reason prob Prompt wet N.R. Prompt non-imperv contam	NIOSH ¥: SCBAF:PD,PP/SAF:PD,PP:ASCBA Escape: GMFOV/SCBAE	Inh Abs Ing Con	In animals: pulm irrit; hematologic effects; liver, kidneys, lung damage, irrit eyes	Eye: Irr immed Skin: Water flush prompt Breath: Resp support Swallow: Medical attention immed	In animals: lungs, eyes, blood, kidneys, liver
Clothing: Goggles: Wash: Change: Remove:	Repeat Reason prob Prompt wet N.R. Prompt non-imperv wet	¥: SCBAF:PD,PP/SAF:PD,PP:ASCBA Escape: GMFOV/SCBAE	Inh Abs Ing Con	Irrit eyes, nose; vomit; kidney damage; para	Eye: Irr immed Skin: Water flush prompt Breath: Resp support Swallow: Medical attention immed	Resp sys, eyes, GI tract
Clothing: Goggles: Wash: Change: Remove:	Repeat Reason prob Prompt wet N.R. Immed wet (flamm)	NIOSH/OSHA 1000 ppm: CCRFOV/PAPROVᵋ 10,000 ppm: GMFOV/SCBAF/SAF/ SA:CFᵋ §: SCBAF:PD,PP/SAF:PD,PP:ASCBA Escape: GMFOV/SCBAE	Inh Ing Con	Irrit eyes, nose, throat; narco; derm	Eye: Irr immed Skin: Water flush prompt Breath: Resp support Swallow: Medical attention immed	Eyes, skin, resp sys

Chemical name, structure/formula, CAS and RTECS Nos., and DOT ID and guide Nos.	Synonyms, trade names, and conversion factors	Exposure limits (TWA unless noted otherwise)	IDLH	Physical description	Chemical and physical properties		Incompatibilities and reactivities	Measurement method (See Table 1)
					MW, BP, SOL Fl.P, IP, Sp.Gr, flammability	VP, FRZ UEL, LEL		
Ethyl acrylate CH₂=CHCOOC₂H₅ 140-88-5 AT0700000 1917 27	Ethyl acrylate (inhibited), Ethyl ester of acrylic acid, Ethyl propenoate 1 ppm = 4.16 mg/m³	NIOSH Ca See Appendix A OSHA 5 ppm (20 mg/m³) ST 25 ppm (100 mg/m³) [skin]	Ca [2000 ppm]	Colorless liquid with an acrid odor.	MW: 100.1 BP: 211°F Sol: 2% Fl.P: 48°F IP: ? Sp.Gr: 0.92 Class IB Flammable Liquid	VP: 29 mm FRZ: -96°F UEL: 14% LEL: 1.4%	Oxidizers, peroxides, polymerizers, strong alkalis, moisture, chlorosulfonic acid [Note: Polymerizes readily unless an inhibitor such as hydroquinone is added.]	Char; CS₂; GC/FID; III [#1450, Esters I]
Ethylamine CH₃CH₂NH₂ 75-04-7 KH2100000 1036 68	Aminoethane, Ethylamine (anhydrous), Monoethylamine 1 ppm = 1.87 mg/m³	NIOSH/OSHA 10 ppm (18 mg/m³)	4000 ppm	Colorless gas or water-white liquid (below 62°F) with an ammonia-like odor. [Note: Shipped as a liquefied compressed gas.]	MW: 45.1 BP: 62°F Sol: Miscible Fl.P: 1°F IP: 8.86 eV Sp.Gr: 0.69 (Liquid) Flammable Gas Class IA Flammable Liquid	VP: >1 atm FRZ: -114°F UEL: 14.0% LEL: 3.5%	Strong acids; strong oxidizers; copper, tin & zinc in presence of moisture; cellulose nitrate	Si gel; H₂SO₄; GC/FID; II(3) [#S144]
Ethyl benzene CH₃CH₂C₆H₅ 100-41-4 DA0700000 1175 26	Ethylbenzol, Phenylethane 1 ppm = 4.41 mg/m³	NIOSH/OSHA 100 ppm (435 mg/m³) ST 125 ppm (545 mg/m³)	2000 ppm	Colorless liquid with an aromatic odor.	MW: 106.2 BP: 277°F Sol: 0.01% Fl.P: 55°F IP: 8.76 eV Sp.Gr: 0.87 Class IB Flammable Liquid	VP(79°F): 10 mm FRZ: -139°F UEL: 6.7% LEL: 1.0%	Strong oxidizers	Char; CS₂; GC/FID; III [#1501, Aromatic Hydro-carbons]
Ethyl bromide CH₃CH₂Br 74-96-4 KH6475000 1891 58	Bromoethane 1 ppm = 4.53 mg/m³	NIOSH See Appendix D OSHA 200 ppm (890 mg/m³) ST 250 ppm (1110 mg/m³)	3500 ppm	Colorless to yellow liquid with an ether-like odor. [Note: A gas above 101°F.]	MW: 109.0 BP: 101°F Sol: 0.9% Fl.P: <4°F IP: 10.29 eV Sp.Gr: 1.46 Class IB Flammable Liquid	VP(70°F): 400 mm FRZ: -182°F UEL: 8.0% LEL: 6.8%	Chemically-active metals such as sodium, potassium, calcium, powdered aluminum, zinc & magnesium	Char; 2-Propanol; GC/FID; III [#1011]

Chemical name, structure/formula, CAS and RTECS Nos., and DOT ID and guide Nos.	Synonyms, trade names, and conversion factors	Exposure limits (TWA unless noted otherwise)	IDLH	Physical description	Chemical and physical properties		Incompatibilities and reactivities	Measurement method (See Table 1)
					MW, BP, SOL Fl.P, IP, Sp.Gr, flammability	VP, FRZ UEL, LEL		
Ethyl butyl ketone CH₃CH₂CO[CH₂]₃CH₃ 106-35-4 MJ5250000 1 ppm = 4.75 mg/m³	Butyl ethyl ketone, 3-Heptanone	NIOSH/OSHA 50 ppm (230 mg/m³)	3000 ppm	Colorless liquid with a powerful, fruity odor.	MW: 114.2 BP: 298°F Sol: 1% Fl.P(oc): 115°F IP: 9.02 eV Sp.Gr: 0.82 Class II Combustible Liquid	VP: 4 mm FRZ: -38°F UEL: ? LEL: ?	Oxidizers, acetaldehyde, perchloric acid	Char; Methanol/CS₂; GC/FID; III [#1301, Ketones II]
Ethyl chloride CH₃CH₂Cl 75-00-3 KH7525000 1037 27	Chloroethane, Hydrochloric ether, Monochloroethane, Muriatic ether 1 ppm = 2.68 mg/m³	NIOSH See Appendix D Handle with caution in the workplace. OSHA 1000 ppm (2600 mg/m³)	20,000 ppm	Colorless gas or liquid (below 54°F) with a pungent, ether-like odor. [Note: Shipped as a liquefied compressed gas.]	MW: 64.5 BP: 54°F Sol: 0.6% Fl.P: NA (Gas) IP: 10.97 eV Sp.Gr: 0.92 (Liquid at 32°F) Flammable Gas Class IA Flammable Liquid	VP: >1 atm FRZ: -218°F UEL: 15.4% LEL: 3.8%	Chemically-active metals such as sodium, potassium, calcium, powdered aluminum, zinc & magnesium; oxidizers; water or steam [Note: Reacts with water to form hydrochloric acid.]	Char(2); CS₂; GC/FID; III [#2519]
Ethylene chlorohydrin CH₂ClCH₂OH 107-07-3 KK0875000 1135 55	2-Chloroethanol, 2-Chloroethyl alcohol, Ethylene chlorhydrin 1 ppm = 3.35 mg/m³	NIOSH/OSHA C 1 ppm (3 mg/m³) [skin]	10 ppm	Colorless liquid with a faint, ether-like odor.	MW: 80.5 BP: 262°F Sol: Miscible Fl.P: 140°F IP: 10.90 eV Sp.Gr: 1.20 Class IIIA Combustible Liquid	VP: 5 mm FRZ: -90°F UEL: 15.9% LEL: 4.9%	Strong oxidizers, strong caustics, water or steam	Char(pet); 2-Propanol/CS₂; GC/FID; III [#2513]
Ethylenediamine NH₂CH₂CH₂NH₂ 107-15-3 KH8575000 1604 29	1,2-Diaminoethane; 1,2-Ethanediamine; Ethylenediamine (anhydrous) 1 ppm = 2.50 mg/m³	NIOSH/OSHA 10 ppm (25 mg/m³)	2000 ppm	Colorless, viscous liquid with an ammonia-like odor. [Note: A solid below 47°F.] [fungicide]	MW: 60.1 BP: 241°F Sol: Miscible Fl.P: 93°F IP: 8.60 eV Sp.Gr: 0.91 Class IC Flammable Liquid	VP: 11 mm FRZ: 47°F UEL: 14.4% LEL: 4.2%	Strong acids & oxidizers, carbon tetrachloride & other chlorinated organic compounds, carbon disulfide [Note: Corrosive to metals.]	XAD-2*; DMF; HPLC/UVD; III [#2540]

Personal protection and sanitation (See Table 3)		Recommendations for respirator selection — maximum concentration for use (MUC) (See Table 4)	Health hazards				
			Route	Symptoms (See Table 5)	First aid (See Table 6)		Target organs (See Table 5)
Clothing: Goggles: Wash: Change: Remove:	Repeat Reason prob Prompt wet N.R. Immed wet (flamm)	NIOSH ¥: SCBAF:PD,PP/SAF:PD,PP:ASCBA Escape: GMFOV/SCBAE	Inh Abs Ing Con	Irrit eyes, resp sys, skin; [carc]	Eye: Skin: Breath: Swallow:	Irr immed Water flush immed Resp support Medical attention immed	Resp sys, eyes, skin
Clothing: Goggles: Wash: Change: Remove: Provide:	Reason prob Any poss Immed contam N.R. Immed wet/Immed non-imperv contam Eyewash, quick drench	NIOSH/OSHA 250 ppm: SA:CF^ε/PAPRS^ε 500 ppm: CCRFS/GMFS/SCBAF/SAF 4000 ppm: SAF:PD,PP §: SCBAF:PD,PP/SAF:PD,PP:ASCBA Escape: GMFS/SCBAE	Inh Abs Ing Con	Irrit eyes; burns skin; resp irrit; derm	Eye: Skin: Breath: Swallow:	Irr immed Water flush immed Resp support Medical attention immed	Resp sys, eyes, skin
Clothing: Goggles: Wash: Change: Remove:	Repeat Reason prob Prompt contam N.R. Immed wet (flamm)	NIOSH/OSHA 1000 ppm: PAPROV*/SA*/SCBA*/CCROV* 2000 ppm: GMFOV/SCBAF/SAF §: SCBAF:PD,PP/SAF:PD,PP:ASCBA Escape: GMFOV/SCBAE	Inh Ing Con	Irrit eyes, muc memb; head; derm; narco, coma	Eye: Skin: Breath: Swallow:	Irr immed Water flush prompt Resp support Medical attention immed	Eyes, upper resp sys, skin, CNS
Clothing: Goggles: Wash: Change: Remove:	Repeat Reason prob Prompt wet N.R. Immed wet (flamm)	OSHA 2000 ppm: SA/SCBA 3500 ppm: SA:CF/SCBAF/SAF §: SCBAF:PD,PP/SAF:PD,PP:ASCBA Escape: GMFOV/SCBAE	Inh Ing Con	Irrit eyes, resp sys, skin; CNS depres; pulm edema; liver, kidney disease; card arrhy; card arrest	Eye: Skin: Breath: Swallow:	Irr immed Soap flush prompt Resp support Medical attention immed	Skin, liver, kidneys, resp sys, CVS, CNS

Personal protection and sanitation (See Table 3)		Recommendations for respirator selection — maximum concentration for use (MUC) (See Table 4)	Health hazards				
			Route	Symptoms (See Table 5)	First aid (See Table 6)		Target organs (See Table 5)
Clothing: Goggles: Wash: Change: Remove:	Repeat Reason prob Prompt contam N.R. Prompt non-imperv contam	NIOSH/OSHA 500 ppm: CCROV*/SA*/SCBA* 1000 ppm: PAPROV*/CCRFOV 1250 ppm: SA:CF* 2500 ppm: GMFOV/SCBAF/SAF 3000 ppm: SAF:PD,PP §: SCBAF:PD,PP/SAF:PD,PP:ASCBA Escape: GMFOV/SCBAE	Inh Ing Con	Irrit eyes, muc memb; head; narco, coma; derm	Eye: Skin: Breath: Swallow:	Irr immed Water flush Resp support Medical attention immed	Eyes, skin, resp sys
Clothing: Goggles: Wash: Change: Remove:	Repeat Reason prob N.R. N.R. Immed wet (flamm)	OSHA 10,000 ppm: SA*/SCBA* 20,000 ppm: SA:CF*/SCBAF/SAF §: SCBAF:PD,PP/SAF:PD,PP:ASCBA Escape: GMFOV/SCBAE	Inh Abs Ing Con	Inco, inebriate; abdom cramps; card arrhy, card arrest; liver, kidney damage	Eye: Skin: Breath: Swallow:	Irr immed Water flush prompt Resp support Medical attention immed	Liver, kidneys, resp sys, CVS
Clothing: Goggles: Wash: Change: Remove: Provide:	Any poss Any poss Immed contam N.R. Immed non-imperv contam Eyewash, quick drench	NIOSH/OSHA 10 ppm: SCBA*/SA* §: SCBAF:PD,PP/SAF:PD,PP:ASCBA Escape: GMFOV/SCBAE	Inh Abs Ing Con	Irrit muc memb; nau, vomit; verti, inco; numb; vis dist; head; thirst; delirium; low BP; collapse, shock, coma	Eye: Skin: Breath: Swallow:	Irr immed Water flush immed Resp support Medical attention immed	Resp sys, liver, kidneys, CNS, skin, CVS
Clothing: Goggles: Wash: Change: Remove: Provide:	Any poss Any poss Immed contam/daily After work if any poss contam Immed wet/Immed non-imperv contam Eyewash (>5%), quick drench	NIOSH/OSHA 250 ppm: SA:CF^ε/PAPRS^ε 500 ppm: CCRFS/GMFS/SCBAF/SAF 2000 ppm: SAF:PD,PP §: SCBAF: PD,PP/SAF:PD,PP:ASCBA Escape: GMFS/SCBAE	Inh Abs Ing Con	Nasal irrit; primary irrit; sens derm, irrit resp sys, asthma; liver, kidney damage	Eye: Skin: Breath: Swallow:	Irr immed Water flush immed Resp support Medical attention immed	Resp sys, liver, kidneys, skin

Chemical name, structure/formula, CAS and RTECS Nos., and DOT ID and guide Nos.	Synonyms, trade names, and conversion factors	Exposure limits (TWA unless noted otherwise)	IDLH	Physical description	Chemical and physical properties		Incompatibilities and reactivities	Measurement method (See Table 1)
					MW, BP, SOL Fl.P, IP, Sp.Gr, flammability	VP, FRZ UEL, LEL		
Ethylene dibromide BrCH$_2$CH$_2$Br 106-93-4 KH9275000 1605 55	1,2-Dibromoethane; Ethylene bromide; Glycol dibromide 1 ppm = 7.81 mg/m^3	NIOSH Ca See Appendix A 0.045 ppm C 0.13 ppm [15-min] OSHA 20 ppm C 30 ppm 50 ppm [5-min max peak]	Ca [400 ppm] ACGIH A2 [skin]	Colorless liquid or solid (below 50°F) with a sweet odor.	MW: 187.9 BP: 268°F Sol: 0.4% Fl.P: NA IP: 9.45 eV Sp.Gr: 2.17 Noncombustible Liquid	VP: 12 mm FRZ: 50°F UEL: NA LEL: NA	Chemically-active metals such as sodium, potassium, calcium, hot aluminum & magnesium; liquid ammonia; strong oxidizers	Char; Benzene/ Methanol; GC/ECD; III [#1008]
Ethylene dichloride ClCH$_2$CH$_2$Cl 107-06-2 KI0525000 1184 26	1,2-Dichloroethane; Ethylene chloride; Glycol dichloride 1 ppm = 4.11 mg/m^3	NIOSH Ca See Appendix A 1 ppm (4 mg/m^3) ST 2 ppm (8 mg/m^3) OSHA 1 ppm (4 mg/m^3) ST 2 ppm (8 mg/m^3)	Ca [1000 ppm]	Colorless liquid with a pleasant, chloroform-like odor. [Note: Decomposes slowly, becomes acidic & darkens in color.]	MW: 99.0 BP: 182°F Sol: 0.9% Fl.P: 56°F IP: 11.05 eV Sp.Gr: 1.24 Class IB Flammable Liquid	VP: 64 mm FRZ: -32°F UEL: 16% LEL: 6.2%	Strong oxidizers & caustics; chemically-active metals such as aluminum or magnesium powder, sodium & potassium; liquid ammonia [Note: Decomposes to vinyl chloride & HCl above 1112°F.]	Char; CS$_2$; GC/FID; III [#1003, Haloge- nated Hydro- carbons]
Ethylene glycol dinitrate O$_2$NOCH$_2$CH$_2$ONO$_2$ 628-96-6 KW5600000 1 ppm = 6.12 mg/m^3	EGDN; 1,2-Ethanediol dinitrate; Ethylene dinitrate; Ethylene nitrate; Glycol dinitrate; Nitroglycol	NIOSH/OSHA ST 0.1 mg/m^3 [skin] [Note: OSHA stayed the start-up date of the PEL on 12/6/89 for the explosives industry until 2/1/90.]	500 mg/m^3	Colorless to yellow, oily, odorless liquid. [Note: An explosive ingredient (60-80%) in dynamite along with nitroglycerine (40-20%).]	MW: 152.1 BP: 387°F Sol: Insoluble Fl.P: 419°F IP: ? Sp. Gr: 1.49 Explosive Liquid	VP: 0.05 mm FRZ: -8°F UEL: ? LEL: ?	Acids, alkalis	Tenax GC; Ethanol; GC/ECD; III [#2507]
Ethyleneimine C$_2$H$_5$N 151-56-4 KX5075000 1185 30 (inhibited)	Aminoethylene, Azirane, Aziridine, Dimethyleneimine, Dimethylenimine, Ethylenimine, Ethylimine 1 ppm = 1.79 mg/m^3	NIOSH Ca See Appendix A OSHA [1910.1012] See Appendix B	Ca [100 ppm]	Colorless liquid with an ammonia-like odor.	MW: 43.1 BP: 133°F Sol: Miscible Fl.P: 12°F IP: 9.20 eV Sp.Gr: 0.83 Class IB Flammable Liquid	VP: 160 mm FRZ: -97°F UEL: 46% LEL: 3.6%	Polymerizes explosively in presence of acids [Note: Explosive silver derivatives may be formed with silver alloys (e.g., silver solder).]	Bub; CHCl$_3$; HPLC/UVD; II(5) [P&CAM #300]

Chemical name, structure/formula, CAS and RTECS Nos., and DOT ID and guide Nos.	Synonyms, trade names, and conversion factors	Exposure limits (TWA unless noted otherwise)	IDLH	Physical description	Chemical and physical properties		Incompatibilities and reactivities	Measurement method (See Table 1)
					MW, BP, SOL Fl.P, IP, Sp.Gr, flammability	VP, FRZ UEL, LEL		
Ethylene oxide C$_2$H$_4$O 75-21-8 KX2450000 1040 69	Dimethylene oxide; 1,2-Epoxy ethane; Oxirane 1 ppm = 1.83 mg/m^3	NIOSH Ca See Appendix A <0.1 ppm (0.18 mg/m^3) C 5 ppm (9 mg/m^3)[10-min/day] OSHA [1910.1047] 1 ppm 5 ppm [15-min Excursion]	Ca [800 ppm] ACGIH A2	Colorless gas or liquid (below 51°F) with an ether-like odor.	MW: 44.1 BP: 51°F Sol: Miscible Fl.P(oc): -20°F IP: 10.56 eV Sp.Gr: 0.82 (Liquid at 50°F) Class IA Flammable Liquid	VP: >1 atm FRZ: -171°F UEL: 100% LEL: 3.0%	Strong acids, alkalis & oxidizers; chlorides of iron, aluminum & tin; oxides of iron & aluminum	Char(pet)*; DMF; GC/ECD; III [#1614]
Ethyl ether C$_2$H$_5$OC$_2$H$_5$ 60-29-7 KI5775000 1155 26	Diethyl ether, Ethyl oxide, Ether, Diethyl oxide, Solvent ether 1 ppm = 3.08 mg/m^3	NIOSH See Appendix D OSHA 400 ppm (1200 mg/m^3) ST 500 ppm (1500 mg/m^3)	19,000 ppm [LEL]	Colorless liquid with a pungent, sweetish odor. [Note: A gas above 84°F.]	MW: 74.1 BP: 94°F Sol: 8% Fl.P: -49°F IP: 9.53 eV Sp.Gr: 0.71 Class IA Flammable Liquid	VP: 440 mm FRZ: -177°F UEL: 36.0% LEL: 1.9%	Strong oxidizers [Note: Tends to form explosive peroxides under influence of air & light.]	Char; Ethyl acetate; GC/FID; III [#1610]
Ethyl formate CH$_3$CH$_2$OCHO 109-94-4 LQ8400000 1190 26	Ethyl ester of formic acid, Ethyl methanoate 1 ppm = 3.08 mg/m^3	NIOSH/OSHA 100 ppm (300 mg/m^3)	8000 ppm	Colorless liquid with a fruity odor.	MW: 74.1 BP: 130°F Sol(64°F): 9% Fl.P: -4°F IP: 10.61 eV Sp.Gr: 0.92 Class IB Flammable Liquid	VP: 200 mm FRZ: -113°F UEL: 16.0% LEL: 2.8%	Nitrates; strong oxidizers, alkalis & acids [Note: Decomposes slowly in water to form ethyl alcohol & formic acid.]	Char; CS$_2$; GC/FID; II(2) [#S36]
Ethyl mercaptan CH$_3$CH$_2$SH 75-08-1 KI9625000 2363 27	Ethanethiol, Ethyl sulfhydrate, Mercaptoethane 1 ppm = 2.58 mg/m^3	NIOSH C 0.5 ppm (1.3 mg/m^3) [15-min] OSHA 0.5 ppm (1 mg/m^3)	2500 ppm	Colorless liquid with a strong skunk-like odor. [Note: A gas above 95°F.]	MW: 62.1 BP: 95°F Sol: 0.7% Fl.P: -55°F IP: 9.29 eV Sp.Gr: 0.84 Class IA Flammable Liquid	VP: 442 mm FRZ: -228°F UEL: 18.0% LEL: 2.8%	Strong oxidizers [Note: Reacts violently with calcium hypochlorite.]	None available

Personal protection and sanitation (See Table 3)		Recommendations for respirator selection — maximum concentration for use (MUC) (See Table 4)	Health hazards			
			Route	Symptoms (See Table 5)	First aid (See Table 6)	Target organs (See Table 5)
Clothing: Goggles: Wash: Change: Remove: Provide:	Repeat Reason prob Immed contam N.R. Immed non-imperv contam Quick drench	NIOSH ¥: SCBAF:PD,PP/SAF:PD,PP:ASCBA Escape: GMFOV/SCBAE	Inh Abs Ing Con	Irrit resp sys, eyes; derm with vesic; [carc]	Eye: Irr immed Skin: Soap wash immed Breath: Resp support Swallow: Medical attention immed	Resp sys, liver, kidneys, skin, eyes
Clothing: Goggles: Wash: Change: Remove:	Repeat Reason prob Prompt contam N.R. Immed wet (flamm)	NIOSH ¥: SCBAF:PD,PP/SAF:PD,PP:ASCBA Escape: GMFOV/SCBAE	Inh Ing Abs Con	CNS depres; nau, vomit; derm; irrit eyes, corneal opacity; [carc]	Eye: Irr immed Skin: Soap wash prompt Breath: Resp support Swallow: Medical attention immed	Kidneys, liver, eyes, skin, CNS
Clothing: Goggles: Wash: Change: Remove: Provide:	Any poss Reason prob Immed contam After work if any poss contam Immed wet (flamm) Quick drench	NIOSH/OSHA 1 mg/m³: SA*/SCBA* 2.5 mg/m³: SA:CF* 5 mg/m³: SCBAF/SAF/SAT:CF* 200 mg/m³: SAF:PD,PP §: SCBAF:PD,PP/SAF:PD,PP:ASCBA Escape: GMFOVHiE/SCBAE	Inh Abs Ing Con	Throb head; dizz; nau, vomit, abdom pain; hypotension, flush; palp; methemoglobinemia; delirium, depres CNS; angina; skin irrit; in animals: anemia; mild liver, kidney damage	Eye: Irr immed Skin: Soap wash immed Breath: Resp support Swallow: Medical attention immed	CVS, blood, skin
Clothing: Goggles: Wash: Change: Remove: Provide:	Any poss Any poss Immed contam/daily After work if any poss contam Immed contam Eyewash, quick drench	NIOSH ¥: SCBAF:PD,PP/SAF:PD,PP:ASCBA Escape: GMFOV/SCBAE	Inh Abs Con Ing	Nau, vomit; head, dizz; pulm edema; liver, kidney damage; eye burns; skin sens; irrit nose, throat; [carc]	Eye: Irr immed (15 min) Skin: Soap wash immed Breath: Resp support Swallow: Medical attention immed	Eyes, lungs, skin, liver, kidneys

Personal protection and sanitation (See Table 3)		Recommendations for respirator selection — maximum concentration for use (MUC) (See Table 4)	Health hazards			
			Route	Symptoms (See Table 5)	First aid (See Table 6)	Target organs (See Table 5)
Clothing: Goggles: Wash: Change: Remove: Provide:	Any poss Reason prob Immed contam N.R. Immed wet (flamm) Eyewash	NIOSH 5 ppm: GMFS†/SCBAF/SAF §: SCBAF:PD,PP/SAF:PD,PP:ASCBA Escape: GMFS†/SCBAE	Inh Ing Con	Irrit eyes, nose, throat; peculiar taste; head; nau, vomit, diarr; dysp, cyan, pulm edema; drow, weak, inco; EKG abnormalities; burns eyes, skin; frostbite; [carc]; in animals: convuls; liver, kidney damage	Eye: Irr immed Skin: Water flush immed Breath: Resp support Swallow: Medical attention immed	Eyes, blood, resp sys, liver, CNS, kidneys
Clothing: Goggles: Wash: Change: Remove:	Repeat Reason prob N.R. N.R. Immed wet (flamm)	OSHA 1000 ppm: CCROV*/PAPROV* 4000 ppm: SA*/SCBA* 10,000 ppm: SA:CF* 19,000 ppm: GMFOV/SCBAF/SAF §: SCBAF:PD,PP/SAF:PD,PP:ASCBA Escape: GMFOV/SCBAE	Inh Ing Con	Dizz; drow; head, excited, narco; nau, vomit; irrit eyes, upper resp sys, skin	Eye: Irr immed Skin: Water wash prompt Breath: Resp support Swallow: Medical attention immed	CNS, skin, resp sys, eyes
Clothing: Goggles: Wash: Change: Remove:	Repeat Reason prob Prompt wet N.R. Any wet immed (flamm)	NIOSH/OSHA 1000 ppm: CCRFOV/PAPROVᵋ 2500 ppm: SA:CFᵋ 5000 ppm: GMFOV/SCBAF/SAF 8000 ppm: SAF:PD,PP §: SCBAF:PD,PP/SAF:PD,PP:ASCBA Escape: GMFOV/SCBAE	Inh Ing Con	Irrit eyes, upper resp sys; narco	Eye: Irr immed Skin: Water flush immed Breath: Resp support Swallow: Medical attention immed	Eyes, resp sys
Clothing: Goggles: Wash: Change: Remove:	Repeat Reason prob Prompt contam N.R. Immed wet (flamm)	NIOSH/OSHA 5 ppm: SA/SCBA/CCROV 12.5 ppm: SA:CF/PAPROV 25 ppm: SCBAF/SAF/GMFOV/PAPRTOV/CCRFOV/SAT:CF 500 ppm: SA:PD,PP 1000 ppm: SAF:PD,PP §: SCBAF:PD,PP/SAF:PD,PP:ASCBA Escape: GMFOV/SCBAE	Inh Ing Con	Head, nau, irrit muc memb; in animals: inco; weak, pulm irrit; liver, kidney damage; cyan	Eye: Irr immed Skin: Soap wash immed Breath: Resp support Swallow: Medical attention immed	Resp sys; in animals: liver, kidneys

Chemical name, structure/formula, CAS and RTECS Nos., and DOT ID and guide Nos.	Synonyms, trade names, and conversion factors	Exposure limits (TWA unless noted otherwise)	IDLH	Physical description	Chemical and physical properties		Incompatibilities and reactivities	Measurement method (See Table 1)
					MW, BP, SOL Fl.P, IP, Sp.Gr, flammability	VP, FRZ UEL, LEL		
N-Ethylmorpholine $C_4H_8ONCH_2CH_3$ 100-74-3 QE4025000	4-Ethylmorpholine	NIOSH/OSHA 5 ppm (23 mg/m³) [skin]	2000 ppm	Colorless liquid with an ammonia-like odor.	MW: 115.2 BP: 281°F SOL: Miscible Fl.P(oc): 90°F IP: ?	VP: 6 mm FRZ: -81°F UEL: ? LEL: ?	Strong acids, strong oxidizers	Si gel; H₂SO₄; GC/FID; II(3) [#S146]
	1 ppm = 4.79 mg/m³				Sp.Gr: 0.90 Class IC Flammable Liquid			
Ethyl silicate $(C_2H_5)_4SiO_4$ 78-10-4 VV9450000	Ethyl orthosilicate, Ethyl silicate (condensed), Tetraethoxysilane, Tetraethyl silicate	NIOSH/OSHA 10 ppm (85 mg/m³)	1000 ppm	Colorless liquid with a sharp, alcohol-like odor.	MW: 208.3 BP: 336°F Sol: Reacts Fl.P: 99°F IP: 9.77 eV	VP: 1 mm FRZ: -117°F UEL: ? LEL: ?	Strong oxidizers, water [Note: Reacts with water to form a silicone adhesive (a milky-white mass).]	XAD-2; CS₂; GC/FID; II(3) [#S264]
1292 29	1 ppm = 8.66 mg/m3				Sp.Gr: 0.93 Class IC Flammable Liquid			
Ferbam $((CH_3)_2NCS_2)_3Fe$ 14484-64-1 NO8750000	tris(Dimethyldithio-carbamate)iron, Ferric dimethyl dithio-carbamate	NIOSH/OSHA 10 mg/m³	N.E.	Dark brown to black, odorless solid. [fungicide]	MW: 416.5 BP: Decomposes Sol: 0.01% Fl.P: ? IP: 7.72 eV	VP: Low MLT: >356°F (Decomposes) UEL: ? LEL: ?	Strong oxidizers	None available
					Combustible Solid			
Ferrovanadium dust FeV 12604-58-9 LK2900000	None	NIOSH/OSHA 1 mg/m³ ST 3 mg/m³	N.E.	Dark, odorless particulate dispersed in air. [Note: Ferro-vanadium metal is an alloy usu-ally containing 50-80% vana-dium.]	MW: 106.8 BP: ? Sol: Insoluble Fl.P: NA IP: NA	VP: 0 mm (approx) MLT: 2696 to 2768°F UEL: NA LEL: NA	Strong oxidizers	Filter; Acid; AA; OSHA [#ID121, #ID125G]
					Noncombustible Solid, but dust may be an explosion hazard.			

Chemical name, structure/formula, CAS and RTECS Nos., and DOT ID and guide Nos.	Synonyms, trade names, and conversion factors	Exposure limits (TWA unless noted otherwise)	IDLH	Physical description	Chemical and physical properties		Incompatibilities and reactivities	Measurement method (See Table 1)
					MW, BP, SOL Fl.P, IP, Sp.Gr, flammability	VP, FRZ UEL, LEL		
Fluorides (as F) 1: NaF 2: Na₃AlF₆ 1: 7681-49-4/ WB0350000 2: 15096-52-3/ WA9625000 1690 54 (NaF)	1: Sodium fluoride, Floridine 2: Cryolite, Cryocide, Cryodust, Sodium aluminum fluoride, Sodium hexafluoroaluminate Synonyms of other compounds vary depending upon the specific compound.	NIOSH/OSHA 2.5 mg/m³	500 mg/m³	1: Odorless white powder or colorless crystals. [Note: Pesticide grade is often dyed blue.] 2: Colorless to dark odorless solid. [Note: Loses color on heating.] [pesticide]	MW: 42.0/209.9 BP: 3099°F/ Decomposes Sol: 4/0.04% Fl.P: NA/NA IP: NA/NA Sp.Gr: 2.78/2.90 Noncombustible Solids	VP: 0/0 mm (approx) MLT: 1819/ 1832°F UEL: NA/NA LEL: NA/NA	Strong oxidizers	None available
Fluorine F₂ 7782-41-4 LM6475000 9192 25 (cryogenic liquid) 1045 20 (compressed)	Fluorine-19	NIOSH/OSHA 0.1 ppm (0.2 mg/m³)	25 ppm	Pale yellow to greenish gas with a pungent, irritating odor.	MW: 38.0 BP: -307°F Sol: Reacts Fl.P: NA IP: 15.70 eV	VP: >1 atm FRZ: -363°F UEL: NA LEL: NA	Water, nitric acid, oxidizers, organic compounds [Note: Reacts violently with all combustible materials, except the metal containers in which it is shipped. Reacts with H₂O to form hydrofluoric acid.]	None available
	1 ppm = 1.58 mg/m³				Nonflammable Gas, but an extremely strong oxidizer.			
Fluorotrichloromethane CCl₃F 75-69-4 PB6125000 1078 12 (gas)	Freon® 11, Monofluorotrichloromethane, Trichlorofluoromethane, Trichloromonofluoromethane, Refrigerant 11	NIOSH/OSHA C 1000 ppm (5600 mg/m³)	10,000 ppm	Colorless to water-white, nearly odorless liquid or gas (above 75°F).	MW: 137.4 BP: 75°F Sol(77°F): 0.1% Fl.P: NA IP: 11.77 eV	VP: 690 mm FRZ: -168°F UEL: NA LEL: NA	Chemically-active metals such as sodium, potassium, calcium, powdered aluminum, zinc, magnesium & lithium shavings; granular barium	Char; CS₂; GC/FID; III [#1006, Trichloro-fluoro-methane]
	1 ppm = 5.71 mg/m³				Sp.Gr: 1.47 (Liquid at 75°F) Noncombustible Liquid			
Formaldehyde HCHO 50-00-0 LP8925000 1198 29 (formalin) 2209 29 (formalin)	1) Gaseous form: Methanal, Methyl aldehyde, Methylene oxide 2) Aqueous solution: Formalin (30 to 50% formaldehyde by weight, which usually contains 6 to 12% methanol)	NIOSH Ca See Appendix A 0.016 ppm C 0.1 ppm [15-min] OSHA [1910.1048] 1 ppm ST 2 ppm	Ca [30 ppm] ACGIH A2	Nearly colorless gas with a pungent, suffocating odor. [Note: Often used in an aqueous solution.]	MW: 30.0 BP: -6/207 to 214°F Sol: Miscible Fl.P: NA(Gas)/ 140-185°F IP: 10.88 eV/? Sp.Gr(77°F): 1.08-1.10 (Formalin) Flammable Gas Class IIIB Combustible Liquid (Formalin)	VP: >1 atm/ 1 mm FRZ: -134°F/? UEL: 73%/? LEL: 7.0%/?	Strong oxidizers, alkalis & acids; phenols; urea [Note: Pure formaldehyde has a tendency to polymerize.]	Filter/Imp(2); none; Vis; III [#3500] [Note: Also XAD-2*; Toluene; GC/FID; #2541.]

Personal protection and sanitation (See Table 3)		Recommendations for respirator selection — maximum concentration for use (MUC) (See Table 4)	Health hazards			
			Route	Symptoms (See Table 5)	First aid (See Table 6)	Target organs (See Table 5)
Clothing: Repeat Goggles: Any poss Wash: Prompt contam Change: N.R. Remove: Immed wet (flamm) Provide: Eyewash (>15%), quick drench		NIOSH/OSHA 50 ppm: SA*/SCBA*/CCROV* 125 ppm: SA:CF*/PAPROV* 250 ppm: SA:CF*/SAF/CCRFOV/ GMFOV/PAPRTOV*/SAT:CF* 2000 ppm: SAF:PD,PP §: SCBAF:PD,PP/SAF:PD,PP:ASCBA Escape: GMFOV/SCBAE	Inh Abs Ing Con	Eye, nose, throat irrit; vis dist; severe eye irrit from splashes	Eye: Irr immed Skin: Water flush prompt Breath: Resp support Swallow: Medical attention immed	Resp sys, eyes, skin
Clothing: Repeat Goggles: Reason prob Wash: Prompt wet Change: N.R. Remove: Immed wet (flamm)		NIOSH/OSHA 100 ppm: SA*/SCBA* 250 ppm: SA:CF* 500 ppm: SCBAF/SAF 1000 ppm: SAF:PD,PP §: SCBAF:PD,PP/SAF:PD,PP:ASCBA Escape: GMFOV/SCBAE	Inh Ing Con	Irrit eyes, nose; in animals: lac; dysp, pulm edema; tremor, narco; liver, kidney damage; anemia	Eye: Irr immed Skin: Soap wash prompt Breath: Resp support Swallow: Medical attention immed	Resp sys, liver, kidneys, blood, skin
Clothing: Repeat Goggles: Reason prob Wash: Prompt contam Change: After work if reason prob contam Remove: Prompt non-imperv contam		NIOSH/OSHA 50 mg/m³: D 100 mg/m³: DXSQ*/SA*/SCBA*/HiE* 250 mg/m³: PAPRD*/SA:CF* 500 mg/m³: HiEF/SCBAF/SAF/ PAPRTHiE*/SAT:CF* 5000 mg/m³: SAF:PD,PP §: SCBAF:PD,PP/SAF:PD,PP:ASCBA Escape: HiEF/SCBAE	Inh Ing Con	Irrit eyes, resp tract; derm; GI dist	Eye: Irr immed Skin: Soap wash prompt Breath: Resp support Swallow: Medical attention immed	Resp sys, skin, GI tract
Clothing: N.R. Goggles: N.R. Wash: N.R. Change: N.R. Remove: N.R.		NIOSH/OSHA 5 mg/m³: DM* 10 mg/m³: D XSQ*/SA*/SCBA* 25 mg/m³: PAPRDM*/SA:CF* 50 mg/m³: HiEF/SCBAF/SAF/ PAPRTHiE*/SAT:CF* 500 mg/m³: SAF:PD,PP §: SCBAF:PD,PP/SAF:PD,PP:ASCBA Escape: HiEF/SCBAE	Inh Con	Irrit eyes, resp sys; in animals: bron, pneuitis	Eye: Irr immed Breath: Resp support	Resp sys, eyes

Personal protection and sanitation (See Table 3)		Recommendations for respirator selection — maximum concentration for use (MUC) (See Table 4)	Health hazards			
			Route	Symptoms (See Table 5)	First aid (See Table 6)	Target organs (See Table 5)
Clothing: Repeat Goggles: Any poss Wash: Prompt contam Change: After work if any poss contam Remove: Promptly non-imperv contam		NIOSH/OSHA 12.5 mg/m³: DM 25 mg/m³: DMXSQ*/SA*/SCBA* 62.5 mg/m³: PAPRDM*†/SA:CF* 125 mg/m3: SCBAF/SAF/HiEF† 500 mg/m3: SAF:PD,PP §: SCBAF:PD,PP/SAF:PD,PP:ASCBA Escape: HiEF†/SCBAE †Note: May need acid gas sorbent	Inh Ing Con	Irrit eyes, resp sys; nau, abdom pain, diarr; excess salv, thirst, sweat; stiff spine; derm; calcification of ligaments of ribs, pelvis	Eye: Irr immed Skin: Soap wash prompt Breath: Fresh air, art resp Swallow: Medical attention immed	Eyes, resp sys, CNS, skeleton, kidneys, skin
Clothing: Any poss Goggles: Any poss Wash: Immed contam Change: N.R. Remove: Immed non-imperv contam Provide: Eyewash, quick drench		NIOSH/OSHA 1 ppm: SA*/SCBA* 2.5 ppm: SA:CF* 5 ppm: SCBAF/SAF 25 ppm: SAF:PD,PP §: SCBAF:PD,PP/SAF:PD,PP:ASCBA Escape: GMFS⁴/SCBAE	Inh Con	Irrit eyes, nose, resp tract; lar spasm, bron spasm; pulm edema; eyes, skin burns; in animals: liver, kidney damage	Eye: Irr immed Skin: Water flush immed Breath: Resp support	Resp sys, eyes, skin; in animals: liver, kidneys
Clothing: Repeat Goggles: Any poss Wash: N.R. Change: N.R. Remove: Prompt non-imperv wet Provide: Eyewash, quick drench		NIOSH/OSHA 10,000 ppm: SA/SCBA §: SCBAF:PD,PP/SAF:PD,PP:ASCBA Escape: GMFOV/SCBAE	Inh Ing Con	Inco, tremors; derm; frostbite; card arrhy, card arrest	Eye: Irr immed Skin: Water flush immed Breath: Resp support Swallow: Medical attention immed	CVS, skin
Clothing: Reason prob Goggles: Any poss Wash: Immed contam Change: N.R. Remove: Immed non-imperv contam Provide: Eyewash, quick drench		NIOSH ¥: SCBAF:PD,PP/SAF:PD,PP:ASCBA Escape: GMFS/SCBAE	Inh Ing Con	Irrit eyes, nose, throat; lac, burns nose, cough; bron spasm, pulm irrit; derm; [carc]	Eye: Irr immed Skin: Water flush prompt Breath: Resp support Swallow: Medical attention immed	Resp sys, eyes, skin

Chemical name, structure/formula, CAS and RTECS Nos., and DOT ID and guide Nos.	Synonyms, trade names, and conversion factors	Exposure limits (TWA unless noted otherwise)	IDLH	Physical description	Chemical and physical properties		Incompatibilities and reactivities	Measurement method (See Table 1)
					MW, BP, SOL Fl.P, IP, Sp.Gr, flammability	VP, FRZ UEL, LEL		
Formic acid HCOOH 64-18-6 LQ4900000 1779 60	Formic acid, 85-95% in aqueous solution; Hydrogencarboxylic acid; Methanoic acid 1 ppm = 1.91 mg/m³	NIOSH/OSHA 5 ppm (9 mg/m³)	30 ppm	Colorless liquid with a pungent, penetrating odor. [Note: Often used in an aqueous solution. The 90% solution freezes at 20°F.]	MW: 46.0 BP: 224°F (90% soln) Sol: Miscible Fl.P(oc): 122°F (90% soln) IP: 11.05 eV Sp.Gr: 1.22 (90% soln) Class IIIA Combustible Liquid	VP: 35 mm FRZ: 20°F (90% soln) UEL: 57% (90% soln) LEL: 18% (90% soln)	Strong oxidizers, strong caustics, concentrated sulfuric acid [Note: Corrosive to metals.]	Filter/ Chrom-103 (2); Water; IC; II(5) [#S173]
Furfural C₄H₃OCHO 98-01-1 LT7000000 1199 29	Fural, 2-Furfuraldehyde, 2-Furancarboxaldehyde, Furfuraldehyde 1 ppm = 3.99 mg/m³	NIOSH See Appendix D OSHA 2 ppm (8 mg/m³) [skin]	250 ppm	Colorless to amber liquid with an almond-like odor. [Note: Darkens in light and air.]	MW: 96.1 BP: 323°F Sol: 8% Fl.P: 140°F IP: 9.21 eV Sp.Gr: 1.16 Class IIIA Combustible Liquid	VP(66°F): 1 mm FRZ: -34°F UEL: 19.3% LEL: 2.1%	Strong acids, oxidizers, strong alkalis	XAD-2*; Toluene; GC/FID; III [#2529]
Furfuryl alcohol C₄H₃OCH₂OH 98-00-0 LU9100000 2874 55	2-Furylmethanol, 2-Hydroxymethylfuran 1 ppm = 4.08 mg/m³	NIOSH/OSHA 10 ppm (40 mg/m³) ST 15 ppm (60 mg/m³) [skin]	250 ppm	Amber liquid with a faint, burning odor.	MW: 98.1 BP: 338°F Sol: Miscible Fl.P: 149°F IP: ? Sp.Gr: 1.13 Class IIIA Combustible Liquid	VP(89°F): 1 mm FRZ: 6°F UEL: 16.3% LEL: 1.8%	Strong oxidizers & acids [Note: Contact with organic acids may lead to polymerization.]	Porapak-Q; Acetone; GC/FID; III [#2505]
Glycidol C₃H₆O₂ 556-52-5 UB4375000 1 ppm = 3.08 mg/m³	2,3-Epoxy-1-propanol; Epoxypropyl alcohol; Hydroxymethyl ethylene oxide; 2-Hydroxymethyl oxiran; 3-Hydroxypropylene oxide	NIOSH/OSHA 25 ppm (75 mg/m³)	500 ppm	Colorless liquid.	MW: 74.1 BP: 320°F (Decomposes) Sol: Miscible Fl.P: 162°F IP: ? Sp.Gr: 1.12 Class IIIA Combustible Liquid	VP(77°F): 0.9 mm FRZ: -49°F UEL: ? LEL: ?	Strong oxidizers, nitrates	Char; THF; GC/FID; III [#1608]

Chemical name, structure/formula, CAS and RTECS Nos., and DOT ID and guide Nos.	Synonyms, trade names, and conversion factors	Exposure limits (TWA unless noted otherwise)	IDLH	Physical description	Chemical and physical properties		Incompatibilities and reactivities	Measurement method (See Table 1)
					MW, BP, SOL Fl.P, IP, Sp.Gr, flammability	VP, FRZ UEL, LEL		
Graphite (natural) C 7782-42-5 MD9659600	Black lead, Mineral carbon, Plumbago, Silver graphite, Stove black	NIOSH/OSHA 2.5 mg/m³ (resp)	N.E.	Steel gray to black, greasy feeling, odorless solid.	MW: 12.0 BP: ? Sol: Insoluble Fl.P: ? IP: NA Sp.Gr: 2.0-2.25 Combustible Solid	VP: 0 mm (approx) MLT: ? UEL: ? LEL: ?	Very strong oxidizers such as fluorine, chlorine trifluoride & potassium peroxide	Filter; none; Grav; III [#0500, Nuisance Dust (total)]
Hafnium and compounds (as Hf) Hf 7440-58-6 (Metal) MG4600000 (Metal 2545 40 (Metal powder, dry) 1326 32 (Metal powder, wet)	Metal: Celtium, Elemental hafnium Synonyms of other compounds vary depending upon the specific compound.	NIOSH/OSHA 0.5 mg/m³	Unknown	Metal: Highly lustrous, ductile, grayish solid.	MW: 178.5 BP: 8316°F Sol: Insoluble Fl.P: NA IP: NA Sp.Gr: 13.31 (Metal) Metal: Explosive in powder form (either dry or with <25% water). Finely divided powder can be ignited by static electricity or EVEN SPONTANEOUSLY.	VP: 0 mm (approx) MLT: 4041°F UEL: NA LEL: NA	Strong oxidizers, chlorine	Filter; Acid; PES; II(5) [#S194]
Heptachlor C₁₀H₅Cl₇ 76-44-8 PC0700000 2761 55	1,4,5,6,7,8,8-Heptachloro-3a,4,7,7a-tetrahydro-4,7-methanoindene	NIOSH Ca See Appendix A 0.5 mg/m³ [skin] OSHA 0.5 mg/m³ [skin]	Ca [700 mg/m³]	White to light tan crystals with a camphor-like odor. [insecticide]	MW: 373.4 BP: 323°F (Decomposes) Sol: Insoluble Fl.P: NA IP: ? Sp.Gr: 1.66 Noncombustible Solid, but may be dissolved in flammable liquids.	VP(77°F): 0.0003 mm MLT: 203°F UEL: NA LEL: NA	Iron, rust	Chrom-102; Toluene; GC/ECD; II(5) [#S287]
n-Heptane CH₃[CH₂]₅CH₃ 142-82-5 MI7700000 1206 27	Heptane, normal-Heptane 1 ppm = 4.17 mg/m³	NIOSH 85 ppm (350 mg/m³) C 440 ppm (1800 mg/m³) [15-min] OSHA 400 ppm (1600 mg/m³) ST 500 ppm (2000 mg/m³)	5000 ppm	Colorless liquid with a gasoline-like odor.	MW: 100.2 BP: 209°F Sol(60°F): 0.005% Fl.P: 25°F IP: 9.90 eV Sp.Gr: 0.68 Class IB Flammable Liquid	VP(72°F): 40 mm FRZ: -131°F UEL: 6.7% LEL: 1.05%	Strong oxidizers	Char; CS₂; GC/FID; III [#1500, Hydrocarbons]

Personal protection and sanitation (See Table 3)	Recommendations for respirator selection — maximum concentration for use (MUC) (See Table 4)	Route	Symptoms (See Table 5)	First aid (See Table 6)	Target organs (See Table 5)
Clothing: Any poss Goggles: Any poss Wash: Immed contam Change: N.R. Remove: Immed non-imperv contam Provide: Eyewash, quick drench	NIOSH/OSHA 30 ppm: SA*/SCBA* §: SCBAF:PD,PP/SAF:PD,PP:ASCBA Escape: GMFOVHiE/SCBAE	Inh Ing Con	Eye irrit, lac; nasal discharge; throat irrit, cough, dysp; nau; skin burns, derm	Eye: Irr immed Skin: Water flush immed Breath: Resp support Swallow: Medical attention immed	Resp sys, skin, kidneys, liver, eyes
Clothing: Repeat Goggles: Reason prob Wash: Prompt contam Change: N.R. Remove: Prompt non-imperv contam	OSHA 20 ppm: CCROV*/SA*/SCBA* 50 ppm: SA:CF*/PAPROV* 100 ppm: CCRFOV/GMFOV/SCBAF/ SAF/PAPRTOV*/SAT:CF* 250 ppm: SAF:PD,PP §: SCBAF:PD,PP/SAF:PD,PP:ASCBA Escape: GMFOV/SCBAE	Inh Abs Ing Con	Irrit eyes, upper resp sys; head; derm	Eye: Irr immed Skin: Water flush prompt Breath: Resp support Swallow: Medical attention immed	Eyes, resp sys, skin
Clothing: Reason prob Goggles: Reason prob Wash: Immed contam Change: N.R. Remove: Immed non-imperv contam Provide: Quick drench	NIOSH/OSHA 100 ppm: PAPROV*/SCBA*/SA*/CCROV* 250 ppm: PAPRTOV*/SA:CF*/CCRFOV/ GMFOV/SCBAF/SAF §: SCBAF:PD,PP/SAF:PD,PP:ASCBA Escape: GMFOV/SCBAE	Inh Abs Ing Con	Dizz; nau, diarr; diuresis; resp, body temperature, depres; vomit; in animals: eye irrit, drow	Eye: Irr immed Skin: Water flush immed Breath: Resp support Swallow: Medical attention immed	Resp sys
Clothing: Repeat Goggles: Reason prob Wash: Prompt contam Change: N.R. Remove: Prompt non-imperv contam	NIOSH/OSHA 250 ppm: SA*/SCBA* 500 ppm: SA:CF*/SCBAF/SAF §: SCBAF:PD,PP/SAF:PD,PP:ASCBA Escape: GMFOV/SCBAE	Inh Ing Con	Irrit eyes, nose, throat, skin; narco	Eye: Irr immed Skin: Water wash prompt Breath: Resp support Swallow: Medical attention immed	Eyes, skin, resp sys, CNS

Personal protection and sanitation (See Table 3)	Recommendations for respirator selection — maximum concentration for use (MUC) (See Table 4)	Route	Symptoms (See Table 5)	First aid (See Table 6)	Target organs (See Table 5)
Clothing: N.R. Goggles: N.R. Wash: N.R. Change: N.R. Remove: N.R.	NIOSH/OSHA 12.5 mg/m³: D 25 mg/m³: DXSQ/SA/SCBA 62.5 mg/m³: PAPRD/SA:CF 125 mg/m³: HiEF/PAPRTHiE/SCBAF/ SAF/SAT:CF 1250 mg/m³: SAF:PD,PP §: SCBAF:PD,PP/SAF:PD,PP:ASCBA Escape: HiEF/SCBAE	Inh Con	Cough, dysp, black sputum, pulm func impairment, lung fib	Eye: Irr immed Breath: Fresh air	Resp sys, CVS
Clothing: Any poss Goggles: Any poss Wash: Prompt contam/daily Change: After work if any poss contam Remove: Prompt non-imperv contam Provide: Eyewash, quick drench	NIOSH/OSHA 2.5 mg/m³: DM 5 mg/m³: DMXSQ/SA/SCBA 12.5 mg/m³: PAPRDM*/SA:CF* 25 mg/m³: HiEF/PAPRTHiE*/SCBAF/ SAF/SAT:CF* 250 mg/m³: SAF:PD,PP §: SCBAF:PD,PP/SAF:PD,PP:ASCBA Escape: HiEF/SCBAE	Inh Ing Con	In animals: irrit eyes, skin, muc memb; liver damage	Eye: Irr immed Skin: Soap wash prompt Breath: Resp support Swallow: Medical attention immed	Eyes, skin, muc memb
Clothing: Reason prob Goggles: Reason prob Wash: Immed contam Change: After work if any poss contam Remove: Immed non-imperv contam Provide: Quick drench	NIOSH ¥: SCBAF:PD,PP/SAF:PD,PP:ASCBA Escape: GMFOVHiE/SCBAE	Inh Abs Ing Con	In animals: tremors, convuls; liver damage; [carc]	Eye: Irr immed Skin: Soap wash immed Breath: Resp support Swallow: Medical attention immed	In animals: CNS, liver
Clothing: Repeat Goggles: Reason prob Wash: Prompt wet Change: N.R. Remove: Immed wet (flamm)	NIOSH 850 ppm: CCROV/SA/SCBA 1000 ppm: PAPROV/CCRFOV 2125 ppm: SA:CF 4250 ppm: GMFOV/SCBAF/SAF/SAT:CF 5000 ppm: SA:PD,PP §: SCBAF:PD,PP/SAF:PD,PP:ASCBA Escape: GMFOV/SCBAE	Inh Ing Con	Li-head, gidd, stupor; no appetite, nau; derm; chemical pneu; unconsciousness	Eye: Irr immed Skin: Soap wash prompt Breath: Resp support Swallow: Medical attention immed	Skin, resp sys, PNS

Chemical name, structure/formula, CAS and RTECS Nos., and DOT ID and guide Nos.	Synonyms, trade names, and conversion factors	Exposure limits (TWA unless noted otherwise)	IDLH	Physical description	Chemical and physical properties		Incompatibilities and reactivities	Measurement method (See Table 1)
					MW, BP, SOL FI.P, IP, Sp.Gr, flammability	VP, FRZ UEL, LEL		

Hexachloroethane
 	
Cl₃CCCl₃
 	
67-72-1
KI4025000
 	
9037 53
	Carbon hexachloride, Perchloroethane	NIOSH
		Ca
		See Appendix A
		1 ppm
		(10 mg/m³)
		[skin]
		OSHA
		1 ppm
		(10 mg/m³)
		[skin]
			Ca
			[300 ppm]
				Colorless crystals with a camphor-like odor.	MW: 236.7
					BP: Sublimes
					Sol(72°F):
					  0.005%
					Fl.P: NA
					IP: 11.22 eV
					
					Sp.Gr: 2.09
					Noncombustible Solid
						VP: 0.2 mm
						MLT: 368°F
						  (Sublimes)
						UEL: NA
						LEL: NA
							Alkalis; metals such as zinc, cadmium, aluminum, hot iron & mercury	Char;
								CS₂;
								GC/FID;
								III
								[#1003, Halogenated Hydrocarbons]

Hexachloronaphthalene C₁₀H₂Cl₆ 1335-87-1 QJ7350000	Halowax® 1014	NIOSH/OSHA 0.2 mg/m³ [skin]	2 mg/m³	White to light-yellow solid with an aromatic odor.	MW: 334.9 BP: 650-730°F Sol: Insoluble Fl.P: NA IP: ? Sp.Gr: 1.78 Noncombustible Solid	VP: <1 mm MLT: 279°F UEL: NA LEL: NA	Strong oxidizers	Filter; Hexane; GC/ECD; II(2) [#S100]
n-Hexane CH₃[CH₂]₄CH₃ 110-54-3 MN9275000 1208 27	Hexane, Hexyl hydride, Normal-hexane 1 ppm = 3.58 mg/m³	NIOSH/OSHA 50 ppm (180 mg/m³)	5000 ppm	Colorless liquid with a gasoline-like odor.	MW: 86.2 BP: 156°F Sol: 0.002% Fl.P: -7°F IP: 10.18 eV Sp.Gr: 0.66 Class IB Flammable Liquid	VP(77°F): 150 mm FRZ: -219°F UEL: 7.5% LEL: 1.1%	Strong oxidizers	Char; CS₂; GC/FID; III [#1500, Hydrocarbons]
2-Hexanone CH₃CO[CH₂]₃CH₃ 591-78-6 MP1400000 1 ppm = 4.17 mg/m³	Butyl methyl ketone, MBK, Methyl butyl ketone, Methyl n-butyl ketone	NIOSH 1 ppm (4 mg/m³) OSHA 5 ppm (20 mg/m³)	5000 ppm	Colorless liquid with an acetone-like odor.	MW: 100.2 BP: 262°F Sol: 2% Fl.P: 77°F IP: 9.34 eV Sp.Gr: 0.81 Class IC Flammable Liquid	VP(77°F): 4 mm FRZ: -71°F UEL: 8% LEL: ?	Strong oxidizers	Char; CS₂; GC/FID; III [#1300, Ketones I]

Chemical name, structure/formula, CAS and RTECS Nos., and DOT ID and guide Nos.	Synonyms, trade names, and conversion factors	Exposure limits (TWA unless noted otherwise)	IDLH	Physical description	Chemical and physical properties		Incompatibilities and reactivities	Measurement method (See Table 1)
					MW, BP, SOL FI.P, IP, Sp.Gr, flammability	VP, FRZ UEL, LEL		

Hexone
 	
CH₃COCH₂CH(CH₃)₂
 	
108-10-1
SA9275000
 	
1245 26
	Isobutyl methyl ketone, Methyl isobutyl ketone, 4-Methyl 2-pentanone, MIBK
	
	1 ppm = 4.17 mg/m³
		NIOSH/OSHA
		50 ppm
		(205 mg/m³)
		ST 75 ppm
		(300 mg/m³)
			3000 ppm	Colorless liquid with a pleasant odor.	MW: 100.2
					BP: 242°F
					Sol: 2%
					Fl.P: 64°F
					IP: 9.30 eV
					
					Sp.Gr: 0.80
					Class IB Flammable Liquid
						VP: 16 mm
						FRZ: -120°F
						UEL(200°F):
						  8.0%
						LEL(200°F):
						  1.2%
							Strong oxidizers, potassium tert-butoxide	Char;
								CS₂;
								GC/FID;
								III
								[#1300, Ketones I]

sec-Hexyl acetate C₈H₁₆O₂ 108-84-9 SA7525000 1233 26	1,3-Dimethylbutyl acetate; Methylisoamyl acetate [CH₃COOCH(CH₃)-CH₂CH(CH₃)₂] 1 ppm = 5.99 mg/m³	NIOSH/OSHA 50 ppm (300 mg/m³)	4000 ppm	Colorless liquid with a mild, pleasant, fruity odor.	MW: 144.2 BP: 297°F Sol: 0.08% Fl.P: 113°F IP: ? Sp.Gr: 0.86 Class II Combustible Liquid	VP: 3 mm FRZ: -83°F UEL: ? LEL: ?	Nitrates; strong oxidizers, alkalis & acids	Char; CS₂; GC/FID; III [#1450, Esters I]
Hydrazine N₂H₄ 302-01-2 MU7175000 2030 59 (<64% soln) 2029 28 (>64% soln)	Diamine, Hydrazine (anhydrous), Hydrazine base 1 ppm = 1.33 mg/m³	NIOSH Ca See Appendix A C 0.03 ppm (0.04 mg/m³) [2-hr] OSHA 0.1 ppm (0.1 mg/m³) [skin]	Ca [80 ppm] ACGIH A2	Colorless liquid with an ammonia-like odor. [Note: A solid below 36°F.]	MW: 32.1 BP: 236°F Sol: Miscible Fl.P: 99°F IP: 8.93 eV Sp.Gr: 1.01 Class IC Flammable Liquid	VP: 10 mm FRZ: 36°F UEL: 98% LEL: 2.9%	Oxidizers, hydrogen peroxide, nitric acid, metallic oxides, acids [Note: Can ignite SPONTANEOUSLY on contact with oxidizers or porous materials such as earth, wood, and cloth.]	Bub; Reagent; Vis; III [#3503]
Hydrogen bromide HBr 10035-10-6 MW3850000 1048 15	Anhydrous hydrogen bromide; Aqueous hydrogen bromide (i.e., Hydrobromic acid) 1 ppm = 3.36 mg/m³	NIOSH/OSHA C 3 ppm (10 mg/m³)	50 ppm	Colorless gas with a sharp, irritating odor. [Note: Often used in an aqueous solution.]	MW: 80.9 BP: -88°F Sol: 49% Fl.P: NA IP: 11.62 eV Nonflammable Gas	VP: >1 atm FRZ: -124°F UEL: NA LEL: NA	Strong oxidizers, strong caustics, metals, moisture [Note: Hydrobromic acid is highly corrosive to metals.]	Si gel; NaHCO₃/ Na₂CO₃; IC; III [#7903, Inorganic Acids]

Personal protection and sanitation (See Table 3)		Recommendations for respirator selection — maximum concentration for use (MUC) (See Table 4)	Health hazards			
			Route	Symptoms (See Table 5)	First aid (See Table 6)	Target organs (See Table 5)
Clothing: Goggles: Wash: Change: Remove:	Repeat N.R. Prompt contam/daily After work if reason prob contam Prompt non-imperv contam	NIOSH ¥: SCBAF:PD,PP/SAF:PD,PP:ASCBA Escape: GMFOV/SCBAE	Inh Abs Ing Con	Irrit eyes; [carc]	Eye: Skin: Breath: Swallow:	Irr immed Soap wash immed Resp support Medical attention immed	Eyes
Clothing: Goggles: Wash: Change: Remove:	Any poss molt/Repeat liq sol/Reason prob fumes Any poss molt/Reason prob liq Prompt contam/daily After work if reason prob contam Immed wet molt/Prompt non-imperv contam liq	NIOSH/OSHA 2 mg/m³: SA*/SCBA* §: SCBAF:PD,PP/SAF:PD,PP:ASCBA Escape: GMFOV/SCBAE	Inh Abs Ing Con	Acne-form derm, nau, conf, jaun, coma	Eye: Skin: Breath: Swallow:	Irr immed Soap wash prompt Resp support Medical attention immed	Liver, skin
Clothing: Goggles: Wash: Change: Remove:	Repeat Reason prob Prompt contam N.R. Immed wet (flamm)	NIOSH/OSHA 500 ppm: SA*/SCBA* 1250 ppm: SA:CF* 2500 ppm: SAT:CF*/SCBAF/SAF 5000 ppm: SAF:PD,PP §: SCBAF:PD,PP/SAF:PD,PP:ASCBA Escape: GMFOV/SCBAE	Inh Ing Con	Li-head; nau, head; numb extremities, musc weak; irrit eyes, nose, derm; chemical pneu; gidd	Eye: Skin: Breath: Swallow:	Irr immed Soap wash immed Resp support Medical attention immed	Skin, eyes, resp sys
Clothing: Goggles: Wash: Change: Remove:	Reason prob Reason prob Prompt contam N.R. Immed wet (flamm)	NIOSH 10 ppm: SA/SCBA 25 ppm: SA:CF 50 ppm: SCBAF/SAF/SAT:CF 2000 ppm: SAF:PD,PP §: SCBAF:PD,PP/SAF:PD,PP:ASCBA Escape: GMFOV/SCBAE	Inh Abs Ing Con	Irrit eyes, nose; peri neur: weak, pares; derm; head; drow	Eye: Skin: Breath: Swallow:	Irr immed Soap wash immed Resp support Medical attention immed	CNS, skin, resp sys

Personal protection and sanitation (See Table 3)		Recommendations for respirator selection — maximum concentration for use (MUC) (See Table 4)	Health hazards			
			Route	Symptoms (See Table 5)	First aid (See Table 6)	Target organs (See Table 5)
Clothing: Goggles: Wash: Change: Remove:	Repeat Reason prob Prompt wet N.R. Immed wet (flamm)	NIOSH/OSHA 500 ppm: CCROV*/SA*/SCBA* 1000 ppm: PAPROV*/CCRFOV 1250 ppm: SA:CF* 2500 ppm: GMFOV/SCBAF/SAF/SAT:CF* 3000 ppm: SAF:PD,PP §: SCBAF:PD,PP/SAF:PD,PP:ASCBA Escape: GMFOV/SCBAE	Inh Ing Con	Irrit eyes, muc memb; head; narco, coma; derm	Eye: Skin: Breath: Swallow:	Irr immed Water flush prompt Resp support Medical attention immed	Resp sys, eyes, skin, CNS
Clothing: Goggles: Wash: Change: Remove:	Repeat Reason prob Prompt wet N.R. Prompt non-imperv wet	NIOSH/OSHA 500 ppm: CCROV*/SA*/SCBA* 1000 ppm: PAPROV*/CCRFOV 1250 ppm: SA:CF* 2500 ppm: GMFOV/SCBAF/SAF 4000 ppm: SAF:PD,PP §: SCBAF:PD,PP/SAF:PD,PP:ASCBA Escape: GMFOV/SCBAE	Inh Ing Con	Head; in animals: irrit eyes, nose, throat; narco	Eye: Skin: Breath: Swallow:	Irr immed Water flush prompt Resp support Medical attention immed	CNS, eyes
Clothing: Goggles: Wash: Change: Remove: Provide:	Any poss Any poss Immed contam N.R. Immed wet (flamm) Eyewash, quick drench	NIOSH ¥: SCBAF:PD,PP/SAF:PD,PP:ASCBA Escape: SCBAE	Inh Abs Ing Con	Irrit eyes, nose, throat; temporary blindness; dizz, nau; derm; burns skin, eyes; [carc]; in animals: bron, pulm edema; liver, kidney damage; convuls	Eye: Skin: Breath: Swallow:	Irr immed Water flush immed Resp support Medical attention immed	CNS, resp sys, skin, eyes
Clothing: Goggles: Wash: Change: Remove: Provide:	Any poss Any poss Immed contam N.R. Immed non-imperv contam Eyewash, quick drench	NIOSH/OSHA 50 ppm: SA:CF£/PAPRAG£/GMFAG/SCBAF/SAF §: SCBAF:PD,PP/SAF:PD,PP:ASCBA Escape: GMFAG/SCBAE	Inh Ing Con	Irrit eyes, nose, throat; skin, eye burns	Eye: Skin: Breath: Swallow:	Irr immed Water flush immed Resp support Medical attention immed	Resp sys, eyes, skin

Chemical name, structure/formula, CAS and RTECS Nos., and DOT ID and guide Nos.	Synonyms, trade names, and conversion factors	Exposure limits (TWA unless noted otherwise)	IDLH	Physical description	Chemical and physical properties		Incompatibilities and reactivities	Measurement method (See Table 1)
					MW, BP, SOL Fl.P, IP, Sp.Gr, flammability	VP, FRZ UEL, LEL		
Hydrogen chloride HCl 7647-01-0 MW4025000 1050 15 (anhydrous)	Anhydrous hydrogen chloride; Aqueous hydrogen chloride (i.e., Hydrochloric acid, Muriatic acid) 1 ppm = 1.52 mg/m³	NIOSH/OSHA C 5 ppm (7 mg/m³)	100 ppm	Colorless to slightly yellow gas with a pungent, irritating odor. [Note: Often used in an aqueous solution.]	MW: 36.5 BP: -121°F Sol(86°F): 67% Fl.P: NA IP: 12.74 eV Nonflammable Gas	VP: >1 atm FRZ: -174°F UEL: NA LEL: NA	Metals, hydroxides, amines, alkalis [Note: Hydrochloric acid is highly corrosive to most metals.]	Si gel; NaHCO₃/ Na₂CO₃; IC; III [#7903, Inorganic Acids]
Hydrogen cyanide HCN 74-90-8 MW6825000 1051 13 1613 57 (5% acid)	Formonitrile, Hydrocyanic acid, Prussic acid 1 ppm = 1.12 mg/m³	NIOSH/OSHA ST 4.7 ppm (5 mg/m³) [skin]	50 ppm	Colorless or pale blue liquid or gas (above 78°F) with a bitter, almond-like odor.	MW: 27.0 Sol: Miscible Fl.P: 0°F IP: 13.60 eV Sp.Gr: 0.69 Class IA Flammable Liquid	VP: 630 mm FRZ: 8°F UEL: 40.0% LEL: 5.6%	Amines, oxidizers, acids, sodium hydroxide, calcium hydroxide, sodium carbonate, water, caustics, ammonia [Note: Can polymerize at 122-140°F.]	Filter/Bub; KOH; ISE; III [#7904, Cyanides] [Note: Also Soda lime Water; Vis #6010.]
Hydrogen fluoride (as F) HF 7664-39-3 MW7875000 1052 15 (anhydrous)	Anhydrous hydrogen fluoride; Aqueous hydrogen fluoride (i.e., Hydrofluoric acid); HF-A 1 ppm = 0.83 mg/m³	NIOSH 3 ppm (2.5 mg/m³) C 6 ppm 5 mg/m³ [15-min] OSHA 3 ppm ST 6 ppm	30 ppm	Colorless gas or fuming liquid (below 67°F) with a strong, irritating odor. [Note: Shipped in cylinders.]	MW: 20.0 BP: 67°F Sol: Miscible Fl.P: NA IP: 15.98 eV Sp.Gr: 1.00 (Liquid at 67°F) Nonflammable Gas	VP: >1 atm FRZ: -118°F UEL: NA LEL: NA	Metals, water or steam [Note: Corrosive to metals. Will attack glass and concrete.]	Si gel; NaHCO₃/ Na₂CO₃; IC; III [#7903, Inorganic Acids]
Hydrogen peroxide H₂O₂ 7722-84-1 MX0900000 2984 60 (8-20% soln) 2014 45 (20-52% soln) 2015 47 (>52% soln)	High-strength hydrogen peroxide, Hydrogen dioxide, Hydrogen peroxide (aqueous), Hydroperoxide, Peroxide 1 ppm = 1.41 mg/m³	NIOSH/OSHA 1 ppm (1.4 mg/m³)	75 ppm	Colorless liquid with a slightly sharp odor. [Note: The pure compound is a crystalline solid below 12°F. Often used in an aqueous solution.]	MW: 34.0 BP: 286°F Sol: Miscible Fl.P: NA IP: 10.54 eV Sp.Gr: 1.39 Noncombustible Liquid, but a powerful oxidizer.	VP(86°F): 5 mm FRZ: 12°F UEL: NA LEL: NA	Oxidizable materials, iron, copper, brass, bronze, chromium, zinc, lead, manganese, silver [Note: Contact with combustible material may result in SPONTANEOUS combustion.]	None available

Chemical name, structure/formula, CAS and RTECS Nos., and DOT ID and guide Nos.	Synonyms, trade names, and conversion factors	Exposure limits (TWA unless noted otherwise)	IDLH	Physical description	Chemical and physical properties		Incompatibilities and reactivities	Measurement method (See Table 1)
					MW, BP, SOL Fl.P, IP, Sp.Gr, flammability	VP, FRZ UEL, LEL		
Hydrogen selenide (as Se) H₂Se 7783-07-5 MX1050000 2202 18	Selenium dihydride, Selenium hydride 1 ppm = 3.37 mg/m³	NIOSH/OSHA 0.05 ppm (0.2 mg/m³)	2 ppm	Colorless gas with an odor resembling decayed horse radish.	MW: 81.0 BP: -42°F Sol(73°F): 0.9% Fl.P: NA (Gas) IP: 9.88 eV Flammable Gas	VP: >1 atm FRZ: -87°F UEL: ? LEL: ?	Strong oxidizers, acids, water, halogenated hydrocarbons	None available
Hydrogen sulfide H₂S 7783-06-4 MX1225000 1053 13	Hydrosulfuric acid, Sewer gas, Sulfuretted hydrogen 1 ppm = 1.42 mg/m³	NIOSH C 10 ppm (15 mg/m³) [10-min] OSHA 10 ppm (14 mg/m³) ST 15 ppm (21 mg/m³)	300 ppm	Colorless gas with a strong odor of rotten eggs. [Note: Sense of smell becomes rapidly fatigued & can NOT be relied upon to warn of the continuous presence of H₂S. Shipped as a liquefied compressed gas.]	MW: 34.1 BP: -77°F Sol: 0.4% Fl.P: NA (Gas) IP: 10.46 eV Flammable Gas	VP: >1 atm FRZ: -122°F UEL: 44.0% LEL: 4.0%	Strong oxidizers, strong nitric acid, metals	Dry tube/ Mol-sieve; Thermal desorp; GC/FID; II(6) [P&CAM #296]
Hydroquinone C₆H₄(OH)₂ 123-31-9 MX3500000 2662 53	1,4-Benzenediol; Dihydroxybenzene; 1,4-Dihydroxybenzene; Quinol	NIOSH C 2 mg/m³ [15-min] OSHA 2 mg/m³	Unknown	Light tan, light gray, or colorless crystals.	MW: 110.1 BP: 545°F Sol: 7% Fl.P: 329°F (Molten) IP: 7.95 eV Sp.Gr: 1.33 Combustible Solid, dust cloud may explode if ignited in an enclosed area.	VP: 0.00001 mm MLT: 338°F UEL: ? LEL: ?	Strong oxidizers, alkalis	Filter; CH₃COOH; HPLC/UVD; III [#5004]
Iodine I₂ 7553-56-2 NN1575000	Iodine crystals, Molecular iodine	NIOSH/OSHA C 0.1 ppm (1 mg/m³)	10 ppm	Violet solid with a sharp, characteristic odor.	MW: 253.8 BP: 365°F Sol: 0.01% Fl.P: NA IP: 9.31 eV Sp.Gr: 4.93 Noncombustible Solid	VP(77°F): 0.3 mm MLT: 236°F UEL: NA LEL: NA	Ammonia, acetylene, acetaldehyde, powdered aluminum, active metals, liquid chlorine	Char*; Na₂CO₃; IC; III [#6005]

Personal protection and sanitation (See Table 3)		Recommendations for respirator selection — maximum concentration for use (MUC) (See Table 4)	Route	Symptoms (See Table 5)	First aid (See Table 6)		Target organs (See Table 5)
Clothing: Goggles: Wash: Change: Remove: Provide:	Any poss pH <3/Repeat pH >3 Any poss Immed contam pH <3/Prompt wet pH >3 N.R. Immed non-imperv contam pH <3/Prompt non-imperv wet pH <3: Eyewash, quick drench	NIOSH/OSHA 50 ppm: SA*/SCBA*/CCRS* 100 ppm: SA:CF*/SCBAF/SAF/GMFS/CCRFS/PAPRS* §: SCBAF:PD,PP/SAF:PD,PP:ASCBA Escape: GMFAG/SCBAE	Inh Ing Con	Inflamm nose, throat, lar; cough, burns throat, choking; burns eyes, skin; derm; in animals: lar spasm; pulm edema	Eye: Skin: Breath: Swallow:	Irr immed Water flush immed Resp support Medical attention immed	Resp sys, skin, eyes
Clothing: Goggles: Wash: Change: Remove: Provide:	Any poss Any poss Immed contam N.R. Immed wet (flamm) Eyewash, quick drench	NIOSH/OSHA 47 ppm: SA/SCBA 50 ppm: SA:CF/SCBAF/SAF §: SCBAF:PD,PP/SAF:PD,PP:ASCBA Escape: GMFS/SCBAE	Inh Abs Ing Con	Asphy and death at high levels; weak, head, conf; nau, vomit; incr rate and depth of respiration or respiration slow and gasping	Eye: Skin: Breath: Swallow:	Irr immed Water flush immed Resp support Medical attention immed	CNS, CVS, liver, kidneys
Clothing: Goggles: Wash: Change: Remove: Provide:	Any poss Any poss Immed contam N.R. Immed non-imperv contam Eyewash, quick drench	NIOSH/OSHA 30 ppm: SA*/SCBA*/PAPRS*/CCRS*/GMFS §: SCBAF:PD,PP/SAF:PD,PP:ASCBA Escape: GMFS/SCBAE	Inh Abs Ing Con	Irrit eyes, nose, throat; pulm edema; skin, eye burns; nasal congestion; bron	Eye: Skin: Breath: Swallow:	Irr immed Water flush immed Resp support Medical attention immed	Eyes, resp sys, skin
Clothing: Goggles: Wash: Change: Remove: Provide:	Any poss Any poss Prompt contam N.R. Immed non-imperv contam Eyewash, quick drench	NIOSH/OSHA 10 ppm: SA*/SCBA* 25 ppm: SA:CF* 50 ppm: SCBAF/SAF 75 ppm: SAF:PD,PP §: SCBAF:PD,PP/SAF:PD,PP:ASCBA Escape: GMFS/SCBAE	Inh Ing Con	Irrit eyes, nose, throat; corneal ulcer; erythema, vesicles on skin; bleaching hair	Eye: Skin: Breath: Swallow:	Irr immed Water flush immed Resp support Medical attention immed	Eyes, skin, resp sys

Personal protection and sanitation (See Table 3)		Recommendations for respirator selection — maximum concentration for use (MUC) (See Table 4)	Route	Symptoms (See Table 5)	First aid (See Table 6)		Target organs (See Table 5)
Clothing: Goggles: Wash: Change: Remove:	N.R. N.R. N.R. N.R. N.R.	NIOSH/OSHA 0.5 ppm: SA/SCBA 1.25 ppm: SA:CF* 2 ppm: SCBAF/SAF §: SCBAF:PD,PP/SAF:PD,PP:ASCBA Escape: GMFS↳/SCBAE	Inh Con	Irrit eyes, nose, throat; nau, vomit, diarr; metallic taste, garlic breath; dizz, lass, ftg; in animals: pneuitis, liver damage	Breath:	Resp support	Resp sys, eyes
Clothing: Goggles: Wash: Change: Remove:	Prevent skin freezing Reason prob N.R. N.R. Immed wet (flamm)	NIOSH/OSHA 100 ppm: SA*/SCBA* 250 ppm: SA:CF* 300 ppm: SCBAF/SAF §: SCBAF:PD,PP/SAF:PD,PP:ASCBA Escape: GMFS/SCBAE	Inh Ing Con	Apnea, coma, convuls; irrit eyes: conj, pain, lac, photo, corneal vesic; irrit resp sys; dizz; head; ftg, irrity; insom; GI dist	Eye: Skin: Breath:	Irr immed Water flush immed Resp support	Resp sys, eyes
Clothing: Goggles: Wash: Change: Remove: Provide:	Repeat Any poss >7%/Reason prob <7% Prompt contam After work if reason prob contam Prompt non-imperv contam >7%: Eyewash	NIOSH/OSHA 50 mg/m³: PAPRD℃ 100 mg/m³: HiEF/SCBAF/SAF/PAPRTHiE℃/SAT:CF℃ 200 mg/m³: SAF:PD,PP §: SCBAF:PD,PP/SAF:PD,PP:ASCBA Escape: HiEF/SCBAE	Inh Ing Con	Irrit eyes: conj; kera; CNS excitement; colored urine, nau, dizz, suffocation; rapid breath; musc twitch, delirium; collapse	Eye: Skin: Breath: Swallow:	Irr immed (15 min) Water flush Fresh air Medical attention immed	Eyes, resp sys, skin, CNS
Clothing: Goggles: Wash: Change: Remove: Provide:	Any poss >7%/Repeat <7% Any poss >7%/Reason prob <7% Immed contam >7%/Prompt contam <7% N.R. Immed non-imperv contam >7%/Prompt non-imperv wet <7% 7%: Eyewash, quick drench	NIOSH/OSHA 1 ppm: SA*/SCBA* 2.5 ppm: SA:CF* 5 ppm: SCBAF/SAF 10 ppm: SAF:PD,PP §: SCBAF:PD,PP/SAF:PD,PP:ASCBA Escape: GMFAGHiE/SCBAE	Inh Ing Con	Irrit eyes, nose; lac; head; tight chest; skin burns, rash; cutaneous hypersensitivity	Eye: Skin: Breath: Swallow:	Irr immed Soap wash immed Resp support Medical attention immed	Resp sys, eyes, skin, CNS, CVS

Chemical name, structure/formula, CAS and RTECS Nos., and DOT ID and guide Nos.	Synonyms, trade names, and conversion factors	Exposure limits (TWA unless noted otherwise)	IDLH	Physical description	Chemical and physical properties		Incompatibilities and reactivities	Measurement method (See Table 1)
					MW, BP, SOL FI.P, IP, Sp.Gr, flammability	VP, FRZ UEL, LEL		
Iron oxide dust and fume (as Fe) Fe$_2$O$_3$ 1309-37-1 NO7525000 1376 37	Ferric oxide	NIOSH 5 mg/m³ OSHA 10 mg/m³	N.E.	Reddish-brown solid. [Note: Exposure to fume may occur during the arc-welding of iron.]	MW: 159.7 BP: ? Sol: Insoluble FI.P: NA IP: NA Sp.Gr: 5.24 Noncombustible Solid	VP: 0 mm (approx) MLT: 2664°F UEL: NA LEL: NA	Calcium hypochlorite	Filter; none; XRF; III [#7200, Welding & Brazing Fume]
Isoamyl acetate CH$_3$COOCH$_2$-CH$_2$CH(CH$_3$)$_2$ 123-92-2 NS9800000 1213 26	Banana oil, Isopentyl acetate, 3-Methyl-1-butanol acetate, 3-Methylbutyl ester of acetic acid, 3-Methylbutyl ethanoate 1 ppm = 5.41 mg/m³	NIOSH/OSHA 100 ppm (525 mg/m³)	3000 ppm	Colorless liquid with a banana-like odor.	MW: 130.2 BP: 288°F Sol: 0.3% FI.P: 77°F IP: ? Sp.Gr: 0.87 Class IC Flammable Liquid	VP: 4 mm FRZ: -109°F UEL: 7.5% LEL(212°F): 1.0%	Nitrates; strong oxidizers, alkalis & acids	Char; CS$_2$; GC/FID; III [#1450, Esters I]
Isoamyl alcohol (primary and secondary) 1: (CH$_3$)$_2$CHCH$_2$CH$_2$OH 2: (CH$_3$)$_2$CHCH(OH)CH$_3$ 1: 123-51-3 2: 528-75-4 1: EL5425000 1105 26	1(primary): Fermentation amyl alcohol, Fusel oil, Isobutyl carbinol, Isopentyl alcohol, 3-Methyl-1-butanol 2(secondary): 3-Methyl-2-butanol 1 ppm = 3.67 mg/m³	NIOSH/OSHA 100 ppm (360 mg/m³) ST 125 ppm (450 mg/m³)	10,000 ppm	Colorless liquids with a disagreeable odor.	MW: 88.2 BP: 270/234°F Sol(57°F): 2%/? FI.P: 109/ 95°F(oc) IP: ?/? Sp.Gr: 0.81 (at 57°F)/0.82 primary: Class II Combustible Liquid secondary: Class IC Flammable Liquid	VP: 28/1 mm FRZ: -179°F/? UEL(212°F): 9.0%/? LEL: 1.2%/?	Strong oxidizers	Char; 2-Propanol/ CS$_2$; GC/FID; III [#1402, Alcohols III]
Isobutyl acetate CH$_3$COOCH$_2$CH(CH$_3$)$_2$ 110-19-0 AI4025000 1213 26	Isobutyl ester of acetic acid, 2-Methylpropyl acetate, 2-Methylpropyl ester of acetic acid, beta-Methylpropyl ethanoate 1 ppm = 4.83 mg/m³	NIOSH/OSHA 150 ppm (700 mg/m³)	7500 ppm	Colorless liquid with a fruity, floral odor.	MW: 116.2 BP: 243°F Sol(77°F): 0.6% FI.P: 64°F IP: 9.97 eV Sp.Gr: 0.87 Class IB Flammable Liquid	VP: 13 mm FRZ: -145°F UEL: 10.5% LEL: 1.3%	Nitrates; strong oxidizers, alkalis & acids	Char; CS$_2$; GC/FID; III [#1450, Esters I]

Chemical name, structure/formula, CAS and RTECS Nos., and DOT ID and guide Nos.	Synonyms, trade names, and conversion factors	Exposure limits (TWA unless noted otherwise)	IDLH	Physical description	Chemical and physical properties		Incompatibilities and reactivities	Measurement method (See Table 1)
					MW, BP, SOL FI.P, IP, Sp.Gr, flammability	VP, FRZ UEL, LEL		
Isobutyl alcohol (CH$_3$)$_2$CHCH$_2$OH 78-83-1 NP9625000 1212 26	IBA, Isobutanol, Isopropylcarbinol, 2-Methyl-1-propanol 1 ppm = 3.08 mg/m³	NIOSH/OSHA 50 ppm (150 mg/m³)	8000 ppm	Colorless, oily liquid with a sweet, musty odor.	MW: 74.1 BP: 227°F Sol: 10% FI.P: 82°F IP: 10.12 eV Sp.Gr: 0.80 Class IC Flammable Liquid	VP: 9 mm FRZ: -162°F UEL(202°F): 10.6% LEL(123°F): 1.7%	Strong oxidizers	Char; 2-Propanol/ CS$_2$; GC/FID; III [#1401, Alcohols II]
Isophorone C$_9$H$_{14}$O 78-59-1 GW7700000	Isoacetophorone; 3,5,5-Trimethyl-2-cyclo-hexenone; 3,5,5-Trimethyl-2-cyclo-hexen-1-one 1 ppm = 5.74 mg/m³	NIOSH/OSHA 4 ppm (23 mg/m³)	800 ppm	Colorless to white liquid with a peppermint-like odor.	MW: 138.2 BP: 419°F Sol: 1% FI.P: 184°F IP: 9.07 eV Sp.Gr: 0.92 Class IIIA Combustible Liquid	VP(77°F): 0.4 mm FRZ: 17°F UEL: 3.8% LEL: 0.8%	Strong oxidizers	Char(pet); CS$_2$; GC/FID; III [#2508]
Isopropyl acetate CH$_3$COOCH(CH$_3$)$_2$ 108-21-4 AI4930000 1220 26	Isopropyl ester of acetic acid, 1-Methylethyl ester of acetic acid, 2-Propyl acetate 1 ppm = 4.25 mg/m³	NIOSH See Appendix D OSHA 250 ppm (950 mg/m³) ST 310 ppm (1185 mg/m³)	16,000 ppm	Colorless liquid with a fruity odor.	MW: 102.2 BP: 194°F Sol: 3% FI.P: 36°F IP: 9.95 eV Sp.Gr: 0.87 Class IB Flammable Liquid	VP: 42 mm FRZ: -92°F UEL: 8% LEL(100°F): 1.8%	Nitrates; strong oxidizers, alkalis & acids	Char; CS$_2$; GC/FID; II(2) [#S50]
Isopropyl alcohol (CH$_3$)$_2$CHOH 67-63-0 NT8050000 1219 26	Dimethyl carbinol, IPA, Isopropanol, 2-Propanol, sec-Propyl alcohol, Rubbing alcohol 1 ppm = 2.50 mg/m³	NIOSH/OSHA 400 ppm (980 mg/m³) ST 500 ppm (1225 mg/m³)	12,000 ppm	Colorless liquid with the odor of rubbing alcohol.	MW: 60.1 BP: 181°F Sol: Miscible FI.P: 53°F IP: 10.10 eV Sp.Gr: 0.79 Class IB Flammable Liquid	VP: 33 mm FRZ: -127°F UEL(200°F): 12.7% LEL: 2.0%	Strong oxidizers, acetaldehyde, chlorine, ethylene oxide, acids, isocyanates	Char; 2-Butanol/ CS$_2$; GC/FID; III [#1400, Alcohols I]

Personal protection and sanitation (See Table 3)		Recommendations for respirator selection — maximum concentration for use (MUC) (See Table 4)	Health hazards			
			Route	Symptoms (See Table 5)	First aid (See Table 6)	Target organs (See Table 5)
Clothing: Goggles: Wash: Change: Remove:	N.R. N.R. N.R. N.R. N.R.	NIOSH 50 mg/m^3: DMFu/SA/SCBA 125 mg/m^3: PAPRDMFu/SA:CF 250 mg/m^3: HiEF/PAPRTHiE/SCBAF/ SAF/SAT:CF 5000 mg/m^3: SA:PD,PP §: SCBAF:PD,PP/SAF:PD,PP:ASCBA Escape: HiEF/SCBAE	Inh	Benign pneumoconiosis with X-ray shadows indistinguishable from fibrotic pneumoconiosis	Breath: Resp support	Resp sys
Clothing: Goggles: Wash: Change: Remove:	Repeat Reason prob Prompt wet N.R. Immed wet (flamm)	NIOSH/OSHA 1000 ppm: CCROV/PAPROV/SA/SCBA 2500 ppm: SA:CF 3000 ppm: GMFOV/SCBAF/SAF §: SCBAF:PD,PP/SAF:PD,PP:ASCBA Escape: GMFOV/SCBAE	Inh Ing Con	Irrit eyes, nose, throat; narco; derm	Eye: Irr immed Skin: Water flush prompt Breath: Resp support Swallow: Medical attention immed	Eyes, skin, resp sys
Clothing: Goggles: Wash: Change: Remove:	Repeat Reason prob Prompt wet N.R. Prompt non-imperv wet	NIOSH/OSHA 1000 ppm: CCRFOV/PAPROV^ε 2500 ppm: SA:CF^ε 5000 ppm: SCBAF/SAF/GMFOV 10,000 ppm: SAF:PD,PP §: SCBAF:PD,PP/SAF:PD,PP:ASCBA Escape: GMFOV/SCBAE	Inh Ing Con	Irrit eyes, nose, throat; narco; head, dizz; dysp, nau, vomit, diarr; skin cracking	Eye: Irr immed Skin: Water flush prompt Breath: Resp support Swallow: Medical attention immed	Eyes, skin, resp sys
Clothing: Goggles: Wash: Change: Remove:	Repeat Reason prob Prompt wet N.R. Immed wet (flamm)	NIOSH/OSHA 1000 ppm: PAPROV^ε/CCRFOV 3750 ppm: SA:CF^ε 7500 ppm: SCBAF/SAF/GMFOV §: SCBAF:PD,PP/SAF:PD,PP:ASCBA Escape: GMFOV/SCBAE	Inh Ing Con	Head; drow; irrit eyes, upper resp sys, skin; anes	Eye: Irr immed Skin: Water flush prompt Breath: Resp support Swallow: Medical attention immed	Skin, eyes, resp sys

Personal protection and sanitation (See Table 3)		Recommendations for respirator selection — maximum concentration for use (MUC) (See Table 4)	Health hazards			
			Route	Symptoms (See Table 5)	First aid (See Table 6)	Target organs (See Table 5)
Clothing: Goggles: Wash: Change: Remove:	Repeat Reason prob Prompt wet N.R. Immed wet (flamm)	NIOSH/OSHA 500 ppm: SA*/SCBA*/CCROV* 1000 ppm: PAPROV*/CCRFOV 1250 ppm: SA:CF* 2500 ppm: GMFOV/SCBAF/SAF 8000 ppm: SAF:PD,PP §: SCBAF:PD,PP/SAF:PD,PP:ASCBA Escape: GMFOV/SCBAE	Inh Ing Con	Irrit eyes, throat; head, drow; skin irrit, cracking	Eye: Irr immed Skin: Water flush prompt Breath: Resp support Swallow: Medical attention immed	Eyes, skin, resp sys
Clothing: Goggles: Wash: Change: Remove: Provide:	Repeat Any poss Prompt wet N.R. Prompt non-imperv wet Eyewash	NIOSH/OSHA 40 ppm: SA*/SCBA*/CCROV* 100 ppm: SA:CF*/PAPROV* 200 ppm: GMFOV/SCBAF/SAF/ PAPRTOV*/SAT:CF* 800 ppm: SAF:PD,PP §: SCBAF:PD,PP/SAF:PD,PP:ASCBA Escape: GMFOV/SCBAE	Inh Ing Con	Irrit eyes, nose, throat; narco; derm	Eye: Irr immed Skin: Soap wash prompt Breath: Resp support Swallow: Medical attention immed	Resp sys, skin
Clothing: Goggles: Wash: Change: Remove:	Repeat Reason prob Prompt wet N.R. Immed wet (flamm)	OSHA 6250 ppm: SA:CF^ε 12,500 ppm: SCBAF/SAF 16,000 ppm: SAF:PD,PP §: SCBAF:PD,PP/SAF:PD,PP:ASCBA Escape: GMFOV/SCBAE	Inh Ing Con	Irrit eyes, nose, skin; derm; narco	Eye: Irr immed Skin: Water flush prompt Breath: Resp support Swallow: Medical attention immed	Eyes, skin, resp sys
Clothing: Goggles: Wash: Change: Remove:	Repeat Reason prob Prompt wet N.R. Immed wet (flamm)	NIOSH/OSHA 1000 ppm: PAPROV^ε/CCRFOV 10,000 ppm: SA:CF^ε 12,000 ppm: GMFOV/SCBAF/SAF §: SCBAF:PD,PP/SAF:PD,PP:ASCBA Escape: GMFOV/SCBAE	Inh Ing Con	Mild irrit eyes, nose, throat; drow, dizz, head; dry cracking skin	Eye: Irr immed Skin: Water flush Breath: Resp support Swallow: Medical attention immed	Eyes, skin, resp sys

Chemical name, structure/formula, CAS and RTECS Nos., and DOT ID and guide Nos.	Synonyms, trade names, and conversion factors	Exposure limits (TWA unless noted otherwise)	IDLH	Physical description	Chemical and physical properties		Incompatibilities and reactivities	Measurement method (See Table 1)
					MW, BP, SOL Fl.P, IP, Sp.Gr, flammability	VP, FRZ UEL, LEL		
Isopropylamine (CH₃)₂CHNH₂ 75-31-0 NT8400000 1221 68	2-Aminopropane, Monoisopropylamine, 2-Propylamine, sec-Propylamine 1 ppm = 2.46 mg/m³	NIOSH See Appendix D OSHA 5 ppm (12 mg/m³) ST 10 ppm (24 mg/m³)	4000 ppm	Colorless liquid with an ammonia-like odor. [Note: A gas above 91°F.]	MW: 59.1 BP: 91°F Sol: Miscible Fl.P(oc): -35°F IP: 8.72 eV Sp.Gr: 0.69 Class IA Flammable Liquid	VP: 460 mm FRZ: -150°F UEL: ? LEL: ?	Strong acids, strong oxidizers	Bub; NaOH; GC/FID; II(3) [#S147]
Isopropyl ether (CH₃)₂CHOCH(CH₃)₂ 108-20-3 TZ5425000 1159 26	Diisopropyl ether, Diisopropyl oxide, 2-Isopropoxy propane 1 ppm = 4.25 mg/m³	NIOSH/OSHA 500 ppm (2100 mg/m³) ACGIH 250 ppm (1050 mg/m³) ST 310 ppm (1320 mg/m³)	10,000 ppm	Colorless liquid with a sharp, sweet, ether-like odor.	MW: 102.2 BP: 154°F Sol: 0.2% Fl.P: -18°F IP: 9.20 eV Sp.Gr: 0.73 Class IB Flammable Liquid	VP: 119 mm FRZ: -76°F UEL: 7.9% LEL: 1.4%	Strong oxidizers, acids [Note: Unstable peroxides may form upon long contact of isopropyl ether with air.]	Char; CS₂; GC/FID; II(3) [#S368]
Isopropyl glycidyl ether C₆H₁₂O₂ 4016-14-2 TZ3500000	1,2-Epoxy-3-isopropoxy-propane; IGE; Isopropoxymethyl oxirane 1 ppm = 4.83 mg/m³	NIOSH C 50 ppm (240 mg/m³) [15-min] OSHA 50 ppm (240 mg/m³) ST 75 ppm (360 mg/m³)	1000 ppm	Colorless liquid.	MW: 116.2 BP: 279°F Sol: 19% Fl.P: 92°F IP: ? Sp.Gr: 0.92 Class IC Flammable Liquid	VP(77°F): 9 mm FRZ: ? UEL: ? LEL: ?	Strong oxidizers, strong caustics	Char; CS₂; GC/FID; II(2) [#S77]
Ketene CH₂=CO 463-51-4 OA7700000 1 ppm = 1.75 mg/m³	Carbomethene, Ethenone, Keto-ethylene	NIOSH/OSHA 0.5 ppm (0.9 mg/m³) ST 1.5 ppm (3 mg/m³)	Unknown	Colorless gas with a penetrating odor.	MW: 42.0 BP: -69°F Sol: Decomposes Fl.P: NA (Gas) IP: 9.61 eV Flammable Gas	VP: >1 atm FRZ: -238°F UEL: ? LEL: ?	Water (decomposes), alcohols, ammonia [Note: Readily polymerizes.]	Bub; FeCl₃; Vis; II(2) [#S92]

Chemical name, structure/formula, CAS and RTECS Nos., and DOT ID and guide Nos.	Synonyms, trade names, and conversion factors	Exposure limits (TWA unless noted otherwise)	IDLH	Physical description	Chemical and physical properties		Incompatibilities and reactivities	Measurement method (See Table 1)
					MW, BP, SOL Fl.P, IP, Sp.Gr, flammability	VP, FRZ UEL, LEL		
Lead (as Pb) Pb 7439-92-1 (Metal) OF7525000 (Metal) 2291 53(soluble cmpds)	Metal: Lead metal, Plumbum [Note: OSHA considers "Lead" to mean metallic Pb, all inorganic Pb cmpds (Pb oxides and Pb salts), and a class of organic Pb cmpds called soaps. All other organic Pb cmpds are excluded from this definition.]	NIOSH 0.100 mg/m³ [Air concentration to be maintained so that worker blood lead remains <0.060 mg/100 g of whole blood.] OSHA [1910.1025] 0.050 mg/m³	700 mg/m³	Metal: A heavy, ductile, soft gray solid.	MW: 207.2 BP: 3164°F Sol: Insoluble Fl.P: NA IP: NA Sp.Gr: 11.34 (Metal) Metal: Noncombustible Solid in bulk form.	VP: 0 mm (approx) MLT: 621°F UEL: NA LEL: NA	Strong oxidizers, hydrogen peroxide, acids	Filter; HNO₃/H₂O₂; AA; III [#7082]
Lindane C₆H₆Cl₆ 58-89-9 GV4900000 2761 55	BHC; gamma-Hexachlorocyclo-hexane; HCH; 1,2,3,4,5,6-Hexachlorocyclo-hexane	NIOSH/OSHA 0.5 mg/m³ [skin]	1000 mg/m³	White to yellow, crystalline powder with a slight musty odor. [pesticide]	MW: 290.8 BP: 614°F Sol: 0.001% Fl.P: NA IP: ? Sp.Gr: 1.85 Noncombustible Solid, but may be dissolved in flammable liquids.	VP: 0.00001 mm MLT: 235°F UEL: NA LEL: NA	Corrosive to metals	Filter/Bub; Isooctane; GC/EConD; III [#5502]
Lithium hydride LiH 7580-67-8 OJ6300000 1414 40 2805 40 (fused solid)	Lithium monohydride	NIOSH/OSHA 0.025 mg/m³	55 mg/m³	Odorless, off-white to gray translucent, crystalline mass or white powder. Sp.Gr: 0.78 Combustible Solid, but can form airborne dust clouds which may explode on contact with flame, heat, or oxidizers.	MW: 7.95 BP: Decomposes Sol: Reacts Fl.P: ? IP: NA	VP: 0 mm (approx) MLT: 1256°F UEL: ? LEL: ?	Strong oxidizers, halogenated hydro-carbons, acids, water [Note: May ignite SPONTANEOUSLY in air and may reignite after fire is extinguished. Reacts with water to form hydrogen & lithium hydroxide.]	None available
L.P.G. C₃H₈/C₃H₆/C₄H₁₀/C₄H₈ 68476-85-7 SE7545000 1075 22	Bottled gas, Compressed petroleum gas, Liquefied hydrocarbon gas, Liquefied petroleum gas, LPG	NIOSH/OSHA 1000 ppm (1800 mg/m³)	19,000 ppm [LEL]	Colorless, noncorrosive, odorless gas when pure. [Note: A foul-smelling odorant is usually added. A fuel mixture of propane, propylene, butanes & butylenes. Shipped as a liquefied compressed gas.]	MW: 42 to 58 BP: >-44°F Sol: ? Fl.P: NA (Gas) IP: 10.95 eV LEL: 2.1% (Propane) 1.6% (Butane) Flammable Gas	VP: >1 atm FRZ: ? UEL: 9.5%) (Propane) 8.4% (Butane)	Strong oxidizers, chlorine dioxide	Combustible gas meter none; none; II(2) [#S93]

Personal protection and sanitation (See Table 3)		Recommendations for respirator selection — maximum concentration for use (MUC) (See Table 4)	Health hazards			
			Route	Symptoms (See Table 5)	First aid (See Table 6)	Target organs (See Table 5)

Clothing:
Goggles:
Wash:
Change:
Remove:
Provide:
	Reason prob
	Any poss
	Immed wet
	N.R.
	Immed wet (flamm)
	Eyewash, quick drench
		OSHA
		125 ppm: SA:CFᴱ/PAPRSᴱ
		250 ppm: CCRFS/GMFS/SCBAF/SAF/
		  PAPRTSᴱ
		4000 ppm: SAF:PD,PP
		§: SCBAF:PD,PP/SAF:PD,PP:ASCBA
		Escape: GMFS/SCBAE
			Inh
			Abs
			Inh
			Con
				Irrit eyes, nose, throat,
				skin; pulm edema; vis
				dist; skin, eye burns;
				derm
					Eye: Irr immed
					Skin: Water flush immed
					Breath: Resp support
					Swallow: Medical attention
					  immed
						Resp sys, skin,
						eyes

Clothing: Goggles: Wash: Change: Remove:	Repeat Reason prob Prompt wet N.R. Immed wet	ACGIH 1000 ppm: PAPROV*/CCROV* 2500 ppm: SA*/SCBA* 6250 ppm: SA:CF* 10,000 ppm: SCBAF/SAF/GMFOV §: SCBAF:PD,PP/SAF:PD,PP:ASCBA Escape: GMFOV/SCBAE	Inh Ing Con	Irrit eyes, nose; resp discomfort; derm; in animals: drow, dizz, unconsciousness, narco	Eye: Irr immed Skin: Soap wash prompt Breath: Resp support Swallow: Medical attention immed	Resp sys, skin
Clothing: Goggles: Wash: Change: Remove:	Repeat Reason prob Prompt contam N.R. Immed wet (flamm)	NIOSH/OSHA 1000 ppm: SA:CFᴱ §: SCBAF:PD,PP/SAF:PD,PP:ASCBA Escape: GMFOV/SCBAE	Inh Ing Con	Irrit eyes, upper resp sys, skin; skin sens	Eye: Irr immed Skin: Soap wash immed Breath: Resp support Swallow: Medical attention immed	Eyes, skin, resp sys
Clothing: Goggles: Wash: Change: Remove:	N.R. N.R. N.R. N.R. N.R.	NIOSH/OSHA 5 ppm: SA*/SCBA* 12.5 ppm: SA:CFᴱ 25 ppm: SCBAF/SAF §: SCBAF:PD,PP/SAF:PD,PP:ASCBA Escape: GMFOV/SCBAE	Inh Con	Irrit eyes, nose, throat, lungs; pulm edema	Breath: Resp support	Resp sys, eyes, skin

Personal protection and sanitation (See Table 3)		Recommendations for respirator selection — maximum concentration for use (MUC) (See Table 4)	Health hazards			
			Route	Symptoms (See Table 5)	First aid (See Table 6)	Target organs (See Table 5)

Clothing:
Goggles:
Wash:
Change:
Remove:
	Repeat
	Reason prob
	Daily
	N.R.
	Prompt non-imperv
	contam
		OSHA
		0.5 mg/m³: SA/HiE/SCBA
		1.25 mg/m³: PAPRHiE/SA:CF
		2.5 mg/m³: HiEF/PAPRTHiE/SCBAF/
		  SAF/SAT:CF
		50 mg/m³: SA:PD,PP
		100 mg/m³: SAF:PD,PP
		§: SCBAF:PD,PP/SAF:PD,PP:ASCBA
		Escape: HiEF/SCBAE
			Inh
			Ing
			Con
				Weak, lass, insom; facial
				pallor, pal eye, anor,
				low-wgt, malnut; constip,
				abdom pain, colic; anemia;
				gingival lead line;
				tremor; para wrist, ankles;
				encephalopathy;
				nephropathy; irrit eyes;
				hypotension
					Eye: Irr immed
					Skin: Soap flush prompt
					Breath: Resp support
					Swallow: Medical attention
					  immed
						GI tract, CNS,
						kidneys, blood,
						gingival tissue

Clothing:
Goggles:
Wash:
Change:
Remove:
Provide:
	Reason prob
	Reason prob
	Immed contam
	After work if reason
	prob contam
	Immed non-imperv contam
	liq
	Quick drench
		NIOSH/OSHA
		5 mg/m³: CCROVDMFu/SA/SCBA
		12.5 mg/m³: PAPROVDMFu*/SA:CF*
		25 mg/m³: CCRFOVHiE/SCBAF/SAF/
		  GMFOVHiE/PAPRTOVHiE*
		1000 mg/m³: SAF:PD,PP
		§: SCBAF:PD,PP/SAF:PD,PP:ASCBA
		Escape: GMFOVHiE/SCBAE
			Inh
			Abs
			Ing
			Con
				Irrit eyes, nose, throat;
				head; nau; clonic convuls;
				resp difficulty; cyan;
				aplastic anemia; skin
				irrit; musc spasm;
				in animals: liver, kidney
				damage
					Eye: Irr immed
					Skin: Soap wash prompt
					Breath: Resp support
					Swallow: Medical attention
					  immed
						Eyes, CNS,
						blood, liver,
						kidneys, skin

Clothing:
Goggles:
Wash:
Change:
Remove:
Provide:
	Any poss sol/Repeat
	air >0.1 mg/m³
	Any poss
	Brush/Immed contam
	After work if any poss
	contam
	Immed contam
	Eyewash/>0.5 mg/m³:
	quick drench
		NIOSH/OSHA
		0.25 mg/m³: SA/SCBA/HiE
		0.625 mg/m³: SA:CF*/PAPRHiE*
		1.25 mg/m³: HiEF/SCBAF/SAF/
		  PAPRTHiE*
		50 mg/m³: SAF:PD,PP
		§: SCBAF:PD,PP/SAF:PD,PP:ASCBA
		Escape: HiEF/SCBAE
			Inh
			Ing
			Con
				Burns eyes, skin; burns
				mouth, esophagus (if
				ingested); nau; musc
				twitches; mental conf;
				blurred vision
					Eye: Irr immed
					Skin: Water flush immed
					Breath: Resp support
					Swallow: Medical attention
					  immed
						Resp sys, skin,
						eyes

Clothing: Goggles: Wash: Change: Remove:	Prevent skin freezing Reason prob N.R. N.R. Immed wet (flamm)	NIOSH/OSHA 10,000 ppm: SA/SCBA 19,000 ppm: SA:CF/SCBAF/SAF/SAT:CF §: SCBAF:PD,PP/SAF:PD,PP:ASCBA Escape: SCBAE	Inh Con	Li-head, drow	Eye: Irr immed Skin: Water flush immed Breath: Resp support	Resp sys, CNS

Chemical name, structure/formula, CAS and RTECS Nos., and DOT ID and guide Nos.	Synonyms, trade names, and conversion factors	Exposure limits (TWA unless noted otherwise)	IDLH	Physical description	Chemical and physical properties		Incompatibilities and reactivities	Measurement method (See Table 1)
					MW, BP, SOL Fl.P, IP, Sp.Gr, flammability	VP, FRZ UEL, LEL		
Magnesium oxide fume MgO 1309-48-4 OM3850000	Magnesia fume	NIOSH See Appendix D OSHA 10 mg/m³	N.E.	Finely divided white particulate dispersed in air. [Note: Exposure may occur when magnesium is burned, thermally cut, or welded upon.]	MW: 40.3 BP: 6512°F Sol(86°F): 0.009% Fl.P: NA IP: NA Sp.Gr: 3.58 Noncombustible Solid	VP: 0 mm (approx) MLT: 5072°F UEL: NA LEL: NA	Chlorine trifluoride, phosphorus penta-chloride	Filter; Acid; ICP; III [#7300, Elements]
Malathion C₁₀H₁₉O₆PS₂ 121-75-5 WM8400000 2783 55	S-[1,2-bis(ethoxycarbonyl) ethyl]0,0-dimethyl-phosphorodithioate; Diethyl (dimethoxyphosphino-thioylthio) succinate	NIOSH/OSHA 10 mg/m³ [skin]	5000 mg/m³	Deep brown to yellow liquid with a garlic-like odor. [Note: A solid below 37°F.] [insecticide]	MW: 330.4 BP: 140°F (Decomposes) Sol: 0.02% Fl.P(oc): >325°F IP: ? Sp.Gr: 1.21 Class IIIB Combustible Liquid, but may be difficult to ignite.	VP: 0.00004 mm FRZ: 37°F UEL: ? LEL: ?	Strong oxidizers, magnesium, alkaline pesticides [Note: Corrosive to metals.]	Filter; Isooctane; GC/FPD; III [#5012]
Maleic anhydride C₄H₂O₃ 108-31-6 ON3675000 2215 60	cis-Butenedioic anhydride; 2,5-Furanedione; Maleic acid anhydride; Toxilic anhydride 1 ppm = 4.08 mg/m³	NIOSH/OSHA 1 mg/m³ (0.25 ppm)	Unknown	Colorless needles, white lumps, or pellets with an irritating, choking odor.	MW: 98.1 BP: 396°F Sol: Reacts Fl.P: 218°F IP: 9.90 eV Sp.Gr: 1.48 Combustible Solid, but may be difficult to ignite.	VP: 0.2 mm MLT: 127°F UEL: 7.1% LEL: 1.4%	Strong oxidizers, water, alkalis, metals, caustics & amines above 150°F [Note: Reacts slowly with water (hydro-lyzes) to form maleic acid.]	Bub; none; HPLC/UVD; II(5) [P&CAM #302]
Manganese compounds (as Mn) Mn 7439-96-5 (Metal) OO9275000 (Metal)	Metal: Colloidal manganese, Manganese-55 Synonyms of other com-pounds vary depending upon the specific com-pound.	NIOSH 1 mg/m³ ST 3 mg/m³ OSHA C 5 mg/m³	N.E.	Metal: A lustrous brittle, silvery solid.	MW: 54.9 BP: 3564°F Sol: Insoluble Fl.P: ? IP: NA Sp.Gr: 7.20 Combustible Solid	VP: 0 mm (approx) MLT: 2271°F UEL: ? LEL: ?	Oxidizers [Note: Will react with water or steam to produce hydrogen.]	Filter; Acid; ICP; III [#7300, Elements]

Chemical name, structure/formula, CAS and RTECS Nos., and DOT ID and guide Nos.	Synonyms, trade names, and conversion factors	Exposure limits (TWA unless noted otherwise)	IDLH	Physical description	Chemical and physical properties		Incompatibilities and reactivities	Measurement method (See Table 1)
					MW, BP, SOL Fl.P, IP, Sp.Gr, flammability	VP, FRZ UEL, LEL		
Mercury vapor Hg 7439-97-6 OV4550000 2809 60	Colloidal mercury, Metallic mercury, Quicksilver	NIOSH/OSHA 0.05 mg/m³ [skin]	28 mg/m³	Silver-white, heavy, odorless liquid.	MW: 200.6 BP: 674°F Sol: Insoluble Fl.P: NA IP: ? Sp.Gr: 13.6 Noncombustible Liquid	VP: 0.0012 mm FRZ: -38°F UEL: NA LEL: NA	Acetylene, ammonia, chlorine dioxide, azides, calcium (amalgam formation), sodium carbide, lithium, rubidium, copper	Hydrar; Acid; AA cold; III [#6009]
Mercury (organo) alkyl compounds (as Hg) Mercury (organo) alkyl	Synonyms vary depending upon the specific (organo) alkyl compound.	NIOSH/OSHA 0.01 mg/m³ ST 0.03 mg/m³ [skin]	10 mg/m³	Appearance and odor vary depending upon the specific (organo) alkyl compound.	Properties vary depending upon the specific (organo) alkyl compound.		Strong oxidizers such as chlorine	None available
Mesityl oxide (CH₃)₂C=CHCOCH₃ 141-79-7 SB4200000 1229 26	Isobutenyl methyl ketone, Isopropylideneacetone, Methyl isobutenyl ketone, 4-Methyl-3-penten-2-one 1 ppm = 4.08 mg/m³	NIOSH 10 ppm (40 mg/m³) OSHA 15 ppm (60 mg/m³) ST 25 ppm (100 mg/m³)	5000 ppm	Oily, colorless to light-yellow liquid with a peppermint- or honey-like odor.	MW: 98.2 BP: 266°F Sol: 3% Fl.P: 87°F IP: 9.08 eV Sp.Gr(59°F): 0.86 Class IC Flammable Liquid	VP: 9 mm FRZ: -52°F UEL: 7.2% LEL: 1.4%	Oxidizers, acids	Char; Methanol/ CS₂; GC/FID; III [#1301, Ketones II]
Methoxychlor Cl₃CCH(C₆H₄OCH₃)₂ 72-43-5 KJ3675000 2761 55	p,p'-Dimethoxydiphenyltri-chloroethane; DMDT; Methoxy-DDT; 2,2-bis(p-Methoxyphenyl)-1,1,1-trichloroethane; 1,1,1-Trichloro-2,2-bis-(p-methoxyphenyl)ethane	NIOSH Ca See Appendix A OSHA 10 mg/m³	Ca [N.E.]	Colorless to light yellow crystals with a slight, fruity odor. [insecticide]	MW: 345.7 BP: Decomposes Sol: Insoluble Fl.P: ? IP: ? Sp.Gr(77°F): 1.41 Combustible Solid, but difficult to burn.	VP: Low MLT: 171-192°F UEL: ? LEL: ?	Oxidizers	Filter; Isooctane; GC/ECD; II(4) [#S371]

Personal protection and sanitation (See Table 3)	Recommendations for respirator selection — maximum concentration for use (MUC) (See Table 4)	Health hazards			
		Route	Symptoms (See Table 5)	First aid (See Table 6)	Target organs (See Table 5)
Clothing: N.R. Goggles: N.R. Wash: N.R. Change: N.R. Remove: N.R.	OSHA 100 mg/m³: DMFu/SA/SCBA 250 mg/m³: SA:CF/PAPRDMFu 500 mg/m³: HiEF/SCBAF/SAF/ PAPRTHiE* 5000 mg/m³: SAF:PD,PP §: SCBAF:PD,PP/SAF:PD,PP:ASCBA Escape: HiEF/SCBAE	Inh Con	Irrit eyes, nose; metal fume fever: cough, chest pain, flu-like fever	Breath: Resp support	Resp sys, eyes
Clothing: Repeat Goggles: Reason prob Wash: Prompt contam Change: N.R. Remove: Prompt non-imperv contam	NIOSH/OSHA 100 mg/m³: SA/SCBA/CCROVDMFu 250 mg/m³: SA:CF*/PAPROVDMFu* 500 mg/m³: SCBAF/SAF/CCRFOVHiE/ GMFOVHiE/PAPRTOVHiE*/ SAT:CF* 5000 mg/m³: SAF:PD,PP §: SCBAF:PD,PP/SAF:PD,PP:ASCBA Escape: GMFOVHiE/SCBAE	Inh Abs Ing Con	Miosis, aching eyes, blurred vision, lac; eye, skin irrit; salv; anor, nau, vomit, abdom cramps, diarr, gidd, conf, ataxia; rhin, head; tight chest, wheez, lar spasm	Eye: Irr immed Skin: Soap wash prompt Breath: Resp support Swallow: Medical attention immed	Resp sys, liver, blood chol, CNS, CVS, GI tract
Clothing: Repeat Goggles: Any poss Wash: Prompt contam Change: N.R. Remove: Prompt non-imperv contam Provide: Eyewash	NIOSH/OSHA 25 mg/m³: SA:CF£ 50 mg/m³: SCBAF/SAF 2000 mg/m³: SAF:PD,PP §: SCBAF:PD,PP/SAF:PD,PP:ASCBA Escape: GMFOVHiE/SCBAE	Inh Ing Con	Conj; photo, double vision; nasal, upper resp irrit; bronchial asthma; derm	Eye: Irr immed Skin: Soap wash immed Breath: Resp support Swallow: Medical attention immed	Eyes, resp sys, skin
Clothing: N.R. Goggles: N.R. Wash: N.R. Change: N.R. Remove: N.R.	NIOSH 10 mg/m³: DMXSQ^/SA/SCBA 25 mg/m³: PAPRDM^/SA:CF 50 mg/m³: HiEF/PAPRTHiE/SCBAF/ SAF/SAT:CF 1000 mg/m³: SA:PD,PP 2000 mg/m³: SAF:PD,PP §: SCBAF:PD,PP/SAF:PD,PP:ASCBA Escape: HiEF/SCBAE	Inh Ing	Parkinson's; asthenia, insom, mental conf; metal fume fever: dry throat, cough, tight chest, dysp, rales, flu-like fever; low-back pain; vomit; mal; ftg	Breath: Resp support Swallow: Medical attention immed	Resp sys, CNS, blood, kidneys

Personal protection and sanitation (See Table 3)	Recommendations for respirator selection — maximum concentration for use (MUC) (See Table 4)	Health hazards			
		Route	Symptoms (See Table 5)	First aid (See Table 6)	Target organs (See Table 5)
Clothing: Repeat Goggles: N.R. Wash: Prompt contam Change: After work if reason prob contam Remove: Prompt non-imperv contam	NIOSH/OSHA 0.5 mg/m³: CCRS†/SA/SCBA 1.25 mg/m³: SA:CF/PAPRS†(canister) 2.5 mg/m³: SCBAF/SAF/SAT:CF/ CCRFS†/GMFS†/ PAPRTS(canister) 28 mg/m³: SA:PD,PP §: SCBAF:PD,PP/SAF:PD,PP:ASCBA Escape: GMFS/SCBAE	Inh Abs Con	Cough, chest pain, dysp, bron pneuitis; tremor, head, ftg, weak; stomatitis, salv; GI dist, anor, low-wgt; prot; irrit eyes, skin	Eye: Irr immed Skin: Soap wash prompt Breath: Resp support Swallow: Medical attention immed	Skin, resp sys, CNS, kidneys, eyes
Clothing: Any poss Goggles: Any poss Wash: Immed contam Change: After work if any poss contam Remove: Immed non-imperv contam Provide: Eyewash, quick drench	NIOSH/OSHA 0.1 mg/m³: SA/SCBA 0.25 mg/m³: SA:CF 0.5 mg/m³: SCBAF/SAF/SAT:CF 10 mg/m³: SA:PD,PP §: SCBAF:PD,PP/SAF:PD,PP:ASCBA Escape: SCBAE	Inh Abs Ing Con	Pares; ataxia, dysarthria; vision, hearing dist; spastic, jerky; dizz; salv; lac; nau, vomit, diarr, constip; skin burns; emotional dist	Eye: Irr immed Skin: Soap wash immed Breath: Resp support Swallow: Medical attention immed	CNS, kidneys, eyes, skin
Clothing: Reason prob Goggles: Reason prob Wash: Immed contam Change: N.R. Remove: Immed wet (flamm) Provide: Quick drench	NIOSH 250 ppm: SA:CF£/PAPROV£ 500 ppm: CCRFOV/GMFOV/SCBAF/SAF 5000 ppm: SAF:PD,PP §: SCBAF:PD,PP/SAF:PD,PP:ASCBA Escape: GMFOV/SCBAE	Inh Ing Con	Irrit eyes, skin, muc memb; narco, coma; in animals: CNS effects	Eye: Irr immed Skin: Water flush immed Breath: Resp support Swallow: Medical attention immed	Eyes, skin, resp sys, CNS
Clothing: Repeat Goggles: N.R. Wash: Prompt contam Change: N.R. Remove: Prompt non-imperv contam	NIOSH ¥: SCBAF:PD,PP/SAF:PD,PP:ASCBA Escape: GMFOVHiE/SCBAE	Inh Ing	None known in humans; in animals: fasc trembling, convuls; kidney, liver damage; [carc]	Skin: Soap wash Breath: Fresh air Swallow: Medical attention immed	None known

Chemical name, structure/formula, CAS and RTECS Nos., and DOT ID and guide Nos.	Synonyms, trade names, and conversion factors	Exposure limits (TWA unless noted otherwise)	IDLH	Physical description	Chemical and physical properties		Incompatibilities and reactivities	Measurement method (See Table 1)
					MW, BP, SOL Fl.P, IP, Sp.Gr, flammability	VP, FRZ UEL, LEL		
Methyl acetate CH₃COOCH₃ 79-20-9 AI9100000	Methyl ester of acetic acid, Methyl ethanoate	NIOSH/OSHA 200 ppm (610 mg/m³) ST 250 ppm (760 mg/m³)	10,000 ppm	Colorless liquid with a fragrant, fruity odor.	MW: 74.1 BP: 135°F Sol: 30% Fl.P: 14°F IP: 10.27 eV	VP: 173 mm FRZ: -145°F UEL: 16% LEL: 3.1%	Nitrates; strong oxidizers, alkalis & acids; water [Note: Reacts slowly with water to form acetic acid & methanol.]	Char; CS₂; GC/FID; II(2) [#S42]
1231 26	1 ppm = 3.08 mg/m³				Sp.Gr: 0.93 Class IB Flammable Liquid			
Methyl acetylene CH₃C≡CH 74-99-7 UK4250000	Allylene, Propine, Propyne, 1-Propyne	NIOSH/OSHA 1000 ppm (1650 mg/m³)	15,000 ppm [LEL]	Colorless gas with a sweet odor. [Note: A fuel that is shipped as a liquefied compressed gas.]	MW: 40.1 BP: -10°F Sol: Insoluble Fl.P: NA (Gas) IP: 10.36 eV	VP: >1 atm FRZ: -153°F UEL: ? LEL: 1.7%	Strong oxidizers such as chlorine, copper alloys [Note: Can decompose explosively at 4.5 to 5.6 atmospheres of pressure.]	Bag; none; GC/FID; II(5) [#S84]
	1 ppm = 1.67 mg/m³				Flammable Gas			
Methyl acetylene-propadiene mixture CH₃C≡CH/CH₂=C=CH₂ UK4920000	MAPP gas, Methyl acetylene-allene mixture, Methyl acetylene-propadiene mixture (stabilized), Propadiene-methyl acetylene, Propyne-allene mixture, Propyne-propadiene mixture	NIOSH/OSHA 1000 ppm (1800 mg/m³) ST 1250 ppm (2250 mg/m³)	15,000 ppm [LEL]	Colorless gas with a strong, characteristic, foul odor. [Note: A fuel that is shipped as a liquefied compressed gas.]	MW: 40.1 BP: -36 to -4°F Sol: Insoluble Fl.P: NA (Gas) IP: ?	VP: >1 atm FRZ: -213°F UEL: ? LEL: ?	Strong oxidizers, copper alloys [Note: Forms explosive compounds at high pressure in contact with alloys containing more than 67% copper.]	Bag; none; GC/FID; II(6) [#S85]
1060 17 (stabilized)	1 ppm = 1.67 mg/m³				Flammable Gas			
Methyl acrylate CH₂=CHCOOCH₃ 96-33-3 AT2800000	Methoxycarbonylethylene, Methyl ester of acrylic acid, Methyl propenoate	NIOSH/OSHA 10 ppm (35 mg/m³) [skin]	1000 ppm	Colorless liquid with an acrid odor.	MW: 86.1 BP: 176°F Sol: 6% Fl.P: 25°F IP: 9.90 eV	VP: 65 mm FRZ: -106°F UEL: 25% LEL: 2.8%	Nitrates, oxidizers such as peroxides, strong alkalis [Note: Polymerizes easily; usually contains an inhibitor such as hydroquinone.]	Char; CS₂; GC/FID; II(2) [#S38]
1919 27 (inhibited)	1 ppm = 3.58 mg/m³				Sp.Gr: 0.96 Class IB Flammable Liquid			

Chemical name, structure/formula, CAS and RTECS Nos., and DOT ID and guide Nos.	Synonyms, trade names, and conversion factors	Exposure limits (TWA unless noted otherwise)	IDLH	Physical description	Chemical and physical properties		Incompatibilities and reactivities	Measurement method (See Table 1)
					MW, BP, SOL Fl.P, IP, Sp.Gr, flammability	VP, FRZ UEL, LEL		
Methylal CH₃OCH₂OCH₃ 109-87-5 PA8750000	Dimethoxymethane, Formal, Formaldehyde dimethylacetal, Methoxymethyl methyl ether, Methylene dimethyl ether	NIOSH/OSHA 1000 ppm (3100 mg/m³)	15,000 ppm [LEL]	Colorless liquid with a chloroform-like odor.	MW: 76.1 BP: 111°F Sol: 33% Fl.P(oc): -26°F IP: 10.00 eV	VP: 330 mm FRZ: -157°F UEL: 13.8% LEL: 2.2%	Strong oxidizers, acids	Char; Hexane; GC/FID; III [#1611]
1234 26	1 ppm = 3.16 mg/m³				Sp.Gr: 0.86 Class IB Flammable Liquid			
Methyl alcohol CH₃OH 67-56-1 PC1400000	Carbinol, Columbian spirits, Methanol, Wood alcohol, Wood spirits	NIOSH/OSHA 200 ppm (260 mg/m³) ST 250 ppm (325 mg/m³) [skin]	25,000 ppm	Colorless liquid with a characteristic pungent odor.	MW: 32.1 BP: 147°F Sol: Miscible Fl.P: 52°F IP: 10.84 eV	VP: 92 mm FRZ: -144°F UEL: 36% LEL: 6.0%	Strong oxidizers	Si gel; Water; GC/FID; III [#2000, Methanol]
1230 28	1 ppm = 1.33 mg/m³				Sp.Gr: 0.79 Class IB Flammable Liquid			
Methylamine CH₃NH₂ 74-89-5 PF6300000	Aminomethane, Anhydrous methylamine, Aqueous methylamine, Monomethylamine	NIOSH/OSHA 10 ppm (12 mg/m³)	100 ppm	Colorless gas with a fish- or ammonia-like odor. [Note: A liquid below 21°F. Shipped as a liquefied compressed gas.]	MW: 31.1 BP: 21°F Sol: Soluble Fl.P: NA (Gas) IP: 8.97 eV	VP: >1 atm FRZ: -136°F UEL: 20.7% LEL: 4.9%	Mercury, strong oxidizers, nitromethane [Note: Corrosive to copper & zinc alloys, aluminum & galvanized surfaces.]	XAD-7; Reagent; HPLC/FLD; OSHA [#40]
1061 19 (anhydrous) 1235 68 (aqueous soln)	1 ppm = 1.29 mg/m³				Sp.Gr: 0.70 (Liquid at 13°F) Flammable Gas			
Methyl (n-amyl) ketone CH₃CO[CH₂]₄CH₃ 110-43-0 MJ5075000	Amyl methyl ketone, n-Amyl methyl ketone, 2-Heptanone	NIOSH/OSHA 100 ppm (465 mg/m³) ACGIH 50 ppm (235 mg/m³)	4000 ppm	Colorless to white liquid with a banana-like, fruity odor.	MW: 114.2 BP: 305°F Sol: 0.4% Fl.P: 102°F IP: 9.33 eV	VP: 3 mm FRZ: -32°F UEL(250°F): 7.9% LEL(151°F): 1.1%	Strong acids, alkalis & oxidizers [Note: Will attack some forms of plastic.]	Char; Methanol/ CS₂; GC/FID; III [#1301, Ketones II]
1110 26	1 ppm = 4.75 mg/m³				Sp.Gr: 0.81 Class II Combustible Liquid			

Personal protection and sanitation (See Table 3)		Recommendations for respirator selection — maximum concentration for use (MUC) (See Table 4)	Health hazards			
			Route	Symptoms (See Table 5)	First aid (See Table 6)	Target organs (See Table 5)
Clothing: Goggles: Wash: Change: Remove:	Repeat Reason prob Prompt wet N.R. Immed wet (flamm)	NIOSH/OSHA 1000 ppm: CCROV*/PAPROV* 2000 ppm: SA*/SCBA* 5000 ppm: SA:CF* 10,000 ppm: GMFOV/SCBAF/SAF §: SCBAF:PD,PP/SAF:PD,PP:ASCBA Escape: GMFOV/SCBAE	Inh Ing Con	Irrit nose, throat; head, drow; optic atrophy	Eye: Irr immed Skin: Water flush prompt Breath: Resp support Swallow: Medical attention immed	Resp sys, skin, eyes
Clothing: Goggles: Wash: Change: Remove:	Prevent skin freezing Reason prob N.R. N.R. Immed wet (flamm)	NIOSH/OSHA 10,000 ppm: SA/SCBA 15,000 ppm: SA:CF/SCBAF/SAF §: SCBAF:PD,PP/SAF:PD,PP:ASCBA Escape: GMFOV/SCBAE	Inh	Hyperexcitability, tremors, anes; irrit resp sys	Breath: Resp support	CNS
Clothing: Goggles: Wash: Change: Remove:	Prevent skin freezing Reason prob N.R. N.R. Immed wet (flamm)	NIOSH/OSHA 10,000 ppm: SA/SCBA 15,000 ppm: SA:CF/SCBAF/SAF §: SCBAF:PD,PP/SAF:PD,PP:ASCBA Escape: GMFS/SCBAE	Inh Con	Frostbite; disorientation, excitement	Eye: Irr immed Skin: Water flush immed Breath: Resp support	CNS, skin, eyes
Clothing: Goggles: Wash: Change: Remove: Provide:	Any poss Reason prob Immed wet N.R. Immed wet (flamm) Quick drench	NIOSH/OSHA 100 ppm: SA*/SCBA* 250 ppm: SA:CF* 500 ppm: SCBAF/SAF 1000 ppm: SAF:PD,PP §: SCBAF:PD,PP/SAF:PD,PP:ASCBA Escape: GMFOV/SCBAE	Inh Abs Con	Irrit eyes, upper resp sys, skin	Eye: Irr immed Skin: Water flush immed Breath: Resp support Swallow: Medical attention immed	Resp sys, eyes, skin

Personal protection and sanitation (See Table 3)		Recommendations for respirator selection — maximum concentration for use (MUC) (See Table 4)	Health hazards			
			Route	Symptoms (See Table 5)	First aid (See Table 6)	Target organs (See Table 5)
Clothing: Goggles: Wash: Change: Remove:	Repeat Reason prob Prompt wet N.R. Immed wet (flamm)	NIOSH/OSHA 10,000 ppm: SA/SCBA 15,000 ppm: SA:CF/SCBAF/SAF §: SCBAF:PD,PP/SAF:PD,PP:ASCBA Escape: GMFOV/SCBAE	Inh Ing Con	Mild irrit eyes, upper resp; anes; irrit skin	Eye: Irr immed Skin: Water flush prompt Breath: Resp support Swallow: Medical attention immed	Skin, resp sys, CNS
Clothing: Goggles: Wash: Change: Remove:	Repeat Reason prob Prompt wet N.R. Immed wet (flamm)	NIOSH/OSHA 2000 ppm: SA/SCBA 5000 ppm: SA:CF 10,000 ppm: SCBAF/SAF/SAT:CF 25,000 ppm: SAF:PD,PP §: SCBAF:PD,PP/SAF:PD,PP:ASCBA Escape: SCBAE	Inh Abs Ing Con	Eye irrit, head, drow, li-head, nau, vomit; vis dist, blindness	Eye: Irr immed Skin: Water flush prompt Breath: Resp support Swallow: Medical attention immed	Eyes, skin, CNS, GI tract
Clothing: Goggles: Wash: Change: Remove: Provide:	Any poss liq Any poss Immed contam N.R. Immed wet (flamm) Eyewash, quick drench	NIOSH/OSHA 100 ppm: SCBAF/SAF/CCRFS/GMFS/PAPRS^ε §: SCBAF:PD,PP/SAF:PD,PP:ASCBA Escape: GMFS/SCBAE	Inh Abs Ing Con	Irrit eyes, resp sys; cough; skin, muc memb burns; derm; conj	Eye: Irr immed Skin: Water flush immed Breath: Resp support Swallow: Medical attention immed	Resp sys, eyes, skin
Clothing: Goggles: Wash: Change: Remove:	Repeat Reason prob Prompt wet N.R. Immed contam	ACGIH 500 ppm: CCROV*/SA*/SCBA* 1250 ppm: SA:CF*/PAPROV* 2500 ppm: CCRFOV/GMFOV/SCBAF/SAF/SAT:CF* 4000 ppm: SAF:PD,PP §: SCBAF:PD,PP/SAF:PD,PP:ASCBA Escape: GMFOV/SCBAE	Inh Ing Con	Irrit eyes, muc memb; head; narco, coma; derm	Eye: Irr immed Skin: Soap wash Breath: Fresh air Swallow: Medical attention immed	Eyes, skin, resp sys, CNS, PNS

Chemical name, structure/formula, CAS and RTECS Nos., and DOT ID and guide Nos.	Synonyms, trade names, and conversion factors	Exposure limits (TWA unless noted otherwise)	IDLH	Physical description	Chemical and physical properties		Incompatibilities and reactivities	Measurement method (See Table 1)
					MW, BP, SOL FI.P, IP, Sp.Gr, flammability	VP, FRZ UEL, LEL		
Methyl bromide CH$_3$Br 74-83-9 PA4900000 1062 55	Bromomethane, Monobromomethane 1 ppm = 3.95 mg/m³	NIOSH Ca See Appendix A Reduce exposure to lowest feasible concentration. OSHA 5 ppm (20 mg/m³) [skin]	Ca [2000 ppm]	Colorless gas with a chloroform-like odor at high concentrations. [Note: A liquid below 38°F. Shipped as a liquefied compressed gas.]	MW: 95.0 BP: 38°F Sol: 2% FI.P: ? IP: 10.54 eV Sp.Gr: 1.73 (Liquid at 32°F) Flammable Gas, but only in presence of a high energy ignition source.	VP: >1 atm FRZ: -137°F UEL: 16.0% LEL: 10%	Aluminum, magnesium, strong oxidizers [Note: Attacks aluminum to form aluminum trimethyl, which is SPONTANE-OUSLY flammable.]	Char(pet) (2); CS$_2$; GC/FID; III [#2520]
Methyl Cellosolve® CH$_3$OCH$_2$CH$_2$OH 109-86-4 KL5775000 1188 26	EGME, Ethylene glycol monomethyl ether, Glycol monomethyl ether, 2-Methoxyethanol 1 ppm = 3.16 mg/m³	NIOSH Reduce exposure to lowest feasible concentration. OSHA 25 ppm (80 mg/m³) [skin]	* [2000 ppm] *Not applicable because of the NIOSH REL.	Colorless liquid with a mild, ether-like odor.	MW: 76.1 BP: 256°F Sol: Miscible FI.P: 102°F IP: 9.60 eV Sp.Gr: 0.96 Class II Combustible Liquid	VP: 6 mm FRZ: -121°F UEL: 1.8% LEL: 14%	Strong oxidizers, caustics	Char; Methanol/ CH$_2$Cl$_2$; GC/FID; III [#1403, Alcohols IV]
Methyl Cellosolve® acetate CH$_3$COOCH$_2$CH$_2$OCH$_3$ 110-49-6 KL5950000 1189 26	EGMEA, Ethylene glycol monomethyl ether acetate, Glycol monomethyl ether acetate, 2-Methoxyethyl acetate 1 ppm = 4.91 mg/m³	NIOSH Reduce exposure to lowest feasible concentration. OSHA 25 ppm (120 mg/m³) [skin]	* [4000 ppm] *Not applicable because of the NIOSH REL.	Colorless liquid with a mild, ether-like odor.	MW: 118.1 BP: 293°F Sol: Miscible FI.P: 120°F IP: ? Sp.Gr: 1.01 Class II Combustible Liquid	VP: 2 mm FRZ: -85°F UEL: 8.2% LEL: 1.7%	Nitrates; strong oxidizers, alkalis & acids	Char; CS$_2$; GC/FID; II(2) [#S39]
Methyl chloride CH$_3$Cl 74-87-3 PA6300000 1063 18	Chloromethane, Monochloromethane 1 ppm = 2.10 mg/m³	NIOSH Ca See Appendix A Reduce exposure to lowest feasible concentration. OSHA 50 ppm (105 mg/m³) ST 100 ppm (210 mg/m³)	Ca [10,000 ppm]	Colorless gas with a faint, sweet odor which is not noticeable at dangerous concentrations. [Note: Shipped as a liquefied compressed gas.]	MW: 50.5 BP: -12°F Sol: 0.5% FI.P: NA (Gas) IP: 11.28 eV Flammable Gas	VP: >1 atm FRZ: -144°F UEL: 17.4% LEL: 8.1%	Chemically-active metals such as potassium, powdered aluminum, zinc & magnesium; water [Note: Reacts with water (hydrolyzes) to form hydrochloric acid.]	Char(2); CH$_2$Cl$_2$; GC/FID; III [#1001]

Chemical name, structure/formula, CAS and RTECS Nos., and DOT ID and guide Nos.	Synonyms, trade names, and conversion factors	Exposure limits (TWA unless noted otherwise)	IDLH	Physical description	Chemical and physical properties		Incompatibilities and reactivities	Measurement method (See Table 1)
					MW, BP, SOL FI.P, IP, Sp.Gr, flammability	VP, FRZ UEL, LEL		
Methyl chloroform CH$_3$CCl$_3$ 71-55-6 KJ2975000 2831 74	Chlorothene; 1,1,1-Trichloroethane; 1,1,1-Trichloroethane (stabilized) 1 ppm = 5.55 mg/m³	NIOSH C 350 ppm (1900 mg/m³) [15-min] OSHA 350 ppm (1900 mg/m³) ST 450 ppm (2450 mg/m³)	1000 ppm	Colorless liquid with a mild, chloroform-like odor.	MW: 133.4 BP: 165°F Sol: 0.4% FI.P: None IP: 11.00 eV Sp.Gr: 1.34 Noncombustible Liquid, however the vapor will burn.	VP: 100 mm FRZ: -23°F UEL: 12.5% LEL: 7.5%	Strong caustics; strong oxidizers; chemically-active metals such as zinc, aluminum, magnesium powders, sodium & potassium; water [Note: Reacts slowly with water to form hydrochloric acid.]	Char; CS$_2$; GC/FID; III [#1003, Halogenated Hydrocarbons]
Methylcyclohexane CH$_3$C$_6$H$_{11}$ 108-87-2 GV6125000 2296 27	Cyclohexylmethane, Hexahydrotoluene 1 ppm = 4.08 mg/m³	NIOSH/OSHA 400 ppm (1600 mg/m³)	10,000 ppm	Colorless liquid with a faint, benzene-like odor.	MW: 98.2 BP: 214°F Sol: Insoluble FI.P: 25°F IP: 9.85 eV Sp.Gr: 0.77 Class IB Flammable Liquid	VP(72°F): 43 mm FRZ: -196°F UEL: 6.7% LEL: 1.2%	Strong oxidizers	Char; CS$_2$; GC/FID; III [#1500, Hydrocarbons]
Methylcyclohexanol CH$_3$C$_6$H$_{10}$OH 25639-42-3 GW0175000 2617 26	Hexahydrocresol, Hexahydromethylphenol 1 ppm = 4.75 mg/m³	NIOSH/OSHA 50 ppm (235 mg/m³)	10,000 ppm	Straw-colored liquid with a weak odor like coconut oil.	MW: 114.2 BP: 311-356°F Sol: 4% FI.P: 154°F IP: ? Sp.Gr: 0.92 Class IIIA Combustible Liquid	VP(86°F): 2 mm FRZ: -58°F UEL: ? LEL: ?	Strong oxidizers	Char; CH$_2$Cl$_2$; GC/FID; II(4) [#S374]
o-Methylcyclohexanone CH$_3$C$_6$H$_9$O 583-60-8 GW1750000 2297 26	2-Methylcyclohexanone 1 ppm = 4.66 mg/m³	NIOSH/OSHA 50 ppm (230 mg/m³) ST 75 ppm (345 mg/m³) [skin]	2500 ppm	Colorless liquid with a weak peppermint-like odor.	MW: 112.2 BP: 325°F Sol: Insoluble FI.P: 118°F IP: ? Sp.Gr: 0.93 Class II Combustible Liquid	VP: 1 mm (approx) FRZ: 7°F UEL: ? LEL: ?	Strong oxidizers	Porapak; Acetone; GC/FID; III [#2521]

Personal protection and sanitation (See Table 3)	Recommendations for respirator selection — maximum concentration for use (MUC) (See Table 4)	Health hazards			
		Route	Symptoms (See Table 5)	First aid (See Table 6)	Target organs (See Table 5)
Clothing: Any poss Goggles: Reason prob Wash: Immed wet Change: N.R. Remove: Immed non-imperv contam Provide: Quick drench	NIOSH ¥: SCBAF:PD,PP/SAF:PD,PP:ASCBA Escape: GMFOV/SCBAE	Inh Abs Ing Con	Head; vis dist; verti; nau, vomit; mal; hand tremor; convuls; dysp; irrit eyes; skin irrit, vesic; [carc]	Eye: Irr immed Skin: Water flush immed Breath: Resp support Swallow: Medical attention immed	CNS, resp sys, skin, eyes
Clothing: Repeat Goggles: Reason prob Wash: Immed contam Change: N.R. Remove: Immed non-imperv contam Provide: Quick drench	NIOSH ¥: SCBAF:PD,PP/SAF:PD,PP:ASCBA Escape: GMFOV/SCBAE	Inh Abs Ing Con	Head, drow, weak; ataxia, tremor, som; anemic pallor, irrit eyes; [carc]	Eye: Irr immed Skin: Water flush prompt Breath: Resp support Swallow: Medical attention immed	CNS, blood, skin, eyes, kidneys
Clothing: Repeat Goggles: Reason prob Wash: Prompt wet Change: N.R. Remove: Prompt non-imperv wet	¥: SCBAF:PD,PP/SAF:PD,PP:ASCBA Escape: GMFOV/SCBAE	Inh Abs Ing Con	Kidney damage; brain damage; eye irrit; in animals: eye, nose, throat irrit; narco	Eye: Irr immed Skin: Water flush prompt Breath: Resp support Swallow: Medical attention immed	Kidneys, brain, CNS, PNS
Clothing: Prevent wet or freezing Goggles: Reason prob Wash: N.R. Change: N.R. Remove: Immed wet (flamm)	NIOSH ¥: SCBAF:PD,PP/SAF:PD,PP:ASCBA Escape: SCBAE	Inh Con	Dizz, nau, vomit; vis dist; stagger; slur speech; convuls, coma; liver, kidney damage; frostbite; [carc]	Eye: Irr immed Skin: Water flush immed Breath: Resp support	CNS, liver, kidneys, skin

Personal protection and sanitation (See Table 3)	Recommendations for respirator selection — maximum concentration for use (MUC) (See Table 4)	Health hazards			
		Route	Symptoms (See Table 5)	First aid (See Table 6)	Target organs (See Table 5)
Clothing: Repeat Goggles: Reason prob Wash: Prompt wet Change: N.R. Remove: Prompt non-imperv wet	NIOSH/OSHA 1000 ppm: SA*/SCBA* §: SCBAF:PD,PP/SAF:PD,PP:ASCBA Escape: GMFOV/SCBAE	Inh Ing Con	Head, lass, CNS depres, poor equi; irrit eyes; derm; card arrhy	Eye: Irr immed Skin: Soap wash prompt Breath: Resp support Swallow: Medical attention immed	Skin, CNS, CVS, eyes
Clothing: Repeat Goggles: Reason prob Wash: Prompt wet Change: N.R. Remove: Immed wet (flamm)	NIOSH/OSHA 4000 ppm: SA/SCBA 10,000 ppm: SA:CF/SCBAF/SAF §: SCBAF:PD,PP/SAF:PD,PP:ASCBA Escape: GMFOV/SCBA	Inh Ing Con	Li-head, drow; skin, nose, throat irrit	Eye: Irr immed Skin: Soap wash prompt Breath: Resp support Swallow: Medical attention immed	Resp sys, skin
Clothing: Repeat Goggles: Reason prob Wash: Prompt contam Change: N.R. Remove: Prompt non-imperv contam	NIOSH/OSHA 500 ppm: SA*/SCBA* 1250 ppm: SA:CF* 2500 ppm: SCBAF:SAF 10,000 ppm: SAF:PD,PP §: SCBAF:PD,PP/SAF:PD,PP:ASCBA Escape: GMFOV/SCBAE	Inh Abs Ing Con	Head; irrit eyes, upper resp sys; in animals: narco; liver, kidney damage	Eye: Irr immed Skin: Soap wash prompt Breath: Resp support Swallow: Medical attention immed	Resp sys, skin, eyes; in animals: CNS, liver, kidneys
Clothing: Repeat Goggles: Reason prob Wash: Prompt contam Change: N.R. Remove: Prompt non-imperv contam	NIOSH/OSHA 500 ppm: SA*/SCBA* 1250 ppm: SA:CF* 2500 ppm: SCBAF:SAF §: SCBAF:PD,PP/SAF:PD,PP:ASCBA Escape: GMFOV/SCBAE	Inh Abs Ing Con	In animals: narco; irrit eyes, muc memb; derm	Eye: Irr immed Skin: Soap wash prompt Breath: Resp support Swallow: Medical attention immed	In animals; resp sys, liver, kidneys, skin

Chemical name, structure/formula, CAS and RTECS Nos., and DOT ID and guide Nos.	Synonyms, trade names, and conversion factors	Exposure limits (TWA unless noted otherwise)	IDLH	Physical description	Chemical and physical properties		Incompatibilities and reactivities	Measurement method (See Table 1)
					MW, BP, SOL Fl.P, IP, Sp.Gr, flammability	VP, FRZ UEL, LEL		
Methylene bisphenyl isocyanate CH₂(C₆H₄NCO)₂ 101-68-8 NQ9350000 2489 53	4,4'-Diphenylmethane diisocyanate; MDI; Methylene bis(4-phenyl isocyanate; Methylene di-p-phenylene ester of isocyanic acid	NIOSH 0.05 mg/m³ (0.005 ppm) C 0.2 mg/m³ (0.020 ppm) [10-min] OSHA C 0.2 mg/m³) (0.02 ppm)	100 mg/m³	White to light-yellow, odorless flakes. [Note: A liquid above 99°F.]	MW: 250.3 BP: 342°F Sol: 0.2% Fl.P(oc): 396°F IP: ? Sp.Gr(122°F): 1.19 Combustible Solid	VP(104°F): 0.001 mm MLT: 99°F UEL: ? LEL: ?	Strong alkalis, acids, alcohol	Bub; Acetylate; HPLC/UVD; III [#5521]
Methylene chloride CH₂Cl₂ 75-09-2 PA8050000 1593 74	Dichloromethane, Methylene dichloride 1 ppm = 3.53 mg/m³	NIOSH Ca See Appendix A Reduce exposure to lowest feasible concentration. OSHA 500 ppm C 1000 ppm 2000 ppm (5-min max peak in any 2 hrs)	Ca [5000 ppm] ACGIH A2, 50 ppm (175 mg/m³)	Colorless liquid with a chloroform-like odor. [Note: A gas above 104°F.]	MW: 84.9 BP: 104°F Sol: 2% Fl.P: ? IP: 11.32 eV Sp.Gr: 1.33 Combustible Liquid	VP: 350 mm FRZ: -139°F UEL: 22% LEL: 14%	Strong oxidizers; caustics; chemically-active metals such as aluminum, magnesium powders, potassium & sodium; concentrated nitric acid	Char(2); CS₂; GC/FID; III [#1005]
Methyl formate HCOOCH₃ 107-31-3 LQ8925000 1243 26	Methyl ester of formic acid, Methyl methanoate 1 ppm = 2.50 mg/m³	NIOSH/OSHA 100 ppm (250 mg/m³) ST 150 ppm (375 mg/m³)	5000 ppm	Colorless liquid with a pleasant odor. [Note: A gas above 89°F.]	MW: 60.1 BP: 89°F Sol: 30% Fl.P: -2°F IP: 10.82 eV Sp.Gr: 0.98 Class IA Flammable Liquid	VP: 476 mm FRZ: -148°F UEL: 23% LEL: 4.5%	Strong oxidizers [Note: Reacts slowly with water to form methanol & formic acid.]	Carbo-B(2); Ethyl acetate; GC/FID; II(5) [#S291]
5-Methyl-3-heptanone CH₃CH₂CO[CH₂]₄CH₃ 541-85-5 MJ7350000 2271 26	Amyl ethyl ketone, Ethyl amyl ketone 1 ppm = 5.33 mg/m³	NIOSH/OSHA 25 ppm (130 mg/m³)	3000 ppm	Colorless liquid with a pungent odor.	MW: 128.2 BP: 315°F Sol: Insoluble Fl.P: 138°F IP: ? Sp.Gr: 0.82 Class II Combustible Liquid	VP: 2 mm FRZ: -70°F UEL: ? LEL: ?	Strong oxidizers	Char; Methanol/ CS₂; GC/FID; III, [#1301, Ketones II]

Chemical name, structure/formula, CAS and RTECS Nos., and DOT ID and guide Nos.	Synonyms, trade names, and conversion factors	Exposure limits (TWA unless noted otherwise)	IDLH	Physical description	Chemical and physical properties		Incompatibilities and reactivities	Measurement method (See Table 1)
					MW, BP, SOL Fl.P, IP, Sp.Gr, flammability	VP, FRZ UEL, LEL		
Methyl hydrazine CH₃NHNH₂ 60-34-4 MV5600000 1244 28	Monomethyl hydrazine, MMH 1 ppm = 1.92 mg/m³	NIOSH Ca See Appendix A C 0.04 ppm (0.08 mg/m³) [2-hr] OSHA C 0.2 ppm (0.35 mg/m³) [skin]	Ca [50 ppm] ACGIH A2	Fuming, colorless liquid with an ammonia-like odor.	MW: 46.1 Sol: Miscible Fl.P: 17°F IP: 8.00 eV Sp.Gr(77°F): 0.87 Class IB Flammable Liquid	VP(77°F): 50 mm FRZ: -62°F UEL: 92% LEL: 2.5%	Oxides of iron; copper; manganese; lead; copper alloys; porous materials such as earth, asbestos, wood & cloth; strong oxidizers such as fluorine & chlorine; nitric acid; hydrogen peroxide	Bub; Pho-acid; Vis; II(3); [#S149]
Methyl iodide CH₃I 74-88-4 PA9450000 2644 55	Iodomethane, Monoiodomethane 1 ppm = 5.90 mg/m³	NIOSH Ca See Appendix A 2 ppm (10 mg/m³) [skin] OSHA 2 ppm (10 mg/m³) [skin]	Ca [800 ppm] ACGIH A2	Colorless liquid with a pungent, ether-like odor. [Note: Turns yellow, red, or brown on exposure to light & moisture.]	MW: 141.9 BP: 109°F Sol: 1% Fl.P: NA IP: 9.54 eV Sp.Gr: 2.28 Noncombustible Liquid	VP(77°F): 400 mm FRZ: -88°F UEL: NA LEL: NA	Strong oxidizers [Note: Decomposes at 518°F.]	Char; Toluene; GC/FID; III [#1014]
Methyl isobutyl carbinol (CH₃)₂CHCH₂CH-(OH)CH₃ 108-11-2 SA7350000 2053 26	Isobutylmethylcarbinol, Methyl amyl alcohol, 4-Methyl-2-pentanol, MIBC 1 ppm = 4.25 mg/m³	NIOSH/OSHA 25 ppm (100 mg/m³) ST 40 ppm (165 mg/m3) [skin]	2000 ppm	Colorless liquid with a mild odor.	MW: 102.2 BP: 271°F Sol: 2% Fl.P: 106°F IP: ? Sp.Gr: 0.81 Class II Combustible Liquid	VP: 3 mm FRZ: -130°F UEL: 5.5% LEL: 1.0%	Strong oxidizers	Char; 2-Propanol/ CS₂; GC/FID; III [#1402, Alcohols III]
Methyl isocyanate CH₃NCO 624-83-9 NQ9450000 2480 30	Methyl ester of isocyanic acid, MIC 1 ppm = 2.37 mg/m³	NIOSH/OSHA 0.02 ppm (0.05 mg/m³) [skin]	20 ppm	Colorless liquid with a sharp, pungent odor.	MW: 57.1 BP: 139°F Sol(59°F): 10% Fl.P: 19°F IP: 10.67 eV Sp.Gr: 0.96 Class IB Flammable Liquid	VP: 348 mm MLT: -49°F UEL: 26% LEL: 5.3%	Water, oxidizers, acids, alkalis, amines, iron, tin, copper	XAD-7; CH₃CN; HPLC/FLD; OSHA [#54]

Personal protection and sanitation (See Table 3)		Recommendations for respirator selection — maximum concentration for use (MUC) (See Table 4)	Route	Symptoms (See Table 5)	First aid (See Table 6)		Target organs (See Table 5)
Clothing: Goggles: Wash: Change: Remove:	Reason prob Any poss Prompt contam After work if reason prob contam Prompt non-imperv contam	NIOSH 2 mg/m³: SA*/SCBA* 5 mg/m³: SA:CF* 10 mg/m³: SCBAF:PD,PP 100 mg/m³: SCBAF:PD,PP §: SCBAF:PD,PP/SAF:PD,PP:ASCBA Escape: GMFOVHiE/SCBAE	Inh Ing Con	Irrit eyes, nose, throat; cough, pulm secretions, chest pain, dysp; asthma	Eye: Skin: Breath: Swallow	Irr immed Soap wash immed Resp support Medical attention immed	Resp sys, eyes
Clothing: Goggles: Wash: Change: Remove:	Repeat Reason prob Prompt wet N.R. Prompt non-imperv wet	NIOSH ¥: SCBAF:PD,PP/SAF:PD,PP:ASCBA Escape: GMFOV/SCBAE	Inh Ing Con	Ftg, weak, sleepiness, li-head; limbs numb, tingle; nau; irrit eyes, skin; [carc]	Eye: Skin: Breath: Swallow:	Irr immed Soap wash prompt Resp support Medical attention immed	Skin, CVS, eyes, CNS
Clothing: Goggles: Wash: Change: Remove:	Repeat Reason prob Prompt wet N.R. Immed wet (flamm)	NIOSH/OSHA 1000 ppm: SA*/SCBA* 2500 ppm: SA:CF* 5000 ppm: SCBAF:SAF §: SCBAF:PD,PP/SAF:PD,PP:ASCBA Escape: GMFOV/SCBAE	Inh Abs Ing Con	Eye, nose irrit; chest oppression, dysp; vis dist; CNS depres; in animals: pulm edema	Eye: Skin: Breath: Swallow:	Irr immed Soap wash immed Resp support Medical attention immed	Eyes, resp sys, CNS
Clothing: Goggles: Wash: Change: Remove:	Any poss Reason prob Prompt wet N.R. Prompt non-imperv wet	NIOSH/OSHA 250 ppm: SA*/SCBA* 625 ppm: SA:CF*/PAPROV* 1000 ppm: CCRFOV 1250 ppm: GMFOV/SCBAF/SAF 3000 ppm: SAF:PD,PP §: SCBAF:PD,PP/SAF:PD,PP:ASCBA Escape: GMFOV/SCBAE	Inh Ing Con	Irrit eyes, muc memb; head; narco, coma; derm	Eye: Skin: Breath: Swallow:	Irr immed Water flush Resp support Medical attention immed	Eyes, skin, resp sys, CNS

Personal protection and sanitation (See Table 3)		Recommendations for respirator selection — maximum concentration for use (MUC) (See Table 4)	Route	Symptoms (See Table 5)	First aid (See Table 6)		Target organs (See Table 5)
Clothing: Goggles: Wash: Change: Remove: Provide:	Any poss Any poss Immed contam N.R. Immed wet (flamm) Eyewash, quick drench	NIOSH ¥: SCBAF:PD,PP/SAF:PD,PP:ASCBA Escape: SCBAE	Inh Abs Ing Con	Irrit eyes, vomit, diarr, resp sys irrit, tremors, ataxia; anoxia, cyan; convuls, [carc]	Eye: Skin: Breath: Swallow:	Irr immed Water flush immed Resp support Medical attention immed	CNS, resp sys, liver, blood, CVS, eyes
Clothing: Goggles: Wash: Change: Remove:	Repeat Reason prob Immed wet N.R. Immed non-imperv contam	NIOSH ¥: SCBAF:PD,PP/SAF:PD,PP:ASCBA Escape: GMFOV/SCBAE	Inh Abs Ing Con	Nau, vomit; verti, ataxia; slurred speech, drow; derm; eye irrit; [carc]	Eye: Skin: Breath: Swallow:	Irr immed Soap flush immed Resp support Medical attention immed	CNS, skin, eyes
Clothing: Goggles: Wash: Change: Remove:	Repeat Reason prob Prompt wet N.R. Immed wet (flamm)	NIOSH/OSHA 625 ppm: SA:CF* 1250 ppm: SCBAF/SAF 2000 ppm: SAF:PD,PP §: SCBAF:PD,PP/SAF:PD,PP:ASCBA Escape: GMFOV/SCBAE	Inh Abs Ing Con	Eye irrit; head, drow; derm	Eye: Skin: Breath: Swallow:	Irr immed Water flush prompt Resp support Medical attention immed	Eyes, skin
Clothing: Goggles: Wash: Change: Remove: Provide:	Any poss Any poss Immed contam N.R. Immed wet (flamm) Eyewash, quick drench	NIOSH/OSHA 0.2 ppm: SA*/SCBA* 0.5 ppm: SA:CF* 1 ppm: SCBAF/SAF 20 ppm: SAF:PD,PP §: SCBAF:PD,PP/SAF:PD,PP:ASCBA Escape: GMFOV/SCBAE	Inh Abs Ing Con	Irrit eyes, nose, throat; cough, secretions, chest pain, dysp; asthma; eye, skin inj; in animals: pulm edema	Eye: Skin: Breath: Swallow:	Irr immed Water flush immed Resp support Medical attention immed	Resp sys, eyes, skin

Chemical name, structure/formula, CAS and RTECS Nos., and DOT ID and guide Nos.	Synonyms, trade names, and conversion factors	Exposure limits (TWA unless noted otherwise)	IDLH	Physical description	Chemical and physical properties		Incompatibilities and reactivities	Measurement method (See Table 1)
					MW, BP, SOL Fl.P, IP, Sp.Gr, flammability	VP, FRZ UEL, LEL		
Methyl mercaptan CH₃SH 74-93-1 PB4375000 1064 13	Mercaptomethane, Methanethiol, Methyl sulfhydrate 1 ppm = 2.00 mg/m³	NIOSH C 0.5 ppm (1 mg/m³) [15-min] OSHA 0.5 ppm (1 mg/m³)	400 ppm	Colorless gas with a disagreeable odor like garlic or rotten cabbage. [Note: A liquid below 43°F. Shipped as a liquefied compressed gas.]	MW: 48.1 BP: 43°F Sol: 2% Fl.P(oc): 0°F (Liquid) IP: 9.44 eV Sp.Gr. 0.90 (Liquid at 32°F) Flammable Gas Class IA Flammable Liquid	VP: >1 atm FRZ: -186°F UEL: 21.8% LEL: 3.9%	Strong oxidizers, bleaches	Filter; HCl/CH₂Cl₂; GC/FPD; OSHA [#26]
Methyl methacrylate CH₂=C(CH₃)COOCH₃ 80-62-6 OZ5075000 1247 27 (inhibited)	Methacrylate monomer, Methyl ester of methacrylic acid, Methyl-2-methyl-2-propenoate 1 ppm = 4.16 mg/m³	NIOSH/OSHA 100 ppm (410 mg/m³)	4000 ppm	Colorless liquid with an acrid, fruity odor.	MW: 100.1 BP: 214°F Sol: Slight Fl.P(oc): 50°F IP: 9.70 eV Sp.Gr. 0.94 Class IB Flammable Liquid	VP(77°F): 40 mm FRZ: -54°F UEL: 8.2% LEL: 1.7%	Nitrates, oxidizers, peroxides, strong alkalis, moisture [Note: May polymerize if subjected to heat, oxidizers, or ultraviolet light. Usually contains an inhibitor such as hydroquinone.]	XAD-2; CS₂; GC/FID; [#2537]
alpha-Methyl styrene C₆H₅C(CH₃)=CH₂ 98-83-9 WL5250000	AMS, Isopropenyl benzene, 1-Methyl-1-phenylethylene, 2-Phenyl propylene 1 ppm = 4.91 mg/m³	NIOSH/OSHA 50 ppm (240 mg/m³) ST 100 ppm (485 mg/m³)	5000 ppm	Colorless liquid with a character-istic odor.	MW: 118.2 BP: 330°F Sol: Insoluble Fl.P: 129°F IP: 8.35 eV Sp.Gr. 0.91 Class II Combustible Liquid	VP: 2 mm FRZ: -10°F UEL: 6.1% LEL: 1.9%	Oxidizers, peroxides, halogens, catalysts for vinyl or ionic polymers; aluminum, iron chloride, copper [Note: Usually con-tains an inhibitor such as tert-butyl catechol.]	Char; CS₂; GC/FID; III [#1501, Aromatic Hydro-carbons]
Mica (containing less than 1% quartz) 12001-26-2 VV8760000	Silicate mica	NIOSH/OSHA 3 mg/m³ (resp)	N.E.	Colorless, odor-less flakes or sheets of hydrous silicates.	MW: 797(approx) BP: ? Sol: Insoluble Fl.P: NA IP: NA Sp.Gr. 2.6-3.2 Noncombustible Solid	VP: 0 mm (approx) MLT: ? UEL: NA LEL: NA	None reported	Filter; none; Grav; III [#0600, Nuisance Dust (resp)]

Chemical name, structure/formula, CAS and RTECS Nos., and DOT ID and guide Nos.	Synonyms, trade names, and conversion factors	Exposure limits (TWA unless noted otherwise)	IDLH	Physical description	Chemical and physical properties		Incompatibilities and reactivities	Measurement method (See Table 1)
					MW, BP, SOL Fl.P, IP, Sp.Gr, flammability	VP, FRZ UEL, LEL		
Molybdenum (soluble compounds as Mo)	Synonyms vary depending upon the specific soluble compound.	NIOSH See Appendix D OSHA 5 mg/m³	N.E.	Appearance and odor vary depending upon the specific soluble compound.	Properties vary depending upon the specific soluble compound.		Alkali metals, sodium, potassium, molten magnesium	Filter; Acid; ICP; III [#7300, Elements]
Molybdenum (insoluble compounds as Mo) Mo (Metal) 7439-93-7 (Metal) QA4680000 (Metal)	Metal: Molybdenum metal Synonyms of other insoluble compounds vary depending upon the specific compound.	NIOSH See Appendix D OSHA 10 mg/m³	N.E.	Metal: Dark gray or black powder with a metallic luster.	MW: 95.9 BP: 8717°F Sol: Insoluble Fl.P: NA IP: NA Sp.Gr. 10.28 (Metal) Flammable in form of dust or powder.	VP: 0 mm (approx) MLT: 4752°F UEL: NA LEL: NA	Strong oxidizers	Filter; Acid; ICP; III [#7300, Elements]
Monomethyl aniline C₆H₅NHCH₃ 100-61-8 BY4550000 2294 57	MA, Methyl aniline, N-Methyl aniline, (Methylamino)benzene, Methylphenyl amine 1 ppm = 4.46 mg/m³	NIOSH/OSHA 0.5 ppm (2 mg/m³) [skin]	100 ppm	Yellow to light-brown liquid with a weak, ammonia-like odor.	MW: 107.2 BP: 384°F Sol: Insoluble Fl.P: 175°F IP: 7.32 eV Sp.Gr. 0.99 Class IIIA Combustible Liquid	VP(97°F): 1 mm FRZ: -71°F UEL: ? LEL: ?	Strong acids, strong oxidizers	Bub; NaOH; GC/FID; II(3) [#S153]
Morpholine C₄H₈ONH 110-91-8 QD6475000 2054 29	Diethylene imidoxide; Diethylene oximide; Tetrahydro-1,4-oxazine; Tetrahydro-p-oxazine 1 ppm = 3.62 mg/m³	NIOSH/OSHA 20 ppm (70 mg/m³) ST 30 ppm (105 mg/m³) [skin]	8000 ppm	Colorless liquid with a weak, ammonia- or fish-like odor. [Note: A solid below 23°F.]	MW: 87.1 BP: 264°F Sol: Miscible Fl.P(oc): 98°F IP: 8.88 eV Sp.Gr. 1.00 Class IC Flammable Liquid	VP: 6 mm FRZ: 23°F UEL: 11.2% LEL: 1.4%	Strong acids, strong oxidizers, metals [Note: Corrosive to metals.]	Si gel; H₂SO₄/NaOH; GC/FID; II(3) [#S150]

Personal protection and sanitation (See Table 3)	Recommendations for respirator selection — maximum concentration for use (MUC) (See Table 4)	Health hazards			
		Route	Symptoms (See Table 5)	First aid (See Table 6)	Target organs (See Table 5)
Clothing: Prevent freezing Goggles: Reason prob Wash: N.R. Change: N.R. Remove: Immed wet (flamm)	NIOSH/OSHA 5 ppm: SA/SCBA/CCROV 12.5 ppm: SA:CF/PAPROV 25 ppm: SCBAF/SAF/CCRFOV/GMFOV/PAPRTOV/SAT:CF 400 ppm: SA:PD,PP §: SCBAF:PD,PP/SAF:PD,PP:ASCBA Escape: GMFOV/SCBAE	Inh Con	Narco; cyan; convuls; pulm irrit	Eye: Irr immed Skin: Water flush immed Breath: Resp support	Resp sys, CNS
Clothing: Repeat Goggles: Reason prob Wash: Prompt wet Change: N.R. Remove: Immed wet (flamm)	NIOSH/OSHA 1000 ppm: PAPROV$^\varepsilon$/CCRFOV 2500 ppm: SA:CF$^\varepsilon$ 4000 ppm: GMFOV/SCBAF/SAF §: SCBAF:PD,PP/SAF:PD,PP:ASCBA Escape: GMFOV/SCBAE	Inh Ing Con	Irrit eyes, nose, throat; derm	Eye: Irr immed Skin: Water flush prompt Breath: Resp support Swallow: Medical attention immed	Eyes, upper resp sys, skin
Clothing: Repeat Goggles: Reason prob Wash: Prompt contam Change: N.R. Remove: Prompt non-imperv contam	NIOSH/OSHA 500 ppm: CCROV*/SA*/SCBA* 1000 ppm: PAPROV*/CCRFOV 1250 ppm: SA:CF* 2500 ppm: GMFOV/SCBAF/SAF 5000 ppm: SAF:PD,PP §: SCBAF:PD,PP/SAF:PD,PP:ASCBA Escape: GMFOV/SCBAE	Inh Ing Con	Irrit eyes, nose, throat; drow; derm	Eye: Irr immed Skin: Water flush prompt Breath: Resp support Swallow: Medical attention immed	Eyes, resp sys, skin
Clothing: N.R. Goggles: N.R. Wash: N.R. Change: N.R. Remove: N.R.	NIOSH/OSHA 15 mg/m³: DM 30 mg/m³: DMXSQ/SA/SCBA 75 mg/m³: PAPRDM/SA:CF 150 mg/m³: HiEF/PAPRTHiE/SCBAF/SAF/SAT:CF 1500 mg/m³: SA:PD,PP §: SCBAF:PD,PP/SAF:PD,PP:ASCBA Escape: HiEF/SCBAE	Inh	Pneumoconiosis, cough, dysp; weak; low-wgt	Eye: Irr immed Breath: Fresh air	Lungs

Personal protection and sanitation (See Table 3)	Recommendations for respirator selection — maximum concentration for use (MUC) (See Table 4)	Health hazards			
		Route	Symptoms (See Table 5)	First aid (See Table 6)	Target organs (See Table 5)
Clothing: Repeat Goggles: Reason prob Wash: Prompt contam Change: N.R. Remove: Prompt non-imperv contam	OSHA 25 mg/m³: DM* 50 mg/m³: DMXSQ*/SA*/SCBA* 125 mg/m³: PAPRDM*/SA:CF* 250 mg/m³: HiEF/PAPRTHiE*/SAT:CF*/SCBAF/SAF 2500 mg/m³: SAF:PD,PP §: SCBAF:PD,PP/SAF:PD,PP:ASCBA Escape: HiEF/SCBAE	Inh Ing Con	In animals: anor; inco; irrit eyes, nose, throat; dysp; anemia	Eye: Irr immed Skin: Water flush Breath: Resp support Swallow: Medical attention immed	Resp sys; in animals: kidneys, blood
Clothing: N.R. Goggles: N.R. Wash: N.R. Change: N.R. Remove: N.R.	OSHA 50 mg/m³: DM^ 100 mg/m³: DMXSQ^/SA/SCBA 250 mg/m³: PAPRDM^/SA:CF 500 mg/m³: HiEF/PAPRTHiE/SAT:CF/SCBAF/SAF 5000 mg/m³: SA:PD,PP §: SCBAF:PD,PP/SAF:PD,PP:ASCBA Escape: HiEF/SCBAE	Inh Ing	In animals: irrit eyes, nose, throat; anor, diarr, low-wgt; listlessness; liver, kidney damage	Breath: Resp support Swallow: Medical attention immed	None known in humans
Clothing: Reason prob Goggles: Reason prob Wash: Immed contam Change: N.R. Remove: Immed non-imperv contam	NIOSH/OSHA 5 ppm: SA/SCBA 12.5 ppm: SA:CF 25 ppm: SCBAF/SAF/SAT:CF 100 ppm: SAF:PD,PP §: SCBAF:PD,PP/SAF:PD,PP:ASCBA Escape: GMFS/SCBAE	Inh Abs Ing Con	Weak, dizz, head; dysp, cyan; methemoglobinemia; pulm edema; liver, kidney damage	Eye: Irr immed Skin: Soap wash immed Breath: Resp support Swallow: Medical attention immed	Resp sys, liver, kidneys, blood
Clothing: Any poss >25%/Repeat <25% Goggles: Any poss Wash: Prompt Contam Change: N.R. Remove: Immed wet (flamm) Provide: 15-25%: Eyewash/ >25%: Quick drench	NIOSH/OSHA 550 ppm: SA:CF$^\varepsilon$PAPROV$^\varepsilon$ 1000 ppm: SCBAF/SAF/CCRFOV 8000 ppm: SAF:PD,PP §: SCBAF:PD,PP/SAF:PD,PP:ASCBA Escape: GMFOV/SCBAE	Inh Abs Ing Con	Vis dist; nose irrit; cough; resp irrit; eye, skin irrit; liver, kidney damage	Eye: Irr immed Skin: Water flush immed Breath: Art resp Swallow: Medical attention immed	Resp sys, eyes, skin

Chemical name, structure/formula, CAS and RTECS Nos., and DOT ID and guide Nos.	Synonyms, trade names, and conversion factors	Exposure limits (TWA unless noted otherwise)	IDLH	Physical description	Chemical and physical properties		Incompatibilities and reactivities	Measurement method (See Table 1)
					MW, BP, SOL Fl.P, IP, Sp.Gr, flammability	VP, FRZ UEL, LEL		
Naphtha (coal tar) 8030-30-6 DE3030000	Crude solvent coal tar naphtha, High solvent naphtha, Naphtha	NIOSH/OSHA 100 ppm (400 mg/m³)	10,000 ppm [LEL]	Reddish-brown, mobile liquid with an aromatic odor.	MW: 110(approx) BP: 320-428°F Sol: Insoluble Fl.P: 100-109°F IP: ?	VP: <5 mm FRZ: ? UEL: ? LEL: ?	Strong oxidizers	Char; CS₂; GC/FID; III [#1550]
2553 27	1 ppm = 4.57 mg/m³ (approx)				Sp.Gr: 0.89-0.97 Class II Combustible Liquid			
Naphthalene C₁₀H₈ 91-20-3 QJ0525000	Naphthalin, Tar camphor, White tar	NIOSH/OSHA 10 ppm (50 mg/m³) ST 15 ppm (75 mg/m³)	500 ppm	Colorless to brown solid with an odor of mothballs. [Note: Shipped as a molten solid.]	MW: 128.2 BP: 424°F Sol: 0.003% Fl.P: 174°F IP: 8.12 eV	VP: 0.08 mm MLT: 176°F UEL: 5.9% LEL: 0.9%	Strong oxidizers, chromic anhydride	Char; CS₂; GC/FID; III [#1501, Aromatic Hydrocarbons]
1334 32					Sp.Gr: 1.15 Combustible Solid, but will take some effort to ignite.			
alpha-Naphthylamine C₁₀H₇NH₂ 134-32-7 QM1400000	1-Aminonaphthalene, 1-Naphthylamine	NIOSH Ca See Appendix A OSHA [1910.1004] See Appendix B	Ca	Colorless crystals with an ammonia-like odor. [Note: Darkens in air to a reddish-purple color.]	MW: 143.2 BP: 573°F Sol: 0.002% Fl.P: 315°F IP: 7.30 eV	VP(220°F): 1 mm MLT: 122°F UEL: ? LEL: ?	Oxidizes in air	Filter/ Si gel; CH₃COOH/ 2-Pro-panol; GC/FID; III [#5518]
2077 55					Sp.Gr(77°F): 1.12 Combustible Solid			
beta-Naphthylamine C₁₀H₇NH₂ 91-59-8 QM2100000	2-Aminonaphthalene, 2-Naphthylamine	NIOSH Ca See Appendix A OSHA [1910.1009] See Appendix B ACGIH A1	Ca	Odorless, white to red crystals with a faint aromatic odor. [Note: Darkens in air to a reddish-purple color.]	MW: 143.2 BP: 583°F Sol: Miscible in hot water Fl.P: 315°F IP: 9.71 eV	VP(226°F): 1 mm MLT: 232°F UEL: ? LEL: ?	None reported	Filter/ Si gel; CH₃COOH/ 2-Pro-panol; GC/FID; III [#5518]
1650 55					Sp.Gr(208°F): 1.06 Combustible Solid			

Chemical name, structure/formula, CAS and RTECS Nos., and DOT ID and guide Nos.	Synonyms, trade names, and conversion factors	Exposure limits (TWA unless noted otherwise)	IDLH	Physical description	Chemical and physical properties		Incompatibilities and reactivities	Measurement method (See Table 1)
					MW, BP, SOL Fl.P, IP, Sp.Gr, flammability	VP, FRZ UEL, LEL		
Nickel carbonyl (as Ni) Ni(CO)₄ 13463-39-3 QR6300000	Nickel tetracarbonyl	NIOSH Ca See Appendix A 0.001 ppm (0.007 mg/m³) OSHA 0.001 ppm (0.007 mg/m³)	Ca [7 ppm]	Colorless to yellow liquid with a musty odor. [Note: A gas above 110°F.]	MW: 170.7 BP: 110°F Sol: 0.05% Fl.P: <-4°F IP: 8.28 eV	VP: 315 mm FRZ: -13°F UEL: ? LEL: 2%	Nitric acid, bromine, chlorine & other oxidizers; flammable materials	Char(low Ni); HNO₃; HGA; III [#6007]
1259 28	1 ppm = 7.10 mg/m³				Sp.Gr(63°F): 1.32 Class IB Flammable Liquid			
Nickel metal and other compounds (as Ni) Ni 7440-02-0 (Metal) QR5950000 (Metal)	Metal: Elemental nickel Synonyms of other compounds vary depending upon the specific compound.	NIOSH Ca See Appendix A 0.015 mg/m³ OSHA 0.1 mg/m³) (soluble cmpds) 1 mg/m³ (metal & insoluble cmpds)	Ca [N.E.]	Metal: Lustrous, silvery solid.	MW: 58.7 BP: 5139°F Sol: Insoluble Fl.P: NA IP: NA	VP: 0 mm (approx) MLT: 2831°F UEL: NA LEL: NA	Strong acids, sulfur, selenium, wood & other combustibles, nickel nitrate	Filter; Acid; ICP; III [#7300, Elements]
					Sp.Gr: 8.90 Noncombustible Solid in bulk form.			
Nicotine C₅H₄NC₄H₇NCH₃ 54-11-5 QS5250000	3-(1-Methyl-2-pyrrolidyl) pyridine	NIOSH/OSHA 0.5 mg/m³ [skin]	35 mg/m³	Pale-yellow to dark brown liquid with a fish-like odor when warm. [insecticide]	MW: 162.2 BP: 482°F (Decomposes) Sol: Miscible Fl.P: 203°F IP: 8.01 eV	VP: 0.08 mm FRZ: -110°F UEL: 4.0% LEL: 0.7%	Strong oxidizers, strong acids	XAD-2; Ethyl acetate; GC/Alk-FID; II(3) [#S293]
1654 55	1 ppm = 6.74 mg/m³				Sp.Gr: 1.01 Class IIIB Combustible Liquid			
Nitric acid HNO₃ 7697-37-2 QU5775000 1760 60 (<40% acid soln) 2031 44 (>40% acid soln) 2032 44 (fuming)	Aqua fortis, Engravers acid, Hydrogen nitrate, Red fuming nitric acid (RFNA), White fuming nitric acid (WFNA)	NIOSH/OSHA 2 ppm (5 mg/m³) ST 4 ppm 10 mg/m³ [Note: Often used in an aqueous solution. Fuming nitric acid is concentrated nitric acid that contains dissolved nitrogen.] 1 ppm = 2.62 mg/m³	100 ppm	Colorless, yellow, or red fuming liquid with an acrid, suffocating odor.	MW: 63.0 BP: 181°F Sol: Miscible Fl.P: NA IP: 11.95 eV Sp.Gr(77°F): 1.50 Noncombustible Liquid, but increases the flammability of combustible materials.	VP: 48 mm FRZ: -44°F UEL: NA LEL: NA	Combustible materials, metallic powders, hydrogen sulfide, carbides, alcohols [Note: Reacts with water to produce heat. Corrosive to metals.]	Si gel; NaHCO₃/ Na₂CO₃; IC; III [#7903, Inorganic Acids]

Personal protection and sanitation (See Table 3)		Recommendations for respirator selection — maximum concentration for use (MUC) (See Table 4)	Health hazards			
			Route	Symptoms (See Table 5)	First aid (See Table 6)	Target organs (See Table 5)
Clothing: Goggles: Wash: Change: Remove:	Repeat Reason prob Prompt wet N.R. Prompt non-imperv wet	NIOSH/OSHA 1000 ppm: PAPROV^ε/CCRFOV 2500 ppm: SA:CF^ε 5000 ppm: GMFOV/SCBAF/SAF 10,000 ppm: SAF:PD,PP §: SCBAF:PD,PP/SAF:PD,PP:ASCBA Escape: GMFOV/SCBAE	Inh Ing Con	Li-head, drow; irrit eyes, nose, skin; derm	Eye: Skin: Breath: Swallow:	Irr immed Soap wash prompt Resp support Medical attention immed	Resp sys, eyes, skin
Clothing: Goggles: Wash: Change: Remove:	Repeat Reason prob Prompt contam After work if reason prob contam Prompt non-imperv contam	NIOSH/OSHA 100 ppm: CCROVDM*/SA*/SCBA* 250 ppm: SA:CF*/PAPROVDM* 500 ppm: CCRFOVHiE/SCBAF/SAF §: SCBAF:PD,PP/SAF:PD,PP:ASCBA Escape: GMFOVHiE/SCBAE	Inh Abs Ing Con	Eye irrit; head; conf, excitement, mal; nau, vomit, abdom pain; irrit bladder; profuse sweat; jaun; hema, hemog, renal shutdown; derm	Eye: Breath: Swallow:	Irr immed Molten: flush immed/ Sol-Liq: soap wash prompt Resp support Medical attention immed	Eyes, blood, liver, kidneys, skin, RBC, CNS
Clothing: Goggles: Wash: Change: Remove: Provide:	Any poss Any poss Immed contam/daily After work if any poss contam Immed contam Eyewash, quick drench	NIOSH ¥: SCBAF:PD,PP/SAF:PD,PP:ASCBA Escape: HiEF/SCBA	Inh Abs Ing Con	Derm; hemorrhagic cystitis; dysp, ataxia, methemoglobinemia; hema; dysuria; [carc]	Eye: Skin: Breath: Swallow:	Irr immed (15 min) Soap wash immed Resp support Medical attention immed	Bladder, skin
Clothing: Goggles: Wash: Change: Remove: Provide:	Any poss Any poss Immed contam/daily After work if any poss contam Immed contam Eyewash, quick drench	NIOSH ¥: SCBAF:PD,PP/SAF:PD,PP:ASCBA Escape: HiEF/SCBA	Inh Abs Con Ing	Derm; hemorrhagic cystitis; dysp; ataxia, methemoglobinemia, hema; dysuria; [carc]	Eye: Skin: Breath: Swallow:	Irr immed (15 min) Soap wash immed Resp support Medical attention immed	Bladder, skin

Personal protection and sanitation (See Table 3)		Recommendations for respirator selection — maximum concentration for use (MUC) (See Table 4)	Health hazards			
			Route	Symptoms (See Table 5)	First aid (See Table 6)	Target organs (See Table 5)
Clothing: Goggles: Wash: Change: Remove:	Repeat N.R. Prompt contam After work if reason prob contam Prompt non-imperv contam	NIOSH ¥: SCBAF:PD,PP/SAF:PD,PP:ASCBA Escape: HiEF/SCBAE	Inh Ing Con	Sens derm, allergic asthma, pneuitis; [carc]	Skin: Breath: Swallow:	Water flush immed Resp support Medical attention immed	Nasal cavities, lungs, skin
Clothing: Goggles: Wash: Change: Remove:	Any poss Any poss Immed wet N.R. Immed wet (flamm)	NIOSH ¥: SCBAF:PD,PP/SAF:PD,PP:ASCBA Escape: GMFS/SCBAE	Inh Ing Con	Head, verti; nau, vomit, epigastric pain; substernal pain; cough, hyperpnea; cyan; weak; leucyt, pneuitis; delirium, convuls; [carc]	Eye: Skin: Breath: Swallow:	Irr immed Soap wash immed Resp support Medical attention immed	Lungs, paranasal sinus, CNS
Clothing: Goggles: Wash: Change: Remove: Provide:	Any poss Any poss Immed contam N.R. Immed non-imperv contam Eyewash, quick drench	NIOSH/OSHA 5 mg/m³: SA/SCBA 12.5 mg/m³: SA:CF* 25 mg/m³: SCBAF/SAF/SAT:CF* 35 mg/m³: SAF:PD,PP §: SCBAF:PD,PP/SAF:PD,PP:ASCBA Escape: GMFOV/SCBAE	Inh Abs Ing Con	Nau, salv, abdom pain, vomit, diarr; head, dizz, disturb hearing, vision; conf, hallu, inco; paroxysmal atrial fibrl; convuls, dysp	Eye: Skin: Breath: Swallow:	Irr immed Water flush immed Resp support Medical attention immed	CNS, CVS, lungs, GI tract
Clothing: Goggles: Wash: Change: Remove: Provide:	Any poss pH <2.5/Repeat pH >2.5 Any poss Immed contam N.R. Immed non-imperv contam pH <2.5: Eyewash, quick drench	NIOSH/OSHA 50 ppm: SA:CF* 100 ppm: SCBAF/SAF/GMFS⁴/CCRFS⁴ §: SCBAF:PD,PP/SAF:PD,PP:ASCBA Escape: GMFS⁴/SCBAE	Inh Ing Con	Irrit eyes, muc memb, skin; delayed pulm edema, pneuitis, bron; dental erosion	Eye: Skin: Breath: Swallow:	Irr immed Water flush immed Resp support Medical attention immed	Eyes, resp sys, skin, teeth

Chemical name, structure/formula, CAS and RTECS Nos., and DOT ID and guide Nos.	Synonyms, trade names, and conversion factors	Exposure limits (TWA unless noted otherwise)	IDLH	Physical description	Chemical and physical properties		Incompatibilities and reactivities	Measurement method (See Table 1)
					MW, BP, SOL Fl.P, IP, Sp.Gr, flammability	VP, FRZ UEL, LEL		
Nitric oxide NO 10102-43-9 QX0525000	Mononitrogen monoxide, Nitrogen monoxide	NIOSH/OSHA 25 ppm (30 mg/m³)	100 ppm	Colorless gas.	MW: 30.0 BP: -241°F Sol: 5% Fl.P: NA IP: 9.27 eV	VP: >1 atm FRZ: -263°F UEL: NA LEL: NA	Combustible materials, ozone, chlorinated hydrocarbons, ammonia, carbon disulfide, metals, fluorine [Note: Reacts with water to form nitric acid.]	Mol-sieve*; Reagent; Vis; II(1) [P&CAM #231]
1660 20	1 ppm = 1.25 mg/m³				Nonflammable Gas, but will accelerate the burning of combustible materials.			
p-Nitroaniline NO₂C₆H₄NH₂ 100-01-6 BY7000000	para-Aminonitrobenzene, 4-Nitroaniline, 4-Nitrobenzenamine, p-Nitrophenylamine, PNA	NIOSH/OSHA 3 mg/m³ [skin]	300 mg/m³	Bright yellow, crystalline powder with a slight ammonia-like odor.	MW: 138.1 BP: 630°F Sol: 0.08% Fl.P: 390°F IP: 8.85 eV	VP(288°F): 1 mm MLT: 295°F UEL: ? LEL: ?	Strong oxidizers, strong reducers [Note: May result in spontaneous heating of organic materials in the presence of moisture.]	Filter; 2-Propanol; HPLC/UVD; II(5) [#S7]
1661 55					Sp.Gr: 1.42 Combustible Solid			
Nitrobenzene C₆H₅NO₂ 98-95-3 DA6475000	Essence of mirbane, Nitrobenzol, Oil of mirbane	NIOSH/OSHA 1 ppm (5 mg/m³) [skin]	200 ppm	Yellow, oily liquid with a pungent odor like paste shoe polish. [Note: A solid below 42°F.]	MW: 123.1 FRZ: 411°F Sol: 0.2% Fl.P: 190°F IP: 9.92 eV	VP(112°F): 1 mm FRZ: 42°F UEL: ? LEL(200°F): 1.8%	Concentrated nitric acid, nitrogen tetroxide, caustics, phosphorus pentachloride, chemically-active metals such as tin or zinc	Si gel; Methanol; GC/FID; III [#2005]
1662 55	1 ppm = 5.12 mg/m³				Sp.Gr: 1.20 Class IIIA Combustible Liquid			
4-Nitrobiphenyl C₆H₅C₆H₄NO₂ 92-93-3 DV5600000	p-Nitrobipheny, p-Nitrodiphenyl, 4-Nitrodiphenyl, 4-Phenylnitrobenzene, p-Phenylnitrobenzene, PNB	NIOSH Ca See Appendix A OSHA [1910.1003] See Appendix B ACGIH A1	Ca	White to yellow, needle-like crystalline solid with a sweetish odor.	MW: 199.2 BP: 644°F Sol: Insoluble Fl.P: 290°F IP: ?	VP: ? MLT: 237°F UEL: ? LEL: ?	Strong reducers	Filter/ Si gel(2); 2-Propanol; GC/FID; II(4) [P&CAM #273]
					Combustible Solid			

Chemical name, structure/formula, CAS and RTECS Nos., and DOT ID and guide Nos.	Synonyms, trade names, and conversion factors	Exposure limits (TWA unless noted otherwise)	IDLH	Physical description	Chemical and physical properties		Incompatibilities and reactivities	Measurement method (See Table 1)
					MW, BP, SOL Fl.P, IP, Sp.Gr, flammability	VP, FRZ UEL, LEL		
Nitroglycerine CH₂NO₃CHNO₃CH₂NO₃ 55-63-0 QX2100000	Glyceryl trinitrate; NG; 1,2,3-Propanetriol trinitrate; Trinitroglycerine	NIOSH/OSHA ST 0.1 mg/m³ [skin] [Note: OSHA stayed the start-up date for the PEL on 12/6/89 for the explosives industry until 2/1/90.]	500 mg/m³	Colorless to pale-yellow, viscous liquid or solid (below 56°F). [Note: An explosive ingredient in dynamite (20-40%) with ethylene glycol dinitrate (80-60%).]	MW: 227.1 BP: Begins to decompose at 122-140°F Sol: 0.1% Fl.P: Explodes IP: ?	VP: 0.0003 mm FRZ: 56°F UEL: ? LEL: ?	Heat, ozone, shock, acids	Tenax GC; Ethanol; GC/ECD; III; [#2507]
3064 26	1 ppm = 9.44 mg/m³				Sp.Gr: 1.60 Class IIIB Combustible Liquid			
Nitromethane CH₃NO₂ 75-52-5 PA9800000	Nitrocarbol	NIOSH See Appendix D OSHA 100 ppm (250 mg/m³)	1000 ppm	Colorless, oily liquid with a disagreeable odor.	MW: 61.0 BP: 214°F Sol: 10% Fl.P: 95°F IP: 11.08 eV	VP: 28 mm FRZ: -20°F UEL: ? LEL: 7.3%	Amines; strong acids, alkalis & oxidizers; hydrocarbons & other combustible materials; metallic oxides [Note: Slowly corrodes steel & copper when wet.]	Chrom-106; Ethyl acetate; GC/FPD; III [#2527]
1261 26	1 ppm = 2.54 mg/m³				Sp.Gr: 1.14 Class IC Flammable Liquid			
1-Nitropropane CH₃CH₂CH₂NO₂ 108-03-2 TZ5075000	Nitropropane, 1-NP	NIOSH/OSHA 25 ppm (90 mg/m³)	2300 ppm	Colorless liquid with a somewhat disagreeable odor.	MW: 89.1 BP: 269°F Sol: 1% Fl.P: 96°F IP: 10.81 eV	VP: 8 mm FRZ: -162°F UEL: ? LEL: 2.2%	Amines; strong acids, alkalis & oxidizers; hydrocarbons & other combustible materials; metal oxides	XAD-4; CS₂; GC/NPD; OSHA [#46]
2608 26	1 ppm = 3.70 mg/m³				Sp.Gr: 1.00 Class IC Flammable Liquid			
2-Nitropropane CH₃CH(NO₂)CH₃ 79-46-9 TZ5250000	Dimethylnitromethane, sec-Nitropropane, iso-Nitropropane, 2-NP	NIOSH Ca See Appendix A Reduce exposure to lowest feasible concentration. OSHA 10 ppm (35 mg/m³)	Ca [2300 ppm] ACGIH C 10 ppm, A2 (35 mg/m³)	Colorless liquid with a pleasant, fruity odor.	MW: 89.1 BP: 249°F Sol: 2% Fl.P: 75°F IP: 10.71 eV	VP: 13 mm FRZ: -135°F UEL: 11.0% LEL: 2.6%	Amines; strong acids, alkalis & oxidizers, metal oxides, combustible materials	Chrom-106; Ethyl acetate; GC/FID; III [#2528]
2608 26	1 ppm = 3.70 mg/m³				Sp.Gr: 0.99 Class IC Flammable Liquid			

Personal protection and sanitation (See Table 3)	Recommendations for respirator selection — maximum concentration for use (MUC) (See Table 4)	Route	Symptoms (See Table 5)	First aid (See Table 6)	Target organs (See Table 5)
Clothing: N.R. Goggles: N.R. Wash: N.R. Change: N.R. Remove: N.R.	NIOSH/OSHA 100 ppm: SA:CF*/PAPRS*ↄ/SA*/ SCBA*/CCRFSↄ/GMFSↄ §: SCBAF:PD,PP/SAF:PD,PP:ASCBA Escape: GMFSↄ/SCBAE	Inh Con	Irrit eyes, nose, throat; drow; unconscious	Breath: Resp support	Resp sys
Clothing: Repeat Goggles: Reason prob Wash: Immed contam/daily Change: After work if reason prob contam Remove: Immed non-imperv contam Provide: Quick drench	NIOSH/OSHA 30 mg/m³: SA*/SCBA* 75 mg/m³: SA:CF* 150 mg/m³: SCBAF/SAF 300 mg/m³: SAF:PD,PP §: SCBAF:PD,PP/SAF:PD,PP:ASCBA Escape: GMFOVDMFu/SCBAE	Inh Abs Ing Con	Cyan, ataxia; tacar, tachypnea; dysp; irrity; vomit, diarr; convuls; resp arrest; anemia; methemoglobinemia	Eye: Irr immed Skin: Water flush immed Breath: Resp support Swallow: Medical attention immed	Blood, heart, lungs, liver
Clothing: Any poss Goggles: Reason prob Wash: N.R. Change: After work if any poss contam Remove: Immed non-imperv contam Provide: Quick drench	NIOSH/OSHA 10 ppm: CCROV*/SA*/SCBA* 25 ppm: SA:CF* 50 ppm: CCRFOV/GMFOV/SCBAF/SAF 200 ppm: SAF:PD,PP §: SCBAF:PD,PP/SAF:PD,PP:ASCBA Escape: GMFOV/SCBAE	Inh Abs Ing Con	Anoxia; irrit eyes; derm; anemia; in animals: liver, kidney damage	Eye: Irr immed Skin: Soap wash immed Breath: Resp support Swallow: Medical attention immed	Blood, liver, kidneys, CVS, skin
Clothing: Any poss Goggles: Any poss Wash: Immed contam/daily Change: After work if any poss contam Remove: Immed contam Provide: Eyewash, quick drench	NIOSH ¥: SCBAF:PD,PP/SAF:PD,PP:ASCBA Escape: HiEF/SCBAE	Inh Abs Ing Con	Head, lethargy, dizz; dysp; ataxia, weak; methemoglobinemia; urinary burning; acute hemorrhagic cystitis; [carc]	Eye: Irr immed Skin: Soap wash immed Breath: Resp support Swallow: Medical attention immed	Bladder, blood

Personal protection and sanitation (See Table 3)	Recommendations for respirator selection — maximum concentration for use (MUC) (See Table 4)	Route	Symptoms (See Table 5)	First aid (See Table 6)	Target organs (See Table 5)
Clothing: N.R. Goggles: N.R. Wash: N.R. Change: N.R. Remove: N.R.	NIOSH/OSHA 100 ppm: SA:CF*/PAPRS*ↄ/SA*/ SCBA*/CCRFSↄ/GMFSↄ §: SCBAF:PD,PP/SAF:PD,PP:ASCBA Escape: GMFSↄ/SCBAE	Inh Con	Irrit eyes, nose, throat; drow; unconscious	Breath: Resp support	Resp sys
Clothing: Repeat Goggles: Reason prob Wash: Immed contam/daily Change: After work if reason prob contam Remove: Immed non-imperv contam Provide: Quick drench	NIOSH/OSHA 30 mg/m³: SA*/SCBA* 75 mg/m³: SA:CF* 150 mg/m³: SCBAF/SAF 300 mg/m³: SAF:PD,PP §: SCBAF:PD,PP/SAF:PD,PP:ASCBA Escape: GMFOVDMFu/SCBAE	Inh Abs Ing Con	Cyan, ataxia; tacar, tachypnea; dysp; irrity; vomit, diarr; convuls; resp arrest; anemia; methemoglobinemia	Eye: Irr immed Skin: Water flush immed Breath: Resp support Swallow: Medical attention immed	Blood, heart, lungs, liver
Clothing: Any poss Goggles: Reason prob Wash: N.R. Change: After work if any poss contam Remove: Immed non-imperv contam Provide: Quick drench	NIOSH/OSHA 10 ppm: CCROV*/SA*/SCBA* 25 ppm: SA:CF*/PAPROV* 50 ppm: CCRFOV/GMFOV/SCBAF/SAF 200 ppm: SAF:PD,PP §: SCBAF:PD,PP/SAF:PD,PP:ASCBA Escape: GMFOV/SCBAE	Inh Abs Ing Con	Anoxia; irrit eyes; derm; anemia; in animals: liver, kidney damage	Eye: Irr immed Skin: Soap wash immed Breath: Resp support Swallow: Medical attention immed	Blood, liver, kidneys, CVS, skin
Clothing: Any poss Goggles: Any poss Wash: Immed contam/daily Change: After work if any poss contam Remove: Immed contam Provide: Eyewash, quick drench	NIOSH ¥: SCBAF:PD,PP/SAF:PD,PP:ASCBA Escape: HiEF/SCBAE	Inh Abs Ing Con	Head, lethargy, dizz; dysp; ataxia, weak; methemoglobinemia; urinary burning; acute hemorrhagic cystitis; [carc]	Eye: Irr immed Skin: Soap wash immed Breath: Resp support Swallow: Medical attention immed	Bladder, blood

Chemical name, structure/formula, CAS and RTECS Nos., and DOT ID and guide Nos.	Synonyms, trade names, and conversion factors	Exposure limits (TWA unless noted otherwise)	IDLH	Physical description	Chemical and physical properties		Incompatibilities and reactivities	Measurement method (See Table 1)
					MW, BP, SOL Fl.P, IP, Sp.Gr, flammability	VP, FRZ UEL, LEL		
Nitroglycerine CH₂NO₃CHNO₃CH₂NO₃ 55-63-0 QX2100000 3064 26	Glyceryl trinitrate; NG; 1,2,3-Propanetriol trinitrate; Trinitroglycerine 1 ppm = 9.44 mg/m³	NIOSH/OSHA ST 0.1 mg/m³ [skin] [Note: OSHA stayed the start-up date for the PEL on 12/6/89 for the explosives industry until 2/1/90.]	500 mg/m³	Colorless to pale-yellow, viscous liquid or solid (below 56°F). [Note: An explosive ingredient in dynamite (20-40%) with ethylene glycol dinitrate (80-60%).]	MW: 227.1 BP: Begins to decompose at 122-140°F Sol: 0.1% Fl.P: Explodes IP: ? Sp.Gr: 1.60 Class IIIB Combustible Liquid	VP: 0.0003 mm FRZ: 56°F UEL: ? LEL: ?	Heat, ozone, shock, acids	Tenax GC; Ethanol; GC/ECD; III; [#2507]
Nitromethane CH₃NO₂ 75-52-5 PA9800000 1261 26	Nitrocarbol 1 ppm = 2.54 mg/m³	NIOSH See Appendix D OSHA 100 ppm (250 mg/m³)	1000 ppm	Colorless, oily liquid with a disagreeable odor.	MW: 61.0 BP: 214°F Sol: 10% Fl.P: 95°F IP: 11.08 eV Sp.Gr: 1.14 Class IC Flammable Liquid	VP: 28 mm FRZ: -20°F UEL: ? LEL: 7.3%	Amines; strong acids, alkalis & oxidizers; hydrocarbons & other combustible materials; metallic oxides [Note: Slowly corrodes steel & copper when wet.]	Chrom-106; Ethyl acetate; GC/FPD; III [#2527]
1-Nitropropane CH₃CH₂CH₂NO₂ 108-03-2 TZ5075000 2608 26	Nitropropane, 1-NP 1 ppm = 3.70 mg/m³	NIOSH/OSHA 25 ppm (90 mg/m³)	2300 ppm	Colorless liquid with a somewhat disagreeable odor.	MW: 89.1 BP: 269°F Sol: 1% Fl.P: 96°F IP: 10.81 eV Sp.Gr: 1.00 Class IC Flammable Liquid	VP: 8 mm FRZ: -162°F UEL: ? LEL: 2.2%	Amines; strong acids, alkalis & oxidizers; hydrocarbons & other combustible materials; metal oxides	XAD-4; CS₂; GC/NPD; OSHA [#46]
2-Nitropropane CH₃CH(NO₂)CH₃ 79-46-9 TZ5250000 2608 26	Dimethylnitromethane, sec-Nitropropane, iso-Nitropropane, 2-NP 1 ppm = 3.70 mg/m³	NIOSH Ca See Appendix A Reduce exposure to lowest feasible concentration. OSHA 10 ppm (35 mg/m³)	Ca [2300 ppm] ACGIH C 10 ppm, A2 (35 mg/m³)	Colorless liquid with a pleasant, fruity odor.	MW: 89.1 BP: 249°F Sol: 2% Fl.P: 75°F IP: 10.71 eV Sp.Gr: 0.99 Class IC Flammable Liquid	VP: 13 mm FRZ: -135°F UEL: 11.0% LEL: 2.6%	Amines; strong acids, alkalis & oxidizers; metal oxides, combustible materials	Chrom-106; Ethyl acetate; GC/FID; III [#2528]

Chemical name, structure/formula, CAS and RTECS Nos., and DOT ID and guide Nos.	Synonyms, trade names, and conversion factors	Exposure limits (TWA unless noted otherwise)	IDLH	Physical description	Chemical and physical properties		Incompatibilities and reactivities	Measurement method (See Table 1)
					MW, BP, SOL Fl.P, IP, Sp.Gr, flammability	VP, FRZ UEL, LEL		
N-Nitrosodimethylamine (CH₃)₂N₂O 62-75-9 IQ0525000 1 ppm = 3.08 mg/m³	Dimethylnitrosamine; N,N-Dimethylnitrosamine; DMNA; N-Methyl-N-nitroso-methanamine; NDMA; N-Nitrosodimethylamine	NIOSH Ca See Appendix A OSHA [1910.1016] See Appendix B ACGIH A2 [skin]	Ca	Yellow, oily liquid with a faint, characteristic odor.	MW: 74.1 BP: 306°F Sol: Soluble Fl.P: ? IP: 8.69 eV Sp.Gr: 1.00 Combustible Liquid	VP: 3 mm FRZ: ? UEL: ? LEL: ?	Strong oxidizers [Note: Should be stored in dark bottles.]	T-Sorb; Methanol/CH₂Cl₂; GC/FID; III [#2522]
Nitrotoluene NO₂C₆H₄CH₃ o: 88-72-2 XT3150000 m: 99-08-1 XT2975000 p: 99-99-0 XT3325000 1664 55	Methyl nitrobenzene, Nitrotoluol o: ortho-Nitrotoluene 2-Nitrotoluene m: meta-Nitrotoluene 3-Nitrotoluene p: para-Nitrotoluene 4-Nitrotoluene 1 ppm = 5.70 mg/m³	NIOSH/OSHA 2 ppm (11 mg/m³) [skin]	200 ppm	Yellow liquids (o-, m-isomers) or crystalline solid (p-isomer) with a weak, aromatic odor.	MW: 137.1 BP: 432/450/460°F Sol: 0.07/0.05/0.04% Fl.P: 223/232/223°F IP: 9.43/9.48/9.50 eV Sp.Gr: 1.16/1.16/1.12 Class IIIB Combustible Liquids	VP(122°F): 1/1/1 mm FRZ/MLT: 25/59/126°F UEL: ? LEL: ?	Strong oxidizers, sulfuric acid	Si gel; Methanol; GC/FID; III [#2005, Nitro-benzenes]
Octachloronaphthalene C₁₀Cl₈ 2234-13-1 QK0250000	Halowax® 1051; 1,2,3,4,5,6,7,8-Octachloronaphthalene; Perchloronaphthalene	NIOSH/OSHA 0.1 mg/m³ ST 0.3 mg/m³ [skin]	Unknown	Waxy, pale-yellow solid with an aromatic odor.	MW: 403.7 BP: 824°F Sol: Insoluble Fl.P: NA IP: ? Sp.Gr: 2.00 Noncombustible Solid	VP: ? MLT: 378°F UEL: NA LEL: NA	Strong oxidizers	Filter; Hexane; GC/FID; II(2) [#S97]
Octane CH₃[CH₂]₆CH₃ 111-65-9 RG8400000 1262 27	n-Octane, normal-Octane 1 ppm = 4.75 mg/m³	NIOSH 75 ppm (350 mg/m³) C 385 ppm (1800 mg/m³) [15-min] OSHA 300 ppm (1450 mg/m³) ST 375 ppm (1800 mg/m³)	5000 ppm	Colorless liquid with a gasoline-like odor.	MW: 114.2 BP: 258°F Sol: Insoluble Fl.P: 56°F IP: 9.82 eV Sp.Gr: 0.70 Class IB Flammable Liquid	VP: 10 mm FRZ: -70°F UEL: 6.5% LEL: 1.0%	Strong oxidizers	Char; CS₂; GC/FID; III [#1500, Hydro-carbons]

Personal protection and sanitation (See Table 3)		Recommendations for respirator selection — maximum concentration for use (MUC) (See Table 4)	Health hazards			
			Route	Symptoms (See Table 5)	First aid (See Table 6)	Target organs (See Table 5)
Clothing: Goggles: Wash: Change: Remove: Provide:	Any poss Reason prob Immed contam After work if any poss contam Immed wet (flamm) Quick drench	NIOSH/OSHA 1 mg/m³: SA*/SCBA* 2.5 mg/m³: SA:CF* 5 mg/m³: SCBAF/SAF/SAT:CF* 200 mg/m³: SAF:PD,PP §: SCBAF:PD,PP/SAF:PD,PP:ASCBA Escape: GMFOVHiE/SCBAE	Inh Abs Ing Con	Throb head; dizz; nau, vomit, abdom pain; hypotension; flush; palp; methemoglobin; delirium, depres CNS; angina; skin irrit	Eye: Skin: Breath: Swallow:	Irr immed Soap wash immed Resp support Medical attention immed	CVS, blood, skin
Clothing: Goggles: Wash: Change: Remove:	Repeat Reason prob Prompt wet N.R. Immed wet (flamm)	OSHA 1000 ppm: SA:CFᶠ/SCBAF/SAF §: SCBAF:PD,PP/SAF:PD,PP:ASCBA Escape: SCBAE	Inh Con Ing	Derm	Eye: Skin: Breath: Swallow:	Irr immed Soap wash prompt Resp support Medical attention immed	Skin
Clothing: Goggles: Wash: Change: Remove:	N.R. Reason prob N.R. N.R. Immed wet (flamm)	NIOSH/OSHA 250 ppm: SA*/SCBA* 625 ppm: SA:CF* 1250 ppm: SCBAF/SAF 2300 ppm: SAF:PD,PP §: SCBAF:PD,PP/SAF:PD,PP:ASCBA Escape: SCBAE	Inh Con Ing	Eye irrit, head, nau, vomit, diarr	Eye: Skin: Breath: Swallow:	Irr immed Soap wash prompt Resp support Medical attention immed	Eyes, CNS
Clothing: Goggles: Wash: Change: Remove:	N.R. Reason prob N.R. N.R. Immed wet (flamm)	NIOSH ¥: SCBAF:PD,PP/SAF:PD,PP:ASCBA Escape: SCBAE	Inh Con Ing	Head, anor, nau, vomit, diarr, irrit resp sys, [carc]	Eye: Skin: Breath: Swallow:	Irr immed Soap wash prompt Resp support Medical attention immed	Resp sys, CNS

Personal protection and sanitation (See Table 3)		Recommendations for respirator selection — maximum concentration for use (MUC) (See Table 4)	Health hazards			
			Route	Symptoms (See Table 5)	First aid (See Table 6)	Target organs (See Table 5)
Clothing: Goggles: Wash: Change: Remove: Provide:	Any poss Any poss Immed contam/daily After work if any poss contam Immed contam Eyewash, quick drench	NIOSH ¥: SCBAF:PD,PP/SAF:PD,PP:ASCBA Escape: HiEF/SCBAE	Inh Abs Ing Con	Nau, vomit, diarr, abdom cramps; head; fever; enlarged liver, jaun; reduced function of liver, kidneys, lungs; [carc]	Eye: Skin: Breath: Swallow:	Irr immed Soap wash immed Resp support Medical attention immed	Liver, kidneys, lungs
Clothing: Goggles: Wash: Change: Remove:	Repeat Reason prob Prompt contam N.R. Prompt non-imperv contam	NIOSH/OSHA 20 ppm: SA*/SCBA* 50 ppm: SA:CF* 100 ppm: SCBAF/SAF/SAT:CF* 200 ppm: SAF:PD,PP §: SCBAF:PD,PP/SAF:PD,PP:ASCBA Escape: GMFOVHiE/SCBAE	Inh Abs Ing Con	Anoxia, cyan; head, weak, dizz, ataxia; dysp; tacar; nau, vomit	Eye: Skin: Breath: Swallow:	Irr immed Soap wash immed Resp support Medical attention immed	Blood, CNS, CVS, skin, GI tract
Clothing: Goggles: Wash: Change: Remove:	Any poss molt/ Reason prob sol-liq Any poss molt/ Reason prob sol-liq Prompt contam/daily After work if reason prob contam Immed non-imperv contam molt/ Prompt non-imperv contam sol-liq	NIOSH/OSHA 1 mg/m³: SA/SCBA §: SCBAF:PD,PP/SAF:PD,PP:ASCBA Escape: GMFOVHiE/SCBAE	Inh Abs Ing Con	Acne-form derm; liver damage, jaun	Eye: Skin: Breath: Swallow:	Irr immed Water flush immed Resp support Medical attention immed	Skin, liver
Clothing: Goggles: Wash: Change: Remove:	Repeat Reason prob Prompt wet N.R. Immed wet (flamm)	NIOSH 750 ppm: SA*/SCBA* 1875 ppm: SA:CF* 3750 ppm: SCBAF/SAF 5000 ppm: SAF:PD,PP §: SCBAF:PD,PP/SAF:PD,PP:ASCBA Escape: GMFOV/SCBAE	Inh Con Ing	Irrit eyes, nose; drow; derm; chemical pneu	Eye: Skin: Breath: Swallow:	Irr immed Soap wash prompt Resp support Medical attention immed	Skin, eyes, resp sys

Chemical name, structure/formula, CAS and RTECS Nos., and DOT ID and guide Nos.	Synonyms, trade names, and conversion factors	Exposure limits (TWA unless noted otherwise)	IDLH	Physical description	Chemical and physical properties		Incompatibilities and reactivities	Measurement method (See Table 1)
					MW, BP, SOL Fl.P, IP, Sp.Gr, flammability	**VP, FRZ UEL, LEL**		
Oil mist (mineral) 8012-95-1 PY8030000	Heavy mineral oil mist, Paraffin oil mist, White mineral oil mist	NIOSH 5 mg/m³ ST 10 mg/m³ OSHA 5 mg/m³	N.E.	Colorless, oily liquid aerosol dispersed in air. [Note: Has an odor like burned lubricating oil.]	MW: Varies BP: 680°F Sol: Insoluble Fl.P(oc): 275-500°F IP: ? Sp.Gr: 0.90 Class IIIB Combustible Liquid	VP: ? FRZ: ? UEL: ? LEL: ?	None reported	Filter; CFC-113; IR; III [#5026]
Osmium tetroxide (as Os) OsO₄ 20816-12-0 RN1140000 2471 55	Osmic acid anhydride, Osmium oxide	NIOSH/OSHA 0.002 mg/m³ (0.0002 ppm) ST 0.006 mg/m³ (0.0006 ppm)	1 mg/m³	Colorless, crystalline solid or pale-yellow mass with an unpleasant, acrid, chlorine-like odor. [Note: A liquid above 105°F.]	MW: 254.2 BP: 266°F Sol(77°F): 6% Fl.P: NA IP: 12.60 eV Sp.Gr: 5.10 Noncombustible Solid	VP(81°F): 11 mm MLT: 105°F UEL: NA LEL: NA	Hydrochloric acid, easily oxidized organic materials [Note: Begins to sublime below BP.]	None available
Oxalic acid HOOCCOOH·2H₂O 144-62-7 RO2450000 2449 54	Ethanedioic acid, Oxalic acid (aqueous), Oxalic acid dihydrate	NIOSH/OSHA 1 mg/m³ ST 2 mg/m³	500 mg/m³	Colorless, odorless powder or granular solid. [Note: The anhydrous form (COOH)₂ is an odorless, white solid.]	MW: 126.1 BP: Sublimes Sol: 14% Fl.P: ? IP: ? Sp.Gr: 1.65 Combustible Solid	VP: <0.001 mm MLT: 373°F (Sublimes) UEL: ? LEL: ?	Strong oxidizers, silver compounds	None available
Oxygen difluoride OF₂ 7783-41-7 RS2100000 2190 20	Difluorine monoxide, Fluorine monoxide, Oxygen fluoride	NIOSH/OSHA C 0.05 ppm (0.1 mg/m³) 1 ppm = 2.24 mg/m³	0.5 ppm	Colorless gas with a peculiar, foul odor.	MW: 54.0 BP: -230°F Sol: 0.02% Fl.P: NA IP: 13.11 eV Nonflammable Gas, but a strong oxidizer.	VP: >1 atm FRZ: -371°F UEL: NA LEL: NA	Combustible materials, chlorine, bromine, iodine, platinum, metal oxides, moist air, hydrogen sulfide, hydrocarbons, water [Note: Reacts very slowly with water to form hydrofluoric acid.]	None available

Chemical name, structure/formula, CAS and RTECS Nos., and DOT ID and guide Nos.	Synonyms, trade names, and conversion factors	Exposure limits (TWA unless noted otherwise)	IDLH	Physical description	Chemical and physical properties		Incompatibilities and reactivities	Measurement method (See Table 1)
					MW, BP, SOL Fl.P, IP, Sp.Gr, flammability	**VP, FRZ UEL, LEL**		
Ozone O₃ 10028-15-6 RS8225000	Triatomic oxygen	NIOSH C 0.1 ppm (0.2 mg/m³) OSHA 0.1 ppm (0.2 mg/m³) ST 0.3 ppm (0.6 mg/m³) 1 ppm = 2.00 mg/m³	10 ppm	Colorless to blue gas with a very pungent odor.	MW: 48.0 BP: -169°F Sol: Insoluble Fl.P: NA IP: 12.52 eV Nonflammable Gas, but a powerful oxidizer.	VP: >1 atm FRZ: -315°F UEL: NA LEL: NA	All oxidizable materials (both organic & inorganic)	Imp; Reagent; Vis; II(1) [P&CAM #154]
Paraquat (CH₃(C₅H₄N)₂CH₃)·2Cl 1910-42-5 DW2275000 2588 53 (solid pesticides)	1,1'-Dimethyl-4,4'-bipyridinium dichloride; N,N'-Dimethyl-4,4'-bipyridinium dichloride; Paraquat chloride; Paraquat dichloride	NIOSH/OSHA 0.1 mg/m³ (resp) [skin]	1.5 mg/m³	Yellow solid with a faint ammonia-like odor. [herbicide] [Note: Paraquat may also be found commercially as a methyl sulfate salt C₁₂H₁₄N₂·2CH₃SO₄.]	MW: 257.2 BP: 347-356°F (Decomposes) Sol: Miscible Fl.P: NA IP: ? Sp.Gr: 1.24 Noncombustible Solid	VP: Low MLT: ? UEL: NA LEL: NA	Strong oxidizers, alkylaryl-sulfonate wetting agents [Note: Corrosive to metals. Decomposes in presence of ultraviolet light.]	Filter; Water; HPLC/UVD; III [#5003]
Parathion (C₂H₅O)₂P(S)OC₆H₄NO₂ 56-38-2 TF4550000 2783 55 (liquid/dry)	O,O-Diethyl-O(p-nitrophenyl) phosphorothioate; Diethyl parathion; Ethyl parathion; Parathion-ethyl	NIOSH 0.05 mg/m³ [skin] OSHA 0.1 mg/m³ [skin] 1 ppm = 12.11 mg/m³	20 mg/m³	Pale-yellow to dark brown liquid with a garlic-like odor. [Note: A solid below 43°F. Pesticide that may be absorbed on a dry carrier.]	MW: 291.3 BP: 707°F Sol: 0.001% Fl.P: ? IP: ? Sp.Gr: 1.27 Combustible Liquid, but may be difficult to ignite.	VP: 0.00004 mm FRZ: 43°F UEL: ? LEL: ?	Strong oxidizers	Filter; Isooctane; GC/FPD; III [#5012]
Pentaborane B₅H₉ 19624-22-7 RY8925000 1380 75	Pentaboron nonahydride, Stable pentaborane	NIOSH/OSHA 0.005 ppm (0.01 mg/m³) ST 0.015 ppm (0.03 mg/m³) 1 ppm = 2.62 mg/m³	3 ppm	Colorless liquid with a pungent odor like sour milk.	MW: 63.1 BP: 140°F Sol: Reacts Fl.P: 86°F IP: 9.90 eV Sp.Gr: 0.62 Class IC Flammable Liquid	VP(77°F): 200 mm FRZ: -52°F UEL: ? LEL: 0.42%	Oxidizers, halogens, water, halogenated hydrocarbons [Note: May ignite SPONTANEOUSLY in moist air. Corrosive to natural rubber. Hydrolyzes slowly with heat in water to form boric acid.]	None available

Personal protection and sanitation (See Table 3)		Recommendations for respirator selection — maximum concentration for use (MUC) (See Table 4)	Health hazards			
			Route	Symptoms (See Table 5)	First aid (See Table 6)	Target organs (See Table 5)
Clothing: Goggles: Wash: Change: Remove:	Repeat N.R. Prompt wet N.R. Prompt non-imperv wet	NIOSH/OSHA 50 mg/m³: SA/SCBA/HiE 125 mg/m³: SA:CF/PAPRHiE 250 mg/m³: HiEF/PAPRTHiE/SAT:CF/ SCBAF/SAF 2500 mg/m³: SA:PD,PP §: SCBAF:PD,PP/SAF:PD,PP:ASCBA Escape: HiEF/SCBAE	Inh	None reported	Skin: Soap wash Breath: Fresh air	Resp sys, skin

Clothing:
Goggles:
Wash:
Change:
 	
Remove:
 	
Provide:
	Reason prob
	Any poss
	Prompt contam
	After work if reason
	  prob contam
	Prompt non-imperv
	  contam
	Eyewash
		NIOSH/OSHA
		0.1 mg/m³: SCBAF/SAF/CCRFSHiE/
		  GMFSHiE
		1 mg/m³: SAF:PD,PP
		§: SCBAF:PD,PP/SAF:PD,PP:ASCBA
		Escape: GMFSHiE/SCBAE
			Inh
			Ing
			Con
				Lac, vis dist, conj; head;
				cough; dysp; derm
					Eye: Irr immed
					Skin: Soap wash immed
					Breath: Resp support
					Swallow: Medical attention
					  immed
						Eyes, resp sys,
						skin

Clothing:
Goggles:
Wash:
Change:
 	
Remove:
	Repeat
	Any poss
	Prompt contam
	After work if reason
	  prob contam
	Prompt non-imperv
	  contam
		NIOSH/OSHA
		25 mg/m³: PAPRDMᴱ/SA:CFᴱ
		50 mg/m³: HiEF/SCBAF/SAF
		500 mg/m³: SAF:PD,PP
		§: SCBAF:PD,PP/SAF:PD,PP:ASCBA
		Escape: HiEF/SCBAE
			Inh
			Abs
			Ing
			Con
				Irrit eyes, muc memb,
				skin; eye burns; local
				pain, cyan; shock,
				collapse, convuls
					Eye: Irr immed
					Skin: Water flush prompt
					Breath: Resp support
					Swallow: Medical attention
					  immed
						Resp sys, skin,
						kidneys, eyes

Clothing: Goggles: Wash: Change: Remove:	N.R. N.R. N.R. N.R. N.R.	NIOSH/OSHA 0.5 ppm: SCBA/SA §: SCBAF:PD,PP/SAF:PD,PP:ASCBA Escape: GMFSᴸ/SCBAE	Inh Con	Intractable head; resp sys irrit, pulm edema; eye, skin burns	Eye: Irr immed Skin: Water flush immed Breath: Resp support	Lungs, eyes

Personal protection and sanitation (See Table 3)		Recommendations for respirator selection — maximum concentration for use (MUC) (See Table 4)	Health hazards			
			Route	Symptoms (See Table 5)	First aid (See Table 6)	Target organs (See Table 5)
Clothing: Goggles: Wash: Change: Remove:	N.R. N.R. N.R. N.R. N.R.	NIOSH/OSHA 1 ppm: SA/SCBA/CCRSᴸ 2.5 ppm: SA:CF/PAPRSᴸ 5 ppm: CCRFSᴸ/GMFSᴸ/SCBAF/SAF/ SAT:CF 10 ppm: SAF:PD,PP §: SCBAF:PD,PP/SAF:PD,PP:ASCBA Escape: GMFSᴸ/SCBAE	Inh	Irrit eyes, muc memb; pulm edema; chronic resp disease	Eye: Medical attention Breath: Fresh air, 100% O₂	Eyes, resp sys

Clothing:
Goggles:
Wash:
Change:
Remove:
Provide:
	Reason prob
	Reason prob
	Immed contam
	N.R.
	Immed non-imperv contam
	Quick drench
		NIOSH/OSHA
		1 ppm: CCROVDMFu*/SA*/SCBA*
		1.5 ppm: CCRFOVDMFu/SA:CF*/
		  SCBAF/SAF/PAPROVDMFu*
		§: SCBAF:PD,PP/SAF:PD,PP:ASCBA
		Escape: GMFOVHiE/SCBAE
			Inh
			Abs
			Ing
			Con
				Irrit eyes, nose, epis;
				derm; fingernail damage;
				irrit GI tract; heart,
				liver, kidney damage;
				acute pulm inflamm
					Eye: Irr immed
					Skin: Water flush immed
					Breath: Resp support
					Swallow: Medical attention
					  immed
						Eyes, resp sys,
						heart, liver,
						kidneys, GI tract

Clothing:
Goggles:
Wash:
Change:
 	
Remove:
Provide:
	Repeat
	Any poss
	Immed contam
	After work if any poss
	  contam
	Immed non-imperv contam
	Eyewash, quick drench
		NIOSH
		0.5 mg/m³: CCROVDMFu/SA/SCBA
		1.25 mg/m³: SA:CF/PAPROVDMFu
		2.5 mg/m³: CCROVHiE/SCBAF/SAF/
		  PAPRTOVHiE/SAT:CF
		20 mg/m³: SA:PD,PP
		§: SCBAF:PD,PP/SAF:PD,PP:ASCBA
		Escape: GMFOVHiE/SCBAE
			Inh
			Abs
			Ing
			Con
				Miosis; rhin; head; tight
				chest, wheez, lar spasm,
				salv, cyan; anor, nau,
				vomit, abdom cramps, diarr;
				sweat; musc fasc, weak,
				para; gidd, conf, ataxia;
				convuls, coma; low BP;
				card irregularities; skin,
				eye irrit
					Eye: Irr immed
					Skin: Soap wash immed
					Breath: Resp support
					Swallow: Medical attention
					  immed
						Resp sys, CNS,
						CVS, eyes, skin,
						blood chol

Clothing: Goggles: Wash: Change: Remove: Provide:	Reason prob Any poss Immed wet N.R. Immed wet (flamm) Eyewash, quick drench	NIOSH/OSHA 0.05 ppm: SA/SCBA 0.125 ppm: SA:CF 0.25 ppm: SCBAF/SAF/SAT:CF 3 ppm: SA:PD,PP §: SCBAF:PD,PP/SAF:PD,PP:ASCBA Escape: GMFS/SCBAE	Inh Abs Ing Con	Dizz, head, drow, li-head; inco, tremor, tonic spasm face, neck, abdom, limbs; convuls; irrit eyes, skin	Eye: Irr immed Skin: Soap wash immed Breath: Resp support Swallow: Medical attention immed	CNS, eyes, skin

Chemical name, structure/formula, CAS and RTECS Nos., and DOT ID and guide Nos.	Synonyms, trade names, and conversion factors	Exposure limits (TWA unless noted otherwise)	IDLH	Physical description	Chemical and physical properties		Incompatibilities and reactivities	Measurement method (See Table 1)
					MW, BP, SOL Fl.P, IP, Sp.Gr, flammability	**VP, FRZ UEL, LEL**		
Pentachloronaphthalene $C_{10}H_3Cl_5$ 1321-64-8 QK0300000	Halowax® 1013; 1,2,3,4,5-Pentachloro-naphthalene	NIOSH/OSHA 0.5 mg/m³ [skin]	Unknown	Pale yellow or white solid or powder with an aromatic odor.	MW: 300.4 BP: 636°F Sol: Insoluble Fl.P: 180°F IP: ? Sp.Gr. 1.73 Combustible Solid	VP: <1 mm MLT: 335°F UEL: ? LEL: ?	Strong oxidizers	Filter/Bub; Isooctane; GC/ECD; II(2) [#S96]
Pentachlorophenol C_6Cl_5OH 87-86-5 SM6300000 2020 53	PCP; Penta; 2,3,4,5,6-Pentachlorophenol	NIOSH/OSHA 0.5 mg/m³ [skin]	150 mg/m³	Colorless to white, crystalline solid with a benzene-like odor. [fungicide]	MW: 266.4 BP: 588°F Sol: 0.001% Fl.P: NA IP: NA Sp.Gr. 1.98 Noncombustible Solid	VP: 0.0001 mm MLT: 374°F UEL: NA LEL: NA	Strong oxidizers	Filter/Bub; Methanol; HPLC/UVD; III [#5512]
n-Pentane $CH_3[CH_2]_3CH_3$ 109-66-0 RZ9450000 1265 27	Pentane, normal-Pentane 1 ppm = 3.00 mg/m³	NIOSH 120 ppm (350 mg/m³) C 610 ppm (1800 mg/m³) [15-min] OSHA 600 ppm (1800 mg/m³) ST 750 ppm (2250 mg/m³)	15,000 ppm [LEL]	Colorless liquid with a gasoline-like odor. [Note: A gas above 97°F. May be utilized as a fuel.]	MW: 72.2 BP: 97°F Sol: 0.04% Fl.P: -57°F IP: 10.34 eV Sp.Gr. 0.63 Class IA Flammable Liquid	VP(65°F): 400 mm FRZ: -202°F UEL: 7.8% LEL: 1.5%	Strong oxidizers	Char; CS₂; GC/FID; III [#1500, Hydro-carbons]
2-Pentanone $CH_3COCH_2CH_2CH_3$ 107-87-9 SA7875000 1249 26	Ethyl acetone, Methyl propyl ketone, MPK 1 ppm = 3.58 mg/m³	NIOSH 150 ppm (530 mg/m³) OSHA 200 ppm (700 mg/m³) ST 250 ppm (875 mg/m³)	5000 ppm	Colorless to water-white liquid with a characteristic acetone-like odor.	MW: 86.1 BP: 215°F Sol: 6% Fl.P: 45°F IP: 9.39 eV Sp.Gr. 0.81 Class IB Flammable Liquid	VP: 16 mm FRZ: -108°F UEL: 8.2% LEL: 1.5%	Oxidizers, bromine trifluoride	Char; CS₂; GC/FID; III [#1300, Ketones I]

Chemical name, structure/formula, CAS and RTECS Nos., and DOT ID and guide Nos.	Synonyms, trade names, and conversion factors	Exposure limits (TWA unless noted otherwise)	IDLH	Physical description	Chemical and physical properties		Incompatibilities and reactivities	Measurement method (See Table 1)
					MW, BP, SOL Fl.P, IP, Sp.Gr, flammability	**VP, FRZ UEL, LEL**		
Perchloromethyl mercaptan Cl_3CSCl 594-42-3 PB0370000 1670 55	PMM, Tricholoromethane sulfenyl chloride, Trichloromethyl sulfur chloride 1 ppm = 7.73 mg/m³	NIOSH/OSHA 0.1 ppm (0.8 mg/m³)	10 ppm	Pale-yellow, oily liquid with an unbearable acrid odor.	MW: 185.9 BP: 297°F (Decomposes) Sol: Insoluble Fl.P: NA IP: ? Sp.Gr. 1.69 Noncombustible Liquid, but will support combustion.	VP: 65 mm FRZ: ? UEL: NA LEL: NA	Alkalis, amines, hot iron, water [Note: Corrosive to most metals.]	None available
Perchloryl fluoride ClO_3F 7616-94-6 SD1925000 1955 15	Chlorine fluoride oxide, Chlorine oxyfluoride, Trioxychlorofluoride 1 ppm = 4.26 mg/m³	NIOSH/OSHA 3 ppm (14 mg/m³) ST 6 ppm (28 mg/m³)	385 ppm	Colorless gas with a characteristic, sweet odor. [Note: Shipped as a liquefied compressed gas.]	MW: 102.5 BP: -52°F Sol: 0.06% Fl.P: NA IP: 13.60 eV Nonflammable Gas, but will support combustion.	VP: >1 atm FRZ: -234°F UEL: NA LEL: NA	Combustibles, strong bases, amines, finely divided metals, oxidizable materials, reducing agents, alcohols	None available
Petroleum distillates (naphtha) 8002-05-9 SE7449000 1255 27	Aliphatic petroleum naphtha, Petroleum naphtha 1 ppm = 4.11 mg/m³ (approx)	NIOSH 350 mg/m³ C 1800 mg/m³ [15-min] OSHA 400 ppm (1600 mg/m³)	10,000 ppm	Colorless liquid with a gasoline- or kerosene-like odor. [Note: A mixture of paraffins (C₅ to C₁₃) that may contain a small amount of aromatic hydrocarbons.]	MW: 99(approx) BP: 86-460°F Sol: Insoluble Fl.P: -40 to -86°F IP: ? Sp.Gr. 0.63-0.66 Flammable Liquid	VP: 40 mm (approx) FRZ: ? UEL: 5.9% LEL: 1.1%	Strong oxidizers	Char; CS₂; GC/FID; III [#1550, Naphthas]
Phenol C_6H_5OH 108-95-2 SJ3325000 1671 55 (solid) 2312 55 (molten) 2821 55 (soln)	Carbolic acid, Hydroxybenzene, Monohydroxy benzene, Phenyl alcohol, Phenyl hydroxide	NIOSH 5 ppm (19 mg/m³) C 15.6 ppm (60 mg/m³) [15-min] [skin] OSHA 5 ppm (19 mg/m³) [skin]	250 ppm	Colorless to light-pink, crystalline solid with a sweet, acrid odor. [Note: Phenol liquefies by mixing with about 8% water.]	MW: 94.1 BP: 359°F Sol(77°F): 9% Fl.P: 175°F IP: 8.50 eV Sp.Gr. 1.06 Combustible Solid	VP: 0.4 mm MLT: 109°F UEL: ? LEL: 1.8%	Strong oxidizers, calcium hypochlorite, aluminum chloride, acids	XAD-7; Methanol; HPLC/UVD; OSHA [#32]

Personal protection and sanitation (See Table 3)		Recommendations for respirator selection — maximum concentration for use (MUC) (See Table 4)	Health hazards			
			Route	Symptoms (See Table 5)	First aid (See Table 6)	Target organs (See Table 5)
Clothing: Goggles: Wash: Change: Remove:	Any poss molt/Repeat sol-liq Any poss molt/Reason prob sol-liq Prompt contam After work if reason prob contam Immed non-imperv contam molt/ Prompt non-imperv contam sol-liq	NIOSH/OSHA 5 mg/m³: SA*/SCBA* §: SCBAF:PD,PP/SAF:PD,PP:ASCBA Escape: GMFOVHiE/SCBAE	Inh Abs Ing Con	Head, ftg, verti, anor; pruritus, acne-form skin eruptions; jaun, liver nec	Eye: Skin: Breath: Swallow:	Irr immed Soap prompt/Molten Flush immed Resp support Medical Attention immed	Skin, liver, CNS
Clothing: Goggles: Wash: Change: Remove: Provide:	Any poss Any poss Immed contam After work if any poss contam Immed non-imperv contam Eyewash, quick drench	NIOSH/OSHA 5 mg/m³: CCROVDMFu*/SA*/SCBA* 12.5 mg/m³: SA:CF*/PAPROVDMFu* 25 mg/m³: CCRFOVHiE/SCBAF/SAF 150 mg/m³: SAF:PD,PP §: SCBAF:PD,PP/SAF:PD,PP:ASCBA Escape: GMFOVHiE/SCBAE	Inh Abs Ing Con	Irrit eyes, nose, throat; sneezing, cough; weak, anor, low-wgt; sweat; head, dizz; nau, vomit; dysp, chest pain; high fever; derm	Eye: Skin: Breath: Swallow:	Irr immed Soap wash immed Resp support Medical attention immed	CVS, resp sys, eyes, liver, kidneys, skin, CNS
Clothing: Goggles: Wash: Change: Remove:	Repeat Reason prob Prompt wet N.R. Immed wet (flamm)	NIOSH 1200 ppm: SA/SCBA 3000 ppm: SA:CF 6000 ppm: SCBAF/SAF/SAT:CF 15,000 ppm: SA:PD,PP §: SCBAF:PD,PP/SAF:PD,PP:ASCBA Escape: GMFOV/SCBAE	Inh Ing Con	Drow; irrit eyes, nose; derm; chemical pneu	Eye: Skin: Breath: Swallow:	Irr immed Water wash prompt Resp support Medical attention immed	Skin, eyes, resp sys
Clothing: Goggles: Wash: Change: Remove:	Repeat Reason prob Prompt wet N.R. Immed wet (flamm)	NIOSH 1000 ppm: CCROV*/PAPROV* 1500 ppm: SA*/SCBA* 3750 ppm: SA:CF* 5000 ppm: GMFOV/SCBAF/SAF §: SCBAF:PD,PP/SAF:PD,PP:ASCBA Escape: GMFOV/SCBAE	Inh Ing Con	Irrit eyes, muc memb; head; derm; narco, coma	Eye: Skin: Breath: Swallow:	Irr immed Water flush Resp support Medical attention immed	Resp sys, eyes, skin, CNS

Personal protection and sanitation (See Table 3)		Recommendations for respirator selection — maximum concentration for use (MUC) (See Table 4)	Health hazards			
			Route	Symptoms (See Table 5)	First aid (See Table 6)	Target organs (See Table 5)
Clothing: Goggles: Wash: Change: Remove:	Repeat Reason prob Prompt wet N.R. Prompt non-imperv wet	NIOSH/OSHA 1 ppm: SA*/SCBA*/CCROV* 2.5 ppm: SA:CF*/PAPROV* 5 ppm: CCRFOV/GMFOV/SCBAF/SAF/SAT:CF*/PAPRTOV* 10 ppm: SAF:PD,PP §: SCBAF:PD,PP/SAF:PD,PP:ASCBA Escape: GMFOV/SCBAE	Inh Abs Ing Con	Lac, eye inflamm; irrit nose, throat; cough; dysp, deep breath pain, coarse rales; vomit; pallor, tacar; acidosis; anuria	Eye: Skin: Breath: Swallow:	Irr immed Soap wash immed Resp support Medical attention immed	Eyes, resp sys, liver, kidneys, skin
Clothing: Goggles: Wash: Change: Remove:	Prevent skin freezing Reason prob N.R. N.R. Immed non-imperv contam	NIOSH/OSHA 30 ppm: SA/SCBA 75 ppm: SA:CF* 150 ppm: SCBAF/SAF 385 ppm: SAF:PD,PP §: SCBAF:PD,PP/SAF:PD,PP:ASCBA Escape: GMFS↓/SCBAE	Inh Con	Resp irrit; skin burns; in animals: methemoglobin; anoxia; cyan; weak, dizz, head; pulm edema; pneuitis	Eye: Breath:	Irr immed Resp support	Resp sys, skin, blood
Clothing: Goggles: Wash: Change: Remove:	Repeat Reason prob Prompt wet N.R. Immed wet (flamm)	NIOSH 850 ppm: SCBA/SA 2125 ppm: SA:CF* 4250 ppm: SCBAF/SAF/SAT:CF* 10,000 ppm: SAF:PD,PP §: SCBAF:PD,PP/SAF:PD,PP:ASCBA Escape: GMFOV/SCBAE	Inh Ing Con	Dizz, drow, head, nau; irrit eyes, nose, throat; dry cracked skin	Eye: Skin: Breath: Swallow:	Irr immed Soap wash prompt Resp support Medical Attention immed	Skin, eyes, resp sys, CNS
Clothing: Goggles: Wash: Change: Remove: Provide:	Any poss Any poss Immed contam After work if any poss contam Immed non-imperv contam Eyewash, quick drench	NIOSH/OSHA 50 ppm: CCROVDM/SA/SCBA 125 ppm: SA:CF/PAPROVDM 250 ppm: SCBAF/SAF/CCRFOVHiE/GMFOVHiE/PAPRTOVHiE §: SCBAF:PD,PP/SAF:PD,PP:ASCBA Escape: GMFOVHiE/SCBAE	Inh Abs Ing Con	Irrit eyes, nose, throat; anor, low-wgt; weak, musc ache, pain; dark urine; cyan; liver, kidney damage; skin burns; derm; ochronosis; tremor, convuls, twitch	Eye: Skin: Breath: Swallow:	Irr immed Soap wash immed Resp support Medical attention immed	Liver, kidneys, skin

Chemical name, structure/formula, CAS and RTECS Nos., and DOT ID and guide Nos.	Synonyms, trade names, and conversion factors	Exposure limits (TWA unless noted otherwise)	IDLH	Physical description	Chemical and physical properties		Incompatibilities and reactivities	Measurement method (See Table 1)
					MW, BP, SOL Fl.P, IP, Sp.Gr, flammability	**VP, FRZ UEL, LEL**		
p-Phenylene diamine $C_6H_4(NH_2)_2$ 106-50-3 SS8050000 1673 53	4-Amino aniline; 1,4-Benzenediamine; p-Diaminobenzene; 1,4-Diaminobenzene	NIOSH/OSHA 0.1 mg/m³ [skin]	Unknown	White to slightly red crystalline solid.	MW: 108.2 BP: 513°F Sol: 5% Fl.P: 312°F IP: 6.89 eV	VP: <1 mm MLT: 295°F UEL: ? LEL: ? Combustible Solid	Strong oxidizers	None available
Phenyl ether (vapor) $C_6H_5OC_6H_5$ 101-84-8 KN8970000	Diphenyl ether, Diphenyl oxide, Phenoxy benzene, Phenyl oxide	NIOSH/OSHA 1 ppm (7 mg/m³)	N.E.	Colorless, crystalline solid or liquid (above 82°F) with a geranium-like odor.	MW: 170.2 BP: 498°F Sol: Insoluble Fl.P: 234°F IP: 8.09 eV Sp.Gr: 1.08	VP(77°F): 0.02 mm MLT: 82°F UEL: 1.5% LEL: 0.8% Class IIIB Combustible Liquid	Strong oxidizers	Char; CS₂; GC/FID; II(2) [#S72]
	1 ppm = 7.08 mg/m³							
Phenyl ether-biphenyl mixture (vapor) $C_6H_5OC_6H_5/C_6H_5C_6H_5$ 8004-13-5 DV1500000	Diphenyl oxide-diphenyl mixture, Dowtherm® A	NIOSH/OSHA 1 ppm (7 mg/m³)	N.E.	Colorless to straw-colored liquid or solid (below 54°F) with a disagreeable, aromatic odor. [Note: A mixture typically contains 75% phenyl ether & 25% biphenyl.]	MW: 166(approx) BP: 495°F Sol: Insoluble Fl.P(oc): 255°F IP: ? Sp.Gr(77°F): 1.06	VP(77°F): 0.08 mm FRZ: 54°F UEL: ? LEL: ? Class IIIB Combustible Liquid	Strong oxidizers	Si gel; Benzene; GC/FID; II(2) [#S73]
	1 ppm = 6.90 mg/m³							
Phenyl glycidyl ether $C_9H_{10}O_2$ 122-60-1 TZ3675000	1,2-Epoxy-3-phenoxy propane; Glycidyl phenyl ether; PGE; Phenyl 2,3-epoxypropyl ether	NIOSH Ca See Appendix A C 1 ppm (6 mg/m³) [15-min] OSHA 1 ppm (6 mg/m³)	Ca [Unknown]	Colorless liquid. [Note: A solid below 38°F.]	MW: 150.1 BP: 473°F Sol: 0.2% Fl.P: 248°F IP: ? Sp.Gr: 1.11	VP: 0.01 mm FRZ: 38°F UEL: ? LEL: ? Class IIIB Combustible Liquid	Strong oxidizers, amines, strong acids, strong bases	Char; CS₂; GC/FID; II(2) [#S74]
	1 ppm = 6.24 mg/m³							

Chemical name, structure/formula, CAS and RTECS Nos., and DOT ID and guide Nos.	Synonyms, trade names, and conversion factors	Exposure limits (TWA unless noted otherwise)	IDLH	Physical description	Chemical and physical properties		Incompatibilities and reactivities	Measurement method (See Table 1)
					MW, BP, SOL Fl.P, IP, Sp.Gr, flammability	**VP, FRZ UEL, LEL**		
Phenylhydrazine $C_6H_5NHNH_2$ 100-63-0 MW8925000 2572 53	Hydrazinobenzene, Monophenylhydrazine	NIOSH Ca See Appendix A C 0.14 ppm (0.6 mg/m³) [2-hr] OSHA 5 ppm (20 mg/m³) ST 10 ppm (45 mg/m³) [skin]	Ca [295 ppm] ACGIH A2	Colorless to pale-yellow liquid or solid (below 67°F) with a faint, aromatic odor.	MW: 108.1 BP: 470°F (Decomposes) Sol: Slight Fl.P: 190°F IP: 7.64 eV Sp.Gr: 1.10	VP(161°F): 1 mm FRZ: 67°F UEL: ? LEL: ? Class IIIA Combustible Liquid	Strong oxidizers, lead dioxide	Bub; Pho-acid; Vis; II(3) [#S160]
	1 ppm = 4.49 mg/m³							
Phosdrin® $C_7H_{13}O_4PO_2$ 7786-34-7 GQ5250000 2783 55	2-Carbomethoxy-1-methylvinyl dimethyl phosphate, Mevinphos	NIOSH/OSHA 0.01 ppm (0.1 mg/m³) ST 0.03 ppm (0.3 mg/m³) [skin]	4 ppm	Pale-yellow to orange liquid with a weak odor. [Note: Insecticide that may be absorbed on a dry carrier.]	MW: 224.2 BP: Decomposes Sol: Miscible Fl.P(oc): 347°F IP: ? Sp.Gr: 1.25	VP: 0.003 mm FRZ: -69°F UEL: ? LEL: ? Class IIIB Combustible Liquid	Strong oxidizers [Note: Corrosive to cast iron, some stainless steels, and brass.]	Chrom-102; Toluene; GC/FPD; III [#2503]
	1 ppm = 9.32 mg/m³							
Phosgene $COCl_2$ 75-44-5 SY5600000 1076 15	Carbon oxychloride, Carbonyl chloride, Chloroformyl chloride	NIOSH 0.1 ppm (0.4 mg/m³) C 0.2 ppm (0.8 mg/m³) [15-min] OSHA 0.1 ppm (0.4 mg/m³)	2 ppm	Colorless gas with a suffocating odor like musty hay. [Note: A fuming liquid below 47°F. Shipped as a liquefied compressed gas.]	MW: 98.9 BP: 47°F Sol: Slight Fl.P: NA IP: 11.55 eV Sp.Gr: 1.43 (Liquid at 32°F)	VP: >1 atm FRZ: -198°F UEL: NA LEL: NA Nonflammable Gas	Moisture [Note: Reacts slowly in water to form hydrochloric acid & carbon dioxide.]	Imp; Reagent; Vis; II(1) [P&CAM #219]
	1 ppm = 4.11 mg/m³							
Phosphine PH_3 7803-51-2 SY7525000 2199 18	Hydrogen phosphide, Phosphorated hydrogen, Phosphorus hydride, Phosphorus trihydride	NIOSH/OSHA 0.3 ppm (0.4 mg/m³) ST 1 ppm (1 mg/m³)	200 ppm	Colorless gas with a fish- or garlic-like odor. [pesticide] [Note: Shipped as a liquefied compressed gas. Pure compound is odorless.]	MW: 34.0 BP: -126°F Sol: Slight Fl.P: NA (Gas) IP: 9.96 eV	VP: >1 atm FRZ: -209°F UEL: ? LEL: ? Flammable Gas	Air, oxidizers, chlorine, acids, moisture, halogenated hydrocarbons, copper [Note: May ignite SPONTANEOUSLY on contact with air.]	Si gel; Reagent; Vis; II(5) [#S332]
	1 ppm = 1.41 mg/m³							

Personal protection and sanitation (See Table 3)		Recommendations for respirator selection — maximum concentration for use (MUC) (See Table 4)	Health hazards					
			Route	Symptoms (See Table 5)	First aid (See Table 6)		Target organs (See Table 5)	
Clothing:	Reason prob	NIOSH/OSHA	Inh	Irrit pharynx, larynx;	Eye:	Irr immed	Resp sys, skin	
Goggles:	Reason prob	2.5 mg/³: SA:CFᶜ	Abs	bronchial asthma; sens	Skin:	Soap wash prompt		
Wash:	Prompt contam/daily	5 mg/m³: SCBAF/SAF	Ing	derm	Breath:	Resp support		
Change:	After work if reason	25 mg/m³: SAF:PD,PP	Con		Swallow:	Medical attention		
	prob contam	§: SCBAF:PD,PP/SAF:PD,PP:ASCBA				immed		
Remove:	Prompt non-imperv contam	Escape: GMFSHiE/SCBAE						
Clothing:	Repeat	NIOSH/OSHA	Inh	Nau; irrit eyes, nose,	Eye:	Irr immed	Eyes, skin, resp	
Goggles:	Reason prob	25 ppm: SA:CFᶜ/PAPROVDMᶜ	Con	skin	Skin:	Soap wash prompt	sys	
Wash:	Prompt contam	50 ppm: CCRFOVHiE/GMFOVHiE/	Ing		Breath:	Resp support		
Change:	N.R.	SCBAF/SAF			Swallow:	Medical attention		
Remove:	Prompt non-imperv contam	100 ppm: SAF:PD,PP §: SCBAF:PD,PP/SAF:PD,PP:ASCBA Escape: GMFOVHiE/SCBAE				immed		
Clothing:	Repeat	NIOSH/OSHA	Inh	Nau; irrit eyes, nose,	Eye:	Irr immed	Eyes, skin, resp	
Goggles:	Reason prob	25 ppm: SA:CFᶜ/PAPROVDMᶜ	Con	skin	Skin:	Soap wash prompt	sys	
Wash:	Prompt contam	50 ppm: CCRFOVHiE/GMFOVHiE/	Ing		Breath:	Resp support		
Change:	N.R.	SCBAF/SAF			Swallow:	Medical attention		
Remove:	Prompt non-imperv contam	100 ppm: SAF:PD,PP §: SCBAF:PD,PP/SAF:PD,PP:ASCBA Escape: GMFOVHiE/SCBAE				immed		
Clothing:	Any poss	NIOSH	Inh	Skin irrit, sens; irrit	Eye:	Irr immed	Skin, eyes, CNS	
Goggles:	Reason prob	¥: SCBAF:PD,PP/SAF:PD,PP:ASCBA	Con	eyes, upper resp sys;	Skin:	Soap wash prompt		
Wash:	Prompt contam	Escape: GMFOV/SCBAE	Ing	narco; [carc]	Breath:	Resp support		
Change:	N.R.				Swallow:	Medical attention		
Remove:	Prompt non-imperv contam					immed		

Personal protection and sanitation (See Table 3)		Recommendations for respirator selection — maximum concentration for use (MUC) (See Table 4)	Health hazards					
			Route	Symptoms (See Table 5)	First aid (See Table 6)		Target organs (See Table 5)	
Clothing:	Any poss	NIOSH	Inh	Skin sens, hemolytic	Eye:	Irr immed	Blood, resp sys,	
Goggles:	Any poss	¥: SCBAF:PD,PP/SAF:PD,PP:ASCBA	Abs	anemia, dysp, cyan; jaun;	Skin:	Soap wash immed	liver, kidneys,	
Wash:	Immed contam/daily	Escape: SCBAE	Ing	kidney damage; vascular	Breath:	Resp support	skin	
Change:	After work if any poss contam		Con	thrombosis; [carc]	Swallow:	Medical attention immed		
Remove:	Immed non-imperv contam							
Provide:	Eyewash, quick drench							
Clothing:	Any poss	NIOSH/OSHA	Inh	Miosis; rhin; head; tight	Eye:	Irr immed	Resp sys, CNS,	
Goggles:	Any poss	0.1 ppm: SA/SCBA	Abs	chest, wheez, lar spasm,	Skin:	Soap wash immed	CVS, skin,	
Wash:	Immed contam	0.25 ppm: SA:CF	Ing	salv, cyan; anor, nau,	Breath:	Resp support	blood chol	
Change:	N.R.	0.5 ppm: SCBAF/SAF/SAT:CF	Con	vomit, abdom cramps, diarr;	Swallow:	Medical attention		
Remove:	Immed non-imperv contam	4 ppm: SA:PD,PP		para; ataxia, convuls;		immed		
Provide:	Eyewash, quick drench	§: SCBAF:PD,PP/SAF:PD,PP:ASCBA Escape: GMFOVHiE/SCBAE		low BP; card irregularities; irrit skin, eyes				
Clothing:	Reason prob	NIOSH/OSHA	Inh	Irrit eyes; dry burning	Eye:	Irr immed	Resp sys, skin,	
Goggles:	Any poss	1 ppm: SA*/SCBA*	Ing	throat; vomit; cough,	Skin:	Water flush immed	eyes	
Wash:	Immed contam	2 ppm: SCBAF/SAF	Con	foamy sputum, dysp,	Breath:	Resp support		
Change:	N.R.	§: SCBAF:PD,PP/SAF:PD,PP:ASCBA		chest pain, cyan; skin	Swallow:	Medical attention		
Remove:	Immed non-imperv contam	Escape: GMFS/SCBAE		burns		immed		
Provide:	Quick drench							
Clothing:	N.R.	NIOSH/OSHA	Inh	Nau, vomit, abdom pain,	Breath:	Resp support	Resp sys	
Goggles:	N.R.	3 ppm: SA/SCBA		diarr; thirst; chest				
Wash:	N.R.	7.5 ppm: SA:CF		pressure, dysp; musc pain,				
Change:	N.R.	15 ppm: SCBAF/SAF/GMFS		chills; stupor or syncope				
Remove:	N.R.	200 ppm: SA:PD,PP §: SCBAF:PD,PP/SAF:PD,PP:ASCBA Escape: GMFS/SCBAE						

Chemical name, structure/formula, CAS and RTECS Nos., and DOT ID and guide Nos.	Synonyms, trade names, and conversion factors	Exposure limits (TWA unless noted otherwise)	IDLH	Physical description	Chemical and physical properties		Incompatibilities and reactivities	Measurement method (See Table 1)
					MW, BP, SOL Fl.P, IP, Sp.Gr, flammability	VP, FRZ UEL, LEL		
Phosphoric acid H₃PO₄ 7664-38-2 TB6300000 1805 60	Metaphosphoric acid, Orthophosphoric acid, Phosphoric acid (aqueous), White phosphoric acid 1 ppm = 4.07 mg/m³	NIOSH/OSHA 1 mg/m³ ST 3 mg/m³	10,000 mg/m³	Thick, colorless, odorless, crystalline solid. [Note: Often used in an aqueous solution.]	MW: 98.0 BP: 415°F Sol: Miscible Fl.P: NA IP: ?. Sp.Gr(77°F): 1.87 (Pure) 1.33 (50% soln) Noncombustible Liquid	VP: 0.03 mm MLT: 108°F UEL: NA LEL: NA	Strong caustics, most metals [Note: Readily reacts with metals to form flammable hydrogen gas.]	Si gel; NaHCO₃/ Na₂CO₃; IC; III [#7903, Inorganic Acids]
Phosphorus (yellow) P₄ 7723-14-0 TH3500000 1381 38	Elemental white phosphorus, White phosphorus	NIOSH/OSHA 0.1 mg/m³	N.E.	White to yellow, soft, waxy solid with acrid fumes in air. [Note: Usually shipped or stored in water.]	MW: 124.0 BP: 536°F Sol: 0.0003% Fl.P: ? IP: ? Sp.Gr: 1.82 Flammable Solid	VP: 0.03 mm MLT: 111°F UEL: ? LEL: ?	Air, oxidizers including elemental sulfur & strong caustics [Note: Ignites SPONTANEOUSLY in moist air.]	Tenax GC; Xylene; GC/FPD; III [#7905]
Phosphorus pentachloride PCl₅ 10026-13-8 TB6125000 1806 39	Pentachlorophosphorus, Phosphoric chloride, Phosphorus perchloride	NIOSH/OSHA 1 mg/m³	200 mg/m³	White to pale-yellow, crystal-line solid with a pungent, unpleasant odor.	MW: 208.3 BP: Sublimes Sol: Reacts Fl.P: NA IP: ? Noncombustible Solid	VP(132°F): 1 mm MLT: 324°F (Sublimes) UEL: NA LEL: NA	Water, magnesium oxide, chemically-active metals such as sodium & potassium, alkalis [Note: Hydrolyzes in water (even in humid air) to form hydrochloric acid and phosphoric acid. Corrosive to metals.]	Filter/Bub; Reagent; Vis; II(5) [#S257]
Phosphorus pentasulfide P₂S₅/P₄S₁₀ 1314-80-3 TH4375000 1340 41	Phosphorus persulfide, Phosphorus sulfide	NIOSH/OSHA 1 mg/m³ ST 3 mg/m³	750 mg/m³	Greenish-gray to yellow, crystal-line solid with an odor of rotten eggs.	MW: 222.3/ 444.6 BP: 957°F Sol: Reacts Fl.P: ? IP: ? Sp.Gr: 2.09 Flammable Solid, which may SPONTANEOUSLY ignite in presence of moisture.	VP(572°F): 1 mm MLT: 550°F UEL: ? LEL: ?	Water, alcohols, strong oxidizers, acids [Note: Reacts with water to form hydrogen sulfide and sulfur dioxide.]	None available

Chemical name, structure/formula, CAS and RTECS Nos., and DOT ID and guide Nos.	Synonyms, trade names, and conversion factors	Exposure limits (TWA unless noted otherwise)	IDLH	Physical description	Chemical and physical properties		Incompatibilities and reactivities	Measurement method (See Table 1)
					MW, BP, SOL Fl.P, IP, Sp.Gr, flammability	VP, FRZ UEL, LEL		
Phosphorus trichloride PCl₃ 7719-12-2 TH3675000 1809 39	Phosphorus chloride 1 ppm = 5.71 mg/m³	NIOSH/OSHA 0.2 ppm (1.5 mg/m³) ST 0.5 ppm (3 mg/m³)	50 ppm	Colorless to yellow, fuming liquid with an odor like hydrochloric acid.	MW: 137.4 BP: 169°F Sol: Reacts Fl.P: NA IP: 9.91 eV Sp.Gr: 1.58 Noncombustible Liquid	VP(70°F): 100 mm FRZ: -170°F UEL: NA LEL: NA	Water, chemically-active metals such as sodium & potassium, aluminum, strong nitric acid, acetic acid, organic matter [Note: Hydrolyzes in water to form hydrochloric acid and phosphoric acid.]	Bub; Reagent; Vis; III [#6402]
Phthalic anhydride C₆H₄(CO)₂O 85-44-9 TI3150000 2214 60	1,2-Benzenedicarboxylic anhydride; PAN; Phthalic acid anhydride	NIOSH/OSHA 6 mg/m³ (1 ppm)	10,000 mg/m³	White solid with a characteristic, acrid odor.	MW: 148.1 BP: 563°F Sol: 0.6% Fl.P: 305°F IP: 10.00 eV Sp.Gr: 1.53 Combustible Solid	VP(206°F): 1 mm MLT: 267°F UEL: 10.5% LEL: 1.7%	Strong oxidizers, water [Note: Converted to phthalic acid in hot water.]	Filter; NH₄OH; HPLC/UVD; II(3) [#S179]
Picric acid C₆H₂OH(NO₂)₃ 88-89-1 TJ7875000 1344 33 (>10% water)	Phenol trinitrate; 2,4,6-Trinitrophenol 1 ppm = 9.52 mg/m³	NIOSH 0.1 mg/m³ ST 0.3 mg/m³ [skin] OSHA 0.1 mg/m³ [skin]	100 mg/m³	Yellow, odorless solid. [Note: Usually used as an aqueous solution.]	MW: 229.1 BP: Explodes above 572°F Sol: 1% Fl.P: 302°F IP: ? Sp.Gr: 1.76 Class IIIB Combustible Liquid	VP(383°F): 1 mm MLT: 252°F UEL: ? LEL: ?	Copper, lead, zinc & other metals; salts; plaster; concrete; ammonia [Note: Corrosive to metals. An explosive mixture results when the aqueous solution crystallizes.]	Filter; Methanol/ Water; HPLC/UVD; II(4) [#S228]
Pindone C₉H₅O₂C(O)C(CH₃)₃ 83-26-1 NK6300000 2472 53	tert-Butyl valone; 1,3-Dioxo-2-pivaloy-lindane; Pival®; Pivalyl; 2-Pivalyl-1,3-indandione	NIOSH/OSHA 0.1 mg/m³	200 mg/m³	Bright yellow powder with almost no odor. [insecticide]	MW: 230.3 BP: Decomposes Sol(77°F): 0.002% Fl.P: ? IP: ? Sp.Gr: 1.06	VP: ? MLT: 230°F UEL: ? LEL: ?	None reported	None available

Personal protection and sanitation (See Table 3)	Recommendations for respirator selection — maximum concentration for use (MUC) (See Table 4)	Health hazards			
		Route	Symptoms (See Table 5)	First aid (See Table 6)	Target organs (See Table 5)
Clothing: Any poss >1.6%/Repeat <1.6% Goggles: Any poss Wash: Immed contam Change: After work if reason prob contam sol Remove: Immed non-imperv contam Provide: >1.6%: Eyewash, quick drench	NIOSH/OSHA 25 mg/m³: SA:CF* 50 mg/m³: SCBAF/SAF/HiEF 2000 mg/m³: SAF:PD,PP §: SCBAF:PD,PP/SAF:PD,PP:ASCBA Escape: HiEF/SCBAE	Inh Ing Con	Irrit upper resp tract, eyes, skin; burns skin, eyes; derm	Eye: Irr immed Skin: Water flush Immed Breath: Resp support Swallow: Medical attention immed	Resp sys, eyes, skin
Clothing: Any poss Goggles: Any poss Wash: Immed contam Change: N.R. Remove: Immed contam Provide: Eyewash, quick drench	NIOSH/OSHA 1 mg/m³: SA 2.5 mg/m³: SA:CF£ 5 mg/m³: SCBAF/SAF 200 mg/m³: SAF:PD,PP §: SCBAF:PD,PP/SAF:PD,PP:ASCBA Escape: SCBAE	Inh Ing Con	Irrit eyes, resp tract; abdom pain, nau, jaun; anemia; cachexia; dental pain, excess salv; jaw pain, swell; burns skin, eyes	Eye: Irr immed Skin: Water flush immed Breath: Resp support Swallow: Medical attention immed	Resp sys, liver, kidneys, jaw, teeth, blood, eyes, skin
Clothing: Any poss Goggles: Any poss Wash: Immed contam Change: After work if reason prob contam Remove: Immed non-imperv contam Provide: Eyewash, quick drench	NIOSH/OSHA 10 mg/m³: SA*/SCBA* 25 mg/m³: SA:CF* 50 mg/m³: SCBAF/SAF 200 mg/m³: SAF:PD,PP §: SCBAF:PD,PP/SAF:PD,PP:ASCBA Escape: GMFOVHiE/SCBAE	Inh Ing Con	Irrit eyes, resp sys; bron; derm	Eye: Irr immed Skin: Water flush immed Breath: Resp support Swallow: Medical attention immed	Resp sys, eyes, skin
Clothing: Repeat Goggles: Reason prob Wash: Prompt contam Change: After work if reason prob contam Remove: Prompt non-imperv contam	NIOSH/OSHA 10 mg/m³: SA*/SCBA* 25 mg/m³: SA:CF* 50 mg/m³: SCBAF/SAF 750 mg/m³: SAF:PD,PP §: SCBAF:PD,PP/SAF:PD,PP:ASCBA Escape: GMFSHiE/SCBAE	Inh Ing Con	Apnea, coma, convuls; irrit eyes, conj pain, lac, photo, kerato-conj, corneal vesic; irrit resp sys; dizz; head; ftg; irrity, insom; GI dist	Eye: Irr immed Skin: Dust off solid; water flush Breath: Resp support Swallow: Medical attention immed	Resp sys, CNS, eyes, skin

Personal protection and sanitation (See Table 3)	Recommendations for respirator selection — maximum concentration for use (MUC) (See Table 4)	Health hazards			
		Route	Symptoms (See Table 5)	First aid (See Table 6)	Target organs (See Table 5)
Clothing: Any poss Goggles: Any poss Wash: Immed contam Change: N.R. Remove: Immed non-imperv contam Provide: Eyewash, quick drench	NIOSH/OSHA 10 ppm: SCBAF/SAF 50 ppm: SAF:PD,PP §: SCBAF:PD,PP/SAF:PD,PP:ASCBA Escape: GMFS↳/SCBAE	Inh Ing Con	Irrit eyes, nose, throat; pulm edema; burns eyes, skin	Eye: Irr immed Skin: Water flush immed Breath: Resp support Swallow: Medical attention immed	Resp sys, eyes, skin
Clothing: Repeat Goggles: Reason prob Wash: Prompt contam Change: After work if reason prob contam Remove: Prompt non-imperv contam	NIOSH/OSHA 30 mg/m³: DM* 60 mg/m³: DMXSQ*/SA*/SCBA* 150 mg/m³: SA:CF*/PAPRDM* 300 mg/m³: SCBAF/SAF/HiEF 10,000 mg/m³: SAF:PD,PP §: SCBAF:PD,PP/SAF:PD,PP:ASCBA Escape: HiEF/SCBAE	Inh Ing Con	Conj; nasal ulcer bleeding, upper resp irrit; bron, bronchial asthma; derm	Eye: Irr immed Skin: Soap wash prompt Breath: Resp support Swallow: Medical attention immed	Resp sys, eyes, skin, liver, kidneys
Clothing: Reason prob Goggles: Reason prob Wash: Prompt contam/daily Change: After work if reason prob contam Remove: Prompt non-imperv contam	NIOSH/OSHA 0.5 mg/m³: DM 1 mg/m³: DMXSQ/SA/SCBA 2.5 mg/m³: PAPRDM/SA:CF 5 mg/m³: HiEF/SCBAF/SAF/PAPRTHiE/SAT:CF 100 mg/m³: SAF:PD,PP §: SCBAF:PD,PP/SAF:PD,PP:ASCBA Escape: HiEF/SCBAE	Inh Abs Con	Irrit eyes; sens derm; yellow-stain hair, skin; weak, myalgia, anuria, polyuria; bitter taste, GI dist; hepatitis, hematuria, album, neph	Eye: Irr immed Skin: Soap wash prompt Breath: Resp support Swallow: Medical attention immed	Kidneys, liver, blood, skin, eyes
Clothing: N.R. Goggles: N.R. Wash: N.R. Change: After work if reason prob contam Remove: N.R.	NIOSH/OSHA 0.5 mg/m³: DM 1 mg/m³: DMXSQ/SA/SCBA 2.5 mg/m³: PAPRDM/SA:CF 5 mg/m³: PAPRTHiE/SCBAF/SAF/HiEF/SAT:CF 200 mg/m: SAF:PD,PP §: SCBAF:PD,PP/SAF:PD,PP:ASCBA Escape: HiEF/SCBAE	Inh	Epis, excess bleeding from minor cuts, bruises; smoky urine, black tarry stools; pain abdom, back	Eye: Irr immed Swallow: Medical attention immed	Blood prothrombin

Chemical name, structure/formula, CAS and RTECS Nos., and DOT ID and guide Nos.	Synonyms, trade names, and conversion factors	Exposure limits (TWA unless noted otherwise)	IDLH	Physical description	Chemical and physical properties		Incompatibilities and reactivities	Measurement method (See Table 1)
					MW, BP, SOL Fl.P, IP, Sp.Gr, flammability	VP, FRZ UEL, LEL		
Platinum (soluble salts as Pt)	Synonyms vary depending upon the specific soluble salt.	NIOSH/OSHA 0.002 mg/m³	N.E.	Appearance and odor vary depending upon the specific soluble salt.	Properties vary depending upon the specific soluble salt.		None reported	Filter; Acid/ Reagent; HGA; II(7) [#S191]
Portland cement

65997-15-1 VV8770000 | Cement, Hydraulic cement, Portland cement silicate | NIOSH/OSHA 10 mg/m³ (total) 5 mg/m³ (resp) | N.E. | Gray, odorless powder. | MW: ? BP: NA Sol: Insoluble Fl.P: NA IP: NA | VP: 0 mm (approx) MLT: NA UEL: NA LEL: NA | None reported | Filter; none; Grav; III [#0500, Nuisance Dust (total)] |
| | | | | | Noncombustible Solid | | | |
| Propane

CH₃CH₂CH₃

74-98-6 TX2275000

1978 22 | Bottled gas, Dimethyl methane, n-Propane, Propyl hydride

1 ppm = 1.83 mg/m³ | NIOSH/OSHA 1000 ppm (1800 mg/m³) | 20,000 ppm [LEL] | Colorless, odorless gas. [Note: A foul-smelling odorant is often added when used for fuel purposes. Shipped as a liquefied compressed gas.] | MW: 44.1 BP: -44°F Sol: 0.01% Fl.P: NA (Gas) IP: 11.07 eV

Flammable Gas | VP: >1 atm FRZ: -306°F UEL: 9.5% LEL: 2.1% | Strong oxidizers | Combustible gas meter none; none; II(2) [#S87] |
| beta-Propiolactone

C₃H₄O₂

57-57-8 RQ7350000 | BPL; Hydroacrylic acid, beta lactone; 3-Hydroxy-beta-lactone; 3-Hydroxy-propionic acid; beta-Lactone; 2-Oxetanone; 3-Propiolacetone

1 ppm = 3.00 mg/m³ | NIOSH Ca See Appendix A

OSHA [1910.1013] See Appendix B

ACGIH 0.5 ppm, A2 (1.5 mg/m³) | Ca | Colorless liquid with a slightly sweet odor. | MW: 72.1 BP: 323°F (Decomposes) Sol: 37% Fl.P: 165°F IP: ?

Sp.Gr: 1.15 Class IIIA Combustible Liquid | VP(77°F): 3 mm FRZ: -28°F UEL: ? LEL: 2.9% | Acetates, halogens, thiocyanates, thiosulfates [Note: May polymerize upon storage.] | None available |

Chemical name, structure/formula, CAS and RTECS Nos., and DOT ID and guide Nos.	Synonyms, trade names, and conversion factors	Exposure limits (TWA unless noted otherwise)	IDLH	Physical description	Chemical and physical properties		Incompatibilities and reactivities	Measurement method (See Table 1)
					MW, BP, SOL Fl.P, IP, Sp.Gr, flammability	VP, FRZ UEL, LEL		
n-Propyl acetate

CH₃COOCH₂CH₂CH₃

109-60-4 AJ3675000

1276 26 | Propylacetate, n-Propyl ester of acetic acid

1 ppm = 4.25 mg/m³ | NIOSH/OSHA 200 ppm (840 mg/m³) ST 250 ppm (1050 mg/m³) | 8000 ppm | Colorless liquid with a mild, fruity odor. | MW: 102.2 BP: 215°F Sol: 2% Fl.P: 55°F IP: 10.04 eV

Sp.Gr: 0.84 Class IB Flammable Liquid | VP(84°F): 40 mm FRZ: -134°F UEL: 8% LEL(100°F): 1.7% | Nitrates; strong oxidizers, alkalis & acids | Char; CS₂; GC/FID; III [#1450, Esters I] |
| n-Propyl alcohol

CH₃CH₂CH₂OH

71-23-8 UH8225000

1274 26 | Ethyl carbinol, 1-Propanol, n-Propanol, Propyl alcohol

1 ppm = 2.50 mg/m³ | NIOSH 200 ppm (500 mg/m³) ST 250 ppm (625 mg/m³) [skin]

OSHA 200 ppm (500 mg/m³) ST 250 ppm (625 mg/m³) | 4000 ppm | Colorless liquid with a mild, alcohol-like odor. | MW: 60.1 BP: 207°F Sol: Miscible Fl.P: 72°F IP: 10.15 eV

Sp.Gr: 0.81 Class IB Flammable Liquid | VP(77°F): 21 mm FRZ: -196°F UEL: 13.7% LEL: 2.2% | Strong oxidizers | Char; 2-Propanol/ CS₂; GC/FID; III [#1401, Alcohols II] |
| Propylene dichloride

CH₃CHClCH₂Cl

78-87-5 TX9625000

1279 27 | Dichloro-1,2-propane; 1,2-Dichloropropane

1 ppm = 4.70 mg/m³ | NIOSH Ca See Appendix A

OSHA 75 ppm (350 mg/m³) ST 110 ppm (510 mg/m³) | Ca [2000 ppm] | Colorless liquid with a chloroform-like odor. [pesticide] | MW: 113.0 BP: 206°F Sol: 0.3% Fl.P: 60°F IP: 10.87 eV

Sp.Gr: 1.16 Class IB Flammable Liquid | VP: 40 mm FRZ: -149°F UEL: 14.5% LEL: 3.4% | Strong oxidizers, strong acids | Char(pet); Acetone/ Cyclo-hexane; GC/ECD; III [#1013, 1,2-Di-chloro-propane] |
| Propylene imine

C₃H₇N

75-55-8 CM8050000

1921 30 (inhibited) | 2-Methylaziridine, 2-Methylethyleneimine, Propyleneimine, Propylene imine (inhibited) Propylenimine

1 ppm = 2.37 mg/m³ | NIOSH Ca See Appendix A 2 ppm (5 mg/m³) [skin] OSHA 2 ppm (5 mg/m³) [skin] | Ca [500 ppm]

ACGIH A2 | Colorless, oily liquid with an ammonia-like odor. | MW: 57.1 BP: 152°F Sol: Miscible Fl.P: 25°F IP: 9.00 eV

Sp.Gr: 0.80 Class IB Flammable Liquid | VP: 112 mm FRZ: -85°F UEL: ? LEL: ? | Acids, strong oxidizers, water, carbonyl compounds, quinones, sulfonyl halides [Note: Subject to violent polymerization in contact with acids. Hydrolyzes in water to form methylethanolamine.] | None available |

Personal protection and sanitation (See Table 3)		Recommendations for respirator selection — maximum concentration for use (MUC) (See Table 4)	Health hazards			
			Route	Symptoms (See Table 5)	First aid (See Table 6)	Target organs (See Table 5)
Clothing: Goggles: Wash: Change: Remove:	Reason prob Reason prob Prompt contam After work if reason prob contam Prompt non-imperv contam	NIOSH/OSHA 0.05 mg/m³: D 0.1 mg/m³: HiEF/SCBAF/SAF 4 mg/m³: SAF:PD,PP §: SCBAF:PD,PP/SAF:PD,PP:ASCBA Escape: HiEF/SCBAE	Inh Ing Con	Irrit eyes, nose; cough, dysp, wheez, cyan; skin sens; lymphocytosis	Eye: Skin: Breath: Swallow:	Irr immed Water flush immed Resp support Medical attention immed	Resp sys, skin, eyes
Clothing: Goggles: Wash: Change: Remove:	Repeat Reason prob Prompt contam N.R. Prompt non-imperv contam	NIOSH/OSHA 50 mg/m³: D 100 mg/m³: SA/SCBA/DXSQ 250 mg/m³: PAPRD/SA:CF 500 mg/m³: SCBAF/SAF/HiEF/SAT:CF/ PAPRTHiE 5000 mg/m³: SA:PD,PP §: SCBAF:PD,PP/SAF:PD,PP:ASCBA Escape: HiEF/SCBAE	Inh Con Ing	Irrit eyes, nose; cough, expectoration; exertional dysp, wheez, chronic bron; derm	Eye: Skin: Breath: Swallow:	Irr immed Soap wash prompt Fresh air Medical attention immed	Resp sys, eyes, skin
Clothing: Goggles: Wash: Change: Remove:	Prevent skin freezing Reason prob N.R. N.R. Immed wet (flamm)	NIOSH/OSHA 10,000 ppm: SA/SCBA 20,000 ppm: SA:CF/SCBAF/SAF §: SCBAF:PD,PP/SAF:PD,PP:ASCBA Escape: SCBAE	Inh Con	Dizz, disorientation, excitation, frostbite	Eye: Skin: Breath:	Irr immed Water flush immed Resp support	CNS
Clothing: Goggles: Wash: Change: Remove: Provide:	Any poss Any poss Immed contam/daily After work if any poss contam Immed contam Eyewash, quick drench	NIOSH ¥: SCBAF:PD,PP/SAF:PD,PP:ASCBA Escape: GMFOV/SCBAE	Inh Abs Ing Con	Skin irrit, blistering, burns; corneal opac; frequent urination; dysuria; hema; [carc]	Eye: Skin: Breath: Swallow:	Irr immed Soap wash immed Resp support Medical attention immed	Kidney, skin, lungs, eyes

Personal protection and sanitation (See Table 3)		Recommendations for respirator selection — maximum concentration for use (MUC) (See Table 4)	Health hazards			
			Route	Symptoms (See Table 5)	First aid (See Table 6)	Target organs (See Table 5)
Clothing: Goggles: Wash: Change: Remove:	Repeat Reason prob Prompt wet N.R. Immed wet (flamm)	NIOSH/OSHA 1000 ppm: PAPROV/CCRFOV 5000 ppm: SA:CF 8000 ppm: GMFOV/SCBAF/SAF §: SCBAF:PD,PP/SAF:PD,PP:ASCBA Escape: GMFOV/SCBAE	Inh Con Ing	Irrit eyes, nose, throat; narco; derm	Eye: Skin: Breath:	Irr immed Water flush prompt Resp support	Resp sys, eyes, skin, CNS
Clothing: Goggles: Wash: Change: Remove:	Repeat Reason prob Prompt wet N.R. Immed wet (flamm)	NIOSH/OSHA 1000 ppm: PAPROV*/CCROV* 2000 ppm: SA*/SCBA* 4000 ppm: SA*/SCBAF/SAF/GMFOV §: SCBAF:PD,PP/SAF:PD,PP:ASCBA Escape: GMFOV/SCBAE	Inh Abs Ing Con	Mild irrit eyes, nose, throat; dry cracking skin; drow, head; ataxia, GI pain; abdom cramps: nau, vomit, diarr	Eye: Skin: Breath: Swallow:	Irr immed Water flush Resp support Medical attention immed	Skin, eyes, resp sys, GI tract
Clothing: Goggles: Wash: Change: Remove:	Repeat Reason prob Prompt wet N.R. Immed wet (flamm)	NIOSH ¥: SCBAF:PD,PP/SAF:PD,PP:ASCBA Escape: GMFOV/SCBAE	Inh Con Ing	Eye irrit; drow, li-head; irrit skin; [carc]; in animals: liver, kidney disease; skin irrit	Eye: Skin: Breath: Swallow:	Irr immed Soap wash prompt Resp support Medical attention immed	Skin, eyes, resp sys, liver, kidneys
Clothing: Goggles: Wash: Change: Remove: Provide:	Reason prob Any poss Immed contam N.R. Immed wet (flamm) Eyewash, quick drench	NIOSH ¥: SCBAF:PD,PP/SAF:PD,PP:ASCBA Escape: GMFS/SCBAE	Inh Abs Ing Con	Eye, skin burns; [carc]	Eye: Skin: Breath: Swallow:	Irr immed Water flush immed Resp support Medical attention immed	Eyes, skin

Chemical name, structure/formula, CAS and RTECS Nos., and DOT ID and guide Nos.	Synonyms, trade names, and conversion factors	Exposure limits (TWA unless noted otherwise)	IDLH	Physical description	Chemical and physical properties		Incompatibilities and reactivities	Measurement method (See Table 1)
					MW, BP, SOL FI.P, IP, Sp.Gr, flammability	VP, FRZ UEL, LEL		
Propylene oxide C_3H_6O 75-56-9 TZ2975000 1280 26 (inhibited)	1,2-Epoxy propane; Methyl ethylene oxide; Methyloxirane; Propene oxide; 1,2-Propylene oxide 1 ppm = 2.42 mg/m³	NIOSH Ca See Appendix A OSHA 20 ppm (50 mg/m³)	Ca [2000 ppm]	Colorless liquid with a benzene-like odor. [Note: A gas above 94°F.]	MW: 58.1 BP: 94°F Sol: 41% FI.P: -35°F IP: 9.81 eV Sp.Gr: 0.83 Class IA Flammable Liquid	VP: 445 mm FRZ: -170°F UEL: 36% LEL: 2.3%	Anhydrous metal chlorides; iron; strong acids, caustics & peroxides [Note: Polymerization may occur due to high temperatures or contamination with alkalis, aqueous acids, amines & acidic alcohols.]	Char; CS₂; GC/FID; III [#1612]
n-Propyl nitrate $CH_3CH_2CH_2NO_3$ 627-13-4 UK0350000 1865 30	Propyl ester of nitric acid 1 ppm = 4.37 mg/m³	NIOSH/OSHA 25 ppm (105 mg/m³) ST 40 ppm (170 mg/m³)	2000 ppm	Colorless to straw-colored liquid with an ether-like odor.	MW: 105.1 BP: 231°F Sol: Insoluble FI.P: 68°F IP: 11.07 eV Sp.Gr: 1.07 Class IB Flammable Liquid	VP: 18 mm FRZ: -148°F UEL: 100% LEL: 2%	Strong oxidizers, combustible materials	Char; CS₂; GC/FID; II(3) [#S227]
Pyrethrum $C_{20}H_{28}O_3$ $C_{21}H_{28}O_3$ $C_{22}H_{28}O_5$ 8003-34-7 UR4200000 9184 31	Cinerin I or II, Jasmolin I or II, Pyrethrin I or II, Pyrethrum I or II	NIOSH/OSHA 5 mg/m³	5000 mg/m³	Brown, viscous oil or solid. [insecticide preparations containing pyrethrins]	MW: 316-372 BP: ? Sol: Insoluble FI.P: 180 to 190°F IP: ? Class IIIA Combustible Liquid	VP: Low MLT: ? UEL: ? LEL: ?	Strong oxidizers	Filter; CH₃CN; HPLC/UVD; III [#5008]
Pyridine C_5H_5N 110-86-1 UR8400000 1282 26	Azabenzene, Azine 1 ppm = 3.29 mg/m³	NIOSH/OSHA 5 ppm (15 mg/m³)	3600 ppm	Colorless to yellow liquid with a nauseating, fish-like odor.	MW: 79.1 BP: 240°F Sol: Miscible FI.P: 68°F IP: 9.27 eV Sp.Gr: 0.98 Class IB Flammable Liquid	VP(77°F): 20 mm FRZ: -44°F UEL: 12.4% LEL: 1.8%	Strong oxidizers, strong acids	Char; CH₂Cl₂; GC/FID; III [#1613]

Chemical name, structure/formula, CAS and RTECS Nos., and DOT ID and guide Nos.	Synonyms, trade names, and conversion factors	Exposure limits (TWA unless noted otherwise)	IDLH	Physical description	Chemical and physical properties		Incompatibilities and reactivities	Measurement method (See Table 1)
					MW, BP, SOL FI.P, IP, Sp.Gr, flammability	VP, FRZ UEL, LEL		
Quinone $C_6H_4O_2$ 106-51-4 DK2625000 2587 55	1,4-Benzoquinone; p-Benzoquinone; 1,4-Cyclohexadiene dioxide; p-Quinone	NIOSH/OSHA 0.4 mg/m³ (0.1 ppm)	300 mg/m³	Pale-yellow solid with an acrid, chlorine-like odor.	MW: 108.1 BP: Sublimes Sol: Slight FI.P: 100 to 200°F IP: 9.68 eV Sp.Gr: 1.32 Combustible Solid	VP(77°F): 0.1 mm MLT: 240°F UEL: ? LEL: ?	Strong oxidizers	XAD-2; Ethanol/ Hexane; HPLC/UVD; II(4) [#S181]
Rhodium (metal fume and insoluble compounds as Rh) Rh 7440-16-6 (Metal) VI9355000 (Metal)	Metal: Elemental rhodium Synonyms of other insoluble compounds vary depending upon the specific compound.	NIOSH/OSHA 0.1 mg/m³	N.E.	Metal: White, hard, ductile, malleable solid with a bluish-gray luster.	MW: 102.9 BP: 6741°F Sol: Insoluble FI.P: NA IP: NA Sp.Gr: 12.41 (Metal) Metal: Noncombustible Solid, but flammable in form of dust or powder.	VP: 0 mm (approx) MLT: 3571°F UEL: NA LEL: NA	None reported	Filter; Acid; AA; II(3) [#S188]
Rhodium (soluble compounds as Rh)	Synonyms vary depending upon the specific soluble compound.	NIOSH/OSHA 0.001 mg/m³	N.E.	Appearance and odor vary depending upon the specific soluble compound.	Properties vary depending upon the specific soluble compound.		None reported	Filter; Acid; HGA; II(3) [#S189]
Ronnel $(CH_3O)_2P(S)OC_6H_2Cl_3$ 299-84-3 TG0525000	O,O-Dimethyl O-(2,4,5-tri-chlorophenyl) phosphoro-thioate; Fenchlorophos	NIOSH/OSHA 10 mg/m³	5000 mg/m³	White to light tan, crystalline solid. [insecticide] [Note: A liquid above 95°F.]	MW: 321.6 BP: Decomposes Sol(77°F): 0.004% FI.P: NA IP: ? Sp.Gr(77°F): 1.49 Noncombustible Solid	VP(77°F): 0.0008 mm MLT: 106°F UEL: NA LEL: NA	Strong oxidizers	Filter/ Chrom-102; Toluene; GC/FID; II(6) [#S299]

Personal protection and sanitation (See Table 3)		Recommendations for respirator selection — maximum concentration for use (MUC) (See Table 4)	Health hazards			
			Route	Symptoms (See Table 5)	First aid (See Table 6)	Target organs (See Table 5)
Clothing: Goggles: Wash: Change: Remove: Provide:	Any poss Reason prob Immed contam N.R. Immed wet (flamm) Quick drench	NIOSH ¥: SCBAF:PD,PP/SAF:PD,PP:ASCBA Escape: GMFS/SCBAE	Inh Con Ing	Irrit eyes, upper resp sys, lungs; skin irrit, blister, burns; [carc]	Eye: Irr immed Skin: Water flush immed Breath: Resp support Swallow: Medical attention immed	Eyes, skin, resp sys
Clothing: Goggles: Wash: Change: Remove:	Repeat Reason prob Prompt contam N.R. Immed non-imperv contam	NIOSH/OSHA 250 ppm: SA/SCBA 625 ppm: SA:CF 1250 ppm: SCBAF/SAF/SAT:CF 2000 ppm: SA:PD,PP §: SCBAF:PD,PP/SAF:PD,PP:ASCBA Escape: GMFS^ʟ/SCBAE	Inh Ing Con	In animals: methemoglobin, anoxia, cyan; dysp, weak, dizz, head; irrit eyes, skin	Eye: Irr immed Skin: Soap wash prompt Breath: Resp support Swallow: Medical attention immed	None known

Clothing:
Goggles:
Wash:
Change:
 	
Remove:
	Repeat
	Reason prob
	Prompt contam
	After work if reason prob contam
	Prompt non-imperv contam
		NIOSH/OSHA
		50 mg/m³: CCROVDMFu*/SA*/SCBA*
		125 mg/m³: PAPROVDMFu*
		250 mg/m³: CCRFOVHiE/SCBAF/SAF/
		            PAPRTOVHiE*
		5000 mg/m³: SAF:PD,PP
		§: SCBAF:PD,PP/SAF:PD,PP:ASCBA
		Escape: GMFOVHiE/SCBAE
			Inh
			Ing
			Con
				Erythema, derm, papules, pruritus, rhin; sneezing; asthma	Eye: Irr immed
					Skin: Soap wash immed
					Breath: Resp support
					Swallow: Medical attention
					            immed
						Resp sys, skin, CNS

Clothing: Goggles: Wash: Change: Remove: Provide:	Reason prob Any poss Immed contam N.R. Immed wet (flamm) Eyewash, quick drench	NIOSH/OSHA 125 ppm: SA:CF[£]/PAPROV[£] 250 ppm: CCRFOV/GMFOV/SCBAF/SAF 3600 ppm: SAF:PD,PP §: SCBAF:PD,PP/SAF:PD,PP:ASCBA Escape: GMFOV/SCBAE	Inh Abs Ing Con	Head, ner, dizz, insomnia; nau, anor; urine frequent; eye irrit; derm; liver, kidney damage	Eye: Irr immed Skin: Water flush immed Breath: Resp support Swallow: Medical attention immed	CNS, liver, kidneys, skin, GI tract

Personal protection and sanitation (See Table 3)		Recommendations for respirator selection — maximum concentration for use (MUC) (See Table 4)	Health hazards			
			Route	Symptoms (See Table 5)	First aid (See Table 6)	Target organs (See Table 5)
Clothing: Goggles: Wash: Change: Remove: Provide:	Reason prob Any poss Immed contam After work if any poss contam Immed non-imperv contam Eyewash, quick drench	NIOSH/OSHA 10 mg/m³: SA/CF[£] 20 mg/m³: SCBAF/SAF 300 mg/m³: SAF:PD,PP §: SCBAF:PD,PP/SAF:PD,PP:ASCBA Escape: GMFOVHiE/SCBAE	Inh Ing Con	Eye irrit, conj; kera; skin irrit	Eye: Irr immed Skin: Soap wash immed Breath: Resp support Swallow: Medical attention immed	Eyes, skin

Clothing:
Goggles:
Wash:
Change:
Remove:
	N.R.
	N.R.
	N.R.
	N.R.
	N.R.
		NIOSH/OSHA
		0.5 mg/m³: DM^
		1 mg/m³: SA/SCBA/DMXSQ^
		2.5 mg/m³: PAPRDMFu/SA:CF
		5 mg/m³: HiEF/SCBAF/SAF/PAPRTHiE/
		            SAT:CF
		100 mg/m³: SA:PD,PP
		200 mg/m³: SAF:PD,PP
		§: SCBAF:PD,PP/SAF:PD,PP:ASCBA
		Escape: HiEF/SCBAE
			Inh	None known	Breath: Resp support
					Swallow: Medical attention
					            immed
						None known

Clothing:
Goggles:
Wash:
Change:
Remove:
	Repeat
	Reason prob
	Prompt contam
	N.R.
	Prompt non-imperv contam
		NIOSH/OSHA
		0.01 mg/m³: HiE*/SA*/SCBA*
		0.025 mg/m³: PAPRHiE*/SA:CF*
		0.05 mg/m³: PAPRTHiE*/HiEF/SCBAF/
		            SAF
		2 mg/m³: SAF:PD,PP
		§: SCBAF:PD,PP/SAF:PD,PP:ASCBA
		Escape: HiEF/SCBAE
			Inh
			Ing
			Con
				In animals: mild eye irrit; CNS damage	Eye: Irr immed
					Skin: Water flush
					Breath: Art Resp
					Swallow: Medical attention
					            immed
						Eyes

Clothing:
Goggles:
Wash:
Change:
 	
Remove:
	Repeat
	Reason prob
	Promptly contam
	After work if reason prob contam
	Prompt non-imperv contam
		NIOSH/OSHA
		100 mg/m³: CCROVDMFu/SA/SCBA
		250 mg/m³: SA:CF/PAPROVDMFu
		500 mg/m³: SCBAF/SAF/CCRFOVHiE/
		            GMFOVHiE/PAPRTOVHiE*
		5000 mg/m³: SAF:PD,PP
		§: SCBAF:PD,PP/SAF:PD,PP:ASCBA
		Escape: GMFOVHiE/SCBAE
			Inh
			Ing
			Con
				In animals: Chol inhibition; eye irrit; liver, kidney damage	Eye: Irr immed
					Skin: Soap wash prompt
					Breath: Resp support
					Swallow: Medical attention
					            immed
						Skin, liver, kidneys, blood plasma

Chemical name, structure/formula, CAS and RTECS Nos., and DOT ID and guide Nos.	Synonyms, trade names, and conversion factors	Exposure limits (TWA unless noted otherwise)	IDLH	Physical description	Chemical and physical properties		Incompatibilities and reactivities	Measurement method (See Table 1)
					MW, BP, SOL Fl.P, IP, Sp.Gr, flammability	VP, FRZ UEL, LEL		
Rotenone $C_{23}H_{22}O_6$ 83-79-4 DJ2800000	1,2,12,12a-Tetrahydro-8,9-dimethoxy-2-(1-methylethenyl)-[1]benzopyrano[3,4-b]furo[2,3-h][1]benzopyran-6(6aH)-one	NIOSH/OSHA 5 mg/m³	Unknown	Colorless to red, odorless, crystalline solid. [insecticide]	MW: 394.4 BP: Decomposes Sol: Insoluble Fl.P: ? IP: ? Sp.Gr: 1.27 Combustible Solid	VP: Low MLT: 330°F UEL: ? LEL: ?	Strong oxidizers, alkalis	Filter; CH₃CN; HPLC/UVD; III [#5007]
Selenium compounds (as Se) Se 7782-49-2 (element) VS7700000 (element) 2658 53 (powder) 2657 53 (disulfide)	Element: Elemental selenium, Selenium alloy Synonyms of other compounds vary depending upon the specific compound.	NIOSH/OSHA 0.2 mg/m³	Unknown	Element: Amorphous or crystalline, red to gray solid. [Note: Occurs as an impurity in most sulfide ores.]	MW: 79.0 BP: 1265°F Sol: Insoluble Fl.P: ? IP: NA Sp.Gr: 4.28 Combustible Solid	VP: 0 mm (approx) MLT: 392°F UEL: ? LEL: ?	Acids, strong oxidizers, chromium trioxide, potassium bromate	Filter; Acid; AA; II(7) [#S190]
Selenium hexafluoride (as Se) SeF₆ 7783-79-1 VS9450000 2194 15	Selenium fluoride	NIOSH/OSHA 0.05 ppm (0.4 mg/m³) 1 ppm = 8.02 mg/m³	5 ppm	Colorless gas.	MW: 193.0 BP: -30°F Sol: Insoluble Fl.P: NA IP: ? Nonflammable Gas	VP: >1 atm FRZ: -59°F UEL: NA LEL: NA	Water [Note: Hydrolyzes very slowly in cold water.]	None available
Silica, amorphous SiO₂ 7631-86-9 VV7310000	Diatomaceous earth, Diatomaceous silica, Diatomite, Silica gel, Silicon dioxide (amorphous)	NIOSH See Appendix D OSHA 6 mg/m³ See Appendix C	N.E.	Transparent to gray, odorless powder. [Note: Amorphous silica is the non-crystalline form of SiO₂.]	MW: 60.1 BP: 4046°F Sol: Insoluble Fl.P: NA IP: NA Sp.Gr: 2.20 Noncombustible Solid	VP: 0 mm (approx) MLT: 3110°F UEL: NA LEL: NA	Fluorine, oxygen difluoride, chlorine trifluoride	Filter; LTA; XRD; III [#7501]

Chemical name, structure/formula, CAS and RTECS Nos., and DOT ID and guide Nos.	Synonyms, trade names, and conversion factors	Exposure limits (TWA unless noted otherwise)	IDLH	Physical description	Chemical and physical properties		Incompatibilities and reactivities	Measurement method (See Table 1)
					MW, BP, SOL Fl.P, IP, Sp.Gr, flammability	VP, FRZ UEL, LEL		
Silica, crystalline (as respirable dust) SiO₂ 14808-60-7 VV7330000	Cristobalite, Quartz, Tridymite, Tripoli	NIOSH Ca See Appendix A 0.05 mg/m³ OSHA 0.05 mg/m³ (cristobalite) 0.05 mg/m³ (tridymite) 0.1 mg/m³ (quartz) 0.1 mg/m³ (tripoli)	Ca [N.E.]	Colorless, odorless solid. [Note: A component of many mineral dusts.]	MW: 60.1 BP: 4046°F Sol: Insoluble Fl.P: NA IP: NA Sp.Gr: 2.66 Noncombustible Solid	VP: 0 mm (approx) MLT: 3110°F UEL: NA LEL: NA	Powerful oxidizers: fluorine, chlorine trifluoride, manganese trioxide, oxygen difluoride, hydrogen peroxide, etc.; acetylene; ammonia	Filter; LTA; XRD; III [#7500]
Silver (metal dust and soluble compounds as Ag) Ag 7440-22-4 (Metal) VW3500000 (Metal)	Metal: Silver metal Synonyms of soluble compounds vary depending upon the specific compound.	NIOSH/OSHA 0.01 mg/m³	N.E.	Metal: White, lustrous solid.	MW: 107.9 BP: 3632°F Sol: Insoluble Fl.P: NA IP: NA Sp.Gr: 10.49 (Metal) Metal: Noncombustible Solid, but flammable in form of dust or powder.	VP: 0 mm (approx) MLT: 1761°F UEL: NA LEL: NA	Acetylene, ammonia, hydrogen peroxide, bromoazide, chlorine trifluoride, ethyleneimine, oxalic acid, tartaric acid	Filter; Acid; ICP; III [#7300, Elements]
Soapstone (containing less than 1% quartz) 3MgO·4SiO₂·H₂O VV8780000	Massive talc, Soapstone silicate, Steatite	NIOSH/OSHA 6 mg/m³ (total) 3 mg/m³ (resp)	N.E.	Odorless, white-gray powder.	MW: 379.3 BP: ? Sol: Insoluble Fl.P: NA IP: NA Sp.Gr: 2.7-2.8 Noncombustible Solid	VP: 0 mm (approx) MLT: ? UEL: NA LEL: NA	None reported	Filter; none; Grav; III [#0500, Nuisance Dust (total)]
Sodium fluoroacetate FCH₂C(O)ONa 62-74-8 AH9100000 2629 53	SFA, Sodium monofluoroacetate	NIOSH/OSHA 0.05 mg/m³ ST 0.15 mg/m³ [skin]	5 mg/m³	Fluffy, colorless to white (sometimes dyed black), odorless powder. [Note: A liquid above 95°F.] [rodenticide]	MW: 100.0 BP: 332°F Sol: Miscible Fl.P: NA IP: ? Noncombustible Solid	VP: Low MLT: 95°F UEL: NA LEL: NA	None reported	Filter; Water; IC; II(5) [#S301]

Personal protection and sanitation (See Table 3)	Recommendations for respirator selection — maximum concentration for use (MUC) (See Table 4)	Route	Symptoms (See Table 5)	First aid (See Table 6)	Target organs (See Table 5)
Clothing: Repeat Goggles: Reason prob Wash: Prompt contam Change: After work if any poss contam Remove: Prompt non-imperv contam	NIOSH/OSHA 5 mg/m³: CCROVDMFu/SA/SCBA 125 mg/m³: SA:CF/PAPROVDMFu 250 mg/m³: SCBAF/SAF/CCRFOVHiE/ GMFOVHiE/SAT:CF/ PAPRTOVHiE 5000 mg/m³: SA:PD,PP §: SCBAF:PD,PP/SAF:PD,PP:ASCBA Escape: GMFOVHiE/SCBAE	Inh Ing Con	Irrit eyes; numb muc memb; nau, vomit, abdom pain; musc tremor, inco, clonic convuls, stupor; pulm irrit; skin irrit	Eye: Irr immed Skin: Soap wash prompt Breath: Resp support Swallow: Medical attention immed	CNS, eyes, resp sys
Recommendations vary depending upon the specific compound.	NIOSH/OSHA 2 mg/m³: DMF^/SA*/SCBA* 5 mg/m³: PAPRDM^*/SA:CF* 10 mg/m³: HiEF/SCBAF/SAF 100 mg/m³: SAF:PD,PP §: SCBAF:PD,PP/SAF:PD,PP:ASCBA Escape: HiEF/SCBAE	Inh Abs Ing Con	Irrit eyes, nose, throat; vis dist; head; chills, fever; dysp, bron; metallic taste, garlic breath, GI dist; derm; skin, eye burns; in animals: anemia; liver, kidney damage	Eye: Irr immed Skin: Soap wash immed Breath: Resp support Swallow: Medical attention immed	Upper resp sys, eyes. skin, liver, kidneys, blood
Clothing: N R Goggles: N R Wash: N R Change: N R Remove: N R	NIOSH/OSHA 0.5 ppm: SA/SCBA 1.25 ppm: SA:CF 2.5 ppm: SCBAF/SAF/SAT:CF 5 ppm: SA:PD,PP §: SCBAF:PD,PP/SAF:PD,PP:ASCBA Escape: GMFS/SCBAE	Inh Con	In animals: pulm irrit, edema	Breath: Resp support	None known in humans
Clothing: N.R. Goggles: N.R. Wash: N.R. Change: N.R. Remove: N.R.	OSHA 30 mg/m³: DM 60 mg/m³: DMXSQ/SA/SCBA 150 mg/m³: PAPRDM/SA:CF 300 mg/m³: HiEF/PAPRTHiE/SCBAF/ SAF/SAT:CF 3000 mg/m³: SA:PD,PP §: SCBAF:PD,PP/SAF:PD,PP:ASCBA Escape: HiEF/SCBAE	Inh	Pneumoconiosis	Eye: Irr immed	Resp sys

Personal protection and sanitation (See Table 3)	Recommendations for respirator selection — maximum concentration for use (MUC) (See Table 4)	Route	Symptoms (See Table 5)	First aid (See Table 6)	Target organs (See Table 5)
Clothing: N.R. Goggles: N.R. Wash: N.R. Change: N.R. Remove: N.R.	NIOSH ¥: SCBAF:PD,PP/SAF:PD,PP:ASCBA Escape: HiEF/SCBAE	Inh	Cough, dysp, wheez; impaired pulm func, progressive symptoms; [carc]	Eye: Irr immed Breath: Fresh air	Resp sys
Clothing: Reason prob Goggles: Any poss Wash: Prompt contam Change: After work if reason prob contam Remove: Prompt non-imperv contam Provide: Eyewash	NIOSH/OSHA 0.25 mg/m³: SA:CF^ε/PAPRHiE^ε 0.5 mg/m³: HiEF/SCBAF/SAF 20 mg/m³: SAF:PD,PP §: SCBAF:PD,PP/SAF:PD,PP:ASCBA Escape: HiEF/SCBAE	Inh Ing Con	Blue-gray eyes, nasal septum, throat, skin; irrit skin, ulceration; GI dist	Eye: Irr immed Skin: Water flush Breath: Resp support Swallow: Medical attention immed	Nasal septum, skin, eyes
Clothing: N R Goggles: N R Wash: N R Change: N R Remove: N R	NIOSH/OSHA 30 mg/m³: DM 60 mg/m³: DMXSQ/SA/SCBA 150 mg/m³: PAPRDM/SA/SCBA 300 mg/m³: HiEF/SCBAF/SAF/ PAPRTHiE*/SAT:CF* 3000 mg/m³: SAF:PD,PP §: SCBAF:PD,PP/SAF:PD,PP:ASCBA Escape: HiEF/SCBAE	Inh Con	Cough, dysp; digital clubbing; cyan; basal crackles, cor pulmonale	Eye: Irr immed Breath: Resp support	Lungs, CVS
Clothing: Any poss Goggles: Reason prob Wash: Immed contam Change: After work if any poss contam Remove: Immed non-imperv contam Provide: Quick drench	NIOSH/OSHA 0.25 mg/m³: DM 0.5 mg/m³: DMXSQ/SA/SCBA 1.25 mg/m³: PAPRDM/SA:CF 2.5 mg/m³: PAPRTHiE/SCBAF/SAF HiEF/SAT:CF 5 mg/m³: SA:PD,PP §: SCBAF:PD,PP/SAF:PD,PP:ASCBA Escape: HiEF/SCBAE	Inh Abs Ing Con	Vomit; appre, auditory halu; facial pares; twitch face musc; pulsus altenans, ectopic heartbeat, tacard, ventfib; pulm edema; nystagmus; convuls	Eye: Irr immed Skin: Water flush immed Breath: Resp support Swallow: Medical attention immed	CVS, lungs, kidneys, CNS

Chemical name, structure/formula, CAS and RTECS Nos., and DOT ID and guide Nos.	Synonyms, trade names, and conversion factors	Exposure limits (TWA unless noted otherwise)	IDLH	Physical description	Chemical and physical properties		Incompatibilities and reactivities	Measurement method (See Table 1)
					MW, BP, SOL Fl.P, IP, Sp.Gr, flammability	VP, FRZ UEL, LEL		
Sodium hydroxide NaOH 1310-73-2 WB4900000 1823 60 (solid) 1824 60 (soln)	Caustic soda, Lye, Soda lye, Sodium hydrate	NIOSH/OSHA C 2 mg/m³	250 mg/m³	Colorless to white, odorless solid (flakes, beads, granular form).	MW: 40.0 BP: 2534°F Sol: 111% Fl.P: NA IP: NA Sp.Gr: 2.13 Noncombustible Solid, but when in contact with water may generate sufficient heat to ignite combustible materials.	VP: 0 mm (approx) MLT: 605°F UEL: NA LEL: NA	Water; acids; flammable liquids; organic halogens; metals such as aluminum, tin & zinc; nitromethane [Note: Corrosive to metals.]	Filter; HCl; Titrate; III [#7401, Alkaline Dusts]
Stibine SbH₃ 7803-52-3 WJ0700000 2676 18	Antimony hydride, Antimony trihydride, Hydrogen antimonide 1 ppm = 5.19 mg/m³	NIOSH/OSHA 0.1 ppm (0.5 mg/m³)	40 ppm	Colorless gas with a disagreeable odor like hydrogen sulfide.	MW: 124.8 BP: -1°F Sol: Slight Fl.P: NA (Gas) IP: 9.51 eV Flammable Gas	VP: >1 atm FRZ: -126°F UEL: ? LEL: ?	Acids, halogenated hydrocarbons, oxidizers, moisture, chlorine, ozone, ammonia	Si gel*; HCl; Vis; III [#6008]
Stoddard solvent 8052-41-3 WJ8925000 1268 27 (petroleum distillate)	Dry cleaning safety solvent, Mineral spirits, Petroleum solvent, Spotting naphtha	NIOSH 350 mg/m³ C 1800 mg/m³ [15-min] OSHA 100 ppm (525 mg/m³)	29,500 mg/m³	Colorless liquid with a kerosene-like odor.	MW: Varies BP: 428-572°F Sol: Insoluble Fl.P: 110°F IP: ? Sp.Gr: 0.78 Class II Combustible Liquid	VP: ? FRZ: ? UEL: ? LEL: ?	Strong oxidizers	Char; CS₂; GC/FID; III [#1550]
Strychnine C₂₁H₂₂N₂O₂ 57-24-9 WL2275000 1692 53	None	NIOSH/OSHA 0.15 mg/m³	3 mg/m³	Colorless to white, odorless, crystalline solid. [pesticide]	MW: 334.4 BP: Decomposes Sol: 0.02% Fl.P: ? IP: ? Sp.Gr: 1.36 Combustible Solid, but difficult to ignite.	VP: Low MLT: 514°F UEL: ? LEL: ?	Strong oxidizers	Filter; Reagent; HPLC/UVD; III [#5016]

Chemical name, structure/formula, CAS and RTECS Nos., and DOT ID and guide Nos.	Synonyms, trade names, and conversion factors	Exposure limits (TWA unless noted otherwise)	IDLH	Physical description	Chemical and physical properties		Incompatibilities and reactivities	Measurement method (See Table 1)
					MW, BP, SOL Fl.P, IP, Sp.Gr, flammability	VP, FRZ UEL, LEL		
Styrene C₆H₅CH=CH₂ 100-42-5 WL3675000 2055 27 (inhibited)	Ethenyl benzene, Phenylethylene, Styrene monomer, Styrol, Vinyl benzene 1 ppm = 4.33 mg/m³	NIOSH/OSHA 50 ppm (215 mg/m³) ST 100 ppm (425 mg/m³)	5000 ppm	Colorless to yellow, oily liquid with a sweet, floral odor.	MW: 104.2 BP: 293°F Sol: Slight Fl.P: 88°F IP: 8.40 eV Sp.Gr: 0.91 Class IC Flammable Liquid	VP: 5 mm FRZ: -23°F UEL: 7.0% LEL: 1.1%	Oxidizers, catalysts for vinyl polymers, peroxides, strong acids, aluminum chloride [Note: May polymerize if contaminated or subjected to heat. Usually contains an inhibitor such as tert-butylcatechol.]	Char; CS₂; GC/FID; III [#1501, Aromatic Hydrocarbons]
Sulfur dioxide SO₂ 7446-09-5 WS4550000 1079 16	Sulfurous acid anhydride, Sulfurous oxide, Sulfur oxide 1 ppm = 2.66 mg/m³	NIOSH/OSHA 2 ppm (5 mg/m³) ST 5 ppm (10 mg/m³)	100 ppm	Colorless gas with a characteristic, irritating, pungent odor. [Note: A liquid below 14°F. Shipped as a liquefied compressed gas.]	MW: 64.1 BP: 14°F Sol: 10% Fl.P: NA IP: 12.30 eV Nonflammable Gas	VP: >1 atm FRZ: -104°F UEL: NA LEL: NA	Powdered and alkali metals such as sodium & potassium, water, ammonia, aluminum [Note: Reacts with water to form sulfuric acid.]	Filters(2); NaHCO₃/ Na₂CO₃; IC; III [#6004]
Sulfuric acid H₂SO₄ 7664-93-9 WS5600000 1830 39 (51-95% acid) 1831 39 (fuming) 1832 39 (spent)	Battery acid, Hydrogen sulfate, Oil of vitriol, Sulfuric acid (aqueous) 1 ppm = 4.08 mg/m³	NIOSH/OSHA 1 mg/m³	80 mg/m³	Colorless to dark-brown, oily, odorless liquid. [Note: Pure compound is a solid below 51°F. Often used in an aqueous solution.]	MW: 98.1 BP: 554°F Sol: Miscible Fl.P: NA IP: ? Sp.Gr: 1.84 (96-98% acid) Noncombustible Liquid, but capable of igniting finely divided combustible materials.	VP(295°F): 1 mm FRZ: 51°F UEL: NA LEL: NA	Organic materials, chlorates, carbides, fulminates, water, powdered metals [Note: Reacts violently with water with evolution of heat. Corrosive to metals.]	Si gel; NaHCO₃/ Na₂CO₃; IC; III [#7903, Inorganic Acids]
Sulfur monochloride S₂Cl₂ 10025-67-9 WS4300000 1828 39	Sulfur chloride, Sulfur subchloride, Thiosulfurous dichloride 1 ppm = 5.61 mg/m³	NIOSH/OSHA C 1 ppm (6 mg/m³)	10 ppm	Light amber to yellow-red, oily liquid with a pungent, nauseating, irritating odor.	MW: 135.0 BP: 280°F Sol: Decomposes Fl.P: 245°F IP: 9.40 eV Sp.Gr: 1.68 Class IIIB Combustible Liquid	VP: 7 mm FRZ: -107°F UEL: ? LEL: ?	Peroxides, oxides of phosphorous, organics, water [Note: Decomposes violently in water to form hydrochloric acid, sulfur dioxide, sulfur, sulfite, thiosulfate, and hydrogen sulfide. Corrosive to metals.]	None available

Personal protection and sanitation (See Table 3)		Recommendations for respirator selection — maximum concentration for use (MUC) (See Table 4)	Health hazards			
			Route	Symptoms (See Table 5)	First aid (See Table 6)	Target organs (See Table 5)
Clothing: Goggles: Wash: Change: Remove: Provide:	Any poss Any poss Immed contam After work if reason prob contam Immed non-imperv contam Eyewash, quick drench	NIOSH/OSHA 50 mg/m³: PAPRDMᵋ/SA:CFᵋ 100 mg/m³: SCBAF/SAF/HiEF 250 mg/m³: SAF:PD,PP §: SCBAF:PD,PP/SAF:PD,PP:ASCBA Escape: HiEF/SCBAE	Inh Ing Con	Irrit nose; pneuitis; burns eyes, skin; temporary loss of hair	Eye: Skin: Breath: Swallow:	Irr immed Water flush immed Resp support Medical attention immed	Eyes, resp sys, skin
Clothing: Goggles: Wash: Change: Remove:	N.R. N.R. N.R. N.R. N.R.	NIOSH/OSHA 1 ppm: SA/SCBA 2.5 ppm: SA:CF 5 ppm: SCBAF/SAF/SAT:CF 40 ppm: SA:PD,PP §: SCBAF:PD,PP/SAF:PD,PP:ASCBA Escape: GMFS/SCBAE	Inh	Head, weak; nau, abdom pain; lumbar pain, hemog, hema, hemolytic anemia; jaun; irrit lung	Breath:	Resp support	Blood, liver kidneys, lungs
Clothing: Goggles: Wash: Change: Remove:	Repeat Reason prob Prompt wet N.R. Prompt non-imperv wet	NIOSH 3500 mg/m³: SA*/SCBA*/CCROV* 5900 mg/m³: PAPROV*/CCRFOV 8750 mg/m³: SA:CF* 17,500 mg/m³: GMFOV/SCBAF/SAF 29,500 mg/m³: SAF:PD,PP §: SCBAF:PD,PP/SAF:PD,PP:ASCBA Escape: GMFOV/SCBAE	Inh Con Ing	Irrit eyes, nose, throat; dizz; derm	Eye: Skin: Breath: Swallow:	Irr immed Soap wash prompt Resp support Medical attention immed	Skin, eyes, resp sys, CNS
Clothing: Goggles: Wash: Change: Remove:	Repeat N.R. Prompt contam After work if any poss contam N.R.	NIOSH/OSHA 0.75 mg/m³: DM 1.5 mg/m³: DMXSQ/SA/SCBA 3 mg/m³: PAPRDM/SA:CF/HiEF/ SCBAF/SAF §: SCBAF:PD,PP/SAF:PD,PP:ASCBA Escape: HiEF/SCBAE	Inh Ing Con	Stiff neck, facial musc; restless, appre, incr acuity of perception; incr reflex excitability; cyan; tetanic convuls with opisthotonos	Eye: Skin: Breath: Swallow:	Irr immed Soap wash prompt Resp support Medical attention immed	CNS

Personal protection and sanitation (See Table 3)		Recommendations for respirator selection — maximum concentration for use (MUC) (See Table 4)	Health hazards			
			Route	Symptoms (See Table 5)	First aid (See Table 6)	Target organs (See Table 5)
Clothing: Goggles: Wash: Change: Remove:	Repeat Reason prob Prompt contam N.R. Immed wet (flamm)	NIOSH/OSHA 500 ppm: CCROV*/SA*/SCBA* 1000 ppm: CCRFOV/PAPROV* 1250 ppm: SA:CF* 2500 ppm: GMFOV/SCBAF/SAF 5000 ppm: SAF:PD,PP §: SCBAF:PD,PP/SAF:PD,PP:ASCBA Escape: GMFOV/SCBAE	Inh Ing Con	Irrit eyes, nose; drow, weak, unsteady gait; narco; defatting derm	Eye: Skin: Breath: Swallow:	Irr immed Water flush Resp support Medical attention immed	CNS, resp sys, eyes, skin
Clothing: Goggles: Wash: Change: Remove: Provide:	Prevent skin freezing Any poss N.R. N.R. Immed wet Eyewash	NIOSH/OSHA 20 ppm: CCRS*/SA*/SCBA* 50 ppm: PAPRS*/SA:CF* 100 ppm: CCRFS/GMFS/PAPRTS*/ SCBAF/SAF/SAT:CF* §: SCBAF:PD,PP/SAF:PD,PP:ASCBA Escape: GMFS/SCBAE	Inh Con	Irrit eyes, nose, throat, rhin; choking, cough; reflex bronchoconstriction; eye, skin burns	Eye: Skin: Breath:	Irr immed Water flush immed Resp support	Resp sys, skin, eyes
Clothing: Goggles: Wash: Change: Remove: Provide:	Any poss >1%/Repeat <1% Any poss Immed contam N.R. Immed non-imperv contam >1%: Eyewash, quick drench	NIOSH/OSHA 25 mg/m³: PAPRAGHiEᵋ/SA:CFᵋ 50 mg/m³: CCRFAGHiE/SCBAF/SAF/ GMFAGHiE 80 mg/m³: SAF:PD,PP §: SCBAF:PD,PP/SAF:PD,PP:ASCBA Escape: GMFAGHiE/SCBAE	Inh Ing Con	Eye, nose, throat irrit; pulm edema, bron; emphy; conj; stomatis; dental erosion; trachbronc; skin, eye burns; derm	Eye: Skin: Breath: Swallow:	Irr immed Water flush immed Resp support Medical attention immed	Resp sys, eyes, skin, teeth
Clothing: Goggles: Wash: Change: Remove: Provide:	Any poss Any poss Immed contam N.R. Immed non-imperv contam Eyewash, quick drench	NIOSH/OSHA 10 ppm: PAPRSᵋ/CCRFS/GMFS/ SCBAF/SAF §: SCBAF:PD,PP/SAF:PD,PP:ASCBA Escape: GMFS/SCBAE	Inh Con Ing	Lac; cough; burn eyes, skin; pulm edema	Eye: Skin: Breath: Swallow:	Irr immed Water flush immed Resp support Medical attention immed	Resp sys, skin, eyes

Chemical name, structure/formula, CAS and RTECS Nos., and DOT ID and guide Nos.	Synonyms, trade names, and conversion factors	Exposure limits (TWA unless noted otherwise)	IDLH	Physical description	Chemical and physical properties		Incompatibilities and reactivities	Measurement method (See Table 1)
					MW, BP, SOL Fl.P, IP, Sp.Gr, flammability	VP, FRZ UEL, LEL		
Sulfur pentafluoride S₂F₁₀ 5714-22-7 WS4480000	Disulfur decafluoride, Sulfur decafluoride	NIOSH/OSHA C 0.01 ppm (0.1 mg/m³)	1 ppm	Colorless liquid or gas (above 84°F) with an odor like sulfur dioxide.	MW: 254.1 BP: 84°F Sol: Insoluble Fl.P: NA IP: ?	VP: 561 mm FRZ: -134°F UEL: NA LEL: NA	None reported	None available
	1 ppm = 10.56 mg/m³				Sp.Gr(32°F): 2.08 Noncombustible Liquid			
Sulfuryl fluoride SO₂F₂ 2699-79-8 WT5075000	Sulfur difluoride dioxide	NIOSH/OSHA 5 ppm (20 mg/m³) ST 10 ppm (40 mg/m³)	1000 ppm	Colorless, odorless gas. [Note: Shipped as a liquefied compressed gas.] [insecticide/ fumigant]	MW: 102.1 BP: -68°F Sol(32°): 0.2% Fl.P: NA IP: 13.04 eV	VP: >1 atm FRZ: -212°F UEL: NA LEL: NA	None reported	None available
2191 15	1 ppm = 4.24 mg/m³				Nonflammable Gas			
2,4,5-T C₆H₂Cl₃OCH₂COOH 93-76-5 AJ8400000	2,4,5-Trichlorophenoxyacetic acid	NIOSH/OSHA 10 mg/m³	Unknown	Colorless to tan, odorless, crystalline solid. [herbicide]	MW: 255.5 BP: Decomposes Sol: 0.03% Fl.P: ? IP: ?	VP: Low MLT: 307°F UEL: ? LEL: ?	None reported	Filter; Methanol; HPLC/UVD; III [#5001]
2765 55					Sp.Gr: 1.80 Combustible Solid, but burns with difficulty.			
Talc (containing no asbestos and less than 1% quartz) Mg₃H₂(SiO₃)₄ 14807-96-6 WW2710000	Hydrous magnesium silicate, Steatite talc	NIOSH/OSHA 2 mg/m³ (resp)	N.E.	Odorless, white powder.	MW: Varies BP: ? Sol: Insoluble Fl.P: NA IP: NA	VP: 0 mm (approx) MLT: 1652 to 1832°F UEL: NA LEL: NA	None reported	Filter; none; Grav; III [#0600, Nuisance Dust (resp)]
					Sp.Gr: 2.70-2.80 Noncombustible Solid			

Chemical name, structure/formula, CAS and RTECS Nos., and DOT ID and guide Nos.	Synonyms, trade names, and conversion factors	Exposure limits (TWA unless noted otherwise)	IDLH	Physical description	Chemical and physical properties		Incompatibilities and reactivities	Measurement method (See Table 1)
					MW, BP, SOL Fl.P, IP, Sp.Gr, flammability	VP, FRZ UEL, LEL		
Tantalum (metal and oxide dust as Ta) Ta 7440-25-7 (Metal) WW5505000 (Metal)	Metal: Tantalum-181 Synonyms of other compounds vary depending upon the specific compound.	NIOSH 5 mg/m³ ST 10 mg/m³ OSHA 5 mg/m³	N.E.	Metal: Steel-blue to gray solid or black powder.	MW: 180.9 BP: 9797°F Sol: Insoluble Fl.P: NA IP: NA	VP: 0 mm (approx) MLT: 5425°F UEL: NA LEL: NA	Strong oxidizers, bromine trifluoride, fluorine	Filter; none; Grav; III [#0500, Nuisance Dust (total)]
					Sp.Gr: 16.65 (Metal) 14.40 (Powder) Noncombustible Solid in bulk form, but powder ignites SPONTANEOUSLY in air.			
TEDP (C₂H₅)₄P₂S₂O₅ 3689-24-5 XN4375000	Dithion, Sulfotepp, Tetraethyl dithiono-pyrophosphate Tetraethyl dithiopyro-phosphate	NIOSH/OSHA 0.2 mg/m³ [skin]	35 mg/m³	Pale-yellow liquid with a garlic-like odor. [Note: A pesticide that may be absorbed on a solid carrier or mixed in a more flammable liquid.]	MW: 322.3 BP: Decomposes Sol: 0.0007% Fl.P: ? IP: ?	VP: 0.0002 mm FRZ: ? UEL: ? LEL: ?	Strong oxidizers, iron [Note: Corrosive to iron.]	None available
1704 55	1 ppm = 13.40 mg/m³				Sp.Gr(77°F): 1.20 Combustible Liquid			
Tellurium and compounds (as Te) Te 13494-80-9 (Metal) WY2625000 (Metal)	Metal: Aurum paradoxum, Metallum problematum Synonyms of other compounds vary depending upon the specific compound.	NIOSH/OSHA 0.1 mg/m³	N.E.	Metal: Odorless, dark-gray to brown, amorphous powder or grayish-white, brittle solid.	MW: 127.6 BP: 1814°F Sol: Insoluble Fl.P: ? IP: NA	VP: 0 mm (approx) MLT: 842°F UEL: ? LEL: ?	Oxidizers, chlorine	Filter; Acid; AA; II(3) [#S204]
					Sp.Gr: 6.24 Combustible Solid			
Tellurium hexafluoride (as Te) TeF₆ 7783-80-4 WY2800000	Tellurium fluoride	NIOSH/OSHA 0.02 ppm (0.2 mg/m³)	1 ppm	Colorless gas with a repulsive odor.	MW: 241.6 BP: Sublimes Sol: Decomposes Fl.P: NA IP: ?	VP: >1 atm FRZ: -36°F (Sublimes) UEL: NA LEL: NA	Water [Note: Hydrolyzes slowly in water to telluric acid.]	Char; NaOH; AA; II(3) [#S187]
2195 15	1 ppm = 10.04 mg/m³				Nonflammable Gas			

Personal protection and sanitation (See Table 3)	Recommendations for respirator selection — maximum concentration for use (MUC) (See Table 4)	Health hazards			
		Route	Symptoms (See Table 5)	First aid (See Table 6)	Target organs (See Table 5)
Clothing: Any poss Goggles: Any poss Wash: N.R. Change: N.R. Remove: Immed non-imperv contam Provide: Eyewash, quick drench	NIOSH/OSHA 0.1 ppm: SA/SCBA 0.25 ppm: SA:CF 0.5 ppm: SCBAF/SAF/SAT:CF 1 ppm: SA:PD,PP §: SCBAF:PD,PP/SAF:PD,PP:ASCBA Escape: GMFAG/SCBAE	Ing Con	In animals: pulm edema, hemorr	Eye: Irr immed Skin: Soap wash immed Breath: Resp support Swallow: Medical attention immed	Resp sys, CNS
Clothing: N.R. Goggles: N.R. Wash: N.R. Change: N.R. Remove: N.R.	NIOSH/OSHA 50 ppm: SA*/SCBA* 125 ppm: SA:CF* 250 ppm: SCBAF/SAF 1000 ppm: SAF:PD,PP §: SCBAF:PD,PP/SAF:PD,PP,ASCBA Escape: GMFS/SCBAE	Inh Con	Conj, rhin, phar; pares; in animals: narco, tremor, convuls; pulm edema; kidney inj	Eye: Irr immed Breath: Resp support	Resp sys, CNS
Clothing: N.R. Goggles: N.R. Wash: N.R. Change: N.R. Remove: N.R.	NIOSH/OSHA 50 mg/m³: DM 100 mg/m³: DMXSQ/SA/SCBA 250 mg/m³: SA:CF/PAPRDM 500 mg/m³: HiEF/SCBAF/SAF/ PAPRTHiE/SAT:CF 5000 mg/m³: SA:PD,PP §: SCBAF:PD,PP/SAF:PD,PP:ASCBA Escape: HiEF/SCBAE	Inh Ing Con	In animals: ataxia; skin irrit, acne-like rash	Eye: Irr immed Skin: Soap wash Breath: Resp support Swallow: Medical attention immed	Skin, liver, GI tract
Clothing: N.R. Goggles: N.R. Wash: N.R. Change: N.R. Remove: N.R.	NIOSH/OSHA 10 mg/m³: DM 20 mg/m³: DMXSQ/SA/SCBA 50 mg/m³: PAPRDM/SA:CF 100 mg/m³: HiEF/PAPRTHiE/SAT:CF SCBAF/SAF 1000 mg/m³: SA:PD,PP §: SCBAF:PD,PP/SAF:PD,PP:ASCBA Escape: HiEF/SCBAE	Inh Con	Fibriotic pneumoconiosis	Eye: Irr immed	Lungs, CVS

Personal protection and sanitation (See Table 3)	Recommendations for respirator selection — maximum concentration for use (MUC) (See Table 4)	Health hazards			
		Route	Symptoms (See Table 5)	First aid (See Table 6)	Target organs (See Table 5)
Clothing: N.R. Goggles: N.R. Wash: N.R. Change: N.R. Remove: N.R.	NIOSH/OSHA 25 mg/m³: DM^ 50 mg/m³: DMXSQ^/DMFu/SA/SCBA 125 mg/m³: PAPRDM^/SA:CF 250 mg/m³: HiEF/PAPRTHiE/SAT:CF SCBAF/SAF 2500 mg/m³: SA:PD,PP §: SCBAF:PD,PP/SAF:PD,PP:ASCBA Escape: HiEF/SCBAE	Inh	In animals: pulm irrit	Breath: Resp support	None known in humans
Clothing: Any poss Goggles: Any poss Wash: Immed contam Change: N.R. Remove: Immed non-imperv contam Provide: Eyewash, quick drench	NIOSH/OSHA 2 mg/m³: SA/SCBA 5 mg/m³: SA:CF 10 mg/m³: SCBAF/SAF 35 mg/m³: SA:PD,PP §: SCBAF:PD,PP/SAF:PD,PP:ASCBA Escape: GMFOVHiE/SCBAE	Inh Abs Ing Con	Eye pain, blurred vision, lac, rhin; head; cyan; anor, nau, vomit, diarr; local sweat, weak, twitch, para, Cheyne-Stokes resp, convuls, low BP, card irregularities; skin, eye irrit	Eye: Irr immed Skin: Soap wash immed Breath: Resp support Swallow: Medical attention immed	CNS, resp sys, CVS
Recommendations vary depending upon the specific compound.	NIOSH/OSHA 0.5 mg/m³: DM^ 1 mg/m³: SA/SCBA/DMXSQ^/DMFu 2.5 mg/m³: PAPRDM^/SA:CF 5 mg/m³: PAPRTHiE/SCBAF/SAF/ SAT:CF/HiEF 50 mg/m3: SA:PD,PP §: SCBAF:PD,PP/SAF:PD,PP:ASCBA Escape: HiEF/SCBAE	Inh Abs Ing Con	Garlic odor on breath, sweat; dry mouth, metal taste; som; anor, nau, no sweat; derm	Eye: Irr immed Skin: Soap wash prompt Breath: Resp support Swallow: Medical attention immed	Skin, CNS
Clothing: N.R. Goggles: N.R. Wash: N.R. Change: N.R. Remove: N.R.	NIOSH/OSHA 0.2 ppm: SA/SCBA 0.5 ppm: SA:CF 1 ppm: SCBAF/SAF/SAT:CF §: SCBAF:PD,PP/SAF:PD,PP:ASCBA Escape: GMFS/SCBAE	Inh	Head; dysp; garlic odor on breath; in animals: pulm edema	Breath: Resp support	Resp sys

Chemical name, structure/formula, CAS and RTECS Nos., and DOT ID and guide Nos.	Synonyms, trade names, and conversion factors	Exposure limits (TWA unless noted otherwise)	IDLH	Physical description	Chemical and physical properties		Incompatibilities and reactivities	Measurement method (See Table 1)
					MW, BP, SOL Fl.P, IP, Sp.Gr, flammability	VP, FRZ UEL, LEL		
TEPP $(C_2H_5)_4P_2O_7$ 107-49-3 UX6825000 2783 55 (dry)	Ethyl pyrophosphate, Tetraethyl pyrophosphate 1 ppm = 12.06 mg/m³	NIOSH/OSHA 0.05 mg/m³ [skin]	10 mg/m³	Colorless to amber liquid with a faint, fruity odor. [Note: A solid below 32°F.] [insecticide]	MW: 290.2 BP: Decomposes Sol: Miscible Fl.P: NA IP: ? Sp.Gr: 1.19 Noncombustible Liquid	VP(86°F): 0.0005 mm FRZ: 32°F UEL: NA LEL: NA	Strong oxidizers, alkalis, water [Note: Hydrolyzes quickly in water to form pyrophosphoric acid.]	Chrom-102 (2); Toluene; GC/FPD; III [#2504, Tetraethyl Pyrophosphate]
Terphenyls $C_6H_5C_6H_4C_6H_5$ 26140-60-3 WZ6450000	Diphenylbenzenes; Phenyl biphenyls; ortho-, meta-, or para-Terphenyl	NIOSH/OSHA C 5 mg/m³ (0.5 ppm)	Unknown	Colorless or light-yellow solid. [Note: Consists of 3 different isomers.]	MW: 230.3 BP: 630-761°F Sol: Insoluble Fl.P(oc): 325 to 405°F IP: 7.99 eV(o) 8.01 eV(m) 7.78 eV(p) Sp.Gr: 1.10-1.23 Combustible Solid	VP: ? MLT: 136 to 415°F UEL: ? LEL: ?	None reported	Filter; CS₂; GC/FID; III [#5021]
1,1,2,2-Tetrachloro-1,2-difluoroethane CCl_2FCCl_2F 76-12-0 KI1420000 1078 12 (refrigerant)	1,2-Difluoro-1,1,2,2-tetrachloroethane; Freon® 112; Halocarbon 112; Refrigerant 112 1 ppm = 8.47 mg/m³	NIOSH/OSHA 500 ppm (4170 mg/m³)	15,000 ppm	Colorless solid or liquid (above 77°F) with a slight, ether-like odor.	MW: 203.8 BP: 199°F Sol(77°F): 0.01% Fl.P: NA IP: 11.30 eV Sp.Gr: 1.65 Noncombustible Solid	VP: 40 mm MLT: 77°F UEL: NA LEL: NA	Chemically-active metals such as potassium, beryllium, powdered aluminum, zinc, magnesium & sodium; acids	Char; CS₂; GC/FID; III [#1016]
1,1,1,2-Tetrachloro-2,2-difluoroethane CCl_3CClF_2 76-11-9 KI1425000 1 ppm = 8.47 mg/m³	2,2-Difluoro-1,1,1,2-tetrachloroethane; Freon® 112a; Halocarbon 112a; Refrigerant 112a	NIOSH/OSHA 500 ppm (4170 mg/m³)	15,000 ppm	Colorless solid with a slight, ether-like odor. [Note: A liquid above 105°F.]	MW: 203.8 BP: 197°F Sol: 0.01% Fl.P: NA IP: ? Noncombustible Solid	VP: ? MLT: 105°F UEL: NA LEL: NA	Chemically-active metals such as potassium, beryllium, powdered aluminum, zinc, magnesium & sodium; acids	Char; CS₂; GC/FID; III [#1016]

Chemical name, structure/formula, CAS and RTECS Nos., and DOT ID and guide Nos.	Synonyms, trade names, and conversion factors	Exposure limits (TWA unless noted otherwise)	IDLH	Physical description	Chemical and physical properties		Incompatibilities and reactivities	Measurement method (See Table 1)
					MW, BP, SOL Fl.P, IP, Sp.Gr, flammability	VP, FRZ UEL, LEL		
1,1,2,2-Tetrachloroethane $CHCl_2CHCl_2$ 79-34-5 KI8575000 1702 55	Acetylene tetrachloride, Symmetrical tetrachloro-ethane 1 ppm = 7.00 mg/m³	NIOSH Ca See Appendix A 1 ppm (7 mg/m³) [skin] OSHA 1 ppm (7 mg/m³) [skin]	Ca [150 ppm]	Colorless to pale-yellow liquid with a pungent, chloroform-like odor.	MW: 167.9 BP: 296°F Sol: 0.3% Fl.P: NA IP: 11.10 eV Sp.Gr(77°F): 1.59 Noncombustible Liquid	VP(86°F): 9 mm FRZ: -33°F UEL: NA LEL: NA	Chemically-active metals, strong caustics, fuming sulfuric acid [Note: Degrades slowly when exposed to air.]	Char(pet); CS₂; GC/FID; III [#1019]
Tetrachloroethylene $Cl_2C=CCl_2$ 127-18-4 KX3850000 1897 74	Perchlorethylene, Perchloroethylene, Perk, Tetrachlorethylene 1 ppm = 6.89 mg/m³	NIOSH Ca See Appendix A Minimize workplace exposure concentrations; limit number of workers exposed. OSHA 25 ppm (170 mg/m³)	Ca [500 ppm]	Colorless liquid with a mild, chloroform-like odor.	MW: 165.8 BP: 250°F Sol(77°F): 0.02% Fl.P: NA IP: 9.32 eV Sp.Gr: 1.62 Noncombustible Liquid	VP: 14 mm FRZ: -2°F UEL: NA LEL: NA	Strong oxidizers; chemically-active metals such as lithium, beryllium & barium; caustic soda; sodium hydroxide; potash	Char; CS₂; GC/FID; III [#1003, Halogenated Hydrocarbons]
Tetrachloronaphthalene $C_{10}H_4Cl_4$ 1335-88-2 QK3700000	Halowax®, Nibren wax, Seekay wax	NIOSH/OSHA 2 mg/m³ [skin]	Unknown	Colorless to pale-yellow solid with an aromatic odor.	MW: 265.9 BP: 593-680°F Sol: Insoluble Fl.P(oc): 410°F IP: ? Sp.Gr: 1.59-1.65 Combustible Solid	VP: <1 mm FRZ: 360°F UEL: ? LEL: ?	Strong oxidizers	Filter/Bub; none; GC/FID; II(2) [#S130]
Tetraethyl lead (as Pb) $Pb(C_2H_5)_4$ 78-00-2 TP4550000 1649 56	Lead tetraethyl, TEL 1 ppm = 13.45 mg/m³	NIOSH/OSHA 0.075 mg/m³ [skin]	40 mg/m³	Colorless liquid (unless dyed red, orange, or blue) with a pleasant, sweet odor. [Note: Main usage is in anti-knock additives for gasoline.]	MW: 323.5 BP: 228°F (Decomposes) Sol: Insoluble Fl.P: 200°F IP: 11.10 eV Sp.Gr: 1.65 Class IIIB Combustible Liquid	VP: 0.2 mm FRZ: -202°F UEL: ? LEL: 1.8%	Strong oxidizers, sulfuryl chloride, rust, potassium permanganate [Note: Decomposes slowly at room temperature and more rapidly at higher temperatures.]	XAD-2; Pentane; GC/PID; III [#2533]

Personal protection and sanitation (See Table 3)	Recommendations for respirator selection — maximum concentration for use (MUC) (See Table 4)	Health hazards			
		Route	Symptoms (See Table 5)	First aid (See Table 6)	Target organs (See Table 5)
Clothing: Any poss Goggles: Any poss Wash: Immed contam Change: N.R. Remove: Immed non-imperv contam Provide: Eyewash, quick drench	NIOSH/OSHA 0.5 mg/m³: SA/SCBA 1.25 mg/m³: SA:CF 2.5 mg/m³: SCBAF/SAF/SAT:CF 10 mg/m³: SA:PD,PP §: SCBAF:PD,PP/SAF:PD,PP:ASCBA Escape: GMFOVHiE/SCBAE	Inh Abs Ing Con	Eye pain, blurred vision, lac, rhin, head, tight chest, cyan; anor, nau, vomit, diarr; weak, twitch, para, Cheyne-Stokes resp, convuls; low BP, card irregularities; local sweating	Eye: Irr immed Skin: Water flush immed Breath: Resp support Swallow: Medical attention immed	CNS, resp sys, CVS, GI tract
Clothing: Any poss molt/Repeat liq-sol Goggles: Any poss molt/Reason prob liq-sol Wash: Prompt contam Change: After work if reason prob contam Remove: Prompt non-imperv contam Provide: Eyewash, quick drench	NIOSH/OSHA 25 mg/m³: DM^£ 50 mg/m³: DMXSQ^£/SA^£/SCBA 125 mg/m³: PAPRDM^£/SA:CF^£ 250 mg/m³: SCBAF/SAF/HiEF 500 mg/m³: SAF:PD,PP §: SCBAF:PD,PP/SAF:PD,PP:ASCBA Escape: HiEF/SCBAE	Inh Ing Con	Irrit eyes, skin; thermal skin burns; head; sore throat	Eye: Irr immed Skin: Water flush immed Breath: Resp support Swallow: Medical attention immed	Skin, resp sys, eyes
Clothing: Repeat Goggles: Reason prob Wash: Prompt wet Change: N.R. Remove: Prompt non-imperv wet	NIOSH/OSHA 5000 ppm: SA/SCBA 12,500 ppm: SA:CF 15,000 ppm: SCBAF/SAF §: SCBAF:PD,PP/SAF:PD,PP:ASCBA Escape: GMFOV/SCBAE	Inh Ing Con	Irrit skin; conj; pulm edema	Eye: Irr immed Skin: Soap wash prompt Breath: Resp support Swallow: Medical attention immed	Lungs, skin
Clothing: Repeat Goggles: Reason prob Wash: Prompt contam Change: N.R. Remove: Prompt non-imperv contam	NIOSH/OSHA 5000 ppm: SA/SCBA 12,500 ppm: SA:CF 15,000 ppm: SCBAF/SAF §: SCBAF:PD,PP/SAF:PD,PP:ASCBA Escape: GMFOV/SCBAE	Inh Ing Con	CNS depres; pulm edema; skin, eye irrit; drow; dysp	Eye: Irr immed Skin: Soap wash prompt Breath: Resp support Swallow: Medical attention immed	Resp sys, skin

Personal protection and sanitation (See Table 3)	Recommendations for respirator selection — maximum concentration for use (MUC) (See Table 4)	Health hazards			
		Route	Symptoms (See Table 5)	First aid (See Table 6)	Target organs (See Table 5)
Clothing: Any poss Goggles: Any poss Wash: Immed contam Change: N.R. Remove: Immed non-imperv contam Provide: Eyewash, quick drench	NIOSH ¥: SCBAF:PD,PP/SAF:PD,PP:ASCBA Escape: GMFOV/SCBAE	Inh Abs Ing Con	Nau, vomit, abdom pain; tremor fingers; jaun, enlarged tend liver; derm; monocy; kidney damage	Eye: Irr immed Skin: Soap wash prompt Breath: Resp support Swallow: Medical attention immed	Liver, kidneys, CNS
Clothing: Repeat Goggles: Reason prob Wash: Prompt contam Change: N.R. Remove: Prompt non-imperv contam	NIOSH ¥: SCBAF:PD,PP/SAF:PD,PP:ASCBA Escape: GMFOV/SCBAE	Inh Ing Con	Irrit eyes, nose, throat; nau; flush face, neck; verti, dizz, inco; head, som; skin eryt; liver damage; [carc]	Eye: Irr immed Skin: Soap wash prompt Breath: Resp support Swallow: Medical attention immed	Liver, kidneys, eyes, upper resp sys, CNS
Clothing: Any poss molt/Repeat liq-sol Goggles: Any poss molt/Reason prob liq-sol Wash: Prompt contam Change: After work if any poss contam Remove: Immed non-imperv contam molt/Prompt non-imperv contam sol	NIOSH/OSHA 20 mg/m³: SCBAF/SAF §: SCBAF:PD,PP/SAF:PD,PP:ASCBA Escape: GMFOVHiE/SCBAE	Inh Abs Ing Con	Acne-form derm; head, ftg, anor, verti; jaun, liver inj	Eye: Irr immed Skin: Soap wash immed Breath: Resp support Swallow: Medical attention immed	Liver, skin
Clothing: Any poss >0.1% Goggles: Reason prob Wash: N.R. Change: After work if any poss contam >0.1% Remove: Immed non-imperv contam (>0.1%) Provide: Quick drench (>0.1%)	NIOSH/OSHA 0.75 mg/m³: SA/SCBA 1.875 mg/m³: SA:CF 3.75 mg/m³: SCBAF/SAF/SAT:CF 40 mg/m³: SA:PD,PP §: SCBAF:PD,PP/SAF:PD,PP:ASCBA Escape: GMFOV/SCBAE	Inh Abs Ing Con	Insom, lass, anxiety; tremor, hyper-reflexia, spastic; bradycardia, hypotension, hypothermia, pallor, nau, anor, low-wgt; disorientation, halu, psychosis, mania, convuls, coma; eye irrit	Eye: Irr immed Skin: Soap wash immed Breath: Resp support Swallow: Medical attention immed	CNS, CVS, kidneys, eyes

Chemical name, structure/formula, CAS and RTECS Nos., and DOT ID and guide Nos.	Synonyms, trade names, and conversion factors	Exposure limits (TWA unless noted otherwise)	IDLH	Physical description	Chemical and physical properties		Incompatibilities and reactivities	Measurement method (See Table 1)
					MW, BP, SOL Fl.P, IP, Sp.Gr, flammability	VP, FRZ UEL, LEL		
Tetrahydrofuran C_4H_8O 109-99-9 LU5950000 2056 26	Diethylene oxide; 1,4-Epoxybutane; Tetramethylene oxide; THF 1 ppm = 3.00 mg/m³	NIOSH/OSHA 200 ppm (590 mg/m³) ST 250 ppm (735 mg/m³)	20,000 ppm [LEL]	Colorless liquid with an ether-like odor.	MW: 72.1 BP: 151°F Sol: Miscible Fl.P: 6°F IP: 9.45 eV Sp.Gr: 0.89 Class IB Flammable Liquid	VP: 132 mm FRZ: -163°F UEL: 11.8% LEL: 2%	Strong oxidizers, lithium-aluminum alloys [Note: Peroxides may accumulate upon prolonged storage in presence of air.]	Char; CS₂; GC/FID; III [#1609]
Tetramethyl lead (as Pb) Pb(CH₃)₄ 75-74-1 TP4725000 1649 56	Lead tetramethyl, TML 1 ppm = 11.11 mg/m³	NIOSH/OSHA 0.075 mg/m³ [skin]	40 mg/m³	Colorless liquid (unless dyed red, orange, or blue) with a fruity odor. [Note: Main usage is in anti-knock additives for gasoline.]	MW: 267.3 BP: 212°F (Decomposes) Sol: Insoluble Fl.P: 100°F IP: 8.50 eV Sp.Gr: 2.00 Class II Combustible Liquid	VP: 23 mm FRZ: -15°F UEL: ? LEL: ?	Strong oxidizers such as sulfuryl chloride or potassium permanganate	XAD-2; Pentane; GC/PID; III [#2534]
Tetramethyl succinonitrile (CH₃)₂C(CN)C(CN)(CH₃)₂ 3333-52-6 WN4025000	Tetramethyl succino-dinitrile, TMSN	NIOSH/OSHA 3 mg/m³ (0.5 ppm) [skin]	5 ppm	Colorless, odorless solid.	MW: 136.2 BP: Sublimes Sol: Insoluble Fl.P: ? IP: ? Sp.Gr: 1.07 Combustible Solid	VP: ? MLT: 338°F (Sublimes) UEL: ? LEL: ?	Strong oxidizers	Char; CS₂; GC/FID; II(3) [#S155]
Tetranitromethane C(NO₂)₄ 509-14-8 PB4025000 1510 47	Tetan, TNM 1 ppm = 8.15 mg/m³	NIOSH/OSHA 1 ppm (8 mg/m³)	5 ppm	Colorless to pale-yellow liquid or solid (below 57°F) with a pungent odor.	MW: 196.0 BP: 259°F Sol: Insoluble Fl.P: ? IP: ? Sp.Gr: 1.62 Combustible Liquid, but difficult to ignite.	VP: 8 mm FRZ: 57°F UEL: ? LEL: ?	Hydrocarbons, alkalis, metals, oxidizers, aluminum, toluene, cotton [Note: May decompose explosively if contaminated with combustible material.]	Imp; none; GC/Alk-FID; II(3) [#S224]

Chemical name, structure/formula, CAS and RTECS Nos., and DOT ID and guide Nos.	Synonyms, trade names, and conversion factors	Exposure limits (TWA unless noted otherwise)	IDLH	Physical description	Chemical and physical properties		Incompatibilities and reactivities	Measurement method (See Table 1)
					MW, BP, SOL Fl.P, IP, Sp.Gr, flammability	VP, FRZ UEL, LEL		
Tetryl (NO₂)₃C₆H₂N(NO₂)CH₃ 479-45-8 BY6300000	N-Methyl-N,2,4,6-tetranitroaniline; Nitramine; 2,4,6-Tetryl; 2,4,6-Trinitrophenyl-N-methylnitramine	NIOSH/OSHA 1.5 mg/m³ [skin]	N.E.	Colorless to yellow, odorless, crystalline solid.	MW: 287.2 BP: 356-374°F (Explodes) Sol: Insoluble Fl.P: Explodes IP: ? Sp.Gr: 1.57 Combustible Solid	VP: Low MLT: 268°F UEL: ? LEL: ?	Oxidizable materials, hydrazine	Filter; Reagent; Vis; II(3) [#S225]
Thallium (soluble compounds as Tl) 7440-28-0 XG3425000 1707 53 (cmpds)	Synonyms vary depending upon the specific soluble compound.	NIOSH/OSHA 0.1 mg/m³ [skin]	20 mg/m³	Appearance and odor vary depending upon the specific soluble compound.	Properties vary depending upon the specific soluble compound.		None reported	Filter; Acid; AA; II(3) [#S306]
Thiram (CH₃)₂NC(S)SSC-(S)N(CH₃)₂ 137-26-8 JO1400000 2771 55	bis(Dimethylthiocarbamoyl) disulfide; Tetramethylthiuram disulfide	NIOSH/OSHA 5 mg/m³	1500 mg/m³	Colorless to yellow, crystalline solid. [Note: Commercial pesticide products may be dyed blue.]	MW: 240.4 BP: Decomposes Sol: Insoluble Fl.P: ? IP: ? Sp.Gr: 1.29 Combustible Solid	VP: Low MLT: 312°F UEL: ? LEL: ?	Strong oxidizers, strong acids, oxidizable materials	Filter; CH₃CN; HPLC/UVD; III [#5005]
Tin (inorganic compounds except oxides, as Sn) Sn 7440-31-5 (Metal) XP7320000 (Metal)	Metal: Metallic tin, Tin flake, Tin powder Synonyms of other inorganic compounds vary depending upon the specific compound.	NIOSH/OSHA 2 mg/m³	400 mg/m³	Metal: Gray to almost silver-white, ductile, malleable, lustrous solid.	MW: 118.7 BP: 4545°F Sol: Insoluble Fl.P: NA IP: NA Sp.Gr: 7.28 (Metal) Metal: Noncombustible Solid, but powdered form may ignite.	VP: 0 mm (approx) MLT: 449°F UEL: NA LEL: NA	Chlorine, turpentine, acids, alkalis	Filter; Acid; AA; I(3) [#S183]

Personal protection and sanitation (See Table 3)		Recommendations for respirator selection — maximum concentration for use (MUC) (See Table 4)	Health hazards			
			Route	Symptoms (See Table 5)	First aid (See Table 6)	Target organs (See Table 5)
Clothing:	Repeat	NIOSH/OSHA	Inh	Irrit eyes, upper resp sys;	Eye: Irr immed	Eyes, skin, resp
Goggles:	Reason prob	1000 ppm: PAPROV^ε/CCRFOV	Con	nau, dizz, head	Skin: Water flush prompt	sys, CNS
Wash:	Prompt wet	5000 ppm: SA:CF^ε	Ing		Breath: Resp support	
Change:	N.R.	10,000 ppm: GMFOV/SCBAF/SAF			Swallow: Medical attention	
Remove:	Immed wet (flamm)	20,000 ppm: SAF:PD,PP			immed	
		§: SCBAF:PD,PP/SAF:PD,PP:ASCBA				
		Escape: GMFOV/SCBAE				
Clothing:	Any poss >1.06%	NIOSH/OSHA	Inh	Insom, bad dreams,	Eye: Irr immed	CNS, CVS, kidneys
Goggles:	Reason prob	0.75 mg/m³: SA/SCBA	Abs	restless, anxious;	Skin: Soap wash immed	
Wash:	N.R.	1.875 mg/m³: SA:CF	Ing	hypotension; nau, anor;	Breath: Resp support	
Change:	N.R.	3.75 mg/m³: SCBAF/SAF/SAT:CF	Con	delirium, mania, convuls;	Swallow: Medical attention	
Remove:	Immed non-imperv contam	40 mg/m³: SA:PD,PP		coma	immed	
Provide:	>1.06%: Quick drench	§: SCBAF:PD,PP/SAF:PD,PP:ASCBA				
		Escape: GMFOV/SCBAE				
Clothing:	Repeat	NIOSH/OSHA	Inh	Head, nau; convuls,	Eye: Irr immed	CNS
Goggles:	Reason prob	30 mg/m³: SA/SCBA	Abs	coma	Skin: Soap wash prompt	
Wash:	Prompt contam	§: SCBAF:PD,PP/SAF:PD,PP:ASCBA	Ing		Breath: Resp support	
Change:	After work if reason prob contam	Escape: GMFOVHiE/SCBAE	Con		Swallow: Medical attention immed	
Remove:	Prompt non-imperv contam					
Clothing:	Reason prob	NIOSH/OSHA	Inh	Irrit eyes, nose, throat;	Eye: Irr immed	Resp sys, eyes,
Goggles:	Any poss	5 ppm: SA:CF^ε/PAPRS^{ε4}/CCRFS^4	Ing	dizz, head; chest pain,	Skin: Soap wash prompt	skin, blood, CNS
Wash:	Prompt contam	GMFS^4/SCBAF/SAF	Con	dysp; methemoglobinuria,	Breath: Resp support	
Change:	After work if reason prob contam	§: SCBAF:PD,PP/SAF:PD,PP:ASCBA		cyan; skin burns	Swallow: Medical attention immed	
Remove:	Immed wet (flamm)	Escape: GMFS^4/SCBAE				
Provide:	Eyewash					

Personal protection and sanitation (See Table 3)		Recommendations for respirator selection — maximum concentration for use (MUC) (See Table 4)	Health hazards			
			Route	Symptoms (See Table 5)	First aid (See Table 6)	Target organs (See Table 5)
Clothing:	Repeat	NIOSH/OSHA	Inh	Sens derm, itch, eryt,	Eye: Irr immed	Resp sys, eyes,
Goggles:	Reason prob	7.5 mg/m³: DM	Abs	edema on nasal folds,	Skin: Soap wash prompt	CNS, skin;
Wash:	Prompt contam/daily	15 mg/m³: DMXSQ*/SA*/SCBA*	Ing	cheeks, neck; kera;	Breath: Resp support	in animals:
Change:	After work if reason prob contam	37.5 mg/m³: PAPRDM*/SA:CF*	Con	sneezing; anemia; ftg;	Swallow: Medical attention	liver, kidneys
Remove:	Prompt non-imperv contam	75 mg/m³: HiEF/SCBAF/SAF		cough, coryza; irrity;	immed	
		3000 mg/m³: SAF:PD,PP		mal, head, lass, insom;		
		§: SCBAF:PD,PP/SAF:PD,PP:ASCBA		nau, vomit		
		Escape: HiEF/SCBAE				
Clothing:	Reason prob	NIOSH/OSHA	Inh	Nau, diarr, abdom pain,	Eye: Irr immed	Eyes, CNS, lungs;
Goggles:	Reason prob	0.5 mg/m³: DM^	Abs	vomit; ptosis, strabismus;	Skin: Water flush prompt	liver, kidneys,
Wash:	Prompt contam	1 mg/m³: DMXSQ^/SA/SCBA	Ing	peri neuritis, tremor;	Breath: Resp support	GI tract, body
Change:	After work if reason prob contam	2.5 mg/m³: PAPRDM^/SA:CF	Con	retster tightness, chest	Swallow: Medical attention	hair
Remove:	Prompt non-imperv contam	5 mg/m³: HiEF/SCBAF/SAF/PAPRTHiE/SAT:CF		pain, pulm edema; sez, chorea, psychosis; liver,	immed	
		20 mg/m³: SAF:PD,PP		kidney damage; alopecia;		
		§: SCBAF:PD,PP/SAF:PD,PP:ASCBA		pares legs		
		Escape: HiEF/SCBAE				
Clothing:	Reason prob	NIOSH/OSHA	Inh	Irrit muc memb; derm;	Eye: Irr immed	Resp sys, skin
Goggles:	Reason prob	50 mg/m³: CCROVDMFu*/SA*/SCBA*	Ing	with ethanol	Skin: Soap wash prompt	
Wash:	Prompt contam	125 mg/m³: PAPROVDMFu*/SA:CF*	Con	consumption: flush, eryt,	Swallow: Medical attention	
Change:	After work if reason prob contam	250 mg/m³: SCBAF/SAF/CCRFOVHiE/PAPRTOVHiE*/GMFOVHiE		pruritus, urticaria, head, nau, vomit, diarr, weak,	immed	
Remove:	Prompt non-imperv contam	1500 mg/m³: SAF:PD,PP		dizz, dysp		
		§: SCBAF:PD,PP/SAF:PD,PP:ASCBA				
		Escape: GMFOVHiE/SCBAE				
Recommendations vary depending upon the specific compound.		NIOSH/OSHA	Inh	Irrit eyes, skin	Eye: Irr immed	Eyes, skin, resp
		10 mg/m³: DM*	Con		Skin: Soap wash immed	sys
		20 mg/m³: DMXSQ^*/SA*/SCBA*			Breath: Resp support	
		50 mg/m³: PAPRDM*/SA:CF*			Swallow: Medical attention	
		100 mg/m³: HiEF/SCBAF/SAF			immed	
		400 mg/m³: SAF:PD,PP				
		§: SCBAF:PD,PP/SAF:PD,PP:ASCBA				
		Escape: HiEF/SCBAE				

Chemical name, structure/formula, CAS and RTECS Nos., and DOT ID and guide Nos.	Synonyms, trade names, and conversion factors	Exposure limits (TWA unless noted otherwise)	IDLH	Physical description	Chemical and physical properties		Incompatibilities and reactivities	Measurement method (See Table 1)
					MW, BP, SOL Fl.P, IP, Sp.Gr, flammability	**VP, FRZ UEL, LEL**		
Tin (organic compounds as Sn)	Synonyms vary depending upon the specific organic compound.	NIOSH/OSHA 0.1 mg/m³ [skin]	Unknown	Appearance and odor vary depending upon the specific organic compound.	Properties vary depending upon the specific organic compound.		Strong oxidizers	Filter/ XAD-2; HPLC; AA; III [#5504]
Titanium dioxide TiO₂ 13463-67-7 XR2275000	Rutile, Titanium oxide, Titanium peroxide	NIOSH Ca See Appendix A OSHA 10 mg/m³	Ca [N.E.]	White, odorless powder.	MW: 79.9 BP: 4532 to 5432°F Sol: Insoluble Fl.P: NA IP: NA Sp.Gr: 4.26 Noncombustible Solid	VP: 0 mm (approx) MLT: 3326 to 3362°F UEL: NA LEL: NA	None reported	Filter; Acid; AA; II(3) [#S385]
Toluene C₆H₅CH₃ 108-88-3 XS5250000 1294 27	Methyl benzene, Methyl benzol, Phenyl methane, Toluol	NIOSH/OSHA 100 ppm (375 mg/m³) ST 150 ppm (560 mg/m³) 1 ppm = 3.83 mg/m³	2000 ppm	Colorless liquid with a sweet, pungent, benzene-like odor.	MW: 92.1 BP: 232°F Sol(61°F): 0.05% Fl.P: 40°F IP: 8.82 eV Sp.Gr: 0.87 Class IB Flammable Liquid	VP(65°F): 20 mm FRZ: -139°F UEL: 7.1% LEL: 1.2%	Strong oxidizers	Char; CS₂; GC/FID; III [#1500, Hydro-carbons]
Toluene-2,4-diisocyanate CH₃C₆H₃(NCO)₂ 584-84-9 CZ6300000 2078 57	TDI; 2,4-TDI; 2,4-Toluene diisocyanate	NIOSH Ca See Appendix A 0.005 ppm (0.04 mg/m³) ST 0.02 ppm (0.15 mg/m³) OSHA 0.005 ppm (0.04 mg/m³) ST 0.02 ppm (0.15 mg/m³) 1 ppm = 7.24 mg/m³	Ca [10 ppm]	Colorless to pale-yellow solid or liquid (above 71°F) with a sharp, pungent odor.	MW: 174.2 BP: 484°F Sol: Insoluble Fl.P: 260°F IP: ? Sp.Gr: 1.22 Class IIIB Combustible Liquid	VP(77°F): 0.01 mm MLT: 71°F UEL: 9.5% LEL: 0.9%	Strong oxidizers, water, acids, bases & amines may cause foam and spatter; alcohols [Note: Reacts slowly with water to form carbon dioxide and polyureas.]	Coated glass wool; Methanol; HPLC/UVD; III [#2535]

Chemical name, structure/formula, CAS and RTECS Nos., and DOT ID and guide Nos.	Synonyms, trade names, and conversion factors	Exposure limits (TWA unless noted otherwise)	IDLH	Physical description	Chemical and physical properties		Incompatibilities and reactivities	Measurement method (See Table 1)
					MW, BP, SOL Fl.P, IP, Sp.Gr, flammability	**VP, FRZ UEL, LEL**		
o-Toluidine CH₃C₆H₄NH₂ 95-53-4 XU2975000 1708 55	o-Aminotoluene, 2-Aminotoluene, 1-Methyl-2-aminobenzene, o-Methylaniline, 2-Methylaniline, ortho-Toluidine	NIOSH Ca See Appendix A 2 ppm (9 mg/m³) [skin] OSHA 5 ppm (22 mg/m³) [skin] 1 ppm = 4.46 mg/m³	Ca [100 ppm] ACGIH A2	Colorless to pale-yellow liquid with an aromatic, aniline-like odor.	MW: 107.2 BP: 392°F Sol: 2% Fl.P: 185°F IP: 7.44 eV Sp.Gr: 1.01 Class IIIA Combustible Liquid	VP: 0.3 mm FRZ: 6°F UEL: ? LEL: ?	Strong oxidizers, nitric acid	Si gel; Ethanol; GC/FID; III [#2002, Aromatic Amines]
Tributyl phosphate (CH₃[CH₂]₃O)₃PO 126-73-8 TC7700000	Butyl phosphate, TBP, Tributyl ester of phosphoric acid, tri-n-Butyl phosphate	NIOSH/OSHA 0.2 ppm (2.5 mg/m³) 1 ppm = 11.07 mg/m³	125 ppm	Colorless to pale-yellow, odorless liquid.	MW: 266.3 BP: 552°F (Decomposes) Sol: 0.6% Fl.P(oc): 295°F IP: ? Sp.Gr: 0.98 Class IIIB Combustible Liquid	VP(351°F): 127 mm FRZ: -112°F UEL: ? LEL: ?	None reported	Filter; Diethyl ether; GC/FPD; II(3) [#S208]
1,1,2-Trichloroethane CHCl₂CH₂Cl 79-00-5 KJ3150000 2831 74	beta-Trichloroethane, Vinyl trichloride	NIOSH Ca See Appendix A 10 ppm (45 mg/m³) [skin] OSHA 10 ppm (45 mg/m³) [skin] 1 ppm = 5.55 mg/m³	Ca [500 ppm]	Colorless liquid with a sweet, chloroform-like odor.	MW: 133.4 BP: 237°F Sol: 0.4% Fl.P: NA IP: 11.00 eV Sp.Gr: 1.44 Noncombustible Liquid	VP: 19 mm FRZ: -34°F UEL: NA LEL: NA	Strong oxidizers & caustics; chemically-active metals such as aluminum, magnesium powders, sodium & potassium	Char; CS₂; GC/FID; III [#1003, Haloge-nated Hydro-carbons]
Trichloroethylene ClCH=CCl₂ 79-01-6 KX4550000 1710 74	Ethylene trichloride, Triclene, Trichloroethene	NIOSH Ca See Appendix A 25 ppm OSHA 50 ppm (270 mg/m³) ST 200 ppm (1080 mg/m³) 1 ppm = 5.46 mg/m³	Ca [1000 ppm]	Colorless liquid (unless dyed blue) with a chloroform-like odor.	MW: 131.4 BP: 189°F Sol(77°F): 0.1% Fl.P: 90°F IP: 9.45 eV Sp.Gr: 1.46 Class IC Flammable Liquid, but burns with difficulty.	VP: 58 mm FRZ: -99°F UEL(77°F): 10.5% LEL(77°F): 8%	Strong caustics & alkalis; chemically-active metals such as barium, lithium, sodium, magnesium, titanium & beryllium	Char; CS₂; GC/FID; III [#1022]

Personal protection and sanitation (See Table 3)	Recommendations for respirator selection — maximum concentration for use (MUC) (See Table 4)	Route	Symptoms (See Table 5)	First aid (See Table 6)	Target organs (See Table 5)
Recommendations vary depending upon the specific compound.	NIOSH/OSHA 1 mg/m³: CCROVDM/SA/SCBA 2.5 mg/m³: SA:CF/PAPROVDM 5 mg/m³: CCRFOVHiE/SCBAF/SAF/ GMFOVHiE/PAPRTOVHiE/ SAT:CF 200 mg/m³: SAF:PD,PP §: SCBAF:PD,PP/SAF:PD,PP:ASCBA Escape: GMFOVHiE/SCBAE	Inh Abs Ing Con	Head, verti; irrit eyes; psycho-neurologic dist; sore throat, cough; abdom pain, vomit; urine retention; paresis, focal anes; skin burns; pruritus; in animals: hemolysis, hepatic nec	Eye: Irr immed Skin: Water flush immed Breath: Resp support Swallow: Medical attention immed	CNS, eyes, liver, urinary tract, skin, blood
Clothing: N.R. Goggles: N.R. Wash: N.R. Change: N.R. Remove: N.R.	NIOSH ¥: SCBAF:PD/PP/SAF:PD,PP:ASCBA Escape: HiEF/SCBAE	Inh	Slight lung fib; [carc]	Breath: Resp support	Lungs
Clothing: Repeat Goggles: Reason prob Wash: Prompt wet Change: N.R. Remove: Immed wet (flamm)	NIOSH/OSHA 1000 ppm: CCROV*/SA*/PAPROV*/ SCBA* 2000 ppm: SA:CF*/SCBAF/SAF/GMFOV §: SCBAF:PD,PP/SAF:PD,PP:ASCBA Escape: GMFOV/SCBAE	Inh Abs Ing Con	Ftg, weak; conf, euph, dizz, head; dilated pupils, lac; ner, musc ftg, insom; pares; derm	Eye: Irr immed Skin: Soap wash prompt Breath: Resp support Swallow: Medical attention immed	CNS, liver, kidneys, skin
Clothing: Repeat Goggles: Any poss Wash: Prompt contam Change: After work if reason prob contam Remove: Prompt non-imperv contam Provide: Eyewash, quick drench	NIOSH ¥: SCBAF:PD,PP/SAF:PD,PP:ASCBA Escape: GMFOV/SCBAE	Inh Ing Con	Irrit nose, throat; choke, paroxysmal cough; chest pain, retster soreness; nau, vomit, abdom pain; bron spasm, pulm edema; dysp, asthma; conj, lac; derm, skin sens; [carc]	Eye: Irr immed Skin: Soap wash immed Breath: Resp support Swallow: Medical attention immed	Resp sys, skin

Personal protection and sanitation (See Table 3)	Recommendations for respirator selection — maximum concentration for use (MUC) (See Table 4)	Route	Symptoms (See Table 5)	First aid (See Table 6)	Target organs (See Table 5)
Clothing: Any poss Goggles: Any poss Wash: Immed contam Change: N.R. Remove: Immed non-imperv contam Provide: Eyewash, quick drench	NIOSH ¥: SCBAF:PD,PP/SAF:PD,PP:ASCBA Escape: GMFOV/SCBAE	Inh Abs Ing Con	Anoxia, head, cyan; weak, dizz, drow; micro hematuria, eye burns; derm; [carc]	Eye: Irr immed Skin: Soap wash immed Breath: Resp support Swallow: Medical attention immed	Blood, kidneys, liver, CVS, skin, eyes
Clothing: Repeat Goggles: Reason prob Wash: Prompt wet Change: N.R. Remove: Prompt non-imperv wet	NIOSH/OSHA 2 ppm: SA/SCBA 5 ppm: SA:CF 10 ppm: SCBAF/SAF 125 ppm: SAF:PD,PP §: SCBAF:PD,PP/SAF:PD,PP:ASCBA Escape: GMFOVHiE/SCBAE	Inh Con Ing	Eyes, resp, skin irrit; head; nau	Eye: Irr immed Skin: Soap wash prompt Breath: Resp support Swallow: Medical attention immed	Resp sys, skin, eyes
Clothing: Repeat Goggles: Reason prob Wash: Prompt contam Change: N.R. Remove: Prompt non-imperv contam	NIOSH ¥: SCBAF:PD,PP/SAF:PD,PP:ASCBA Escape: GMFOV/SCBAE	Inh Abs Ing Con	Irrit nose, eyes; CNS depres; liver, kidney damage; [carc]	Eye: Irr immed Skin: Soap wash prompt Breath: Resp support Swallow: Medical attention immed	CNS, eyes, nose, liver, kidneys
Clothing: Repeat Goggles: Reason prob Wash: Prompt wet Change: N.R. Remove: Prompt non-imperv wet	NIOSH ¥: SCBAF:PD,PP/SAF:PD,PP:ASCBA Escape: GMFOV/SCBAE	Inh Ing Con	Head, verti; vis dist, tremors, som, nau, vomit; irrit eyes; derm; card arrhy, pares; [carc]	Eye: Irr immed Skin: Soap wash prompt Breath: Resp support Swallow: Medical attention immed	Resp sys, heart, liver, kidneys, CNS, skin

Chemical name, structure/formula, CAS and RTECS Nos., and DOT ID and guide Nos.	Synonyms, trade names, and conversion factors	Exposure limits (TWA unless noted otherwise)	IDLH	Physical description	Chemical and physical properties		Incompatibilities and reactivities	Measurement method (See Table 1)
					MW, BP, SOL Fl.P, IP, Sp.Gr, flammability	VP, FRZ UEL, LEL		
Trichloronaphthalene $C_{10}H_5Cl_3$ 1321-65-9 QK4025000	Halowax®, Nibren wax, Seekay wax	NIOSH/OSHA 5 mg/m³ [skin]	Unknown	Colorless to pale-yellow solid with an aromatic odor.	MW: 231.5 BP: 579-669°F Sol: Insoluble Fl.P(oc): 392°F IP: ? Sp.Gr: 1.58 Combustible Solid	VP: <1 mm MLT: 199°F UEL: ? LEL: ?	Strong oxidizers	Filter/Bub; none; GC/FID; II(2) [#S128]
1,2,3-Trichloropropane $CH_2ClCHClCH_2Cl$ 96-18-4 TZ9275000	Allyl trichloride, Glycerol trichlorohydrin, Glyceryl trichlorohydrin, Trichlorohydrin	NIOSH Ca See Appendix A 10 ppm (60 mg/m³) [skin] OSHA 10 ppm (60 mg/m³)	Ca [1000 ppm]	Colorless liquid with a chloroform-like odor.	MW: 147.4 BP: 314°F Sol: 0.1% Fl.P(oc): 180°F IP: ? Sp.Gr: 1.39 Class IIIA Combustible Liquid	VP(115°F): 10 mm FRZ: 6°F UEL: ? LEL: ?	Chemically-active metals, strong caustics & oxidizers	Char; CS₂; GC/FID; III [#1003, Haloge- nated Hydro- carbons]
	1 ppm = 6.13 mg/m³							
1,1,2-Trichloro-1,2,2-trifluoroethane CCl_2FCClF_2 76-13-1 KJ4000000	Chlorofluorocarbon-113, CFC-113, Freon® 113, Genetron® 113, Halocarbon 113, Refrigerant 113, TTE	NIOSH/OSHA 1000 ppm (7600 mg/m³) ST 1250 ppm (9500 mg/m³)	4500 ppm	Colorless to water-white liquid with an odor like carbon tetrachloride at high concentra-tions. [Note: A gas above 118°F.]	MW: 187.4 BP: 118°F Sol(77°F): 0.02% Fl.P: ? IP: 11.99 eV Sp.Gr(77°F): 1.56 Noncombustible Liquid at ordinary temperatures, but the gas will ignite and burn weakly at 1256°F	VP: 285 mm FRZ: -31°F UEL: ? LEL: ?	Chemically-active metals such as calcium, powdered aluminum, zinc, magnesium & beryllium [Note: Decomposes if in contact with alloys containing >2% magnesium.]	Char; CS₂; GC/FID; III [#1020]
	1 ppm = 7.79 mg/m³							
Triethylamine $(C_2H_5)_3N$ 121-44-8 YEO175000 1296 68	TEA	NIOSH See Appendix D OSHA 10 ppm (40 mg/m³) ST 15 ppm (60 mg/m³)	1000 ppm	Colorless liquid with a strong, ammonia-like odor.	MW: 101.2 BP: 193°F Sol: 2% Fl.P(oc): 16°F IP: 7.50 eV Sp.Gr: 0.73 Class IB Flammable Liquid	VP: 54 mm FRZ: -175°F UEL: 8.0% LEL: 1.2%	Strong oxidizers, strong acids	Bub; NaOH; GC/FID; II(3) [#S152]
	1 ppm = 4.21 mg/m³							

Chemical name, structure/formula, CAS and RTECS Nos., and DOT ID and guide Nos.	Synonyms, trade names, and conversion factors	Exposure limits (TWA unless noted otherwise)	IDLH	Physical description	Chemical and physical properties		Incompatibilities and reactivities	Measurement method (See Table 1)
					MW, BP, SOL Fl.P, IP, Sp.Gr, flammability	VP, FRZ UEL, LEL		
Trifluorobromomethane $CBrF_3$ 75-63-8 PA5425000 1009 12	Bromotrifluoromethane, Fluorocarbon 1301, Freon® 13B1, Halocarbon 13B1, Halon® 1301, Monobromotrifluoromethane, Refrigerant 13B1, Trifluoromonobromomethane	NIOSH/OSHA 1000 ppm (6100 mg/m³)	50,000 ppm	Colorless, odorless gas. [Note: Shipped as a liquefied compressed gas.]	MW: 148.9 BP: -72°F Sol: 0.03% Fl.P: NA IP: 11.78 eV Nonflammable Gas	VP: >1 atm FRZ: -267°F UEL: NA LEL: NA	Chemically-active metals such as calcium, powdered aluminum, zinc & magnesium	Char(2); CH₂Cl₂; GC/FID; III [#1017]
	1 ppm = 6.19 mg/m³							
2,4,6-Trinitrotoluene $CH_3C_6H_2(NO_2)_3$ 118-96-7 XU0175000 1356 33 (wet)	1-Methyl-2,4,6-trinitrobenzene; TNT; Trinitrotoluene; sym-Trinitrotoluene; Trinitrotoluol	NIOSH/OSHA 0.5 mg/m³ [skin]	N.E.	Colorless to pale-yellow, odorless solid or crushed flakes. [Note: Classified as a high explosive.]	MW: 227.1 BP: 464°F (Explodes) Sol(77°F): 0.01% Fl.P: ? (Explodes) IP: 10.59 eV Sp.Gr: 1.65 Combustible Solid	VP(180°F): 0.05 mm MLT: 176°F UEL: ? LEL: ?	Strong oxidizers, ammonia, strong alkalis, combustible materials, heat [Note: Rapid heating will result in detonation.]	Tenax GC; Acetone; GC/TEA; OSHA [#44]
Triorthocresyl phosphate $(CH_3C_6H_4O)_3PO$ 78-30-8 TD0350000 2574 55	TCP, TOCP, Tri-o-cresyl ester of phosphoric acid, Tri-o-cresyl phosphate, o-Trinitrotoluol	NIOSH/OSHA 0.1 mg/m³ [skin]	40 mg/m³	Colorless to pale-yellow, odorless liquid or solid (below 52°F).	MW: 368.4 BP: 770°F (Decomposes) Sol: Slight Fl.P: 437°F IP: ? Sp.Gr: 1.20 Class IIIB Combustible Liquid	VP(77°F): 0.00002 mm FRZ: 52°F UEL: ? LEL: ?	Oxidizers	Filter; Diethyl ether; GC/FPD; II(3) [#S209]
	1 ppm = 15.31 mg/m³							
Triphenyl phosphate $(C_6H_5O)_3PO$ 115-86-6 TC8400000	Phenyl phosphate, TPP, Triphenyl ester of phosphoric acid	NIOSH/OSHA 3 mg/m³	N.E.	Colorless, crystalline powder with a phenol-like odor.	MW: 326.3 BP: 776°F Sol: Insoluble Fl.P: 428°F IP: ? Sp.Gr: 1.29 Combustible Solid	VP(380°F): 1 mm MLT: 120°F UEL: ? LEL: ?	None reported	Filter; Diethyl ether; GC/FPD; II(3) [#S210]

Personal protection and sanitation (See Table 3)	Recommendations for respirator selection — maximum concentration for use (MUC) (See Table 4)	Health hazards			
		Route	Symptoms (See Table 5)	First aid (See Table 6)	Target organs (See Table 5)
Clothing: Any poss molt-fume/Repeat liq-sol Goggles: Any poss molt/Reason prob liq-sol Wash: Prompt contam Change: After work if reason prob contam sol Remove: Immed non-imperv contam molt/Prompt non-imperv contam sol	NIOSH/OSHA 50 mg/m³: SCBAF/SAF §: SCBAF:PD,PP/SAF:PD,PP:ASCBA Escape: GMFOVHiE/SCBAE	Inh Abs Ing Con	Acne-form derm; anor, nau; verti; jaun, liver inj	Eye: Irr immed Skin: Soap wash prompt Breath: Resp support Swallow: Medical attention immed	Skin, liver
Clothing: Repeat Goggles: Any poss Wash: Immed contam Change: N.R. Remove: Immed non-imperv contam Provide: Eyewash, quick drench	NIOSH ¥: SCBAF:PD,PP/SAF:PD,PP:ASCBA Escape: GMFOV/SCBAE	Inh Abs Ing Con	Irrit eyes, throat; CNS depres; liver inj; skin irrit; [carc]	Eye: Irr immed Skin: Soap wash prompt Breath: Resp support Swallow: Medical attention immed	Eyes, resp sys, skin, CNS, liver
Clothing: Repeat Goggles: Any poss Wash: Prompt wet Change: N.R. Remove: Prompt non-imperv wet	NIOSH/OSHA 4500 ppm: SA/SCBA §: SCBAF:PD,PP/SAF:PD,PP:ASCBA Escape: GMFOV/SCBAE	Inh Ing Con	Irrit throat, drow, derm; in animals: card arrhy	Eye: Irr immed Skin: Soap wash prompt Breath: Resp support Swallow: Medical attention immed	Skin, heart
Clothing: Repeat Goggles: Any poss Wash: Immed contam Change: N.R. Remove: Immed wet (flamm) Provide: >1%: Eyewash, quick drench	OSHA 250 ppm: SA:CF$^\varepsilon$ 500 ppm: SCBAF/SAF 1000 ppm: SAF:PD,PP §: SCBAF:PD,PP/SAF:PD,PP:ASCBA Escape: GMFS/SCBAE	Inh Ing Abs Con	Irrit eyes, resp sys, skin	Eye: Irr immed Skin: Soap wash immed Breath: Resp support Swallow: Medical attention immed	Resp sys, eyes, skin

Personal protection and sanitation (See Table 3)	Recommendations for respirator selection — maximum concentration for use (MUC) (See Table 4)	Health hazards			
		Route	Symptoms (See Table 5)	First aid (See Table 6)	Target organs (See Table 5)
Clothing: N.R. Goggles: N.R. Wash: N.R. Change: N.R. Remove: N.R.	NIOSH/OSHA 10,000 ppm: SA/SCBA 25,000 ppm: SA:CF 50,000 ppm: SCBAF/SAF/SAT:CF §: SCBAF:PD,PP/SAF:PD,PP:ASCBA Escape: GMFOV/SCBAE	Inh Con (liq)	Li-head; card arrhy	Breath: Resp support	Heart, CNS
Clothing: Repeat Goggles: Reason prob Wash: Prompt contam/daily Change: After work if reason prob contam Remove: Prompt non-imperv contam	NIOSH/OSHA 5 mg/m³: SA*/SCBA* 12.5 mg/m³: SA:CF* 25 mg/m³: SCBAF/SAF 1000 mg/m³: SAF:PD,PP §: SCBAF:PD,PP/SAF:PD,PP:ASCBA Escape: GMFOVHiE/SCBAE	Inh Abs Ing Con	Liver damage, jaun; cyan; sneezing; cough, sore throat; peri neur, musc pain; kidney damage; cataract; sens derm; leucyt; anemia; card irregularities	Eye: Irr immed Skin: Soap wash prompt Breath: Resp support Swallow: Medical attention immed	Blood, liver, eyes, CVS, CNS, kidneys, skin
Clothing: Repeat Goggles: N.R. Wash: Prompt contam Change: N.R. Remove: Prompt non-imperv contam	NIOSH/OSHA 0.5 mg/m³: DM 1 mg/m³: DMXSQ/SA/SCBA 2.5 mg/m³: PAPRDM/SA:CF 5 mg/m³: HiEF/PAPRTHiE/SCBAF/SAF/SAT:CF 40 mg/m³: SA:PD,PP §: SCBAF:PD,PP/SAF:PD,PP:ASCBA Escape: HiEF/SCBAE	Inh Abs Ing Con	GI dist; peri neur; cramps in calves, pares in feet or hands; weak feet, wrist drop, para	Eye: Irr immed Skin: Soap wash immed Breath: Resp support Swallow: Medical attention immed	PNS, CNS
Clothing: N.R. Goggles: N.R. Wash: N.R. Change: N.R. Remove: N.R.	NIOSH/OSHA 15 mg/m³: D 30 mg/m³: DXSQ/SA/SCBA/HiE 75 mg/m³: PAPRDM/SA:CF 150 mg/m³: HiEF/PAPRTHiE/SAT:CF SCBAF/SAF 1500 mg/m³: SA:PD,PP §: SCBAF:PD,PP/SAF:PD,PP:ASCBA Escape: HiEF/SCBAE	Inh Ing	Minor changes in blood enzymes; in animals: musc weak, para	Breath: Resp support Swallow: Medical attention immed	Blood

Chemical name, structure/formula, CAS and RTECS Nos., and DOT ID and guide Nos.	Synonyms, trade names, and conversion factors	Exposure limits (TWA unless noted otherwise)	IDLH	Physical description	Chemical and physical properties		Incompatibilities and reactivities	Measurement method (See Table 1)
					MW, BP, SOL Fl.P, IP, Sp.Gr, flammability	VP, FRZ UEL, LEL		
Turpentine C₁₀H₁₆ (approx) 8006-64-2 YO8400000 1299 27	Gumspirits, Gum turpentine, Spirits of turpentine, Steam distilled turpentine, Sulfate wood turpentine, Turps, Wood turpentine 1 ppm = 5.65 mg/m³	NIOSH/OSHA 100 ppm (560 mg/m³)	1500 ppm	Colorless liquid with a characteristic odor.	MW: 136(approx) BP: 309-338°F Sol: Insoluble Fl.P: 95°F IP: ? Sp.Gr: 0.86 Class IC Flammable Liquid	VP(77°F): 5 mm FRZ: -58 to -76°F UEL: ? LEL: 0.8%	Strong oxidizers, chlorine, chromic anhydride, stannic chloride	Char; CS₂; GC/FID; III [#1551]
Uranium (insoluble compounds as U) U 7440-61-1 (Metal) YR3490000 (Metal) 2979 65 (Metal) 9175 65 (Metal)	Metal: Uranium I Synonyms of other insoluble compounds vary depending upon the specific compound.	NIOSH Ca See Appendix A 0.2 mg/m³ ST 0.6 mg/m³ OSHA 0.2 mg/m³ ST 0.6 mg/m³	Ca [30 mg/m³]	Metal: Silver-white, malleable, ductile, lustrous solid. [Note: Weakly radioactive.]	MW: 238.0 BP: 6895°F Sol: Insoluble Fl.P: ? IP: NA Sp.Gr(77°F): 19.05 (Metal) Metal: Combustible Solid, especially turnings and powder.	VP: 0 mm MLT: 2097°F UEL: ? LEL: ?	Carbon dioxide, carbon tetrachloride, nitric acid, fluorine [Note: Complete coverage of uranium metal scrap with oil is essential for prevention of fire.]	None available
Uranium (soluble compounds as U)	Synonyms vary depending upon the specific soluble compound.	NIOSH Ca See Appendix A 0.05 mg/m³ OSHA 0.05 mg/m³	Ca [20 mg/m³]	Appearance and odor vary depending upon the specific soluble compound.	Properties vary depending upon the specific soluble compound.		Uranyl nitrates: combustibles. Uranium hexafluoride: water.	None available
Vanadium pentoxide (respirable dust as V₂O₅) V₂O₅ 1314-62-1 YW2450000 2862 55	Divanadium pentoxide, Vanadic anhydride, Vanadium oxide, Vanadium pentaoxide	NIOSH C 0.05 mg/m³ [15-min] OSHA 0.05 mg/m³	70 mg/m³	Yellow-orange powder or dark gray, odorless flakes dispersed in air.	MW: 181.9 BP: 3182°F (Decomposes) Sol: 0.8% Fl.P: NA IP: NA Sp.Gr: 3.36 Noncombustible Solid, but may increase intensity of fire when in contact with combustible materials.	VP: 0 mm (approx) MLT: 1274°F UEL: NA LEL: NA	Lithium, chlorine trifluoride	Filter; THF; XRD; III [#7504]

Chemical name, structure/formula, CAS and RTECS Nos., and DOT ID and guide Nos.	Synonyms, trade names, and conversion factors	Exposure limits (TWA unless noted otherwise)	IDLH	Physical description	Chemical and physical properties		Incompatibilities and reactivities	Measurement method (See Table 1)
					MW, BP, SOL Fl.P, IP, Sp.Gr, flammability	VP, FRZ UEL, LEL		
Xylenes (o-, m-, p- isomers) C₆H₄(CH₃)₂ 1330-20-7 ZE2100000 1307 27	o: 1,2-Dimethylbenzene; o-Xylol, m: 1,3-Dimethylbenzene; m-Xylol, p: 1,4-Dimethylbenzene; p-Xylol 1 ppm = 4.41 mg/m³	NIOSH/OSHA 100 ppm (435 mg/m³) ST 150 ppm (655 mg/m³)	1000 ppm	Colorless liquids with an aromatic odor. [Note: Pure p-xylene is a solid below 56°F.]	MW: 106.2 BP: 292/269/ 281°F Sol: Insoluble Fl.P: 63/84/ 81°F IP: 8.56/8.56/8.44 eV Sp.Gr: 0.88/0.86/0.86 Class IB Flammable Liquid (o) Class IC Flammable Liquids (m,p)	VP: 7/9/9 mm FRZ: -13/ -54/56°F UEL: 7.0/7.0/7.0% LEL: 1.1/1.0/1.1%	Strong oxidizers	Char; CS₂; GC/FID; III [#1501, Aromatic Hydro-carbons]
Xylidine (CH₃)₂C₆H₃NH₂ 1300-73-8 ZE8575000 1711 55	Aminodimethylbenzene, Aminoxylene, Dimethylaminobenzene, Dimethylaniline, Xylidine isomers, Xylidine (mixed o-, m-, p-) 1 ppm = 5.04 mg/m³	NIOSH/OSHA 2 ppm (10 mg/m³) [skin]	150 ppm	Pale-yellow to brown liquid with a weak, aromatic, amine-like odor.	MW: 121.2 BP: 415-439°F Sol: Slight Fl.P: 206°F IP: ? Sp.Gr: 0.98 Class IIIB Combustible Liquid	VP: <1 mm FRZ: ? UEL: ? LEL: 1.0%	Strong oxidizers, hypochlorite bleaches	Si gel; Ethanol; GC/FID; III [#2002]
Yttrium compounds (as Y) Y 7440-65-5 (Metal) ZG2980000 (Metal)	Metal: Yttrium metal Synonyms of other compounds vary depending upon the specific compound.	NIOSH/OSHA 1 mg/m³	N.E.	Metal: Dark gray to black solid.	MW: 88.9 BP: 5301°F Sol: ? Fl.P: NA IP: NA Sp.Gr: 4.47 Noncombustible Solid in bulk form.	VP: 0 mm (approx) MLT: 2732°F UEL: NA LEL: NA	Oxidizers	Filter; Acid; ICP; III [#7300, Elements]
Zinc chloride fume ZnCl₂ 7646-85-7 ZH1400000	None	NIOSH/OSHA 1 mg/m³ ST 2 mg/m³	4800 mg/m³	White particulate dispersed in air.	MW: 136.3 BP: 1350°F Sol(77°F): 432% Fl.P: NA IP: NA Sp.Gr(77°F): 2.91 Noncombustible Solid	VP: 0 mm (approx) MLT: 554°F UEL: NA LEL: NA	Potassium	None available

Personal protection and sanitation (See Table 3)		Recommendations for respirator selection — maximum concentration for use (MUC) (See Table 4)	Health hazards			
			Route	Symptoms (See Table 5)	First aid (See Table 6)	Target organs (See Table 5)
Clothing:	Repeat	NIOSH/OSHA	Inh	Irrit eyes, nose, throat;	Eye: Irr immed	Skin, eyes,
Goggles:	Reason prob	1000 ppm: CCRFOV/PAPROV$^£$	Abs	head, verti; hema, album;	Skin: Soap wash prompt	kidneys, resp sys
Wash:	Prompt contam	1500 ppm: SA:CF$^£$/GMFOV/SCBAF/SAF	Ing	skin irrit, sens	Breath: Resp support	
Change:	N.R.	§: SCBAF:PD,PP/SAF:PD,PP:ASCBA	Con		Swallow: Medical attention	
Remove:	Immed wet (flamm)	Escape: GMFOV/SCBAE			immed	
Clothing:	Repeat	NIOSH	Inh	Derm; [carc];	Eye: Irr immed	Skin, bone
Goggles:	Any poss	¥: SCBAF:PD,PP/SAF:PD,PP:ASCBA	Con	in animals: lung, lymph	Skin: Soap wash prompt	marrow,
Wash:	Prompt contam	Escape: HiEF/SCBAE	Ing	node damage; derm	Breath: Resp support	lymphatics
Change:	After work if reason prob contam				Swallow: Medical attention immed	
Remove:	Prompt non-imperv contam					
Provide:	Eyewash					
Clothing:	Any poss UF$_6$/Repeat	NIOSH	Inh	Lac, conj; short breath,	Eye: Irr immed	Resp sys, blood,
Goggles:	Any poss	¥: SCBAF:PD,PP/SAF:PD,PP:ASCBA	Con	cough, chest rales; nau,	Skin: Water flush immed	liver, kidneys,
Wash:	Prompt contam/UF$_6$: Immed contam and daily	Escape(Halides): GMFAGHiE/SCBAE	Ing	vomit; skin burns; RBC,	Breath: Resp support	lymphatics, skin,
Change:	After work if reason prob contam/UF$_6$: After work if any poss contam	Escape(Non-halides): HiEF/SCBAE		casts in urine; album; high BUN; [carc]	Swallow: Medical attention immed	bone marrow
Remove:	Prompt non-imperv contam/ UF$_6$: Immed non-imperv contam					
Provide:	Eyewash (UF$_6$),quick drench					
Clothing:	Repeat	NIOSH/OSHA	Inh	Irrit eyes; green tongue,	Eye: Irr immed	Resp sys, skin,
Goggles:	Reason prob	0.5 mg/m^3: HiE*/SA*/SCBA*	Ing	metallic taste, eczema;	Skin: Soap wash prompt	eyes
Wash:	Prompt contam	1.25 mg/m^3: SA:CF*/PAPRHiE*	Con	cough; fine rales, wheez,	Breath: Resp support	
Change:	N.R.	2.5 mg/m^3: HiEF/SCBAF/SAF/ PAPRTHiE*		bron, dysp; irrit throat	Swallow: Medical attention immed	
Remove:	Prompt non-imperv contam	70 mg/m^3: SAF:PD,PP				
		§: SCBAF:PD,PP/SAF:PD,PP:ASCBA				
		Escape: HiEF/SCBAE				

Personal protection and sanitation (See Table 3)		Recommendations for respirator selection — maximum concentration for use (MUC) (See Table 4)	Health hazards			
			Route	Symptoms (See Table 5)	First aid (See Table 6)	Target organs (See Table 5)
Clothing:	Repeat	NIOSH/OSHA	Inh	Dizz, excitement, drow,	Eye: Irr immed	CNS, eyes, GI
Goggles:	Reason prob	1000 ppm: CCROV*/PAPROV*/SA*/ SCBA*	Abs	inco, staggering gait;	Skin: Soap wash prompt	tract, blood,
Wash:	Prompt contam	§: SCBAF:PD,PP/SAF:PD,PP:ASCBA	Ing	irrit eyes, nose, throat;	Breath: Resp support	liver, kidneys,
Change:	N.R.	Escape: GMFOV/SCBAE	Con	corneal vacuolization;	Swallow: Medical attention immed	skin
Remove:	Immed wet (flamm)			anor, nau, vomit, abdom pain; derm		
Clothing:	Any poss	NIOSH/OSHA	Inh	Anoxia, cyan; lung, liver,	Eye: Irr immed	Blood, lungs,
Goggles:	Any poss	20 ppm: CCROV/SA/SCBA	Abs	kidney damage	Skin: Soap wash immed	liver, kidneys,
Wash:	Immed contam	50 ppm: SA:CF/PAPROV	Ing		Breath: Resp support	CVS
Change:	N.R.	100 ppm: CCRFOV/PAPRTOV/GMFOV/ SCBAF/SAF	Con		Swallow: Medical attention immed	
Remove:	Immed non-imperv contam	150 ppm: SA:PD,PP				
Provide:	Eyewash, quick drench	§: SCBAF:PD,PP/SAF:PD,PP:ASCBA				
		Escape: GMFOV/SCBAE				
Clothing:	N.R.	NIOSH/OSHA	Inh	Irrit eyes;	Eye: Irr immed	Eyes, lungs
Goggles:	Any poss	5 mg/m^3: DM	Ing	in animals: pulm irrit;	Skin: Soap wash prompt	
Wash:	N.R.	10 mg/m^3: DMXSQ/SA/SCBA	Con	eye inj; possible liver	Breath: Resp support	
Change:	N.R.	25 mg/m^3: PAPRDM/SA:CF		damage	Swallow: Medical attention immed	
Remove:	N.R.	50 mg/m^3: HiEF/PAPRTHiE/SAT:CF/ SCBAF/SAF				
Provide:	Eyewash	500 mg/m^3: SA:PD,PP				
		§: SCBAF:PD,PP/SAF:PD,PP:ASCBA				
		Escape: HiEF/SCBAE				
Clothing:	N.R.	NIOSH/OSHA	Inh	Conj; irrit nose, throat;	Breath: Resp support	Resp sys, skin,
Goggles:	N.R.	10 mg/m^3: DMFu*/SA*/SCBA*	Con	cough, copious sputum;		eyes
Wash:	N.R.	25 mg/m^3: PAPRDMFu*/SA:CF*		dysp, chest pain, pulm		
Change:	N.R.	50 mg/m^3: HiEF/PAPRTHiE*/SCBAF/SAF		edema, broncopneu; pulm		
Remove:	N.R.	2000 mg/m^3: SAF:PD,PP		fib, cor pulmonale; fever;		
		§: SCBAF:PD,PP/SAF:PD,PP:ASCBA		cyan; tachypnea; burn		
		Escape: HiEF/SCBAE		skin; irrit skin, eyes		

Chemical name, structure/formula, CAS and RTECS Nos., and DOT ID and guide Nos.	Synonyms, trade names, and conversion factors	Exposure limits (TWA unless noted otherwise)	IDLH	Physical description	Chemical and physical properties		Incompatibilities and reactivities	Measurement method (See Table 1)
					MW, BP, SOL FI.P, IP, Sp.Gr, flammability	VP, FRZ UEL, LEL		
Vanadium pentoxide (respirable fume as V₂O₅) V₂O₅ 1314-62-1 YW2460000 2862 55	Divanadium pentoxide, Vanadic anhydride, Vanadium oxide, Vanadium pentaoxide	NIOSH C 0.05 mg/m³ [15-min] OSHA 0.05 mg/m³	70 mg/m³	Finely divided particulate dispersed in air.	MW: 181.9 BP: 3182°F (Decomposes) Sol: 0.1% Fl.P: NA IP: NA Sp.Gr: 3.36 Noncombustible Solid	VP: 0 mm (approx) MLT: 1274°F UEL: NA LEL: NA	Lithium, chlorine trifluoride	Filter; THF; XRD; III [#7504]
Vinyl chloride CH₂=CHCl 75-01-4 KU9625000 1086 17	Chloroethene, Chloroethylene, Ethylene monochloride, Monochloroethene, Monochloroethylene, VC, Vinyl chloride monomer (VCM) 1 ppm = 2.60 mg/m³	NIOSH Ca See Appendix A Lowest reliably detectable concentration. OSHA [1910.1017] 1 ppm C 5 ppm [15-min]	Ca ACGIH A1	Colorless gas or liquid (below 56°F) with a pleasant odor at high concentrations. [Note: Shipped as a liquefied compressed gas.]	MW: 62.5 BP: 7°F Sol(77°F): 0.1% Fl.P: NA (Gas) IP: 9.99 eV Flammable Gas	VP: >1 atm FRZ: -256°F UEL: 33.0% LEL: 3.6%	Copper, oxidizers, aluminum, peroxides, iron, steel [Note: Polymerizes in air, sunlight, or heat unless stabilized by inhibitors such as phenol. Attacks iron & steel in presence of moisture.]	Char(2); CS₂; GC/FID; III [#1007]
Vinyl toluene CH₂=CHC₆H₄CH₃ 25013-15-4 WL5075000 2618 27	Ethenylmethylbenzene, Methylstyrene, Tolyethylene 1 ppm = 4.91 mg/m³	NIOSH/OSHA 100 ppm (480 mg/m³) ACGIH 50 ppm (240 mg/m³)	5000 ppm	Colorless liquid with a strong, disagreeable odor.	MW: 118.2 BP: 339°F Sol: 0.009% Fl.P: 120°F IP: 8.20 eV Sp.Gr: 0.89 Class II Combustible Liquid	VP: 1 mm FRZ: -106°F UEL: 11.0% LEL: 0.8%	Oxidizers, peroxides, strong acids, iron or aluminum salts [Note: Usually inhibited with tert-butyl catechol to prevent polymerization.]	Char; CS₂; GC/FID; III [#1501, Aromatic Hydrocarbons]
Warfarin C₁₉H₁₆O₄ 81-81-2 GN4550000 3027 55 (Coumarin derivative pesticide)	3-(alpha-Acetonyl)-benzyl-4-hydroxycoumarin, 4-Hydroxy-3-(3-oxo-1-phenyl butyl)-2H-1-benzopyran-2-one, WARF	NIOSH/OSHA 0.1 mg/m³	350 mg/m³	Colorless, odorless, crystalline powder. [pesticide]	MW: 308.3 BP: Decomposes Sol: 0.002% Fl.P: ? IP: ? Combustible Solid	VP(71°F): 0.09 mm MLT: 322°F UEL: ? LEL: ?	Strong oxidizers	Filter; Methanol; HPLC/UVD; III [#5002]

Chemical name, structure/formula, CAS and RTECS Nos., and DOT ID and guide Nos.	Synonyms, trade names, and conversion factors	Exposure limits (TWA unless noted otherwise)	IDLH	Physical description	Chemical and physical properties		Incompatibilities and reactivities	Measurement method (See Table 1)
					MW, BP, SOL FI.P, IP, Sp.Gr, flammability	VP, FRZ UEL, LEL		
Zinc oxide fume ZnO 1314-13-2 ZH4810000	None	NIOSH/OSHA 5 mg/m³ ST 10 mg/m³	N.E.	Fine white, odorless particulate dispersed in air.	MW: 81.4 BP: ? Sol: Insoluble Fl.P: NA IP: NA Sp.Gr: 5.61 Noncombustible Solid	VP: 0 mm (approx) MLT: 3587°F UEL: NA LEL: NA	Chlorinated rubber (at 419°F)	Filter; none; XRD; III [#7502]
Zirconium compounds (as Zr) Zr 7440-67-7 (Metal) ZH7070000 (Metal) 2008 37 (metal powder, dry) 1358 32 (metal powder, wet)	Metal: Zirconium metal Synonyms of other compounds vary depending upon the specific compound.	NIOSH 5 mg/m³ ST 10 mg/m³ [Zr compounds excluding Zirconium tetrachloride.] OSHA 5 mg/m³ ST 10 mg/m³	500 mg/m³	Metal: Soft, malleable, ductile, solid or gray to gold amorphous powder.	MW: 91.2 BP: 6471°F Sol: Insoluble Fl.P: NA IP: NA Sp.Gr: 6.51 Metal: Combustible, but solid form is difficult to ignite; however, powder form may ignite SPONTANEOUSLY and can continue burning under water.	VP: 0 mm (approx) MLT: 3375°F UEL: NA LEL: NA	Potassium nitrate, oxidizers [Note: Fine powder may be stored completely immersed in water.]	Filter Acid; ICP; III [#7300, Elements]

Personal protection and sanitation (See Table 3)	Recommendations for respirator selection — maximum concentration for use (MUC) (See Table 4)	Route	Symptoms (See Table 5)	First aid (See Table 6)	Target organs (See Table 5)
Clothing: N.R. Goggles: N.R. Wash: N.R. Change: N.R. Remove: N.R.	NIOSH/OSHA 0.5 mg/m³: HiE*/SA*/SCBA* 1.25 mg/m³: SA:CF*/PAPRHiE* 2.5 mg/m³: HiEF/SCBAF/SAF/ PAPRTHiE* 70 mg/m³: SAF:PD,PP §: SCBAF:PD,PP/SAF:PD,PP:ASCBA Escape: HiEF/SCBAE	Inh Con	Irrit eye; green tongue, metallic taste; irrit throat, cough, fine rales, wheez, bron, dysp; eczema	Breath: Resp support	Resp sys, skin, eyes
Clothing: Any poss Goggles: Any poss Wash: Immed contam/daily Change: After work if any poss contam Remove: Immed contam Provide: Eyewash, quick drench	NIOSH ¥: SCBAF:PD,PP/SAF:PD,PP:ASCBA Escape: GMFS/SCBAE	Inh	Weak; abdom pain, GI bleeding; hepatomegaly; pallor or cyan of extremities; [carc]	Breath: Resp support	Liver, CNS, blood, resp sys, lymphatic sys
Clothing: Repeat Goggles: Reason prob Wash: Prompt contam Change: N.R. Remove: Prompt non-imperv contam	ACGIH 500 ppm: CCROV*/SA*/SCBA* 1000 ppm: PAPROV*/CCRFOV 1250 ppm: SA:CF* 2500 ppm: GMFOV/SCBAF/SAF 5000 ppm: SAF:PD,PP §: SCBAF:PD,PP/SAF:PD,PP:ASCBA Escape: GMFOV/SCBAE	Inh Con Ing	Irrit eyes, skin, upper resp sys; drow	Eye: Irr immed Skin: Soap flush prompt Breath: Resp support Swallow: Medical attention immed	Eyes, skin, resp sys
Clothing: Repeat Goggles: N.R. Wash: Prompt contam Change: After work if any poss contam Remove: Prompt non-imperv contam	NIOSH/OSHA 0.5 mg/m³: DM 1 mg/m³: DMXSQ/SA/SCBA 2.5 mg/m³: PAPRDM/SA:CF 5 mg/m³: HiEF/PAPRTHiE/SCBAF/SAF/ SAT:CF 100 mg/m³: SA:PD,PP 200 mg/m³: SAF:PD,PP §: SCBAF:PD,PP/SAF:PD,PP:ASCBA Escape: HiEF/SCBAE	Inh Abs Ing Con	Hema, back pain; hematoma arms, legs; epis, bleeding lips, muc memb hemorr; abdom pain, vomit, fecal blood; petechial rash; abnormal hematologic indices	Eye: Irr immed Skin: Soap wash prompt Breath: Resp support Swallow: Medical attention immed	Blood, CVS

Personal protection and sanitation (See Table 3)	Recommendations for respirator selection — maximum concentration for use (MUC) (See Table 4)	Route	Symptoms (See Table 5)	First aid (See Table 6)	Target organs (See Table 5)
Clothing: N.R. Goggles: N.R. Wash: N.R. Change: N.R. Remove: N.R.	NIOSH/OSHA 50 mg/m³: DMFu/SA/SCBA 125 mg/m³: PAPRDMFu/SA:CF 250 mg/m³: HiEF/PAPRTHiE/SAT:CF SCBAF/SAF 2500 mg/m³: SA:PD,PP §: SCBAF:PD,PP/SAF:PD,PP:ASCBA Escape: HiEF/SCBAE	Inh	Sweet, metallic taste; dry throat, cough; chills, fever; tight chest, dysp, rales, reduced pulm func; head; blurred vision; musc cramps, low back pain; nau, vomit; ftg, lass, mal	Breath: Resp support	Resp sys
Recommendations vary depending upon the specific compound.	NIOSH/OSHA 25 mg/m³: DM 50 mg/m³: DMXSQ/SA/SCBA 125 mg/m³: PAPRDM/SA:CF 250 mg/m³: HiEF/PAPRTHiE/SAT:CF/ SCBAF/SAF 500 mg/m³: SA:PD,PP §: SCBAF:PD,PP/SAF:PD,PP:ASCBA Escape: HiEF/SCBAE	Inh Con	Skin granulomas; in animals: X-ray evidence of retention in lungs; irrit skin, muc memb	Eye: Irr immed Skin: Soap wash Breath: Resp support Swallow: Medical attention immed	Resp sys, skin

NIOSH OCCUPATIONAL CARCINOGENS

NIOSH has identified numerous chemicals that should be treated as occupational carcinogens even though OSHA has not identified them as such. In determining their carcinogenicity, NIOSH uses a classification outlined in 29 CFR 1990.103, which states in part:

> Potential occupational carcinogen means any substance, or combination or mixture of substances, which causes an increased incidence of benign and/or malignant neoplasms, or a substantial decrease in the latency period between exposure and onset of neoplasms in humans or in one or more experimental mammalian species as the result of any oral, respiratory or dermal exposure, or any other exposure which results in the induction of tumors at a site other than the site of administration. This definition also includes any substance which is metabolized into one or more potential occupational carcinogens by mammals.

NIOSH has not identified thresholds for carcinogens that will protect 100% of the population. NIOSH usually recommends that occupational exposures to carcinogens be limited to the lowest feasible concentration. To ensure maximum protection from carcinogens through the use of respiratory protection, only the most reliable and protective respirators are recommended. These include (1) a self-contained breathing apparatus (SCBA) that has a full facepiece and is operated in a positive-pressure mode, or (2) a supplied-air respirator that has a full facepiece and is operated in a pressure-demand or other positive-pressure mode in combination with an auxiliary SCBA operated in a pressure-demand or other positive-pressure mode.

THIRTEEN OSHA-REGULATED CARCINOGENS

Without establishing PELs, OSHA promulgated standards in 1974 to regulate the industrial use of 13 chemicals identified as occupational carcinogens (2-acetylaminofluorene, 4-aminodiphenyl, benzidine, bis-chloromethyl ether, 3,3'-dichlorobenzidine, 4-dimethylaminoazobenzene, ethyleneimine, methyl chloromethyl ether, alpha-naphthylamine, beta-naphthylamine, 4-nitrobiphenyl, N-nitrosodimethylamine, and beta-propiolactone). Exposures of workers to these 13 chemicals are to be controlled through the required use of engineering controls, work practices, and personal protective equipment, including respirators. See 29 CFR 1910.1003–1910.1016 for specific details of these requirements.

Respirator selections in the *Pocket Guide* are based on NIOSH policy, which considers the 13 chemicals to be potential occupational carcinogens.

SUPPLEMENTARY EXPOSURE LIMITS

Asbestos

NIOSH considers asbestos to be a carcinogen and recommends that exposures be reduced to the lowest possible concentration. For asbestos fibers >5 micrometers in length, NIOSH recommends an REL of 100,000 fibers per cubic meter of air (100,000 fibers/m³), which is equal to 0.1 fiber per cubic centimeter of air (0.1 fiber/cm³), as determined by a 400-liter air sample and NIOSH Analytical Method #7400.

The OSHA PEL for asbestos fibers is an 8-hour TWA airborne concentration of 0.2 fiber (longer than 5 micrometers and having a length-to-diameter ratio of at least 3 to 1) per cubic centimeter of air (0.2 fiber/cm³), as determined by the membrane filter method at approximately 400X magnification with phase contrast illumination. An "action level" of 0.1 fiber/cm³ as an 8-hour TWA was established as the concentration above which employers must initiate compliance activities such as worker training and medical surveillance.

ACGIH designates asbestos as a confirmed human carcinogen (A1) with the following TLVs: amosite, 0.5 fiber/cm³; chrysotile, 2 fibers/cm³; crocidolite, 0.2 fiber/cm³; and other forms, 2 fibers/cm³. ACGIH considers "fibers" to be those particles >5 micrometers in length with an aspect ratio equal to or greater than 3:1 as determined by the membrane filter method at 400 to 450X magnification (4-mm objective) with phase contrast illumination.

Chromic Acid and Chromates (as CrO₃), Chromium(II) and Chromium(III) Compounds (as Cr), and Chromium Metal (as Cr)

The NIOSH REL (10-hour TWA) is 0.001 mg Cr(VI)/m³ for all hexavalent chromium [Cr(VI)] compounds. NIOSH considers all Cr(VI) compounds (including chromic acid) to be potential occupational carcinogens. The NIOSH REL (8-hour TWA) is 0.5 mg Cr/m³ for chromium metal and chromium(II) and chromium(III) compounds.

The OSHA PEL is 0.1 mg CrO₃/m³ (ceiling) for chromic acid and chromates; 0.5 mg Cr/m³ (8-hour TWA) for chromium(II) and chromium(III) compounds; and 1 mg Cr/m³ (8-hour TWA) for chromium metal.

ACGIH has various TLVs for chromium compounds. The 8-hour TWA is 0.5 mg Cr/m³ for chromium metal, for chromium(II) compounds (e.g., chromous chloride and chromous sulfate), and for chromium(III) compounds (e.g., chromic oxide, chromic sulfate, chromic chloride, chromic potassium sulfate, and chromite ore). The 8-hour TWA is 0.05 mg Cr(VI)/m³ for water-soluble Cr(VI) compounds, which include chromic acid and its anhydride, and the monochromates and dichromates of sodium,

SUPPLEMENTARY EXPOSURE LIMITS (Continued)

potassium, ammonium, lithium, cesium, and rubidium. The 8-hour TWA is 0.05 mg Cr/m³ for certain water-insoluble Cr(VI) compounds that are designated as confirmed human carcinogens (A1) (including zinc chromate, calcium chromate, lead chromate, barium chromate, strontium chromate, and sintered chromium trioxide).

Coal Tar Pitch Volatiles

The OSHA PEL (8-hour TWA) for coal tar pitch volatiles is 0.2 mg/m³ (benzene-soluble fraction). OSHA defines "coal tar pitch volatiles" as the fused polycyclic hydrocarbons that volatilize from the distillation residues of coal, petroleum, wood, and other organic matter such as anthracene, benzo(a)pyrene (BaP), phenanthrene, acridine, chrysene, pyrene, etc. Asphalt is not covered under the OSHA standard for coal tar pitch volatiles.

NIOSH considers coal tar products (i.e., coal tar, coal tar pitch, or creosote) to be carcinogenic (Ca); the NIOSH REL (10-hour TWA) for coal tar products is 0.1 mg/m³ (cyclohexane-extractable fraction).

ACGIH designates coal tar pitch volatiles as a confirmed human carcinogen (A1) with an 8-hour TWA of 0.2 mg/m³ (benzene-solubles).

Cotton Dust (raw)

The NIOSH REL for cotton dust is 0.200 mg/m³ (lint-free cotton dust).

As found in the OSHA Table Z-1-A (29 CFR 1910.1000), the PEL for cotton dust (raw) is 1 mg/m³ for the cotton waste processing operations of waste recycling (sorting, blending, cleaning, and willowing) and garretting. PELs for other sectors (as found in 29 CFR 1910.1043) are 0.200 mg/m³ for yarn manufacturing and cotton washing operations, 0.500 mg/m³ for textile mill waste house operations or for dust from "lower grade washed cotton" used during yarn manufacturing, and 0.750 mg/m³ for textile slashing and weaving operations. The OSHA standard in 29 CFR 1910.1043 does not apply to cotton harvesting, ginning, or the handling and processing of woven or knitted materials and washed cotton. All PELs for cotton dust are mean concentrations of lint-free, respirable cotton dust collected by the vertical elutriator or an equivalent method and averaged over an 8-hour period.

The ACGIH TLV for cotton dust (raw) is 0.200 mg/m³ (lint-free dust) as an 8-hour TWA measured by the vertical elutriator cotton-dust sampler.

SUBSTANCES WITH NO ESTABLISHED RELs

After reviewing available published literature, NIOSH provided comments to OSHA on August 1, 1988, regarding the "Proposed Rule on Air Contaminants" (29 CFR 1910, Docket No. H-020). In these comments, NIOSH questioned whether the PELs proposed for the following substances included in the *Pocket Guide* were adequate to protect workers from recognized health hazards: acetylene tetrabromide, chlorobenzene, ethyl bromide, ethyl chloride, ethyl ether, furfural, isopropyl acetate, isopropylamine, molybdenum (soluble compounds as Mo), nitromethane, and silica (amorphous). At that time, NIOSH also conducted a limited evaluation of the literature on magnesium oxide fume and molybdenum (insoluble compounds as Mo) and concluded that the documentation cited by OSHA was inadequate to support the proposed PELs of 10 mg/m³.

Appendix L
TOXICOLOGY REVIEW

This abstract of Chapter 7 (Industrial Toxicology) from the NIOSH textbook, *The Industrial Environment: Its Evaluation and Control* is presented in this appendix to aquaint the hazardous waste worker with toxicology terminology and concepts and to meet the requirements of OSHA 1910.120 HAZWOPER.

INTRODUCTION

Toxicology is the study of the nature and action of poisons. Since about 1900, there has been increasing social concern for the health of workers exposed to a diversity of chemicals. This has led to intensive investigations of the toxicity of these materials in order that proper precautions may be taken in their use. This is the area of industrial toxicology that will be briefly reviewed in this appendix.

French hatters of the 17th century discovered that mercuric nitrate aided greatly in the felting of fur. Such use led to chronic mercury poisoning so widespread among members of that trade that the expression "mad as a hatter" entered our folk language. Exposure to other hazardous substances is an outgrowth of modern technology. In addition to newly developed chemicals, many materials first synthesized in the late 19th century have found widespread industrial use.

Toxicological research now has its place in assessing the safety of new chemicals prior to the extension of their use beyond exploratory stages. Information on the qualitative and quantitative actions of a chemical in the body can be used to predict tentative safe levels of exposure as well as to predict the signs and symptoms to be watched for as indicative of excessive exposure. An example of the mechanism of action of the chemical can hopefully lead to rational rather than symptomatic therapy in the event of damage from excessive exposure. Both in the application of newer refined research techniques of toxicology and in his or her communication of knowledge vital to the public health, the toxicologist considers old as well as new hazardous substances. Most people are careful in handling a new chemical whether or not they have been warned specifically of its possible toxicity. Despite the potential hazards of thousands of new chemicals each year most injuries from chemicals are due to those that have been familiar for a generation or more. It is important for the perspective of the toxicologist that she keep this fact well to the forefront of her mind. She must not neglect talking about the hazards of the old standbys, lead, benzene and chlorinated hydrocarbons, just because this week she discovered the horrifying action of something brand new.

DISCIPLINES INVOLVED IN INDUSTRIAL TOXICOLOGY

In order to assess the potential hazard of a substance to the heatlh of workers industrial toxicology draws on the expertise of many disciplines.

Chemistry: The chemical properties of a compound can often be one of the main factors in its toxicity. The vapor pressure indicates whether or not a given substance has the potential to pose a hazard from inhalation. The solubility of a substance in aqueous and lipid media is a guiding factor in determining the rate of uptake and excretion of inhaled substances. The toxicologist needs to determine the concentration of toxic agents in air and in body organs and fluids. It is important to know if a substance is, for example, taken up by the liver, stored in the bones, or rapidly excreted. For this, analytical methods are needed that are both sensitive and specific.

Biochemistry: The toxicologist needs knowledge of the pathways of metabolism of foreign compounds in the body. Such information can serve as the basis for monitoring the exposure of workers, as for example the assessment of benzene exposure by the analysis of phenol in the urine. Differences in metabolic pathways among animal species form one basis for selective toxicity. Such knowledge is useful, for example, in developing compounds that will be maximally toxic to insects and minimally so to other species. Such knowledge can also serve as a guide in the choice of a species of experimental animal with a metabolic pathway similar to that of man for studies that will be extrapolated to predict safe levels for human exposure. Rational therapy for injury from toxic chemicals has as its basis an understanding of the biochemical changes they produce. In the important area of joint toxic action, understanding comes from *elucidation* of biochemical action. If, for example, Compound A induces enzymes that serve to detoxify Compound B, the response to the combination may be less than additive. On the other hand, if Compound A should act to inhibit the enzyme that serves to detoxify Compound B then the response may be more than additive.

Physiology: The toxicologist needs to know something of the normal functioning of organ systems. Toxicology searches for means to detect reversible physiological changes produced by concentrations of toxic substances too low to produce irreversible histological damage or death in experimental animals. Measurement of increases in pulmonary flow-resistance has proved a sensitive tool for assessing the response to irritant gases and aerosols. Tests of pulmonary function can be used to assess response of workers to industrial environments. Renal clearance and other kidney function tests can serve to detect renal damage. The effects of exercise or nonspecific stress on the degree of response to toxic chemicals is another important research area in toxicology.

Pathology: The toxicologist is concerned with gross and *histological* damage caused by toxic substances. Most toxicological studies include a pathological examination that may include examination of subcellular structure by electron microscopy.

Immunology and Immunochemistry: It is recognized that immunology and immunochemistry constitute an important area for investigation. The response to many chemicals, especially inhaled products of biologic origin, has as its basis the immune reaction.

Physics and Engineering: The toxicologist who is concerned with inhalation as the route of exposure needs some knowledge of physics and engineering in order to establish controlled concentrations of the substances he studies. If the toxic materials are to be administered as airborne particles, knowledge is needed of methods of generation of aerosols and methods of sampling and sizing appropriate to the material studied.

An understanding of the factors governing penetration, deposition, retention, and clearance of particulate material from the respiratory tract requires knowledge of both the physical laws governing aerosol behavior and the anatomy and physiology of the respiratory tract. The interest in prolonged exposure to closed atmospheres encountered in manned space travel or deep sea exploration led to experimental studies involving round-the-clock exposures of experimental animals for long periods.

Statistics: The toxicologist relies heavily on statistics, as the calculation of the LD_{50} (Lethal Dose—50% probable) is a statistical calculation.

Communication: The ultimate aim of the toxicologist is the prevention of damage to people and the environment by toxic agents. One important function is the distribution of information in such terms that the people in need of the information will understand it. The toxicologist is called on to make value judgments in extrapolation of her findings in order to advise governmental agencies and others faced with the problem of setting safe levels, be they air pollution standards or Threshold Limit Values for industrial exposure or tolerance levels of pesticide residues in food. They are value judgments and as such should be subject to frequent review as new knowledge and experience accumulate.

DOSE-RESPONSE RELATIONSHIPS

Experimental toxicology is in essence biological assay with the concept of a dose-response relationship as its unifying theme. The potential toxicity (harmful action) inherent in a substance is manifest only when that substance comes in contact with a living biological system. A chemical normally thought of as harmless will evoke a toxic response if added to a biological system in sufficient amount. For example, the inadvertent inclusion of large amounts of sodium chloride in feeding formulae in a hospital nursery led to infant mortality. Conversely, for a chemical normally thought of as toxic there is a minimal concentration that will produce no toxic effect if added to a biological system. The toxic potency of a chemical is thus ultimately defined by the relationship between the dose (the amount) of the chemical and the response that is produced in a biological system.

In preliminary toxicity testing, death of the animals is the response most commonly chosen. Given a compound with no known toxicity data, the initial step is one of range finding. A dose is administered and, depending on the outcome, is increased or decreased until a critical range is found over which, at the upper end, all animals die and, at the lower end, all animals survive. Between these extremes is the range in which the toxicologist accumulates data that enable him to prepare a dose-response curve relating percent mortality to dose administered.

From the dose-response curve, the dose that will produce death in 50% of the animals may be calculated. This value is commonly abbreviated as LD_{50}. It is a statistically obtained value representing the best estimation that can be made from the experimental data at hand. The dose is expressed as amount per unit of body weight. The value should be accompanied by an indication of the species of experimental animal used, the route of administration of the compound, the vehicle used to dissolve or suspend the material if applicable, and the time period over which the animals were observed. For example, a publication might state "For rats,

the 24 hr. ip LD_{50} for 'X' in corn oil was 66 mg/kg (95% confidence limits 59–74).'' This would indicate to the reader that the material was given to rats as an intraperitoneal injection of compound X dissolved or suspended in corn oil and that the investigator had limited her mortality count to 24 hours after administering the compound. If the experiment has involved inhalation as the route of exposure, the dose to the animals is expressed as parts per million, mg/m^3, or some other appropriate expression of concentration of the material in the air of the exposure chamber, and the length of exposure time is specified. In this case the term LC_{50} is used to designate the concentration in air that may be expected to kill 50% of the animals exposed for the specified length of time.

The simple determination of the LD_{50} for an unknown compound provides an initial comparative index for the location of the compound in the overall spectrum of toxic potency. Table L.1 shows an attempt at utilizing LD_{50} and LC_{50} values to set up an approximate classification of toxic substances.

Over and above the specific LD_{50} value, the slope of the dose-response curve provides useful information. It suggests an index of the margin of safety, that is the magnitude of the range of doses involved in going from a noneffective dose to a lethal dose. It is obvious that if the dose-response curve is very steep, this margin of safety is slight. Another situation may arise in which one compound would be rated as more toxic than a second compound if the LD_{50} values alone were compared, but the reverse assessment of relative toxicity would be reached if the comparison was made of the LD_5 values for the two compounds because the dose-response curve for the second compound had a more gradual slope. It should thus be apparent that the slope of the dose-response curves may be of considerable significance with respect to establishing relative toxicities of compounds.

By similar experiments dose-response curves may be obtained using a *criterion* other than mortality as the response and an ED_{50} value is obtained. This is the dose that produced the chosen response in 50% of the treated animals. When the study of a toxic substance progresses to the point at which its action on the organism may be studied as graded response in groups of animals, dose-response curves of a slightly different sort are generally used. One might see, for example, a dose-response curve relating the degree of depression of brain choline esterase to the dose of an organic phosphorus ester or a dose-response curve relating the increase in pulmonary flow-resistance to the concentration of sulfur dioxide inhaled.

TABLE L.1 Toxicity Classes

Toxicity Rating	Descriptive Term	LD_{50}-Wt/kg Single Oral Dose Rats	4 hr Inhalation LC_{50}-PPM Rats
1	Extremely toxic	1 mg or less	<10
2	Highly toxic	1–50 mg	10–100
3	Moderately toxic	50–500 mg	100–1,000
4	Slightly toxic	0.5–5 g	1,000–10,000
5	Practically nontoxic	5–15 g	10,000–100,000
6	Relatively harmless	15 g or more	>100,000

ROUTES OF EXPOSURE

Toxic chemicals can enter the body by various routes. The most important route of exposure in industry is inhalation. Next in importance is contact with skin and eyes. The response to a given dose of toxic agent may vary markedly depending on the route of entry. *The intensity of toxic action is a function of the concentration of the toxic agent that reaches the site of action.* The route of exposure can have an influence on the concentration reaching the site of action.

Parenteral: Aside from the obvious use in administration of drugs, injection is considered mainly as a route of exposure of experimental animals. In the case of injection, the dose administered is known with accuracy. Intravenous (iv) injection introduces the material directly into the circulation, hence comparison of the degree of response to iv injection with the response to the dose administered by another route can provide information on the rate of uptake of the material by the alternate route. When a material is administered by injection, the highest concentration of the toxic material in the body occurs at the time of entrance. The organism receives the initial impact at the maximal concentration without opportunity for a gradual reaction, whereas if the concentration is built up more gradually by some other route of exposure, the organism may have time to develop some resistance or physiological adjustment that could produce a modified response. In experimental studies intraperitoneal (ip) injection of the material into the abdominal fluid is a frequently used technique. The major venous blood circulation from the abdominal contents proceeds via the portal circulation to the liver. A material administered by ip injection is subject to the special metabolic transformation mechanisms of the liver, as well as the possibility of excretion via the bile before it reaches the general circulation. If the LD_{50} of a compound given by ip injection was much higher (i.e., the toxicity is lower) than the LD_{50} by iv injection, this fact would suggest that the material was being detoxified by the liver or that the bile was a major route of excretion of the material. If the values for LD_{50} were very similar for ip and iv injection, it would suggest that neither of these factors played a major role in the handling of that particular compound by that particular species of animal.

Oral: Ingestion occurs as a route of exposure of workers through eating or smoking with contaminated hands or in contaminated work areas. Ingestion of inhaled material also occurs. One mechanism for the clearance of particles from the respiratory tract is the carrying up of the particles by the action of the ciliated lining of the respiratory tract. These particles are then swallowed and absorption of the material may occur from the gastrointestinal tract. This situation is most likely to occur with larger size particles (2μ and up) although smaller particles deposited in the alveoli may be carried by phagocytes to the upward moving mucous carpet and eventually be swallowed.

In experimental work, compounds may be administered orally as either a single or multiple dose given by stomach tube or the material may be incorporated in the diet or drinking water for periods varying from several weeks or months up to several years or the lifetime of the animals. In either case, the dose the animals actually receive may be ascertained with considerable accuracy. Except in the case of a substance that has a corrosive action or in some way damages the lining of the gastrointestinal tract, the response to a substance administered orally will depend on how readily it is absorbed from the gut. Uranium, for example, is capable of producing kidney damage, but is poorly absorbed from the gut and so oral administration produces only low concentrations at the site of action. On the other hand, ethyl alcohol, which has as a target organ the central nervous system, is very rapidly absorbed and within an hour 90% of an ingested dose has been absorbed.

The epithelium of the gastrointestinal tract is poorly permeable to the ionized form of organic compounds. Absorption of such materials generally occurs by diffusion of the lipid-soluble nonionized form. Weak acids that are predominantly nonionized in the high acidity (pH 1.4) of gastric juice are absorbed from the stomach. The surface of the intestinal mucosa has a pH of 5.3. At this higher pH weak bases are less ionized and more readily absorbed. The pH of a compound thus becomes an important factor in predicting absorption from the gastrointestinal tract.

Inhalation: Inhalation exposures are of prime importance to the industrial toxicologist. The dose actually received and retained by the animals is not known with the same accuracy as when a compound is given by the routes previously discussed. This depends on the ventilation rate of the individual. In the case of a gas, it is influenced by solubility and in the case of an aerosol by particle size. The concentration and time of exposure can be measured and this gives a working estimate of the exposure. Two techniques are sometimes utilized in an attempt to determine the dose with precision and still administer the compound via the lung. One is intratracheal injection, which may be used in some experiments in which it is desirable to deliver a known amount of particulate material into the lung. The other is so-called precision gassing. In this technique the animal or experimental subject breathes through a valve system and the volume of exhaled air and the concentration of toxic material in it are determined. A comparison of these data with the concentration in the atmosphere of the exposure chamber gives an indication of the dose retained.

Cutaneous: Cutaneous exposure ranks first in the production of occupational disease, but not necessarily first in severity. The skin and its associated film of lipids and sweat may act as an effective barrier. The chemical may react with the skin surface and cause primary irritation. The agent may penetrate the skin and cause sensitization to repeated exposure. The material may penetrate the skin in an amount sufficient to cause systemic poisoning. In assessing the toxicity of a compound by skin application, a known amount of the material to be studied is placed on the clipped skin of the animal and held in place with a rubber cuff. Some materials such as acids, alkalis, and many organic solvents are primary skin irritants and produce skin damage on initial contact. Other materials are sensitizing agents. The initial contact produces no irritant response, but may render the individual sensitive and dermatitis may result from future contact. Ethyleneamines and the catechols in the well-known members of the Rhus family (poison ivy and poison oak) are examples of such agents. The physiochemical properties of a material are the main determinant of whether or not a material will be absorbed through the skin. Among the important factors are pH, extent of ionization, water and lipid solubility, and molecular size. Some compounds, such as phenol and phenolic derivatives, can readily penetrate the skin in amounts sufficient to produce systemic intoxication. If the skin is damaged, the normal protective barrier to absorption of chemicals is lessened and penetration may occur. An example of this is a description of cases of mild lead intoxication that occurred in an operation that involved an inorganic lead salt and also a cutting oil. Inorganic lead salts would not be absorbed through intact skin, but the dermatitis produced by the cutting oil permitted increased absorption.

Ocular: The assessment of possible damage resulting from the exposure of the eyes to toxic chemicals should also be considered. The effects of accidental contamination of the eye can vary from minor irritation to complete loss of vision. In addition to the accidental splashing of substances into the eyes, some mists, vapors, and gases produce varying degrees of eye irritation, either acute or chronic. In some instances a chemical that does no damage to the eye can be absorbed in sufficient amount to cause systemic poisoning.

CRITERIA OF RESPONSE

After the toxic material has been administered by one of the routes of exposure discussed previously, there are various criteria to evaluate the response. These criteria are oriented whenever possible towards elucidating the mechanisms of action of the material.

Mortality: As has been indicated, the LD_{50} of a substance serves as an initial test to place the compound appropriately in the spectrum of toxic agents. Mortality is also a criterion of response in long-term chronic studies. In such studies, the investigator must be certain that the mortality observed was due to the chronic low level of the material she is studying; hence, an adequate control group of untreated animals subject to otherwise identical conditions is maintained for the duration of the experiment.

Pathology: By examination of both gross and microscopic pathology of the organs of animals exposed, it is possible to get an idea of the site of action of the toxic agent, the mode of action, and the cause of death. Pathological changes are usually observed at dose levels that are below those needed to produce the death of animals. The liver and the kidney are organs particularly sensitive to the action of a variety of toxic agents. In some instances the pathological lesion is typical of the specific toxic agents, for example, the silicotic nodules in the lungs produced by inhalation of free silica or the pattern of liver damage resulting from exposure to carbon tetrachloride and some other hepatotoxins. In other cases the damage may be more diffuse and less specific in nature.

Growth: In chronic studies the effect of the toxic agent on the growth rate of the animals is another criterion of response. Levels of the compound that do not produce death or overt pathology may result in a diminished rate of growth. A record is also made of the food intake. This will indicate whether diminished growth results from lessened food intake or from less efficient use of food ingested. It sometimes happens that when an agent is administered by incorporation into the diet, especially at high levels, the food is unpalatable to the animals and they simply refuse to eat it.

Organ Weight: The weight of various organs, or more specifically the ratio of organ weight to body weight, may be used as a criterion of response. In some instances such alterations are specific and explicable, as for example the increase of lung weight to body weight ratio as a measure of the edema produced by irritants such as ozone or oxides of nitrogen. In other instances the increase is a less specific general hypertrophy of the organ, especially of the liver and kidney.

Physiological Function Tests: Physiological function tests are useful criteria of response both in experimental studies and in assessing the response of exposed workers. They can be especially useful in chronic studies in that they do not necessitate the killing of the animal and can, if desired, be done at regular intervals throughout the period of study. Tests in common clinical use such as brom-

sulphalein retention, thymol turbidity, or serum alkaline phosphatase may be used to assess the effect of an agent on liver function. The examination of the renal clearance of various substances helps give an indication of localization of kidney damage. The ability of the kidney (especially in the rat) to produce a concentrated urine may be measured by the osmolality of the urine produced. This has been suggested for the evaluation of alterations in kidney function. Alterations may be detected following inhalation of materials such as chlorotrifluoroethylene at levels of reversible response. In some instances measurement of blood pressure has proved a sensitive means of evaluating response. Various tests of pulmonary function have been used to evaluate the response of both experimental animals and exposed workers. These tests include relatively simple tests that are suitable for use in field surveys as well as more complex methods possible only under laboratory conditions. Simple tests include such measurements as peak expiratory flowrate (PEFR), forced vital capacity (FVC), and 1-second forced expiratory volume ($FEV_{1.0}$). More complex procedures include the measurement of pulmonary mechanics (flow-resistance and compliance).

Biochemical Studies: The study of biochemical response to toxic agents leads in many instances to an understanding of the mechanism of action. Tests of toxicity developed in animals should be oriented to determination of early response from exposures that are applicable to the industrial scene. Many toxic agents act by inhibiting the action of specific enzymes. This action may be studied *in vitro* and *in vivo*. In the first case, the toxic agent is added to tissue slices or tissue homogenate from normal animals and the degree of inhibition of enzymatic activity is measured by an appropriate technique. In the second case, the toxic agent is administered to the animals; after the desired interval the animals are killed and the degree of enzyme inhibition is measured in the appropriate tissues. A judicious combination of *in vivo* and *in vitro* studies is especially useful when biotransformation to a toxic compound is involved. The classic example of this is the toxicity of fluoroacetate. This material, which was extremely toxic when administered to animals of various species, did not inhibit any known enzymes *in vitro*. Fluoroacetate entered the carboxylic acid cycle of metabolism as if it were acetic acid. The product formed was fluorocitrate, which was a potent inhibitor of the enzyme aconitase. Biological conversion in the living animal had resulted in the formation of a highly toxic compound. The term *lethal synthesis* describes such a transformation.

DETOXICATION MECHANISMS

The term *biotransformation* is in many ways preferable to *detoxication,* for in many instances the toxic element may be the metabolic product rather than the compound administered. There are some instances, of course, such as the conversion of cyanide to thiocyanate, that are truly detoxication in the strict sense.

Tests for the level of metabolites of toxic agents in the urine have found wide use in industrial toxicology as a means of evaluating exposure of workers. These are commonly referred to as biologic threshold limits, that serve as biologic counterparts to the TLV's. The presence of the metabolic product does not of necessity imply poisoning; indeed, the opposite is more commonly the case. Normal values have been established and an increase above these levels indicates that exposure has occurred and thus provides a valuable screening mechanism for the prevention of injury from continued or excessive exposure. Table L.2 lists some of these

TABLE L.2 Metabolic Products Useful As Indices Of Exposure

Product in Urine	Toxic Agents
Organic Sulfate	Benzene Phenol Aniline
Hippuric Acid	Toluene Ethyl benzene
Thiocyanate	Cyanate Nitriles
Glucuronates	Phenol Benzene Terpenes
Formic Acid	Methyl alcohol
2,6, dinitro-4-amino toluene	TNT
p-nitrophenol	Parathion
p-aminophenol	Aniline

metabolic products which have been used to evaluate exposure as well as the agents for which they may be used.

There are other instances in which a biochemical alteration produced by the toxic agent is useful as a criterion of evaluating exposure. Lead, for example, interferes in porphyrin metabolism and increased levels of delta-aminolevulinic acid may be detected in the urine following lead exposure. Levels of plasma choline esterase may be used to evaluate exposure to organic phosphorus insecticides. Levels of carboxyhemoglobin provide a means of assessing exposure to carbon monoxide. Levels of methemoglobin can be used to evaluate exposure to nitrobenzene or aniline. Hemolysis of red cells is observed in exposure to arsine. Analysis of blood, urine, hair, or nails for various metals is also used to evaluate exposure, though whether these would be termed *biochemical tests* depends somewhat on whether you are speaking with an engineer or a biochemist.

The use of biologic threshold limit values provides a valuable adjunct to the TLV's, which are based on air analysis. The analysis of blood, urine, hair, or exhaled air for a toxic material per se (e.g., Pb, As) or for a metabolite of the toxic agent (e.g., thiocyanate, phenol) gives an indication of the exposure of an individual worker. These tests represent a very practical application of data from experimental toxicology. Research in industrial toxicology is often oriented toward the search for a test suitable for use as a biologic threshold that will indicate exposure at a level below which damage occurs.

Behavioral Studies: When any toxic agent is administered to experimental animals, the experienced investigator is alert for any signs of abnormal behavior. Such things as altered gait, bizarre positions, aggressive behavior, increased or decreased activity, or tremors or convulsions can suggest possible sites of action or mechanisms of action. The ability of an animal to maintain its balance on a rolling bar is a frequently used test of coordination. The loss of learned conditioned reflexes has also been used and by judicious combination of these tests it is possible to determine, for example, that the neurological response to methyl cellosolve differs from the response to ethyl alcohol. Ability to solve problems or make perceptual distinctions has been used on human subjects, especially in an effort to determine the possible effects of low

levels of carbon monoxide and other agents that might be expected to interfere with efficient performance of necessary tasks, thus creating a subtle hazard. Effects on neurological variables, such as dark adaptation of the eye, have been used in determining threshold limit values.

Reproductive Effects: It is possible that a level of a toxic material can have an effect on either male or female animals that will result in decreased fertility. In fertility studies the chemical is given to males and females in daily doses for the full cycle of oogenesis and spermatogenesis prior to mating. If gestation is established, the fetuses are removed by caesarean section one day prior to delivery. The litter size and viability are compared with untreated groups. The young are then studied to determine possible effects on survival, growth rate, and maturation. The tests may be repeated through a second and third litter of the treated animals. If it is considered necessary the test may be extended through the second and third generation.

Teratogenic Effects: Chemicals administered to the pregnant animal may, under certain conditions, produce malformations of the fetus without inducing damage to the mother or killing the fetus. The experience with the birth of many infants with limb anomalies resulting from the use of thalidomide by the mothers during pregnancy alerted the toxicologists to the need for more rigid testing in this difficult area. Another example of human experience in recent times was the teratogenic effect of methyl mercury as demonstrated in the incidents of poisoning in Minamata Bay, Japan. The study of the teratogenic potential poses a very complex toxicological problem. The susceptibility of various species of animals varies greatly in the area of teratogenic effects. The timing of the dose is very critical as a chemical may produce severe malformations of one sort if it reaches the embryo at one period of development and either no malformations or malformations of a completely different character if it is administered at a later or earlier period of embryogenesis.

Carcinogenicity: The study of the carcinogenic effects of a toxic chemical is a complex experimental problem. Such testing involves the use of sizable groups of animals observed over a period of two years in rats or four to five years in dogs because of the long latent period required for the development of cancer. Efforts to shorten the time lag have led to the use of aging animals. This may reduce the lag period one third to one fourth. Various strains of inbred mice or hamsters are frequently used in such experiments. Quite frequently materials are screened by painting on the skin of experimental animals, especially mice. Industrial experience down through the years has made plain the hazard of cancer from exposure to various chemicals. Among these are many of the polynuclear hydrocarbons, beta-naphylamine, which produces bladder cancer, chromates, and nickel carbonyl, which produce lung cancer.

FACTORS INFLUENCING INTENSITY OF TOXIC ACTION

Rate of Entry and Route of Exposure: The degree to which a biological system responds to the action of a toxic agent is in many cases markedly influenced by the rate and route of exposure. It has already been indicated that when a substance is administered as an iv injection, the material has maximum opportunity to be carried by the blood stream throughout the body, whereas other routes of exposure interpose a barrier to distribution of the material. The effectiveness of this barrier will govern the intensity of toxic action of a given amount of toxic agent administered by various routes.

Lead, for example, is toxic both by ingestion and by inhalation. An equivalent dose, however, is more readily absorbed from the respiratory tract than from the gastrointestinal tract, and hence produces a greater response.

There is frequently a difference in intensity of response and sometimes a difference even in the nature of the response between the acute and chronic toxicity of a material. If a material is taken into the body at a rate sufficiently slow that the rate of excretion and/or detoxification keeps pace with the intake, it is possible that no toxic response will result even though the same total amount of material taken in at a faster rate would result in a concentration of the agent at the site of action sufficient to produce a toxic response. Information of this sort enters into the concept of a threshold limit for safe exposure. Hydrogen sulfide is a good example of a substance that is rapidly lethal at high concentrations as evidenced by the many accidental deaths it has caused. It has an acute action on the nervous system with rapid production of respiratory paralysis unless the victim is promptly removed to fresh air and revived with appropriate artificial respiration. On the other hand, hydrogen sulfide is rapidly oxidized in the plasma to nontoxic substances, and many times the lethal dose produces relatively little effect if administered slowly. Benzene is a good example of a material that differs in the nature of response depending on whether the exposure is an acute one to a high concentration or a chronic exposure to a lower level. If one used as criteria the 4 hr LC_{50} for rats of 16,000 ppm, which has been reported for benzene, one would conclude (from Table L.1) that this material would be ''practically nontoxic,'' which, of course, is contrary to fact. The mechanism of acute death is narcosis. Chronic exposure to low levels of benzene on the other hand produces damage to the blood-forming tissue of the bone marrow, and chronic benzene intoxication may appear even many years after the actual exposure to benzene has ceased.

Age: It is well known that, in general, infants and the newborn are more sensitive to many toxic agents than are adults of the same species, but this has relatively little bearing on a discussion of industrial toxicology. Older persons or older animals are also often more sensitive to toxic action than are younger adults. With aging comes a diminished reserve capacity in the face of toxic stress. This reserve capacity may be either functional or anatomical. The excess mortality in the older age groups during and immediately following the well-known acute air pollution incidents is a case in point. There is experimental evidence from electron microscope studies that younger animals exposed to pollutants have a capacity to repair lung damage that was lost in older animals.

State of Health: Pre-existing disease can result in greater sensitivity to toxic agents. In the case of specific diseases that would contraindicate exposure to specific toxic agents, pre-placement medical examination can prevent possible hazardous exposure. For example, an individual with some degree of pre-existing methemoglobinemia would not be placed in a work situation involving exposure to nitrobenzene. Since it is known that the uptake of manganese parallels the uptake of iron, it would be unwise to employ a person with known iron deficiency anemia as a manganese miner. It has been shown that viral agents will increase the sensitivity of animals to exposure to oxidizing type air pollutants. Nutritional status also affects response to toxic agents.

Previous Exposure: Previous exposure to a toxic agent can lead to either tolerance, increased sensitivity, or make no difference in the degree of response. Some toxic agents function by sensitization and the initial exposures produce no observable response, but subsequent exposures will do so. In these cases the individuals who are thus sensitized must be removed from exposure. In other instances, if an individual is re-exposed to a substance before complete reversal of the change produced by a previous exposure, the effect may be more dangerous. A case in point would be an exposure to an organophosphorus insecticide that would lower the level of acetylcholine esterase. Given time, the level will be restored to normal. If another exposure occurs prior to this, the enzyme activity may be further reduced to dangerous levels. Previous exposure to low levels of a substance may in some cases protect against subsequent exposure to levels of a toxic agent that would be damaging if given initially. This may come about through the induction of enzymes that detoxify the compound or by other mechanisms often not completely understood. It has been shown, for example, that exposure of mice to low levels of ozone will prevent death from pulmonary edema in subsequent high exposures. There is also a considerable cross tolerance among the oxidizing irritants such as ozone and hydrogen peroxide, an exposure to low levels of the one protecting against high levels of the other.

Environmental Factors: Physical factors can also affect the response to toxic agents. In industries such as smelting or steel making, high temperatures are encountered. Pressures different than normal ambient atmospheric pressure can be encountered in caissons or tunnel construction.

Host Factors: For many toxic agents the response varies with the species of animal. There are often differences in the response of males and females to the same agent. Hereditary factors also can be of importance. Genetic defects in metabolism may render certain individuals more sensitive to a given toxic agent.

CLASSIFICATION OF TOXIC MATERIALS

Within the scope of this review it is not possible to discuss the specific toxic action of a variety of materials, although where possible specific information has been used to illustrate the principles discussed. Toxic agents may be classified in several ways. No one of these is of itself completely satsifactory. A toxic agent may have its action on the organ with which it comes into first contact. Let us assume for the moment that the agent is inhaled. Materials such as irritant gases or acid mists produce a more or less rapid response from the respiratory tract when present in sufficient concentration. Other agents, such as silica or asbestos, also damage the lungs but the response is seen only after lengthy exposure. Other toxic agents may have no effect on the organ through which they enter the body, but exert what is called systemic toxic action when they have been absorbed and translocated to the site of biological action. Examples of such agents would be mercury vapor, manganese, lead, chlorinated hydrocarbons, and many others that are readily absorbed through the lungs, but produce typical toxic symptoms only in other organ systems.

Physical Classifications: This type of classification is an attempt to base the discussion of toxic agents on the form in which they are present in the air. These are discussed as gases and vapors or as aerosols.

Gases and Vapors: In common industrial hygiene usage the term *gas* is usually applied to a substance that is in the gaseous state at room temperature and pressure and the term *vapor* is applied to the gaseous phase of a material that is ordinarily a solid or a liquid at room temperature and pressure. In considering the toxicity of a gas or vapor, the solubility of the material is of the utmost importance. If the material is an irritant gas, solubility in aqueous media will determine the amount of material that reaches the lung and hence its site of action. A highly soluble gas, such as ammonia, is taken up readily by the mucous membranes of the nose and upper respiratory tract. Sensory response to irritation in these areas provides the individual with warning of the presence of an irritant gas. On the other hand, a relatively insoluble gas such as nitrogen dioxide is not scrubbed out by the upper respiratory tract, but penetrates readily to the lung. Amounts sufficient to lead to pulmonary edema and death may be inhaled by an individual who is not at the time aware of the hazard. The solubility coefficient of a gas or vapor in blood is one of the factors determining rate of uptake and saturation of the body. With a very soluble gas, saturation of the body is slow, is largely dependent on ventilation of the lungs, and is only slightly influenced by changes in circulation. In the case of a slightly soluble gas, saturation is rapid, depends chiefly on the rate of circulation, and is little influenced by the rate of breathing. If the vapor has a high fat solubility, it tends to accumulate in the fatty tissues that it reaches carried in the blood. Since fatty tissue often has a meager blood supply, complete saturation of the fatty tissue may take a longer period. It is also of importance whether the vapor or gas is one that is readily metabolized. Conversion to a metabolite would tend to lower the concentration in the blood and shift the equilibrium toward increased uptake. It is also of importance whether such metabolic products are toxic.

Aerosols: An aerosol is composed of solid or liquid particles of microscopic size dispersed in a gaseous medium (for our purposes, air). Special terms are used for indicating certain types of particles. Some of these are: *dust,* a dispersion of solid particles usually resulting from the fracture of larger masses of material such as in drilling, crushing, or grinding operations; *mist,* a dispersion of liquid particles, many of which are visible; *fog,* visible aerosols of a liquid formed by condensation; *fume,* an aerosol of solid particles formed by condensation of vaporized materials; *smoke,* aerosols resulting from incomplete combustion that consist mainly of carbon and other combustible materials. The toxic response to an aerosol depends obviously on the nature of the material, which may have as a target organ the respiratory system or may be a systemic toxic agent acting elsewhere in the body. In either case, the toxic potential of a given material dispersed as an aerosol is only partially described by a statement of the concentration of the material in terms of weight per unit volume or number of particles per unit volume. For a proper assessment of the toxic hazard, it is necessary to have information also on the particle size distribution of the material. Understanding of this fact has led to the development of instruments that sample only particles in the respirable size range.

The particle size of an aerosol is the key factor in determining its site of deposition in the respiratory tract and as a sequel to this, the clearance mechanisms that will be available for its subsequent removal. The deposition of an aerosol in the respiratory tract depends on the physical forces of impaction, settling, and diffusion or Brownian movement that apply to the removal of any aerosol from the atmosphere, as well as on anatomical and physiological factors such as the geometry of the lungs and the air-flow rates and patterns occurring during the respiratory cycle.

In the limited space available only one point will be emphasized here, namely, the toxicological importance of particles below 1 μm in size. Aerosols in the range of 0.2–0.4 μm tend to be fairly stable in the atmosphere. This comes about because they are too small to be effectively removed by forces of settling or impaction and too big to be effectively removed by diffusion. Since these are the forces that lead to deposition in the respiratory tract, it has been predicted theoretically and confirmed experimentally that a lesser *percentage* of these particles is deposited in the respiratory tract. On the other hand, since they are stable in the atmosphere, there are large numbers of them present to be inhaled, and to dismiss this size range as of minimal importance is an error in toxicological thinking that should be corrected whenever it is encountered. Aerosols in the size range below 0.1 μm are also of major toxicological importance. The percentage deposition of these extremely small particles is as great as for 1 μm particles and this deposition is alveolar. Particles in the submicron range also appear to have greater potential for interaction with irritant gases, a fact that is of importance in air pollution toxicology.

Chemical Classification: Toxic compounds may be classified according to their chemical nature.

Physiological Classification: Such classification attempts to frame the discussion of toxic materials according to their biological action.

Irritants: The basis of classifying these materials is their ability to cause inflammation of mucous membranes with which they come in contact. While many irritants are strong acids or alkalis familiar as corrosive to nonliving things such as lab coats or bench tops, bear in mind that inflammation is the reaction of a living tissue and is distinct from chemical corrosion. The inflammation of tissue results from concentrations far below those needed to produce corrosion. As was indicated earlier in discussing gases and vapors, solubility is an important factor in determining the site of irritant action in the respiratory tract. Highly soluble materials such as ammonia, alkaline dusts and mists, hydrogen chloride, and hydrogen fluoride affect mainly the upper respiratory tract. Other materials of intermediate solubility such as the halogens, ozone, diethyl or dimethyl sulfate, and phosphorus chlorides affect both the upper respiratory tract and the pulmonary tissue. Insoluble materials, such as nitrogen dioxide, arsenic trichloride, or phosgene, affect primarily the lung. There are exceptions to the statement that solubility serves to indicate site of action. One such is ethyl ether and other insoluble compounds that are readily absorbed unaltered from the alveoli and hence do not accumulate in that area. In the upper respiratory passages and bronchi where the material is not readily absorbed, it can accumulate in concentrations sufficient to produce irritation. Another exception is in materials such as bromobenzyl cyanide that is a vapor from a liquid boiling well above room temperature. It is taken up by the eyes and skin as a mist. In initial action, then, it is a powerful lachrymator and upper respiratory irritant, rather than producing a primarily alveolar reaction as would be predicted from its low solubility.

Irritants can also cause changes in the mechanics of respiration such as increased pulmonary flow-resistance or decreased compliance (a measure of elastic behavior of the lungs). One group of irritants, among which are sulfur dioxide, acetic acid, formaldehyde, formic acid, sulfuric acid, acrolein, and iodine, produce a pattern in which the flow-resistance is increased, the compliance is decreased only slightly, and at higher concentrations the frequency of breathing is decreased. Another group, among which are ozone and oxides of nitrogen, has little effect on resistance, produces a decrease in compliance and an increase in respiratory rate. There is evidence that in the case of irritant aerosol, the irritant potency of a given material tends to increase with decreasing particle size as assessed by the increase in flow-resistance. Following respiratory mechanics measurements in cats exposed to irritant aerosols, the histologic sections prepared after rapid freezing of the lungs showed the anatomical sites of constriction. Long-term chronic lung impairment may be caused by irritants either as sequelae to a single very severe exposure or as the result of chronic exposure to low concentrations of the irritant. There is evidence in experimental animals that long-term exposure to respiratory irritants can lead to increased mucous secretion and a condition resembling the pathology of human chronic bronchitis without the intermediary of infection. The epidemiological assessment of this factor in the health of residents of polluted urban atmospheres is currently a vital area of research.

Irritants are usually further subdivided into primary and secondary irritants. A primary irritant is a material that for all practical purposes exerts no systemic toxic action either because the products formed on the tissues of the respiratory tract are non-toxic or because the irritant action is far in excess of any systemic toxic action. Examples of the first type would be hydrochloric acid or sulfuric acid. Examples of the second type would be materials such as Lewisite or mustard gas, which would be quite toxic on absorption but death from the irritation would result before sufficient amounts to produce systemic poisoning would be absorbed. Secondary irritants are materials that do produce irritant action on mucous membranes, but this effect is overshadowed by systemic effects resulting from absorption. Examples of materials in this category are hydrogen sulfide and many of the aromatic hydrocarbons and other organic compounds. The direct contact of liquid aromatic hydrocarbons with the lung can cause chemical pneumonitis with pulmonary edema, hemorrhage, and tissue necrosis. It is for this reason that in the case of accidental ingestion of these materials the induction of vomiting is contraindicated because of possible aspiration of the hydrocarbon into the lungs.

Asphyxiants: The basis of classifying these materials is their ability to deprive the tissue of oxygen. In the case of severe pulmonary edema caused by an irritant such as nitrogen dioxide or laryngeal spasm caused by a sudden severe exposure to sulfuric acid mist, the death is from asphyxia, but this results from the primary irritant action. The materials we classify here as asphyxiants do not damage the lung. Simple asphyxiants are physiologically inert gases that act when they are present in the atmosphere in sufficient quantity to exclude an adequate oxygen supply. Among these are such substances as nitrogen, nitrous oxide, carbon dioxide, hydrogen, helium, and the aliphatic hydrocarbons such as methane and ethane. All of these gases are not chemically unreactive and among them are many materials that pose a major hazard of fire and explosion. Chemical asphyxiants are materials that have as their specific toxic action rendering the body incapable of utilizing an adequate oxygen supply. They are thus active in concentrations far below the level needed for damage from the simple asphyxiants. The two classic examples of chemical asphyxiants are carbon monoxide and cyanides. Carbon monoxide interferes with the transport of oxygen to the tissues by its affinity for hemoglobin. The carboxy-hemoglobin thus formed is unavailable for the transport of oxygen.

Over and above the familiar lethal effects, there is concern about how low level exposures will affect performance of such tasks as automobile driving and so on. In the case of cyanide, there is no interference with the transport of oxygen to the tissues. Cyanide transported to the tissues forms a stable complex with the ferric iron of ferric cytochrome oxidase resulting in inhibition of enzyme action. Since aerobic metabolism is dependent on this enzyme system, the tissues are unable to utilize the supply of oxygen, and tissue hypoxia results. Therapy is directed toward the formation of an inactive complex before the cyanide has a chance to react with the cytochrome. Cyanide will complex with methemoglobin so nitrite is injected to promote the formation of methemoglobin. Thiosulfate is also given as this provides the sulfate needed to promote the enzymatic conversion of cyanide to the less toxic thiocyanate.

Primary Anesthetics: The main toxic action of these materials is their depressant effect on the central nervous system, particularly the brain. The degree of anesthetic effect depends on the effective concentration in the brain as well as on the specific pharmacologic action. Thus, the effectiveness is a balance between solubility (which decreases) and pharmacological potency (which increases) as one moves up a homologous series of compounds of increasing chain length. The anesthetic potency of the simple alcohols also rises with increasing number of carbon atoms through amyl alcohol, which is the most powerful of the series. The presence of multiple hydroxyl groups diminishes potency. The presence of carboxyl groups tends to prevent anesthetic activity, which is slightly restored in the case of an ester. Thus acetic acid is not anesthetic, but ethyl acetate is mildly so. The substitution of a halogen for a hydrogen of the fatty hydrocarbons greatly increases the anesthetic action, but confers toxicity to other organ systems, which outweighs the anesthetic action.

Hepatotoxic Agents: These are materials that have as their main toxic action the production of liver damage. Carbon tetrachloride produces severe diffuse central necrosis of the liver. Tetrachloroethane is probably the most toxic of the chlorinated hydrocarbons and produces acute yellow atrophy of the liver. Nitrosamines are capable of producing severe liver damage. There are many compounds of plant origin such as some of the toxic components of the mushroom Amanita phalloides, alkaloids from Senecio, and aflatoxins that are capable of producing severe liver damage and in some instances are powerful hepatocarcinogens.

Nephrotoxic Agents: These are materials that have as their main toxic action the production of kidney damage. Some of the halogenated hydrocarbons produce damage to the kidney as well as to the liver. Uranium produces kidney damage, mostly limited to the distal third of the proximal convoluted tubule.

Neurotoxic Agents: These are materials that in one way or another produce their main toxic symptoms on the nervous system. Among them are metals such as manganese, mercury, and thallium. The central nervous system seems particularly sensitive to organometallic compounds, and neurological damage results from such materials as methylmercury and tetraethyl lead. Trialkyl tin compounds may cause edema of the central nervous system. Carbon disulfide acts mainly on the nervous system. The organic phosphorus insecticides lead to an accumulation of acetyl choline because of the inhibition of the enzyme that would normally remove it and hence cause their main symptoms in the nervous system.

Agents That Act on the Blood or Hematopoietic System: Some toxic agents such as nitrites, aniline, and toluidine convert hemoglobin to methemoglobin. Nitrobenzene forms methemoglobin and also lowers the blood pressure. Arsine produces hemolysis of the red blood cells. Benzene damages the hematopoietic cells of the bone marrow.

Agents That Damage the Lung: In this category are materials that produce damage of the pulmonary tissue but not by immediate irritant action. Fibrotic changes are produced by materials such as free silica, which produces the typical silicotic nodule. Asbestos also produces a typical damage to lung tissue and there is newly aroused interest in this subject from the point of view of possible effects of low level exposure of individuals who are not asbestos workers. Other dusts, such as coal dust, can produce pneumoconiosis that, with or without tuberculosis superimposed, has been of long concern in mining. Many dusts of organic origin such as those arising in the processing of cotton or wood can cause pathology of the lungs and/or alterations in lung function. The proteolytic enzymes added to laundry products are an occupational hazard of current interest. Toluenediisocyanate (TDI) is another material which can cause impaired lung function at very low concentrations and there is evidence of chronic as well as acute effects.

INDEX

Note: Page numbers in italics refer to figures; page numbers followed by t indicate tables.